# Stem Cell Labeling for Delivery and Tracking Using Noninvasive Imaging

# Series in Medical Physics and Biomedical Engineering

Series Editors: John G Webster, Slavik Tabakov, Kwan-Hoong Ng

*Other recent books in the series:*

**Practical Biomedical Signal Analysis using MATLAB®**
K J Blinowska and J Żygierewicz (Eds.)

**Physics for Diagnostic Radiology, Third Edition**
P P Dendy and B Heaton (Eds.)

**Nuclear Medicine Physics**
J J Pedroso de Lima (Ed.)

**Handbook of Photonics for Biomedical Science**
Valery V Tuchin (Ed.)

**Handbook of Anatomical Models for Radiation Dosimetry**
Xie George Xu and Keith F Eckerman (Eds.)

**Fundamentals of MRI: An Interactive Learning Approach**
Elizabeth Berry and Andrew J Bulpitt

**Handbook of Optical Sensing of Glucose in Biological Fluids and Tissues**
Valery V Tuchin (Ed.)

**Intelligent and Adaptive Systems in Medicine**
Oliver C L Haas and Keith J Burnham

**A Introduction to Radiation Protection in Medicine**
Jamie V Trapp and Tomas Kron (Eds.)

**A Practical Approach to Medical Image Processing**
Elizabeth Berry

**Biomolecular Action of Ionizing Radiation**
Shirley Lehnert

**An Introduction to Rehabilitation Engineering**
R A Cooper, H Ohnabe, and D A Hobson

**The Physics of Modern Brachytherapy for Oncology**
D Baltas, N Zamboglou, and L Sakelliou

**Electrical Impedance Tomography**
D Holder (Ed.)

Series in Medical Physics and Biomedical Engineering

# Stem Cell Labeling for Delivery and Tracking Using Noninvasive Imaging

*Edited by*

## Dara L. Kraitchman
*Johns Hopkins University*
*Baltimore, Maryland*

## Joseph C. Wu
*Stanford University School of Medicine*
*Stanford, California*

**CRC Press**
Taylor & Francis Group
Boca Raton  London  New York

CRC Press is an imprint of the
Taylor & Francis Group, an **informa** business

A TAYLOR & FRANCIS BOOK

CRC Press
Taylor & Francis Group
6000 Broken Sound Parkway NW, Suite 300
Boca Raton, FL 33487-2742

First issued in paperback 2019

© 2012 by Taylor & Francis Group, LLC
CRC Press is an imprint of Taylor & Francis Group, an Informa business

No claim to original U.S. Government works

ISBN-13: 978-1-4398-0751-4 (hbk)
ISBN-13: 978-0-367-86445-3 (pbk)

### Library of Congress Cataloging-in-Publication Data

Stem cell labeling for delivery and tracking using noninvasive imaging / editors, Dara L. Kraitchman, Joseph C. Wu.
    p. ; cm. -- (Series in medical physics and biomedical engineering)
    Includes bibliographical references and index.
    ISBN 978-1-4398-0751-4 (hardcover : alk. paper)
    1. Stem cells. 2. Molecular probes. 3. Radioisotope scanning. I. Kraitchman, Dara L. II. Wu, Joseph Ching-Ming, 1971- III. Series: Series in medical physics and biomedical engineering.
    [DNLM: 1. Stem Cells--radionuclide imaging. 2. Cell Tracking. 3. Magnetic Resonance Imaging--methods. 4. Staining and Labeling. QU 325]

QH588.S83S73884 2012
616.02'774--dc23
    2011027660

**Visit the Taylor & Francis Web site at**
**http://www.taylorandfrancis.com**

**and the CRC Press Web site at**
**http://www.crcpress.com**

# Contents

# *About the Series*

The *Series in Medical Physics and Biomedical Engineering* describes the applications of physical sciences, engineering and mathematics in medicine and clinical research.

The series seeks (but is not restricted to) publications in the following topics:

- Artificial Organs
- Assistive Technology
- Bioinformatics
- Bioinstrumentation
- Biomaterials
- Biomechanics
- Biomedical Engineering
- Clinical Engineering
- Imaging
- Implants
- Medical Computing and Mathematics
- Medical/Surgical Devices

- Patient Monitoring
- Physiological Measurement
- Prosthetics
- Radiation Protection, Health Physics and Dosimetry
- Regulatory Issues
- Rehabilitation Engineering
- Sports Medicine
- Systems Physiology
- Telemedicine
- Tissue Engineering
- Treatment

The *Series in Medical Physics and Biomedical Engineering* is an international series that meets the need for up-to-date texts in this rapidly developing field. Books in the series range in level from introductory graduate textbooks and practical handbooks to more advanced expositions of current research.

The **Series in Medical Physics and Biomedical Engineering** is the official book series of the International Organization for Medical Physics.

## The International Organization for Medical Physics

The International Organization for Medical Physics (IOMP), founded in 1963, is a scientific, educational, and professional organization of 76 national adhering organisations, more than 16,500 individual members, several Corporate Members and four international Regional Organizations.

IOMP is administered by a Council, which includes delegates from each of the Adhering National Organizations. Regular meetings of Council are held electronically as well as every three years at the World Congress on Medical Physics and Biomedical Engineering. The President and other Officers form the Executive Committee and there are also committees covering the main areas of activity, including Education and Training, Scientific, Professional Relations and Publications.

## Objectives

- To contribute to the advancement of medical physics in all its aspects.
- To organize international cooperation in medical physics, especially in developing countries.
- To encourage and advise on the formation of national organizations of medical physics in those countries which lack such organizations.

## Activities

Official journals of the IOMP are *Physics in Medicine and Biology, Medical Physics* and *Physiological Measurement*. The IOMP publishes a bulletin *Medical Physics World* twice a year which is distributed to all members.

A World Congress on Medical Physics and Biomedical Engineering is held every three years in co-operation with IFMBE through the International Union for Physics and Engineering Sciences in Medicine (IUPESM). A regionally based International Conference on Medical Physics is held between World Congresses. IOMP also sponsors international conferences, workshops and courses. IOMP representatives contribute to various international committees and working groups.

The IOMP has several programmes to assist medical physicists in developing countries. The joint IOMP Library programme supports 69 active libraries in 42 developing countries and the Used Equipment Programme coordinates equipment donations. The Travel Assistance Programme provides a limited number of grants to enable physicists to attend the World Congresses.

The IOMP website is being developed to include a scientific database of international standards in medical physics and a virtual education and resource centre.

Information on the activities of the IOMP can be found on its web site at www.iomp.org.

# Contributors

**Ali S. Arbab, MD, PhD**
Department of Radiology
Henry Ford Hospital
Wayne State University School of
Medicine
Detroit, Michigan

**Aravind Arepally, MD, FSIR**
Piedmont Healthcare
Radiology Associates of Atlanta
Atlanta, Georgia

**Megan T. Baldridge, PhD**
Department of Molecular and
Human Genetics
Baylor College of Medicine
Houston, Texas

**Ernst Jan Bos, MD**
Department of Cardiothoracic
Surgery
Stanford University School of
Medicine
Stanford, California

and

Department of Surgery
Leiden University Medical Center
Leiden, the Netherlands

**Peter R. Brink, PhD**
Department of Physiology and
Biophysics
Stony Brook University
Stony Brook, New York

**Jeff W. M. Bulte, PhD**
Russell H. Morgan Department
of Radiology and Radiological
Science
Johns Hopkins University School of
Medicine
Baltimore, Maryland

**Jochen Cammin, PhD**
Russell H. Morgan Department
of Radiology and Radiological
Science
Johns Hopkins University School of
Medicine
Baltimore, Maryland

**John A. Carrino, MD, MPH**
Russell H. Morgan Department
of Radiology and Radiological
Science
Johns Hopkins University School of
Medicine
Baltimore, Maryland

**Ira S. Cohen, MD, PhD**
Department of Physiology and
Biophysics Institute for Molecular
Cardiology
Stony Brook University
Stony Brook, New York

**Jason M. Criscione**
Department of Biomedical
Engineering
Yale University
New Haven, Connecticut

**Marcel M. Daadi, PhD**
Department of Neurosurgery
Stanford University School of
    Medicine
Stanford, California

**Richard P. Davis, PhD**
Department of Anatomy and
    Embryology
Leiden University Medical Centre
Leiden, the Netherlands

**Tarek M. Fahmy, PhD**
Departments of Biomedical and
    Chemical Engineering
Yale University
New Haven, Connecticut

**Paula J. Foster, PhD**
Robarts Research Institute
University of Western Ontario
London, Ontario, Canada

**Joseph A. Frank, MS, MD**
Radiology and Imaging Sciences
    Clinical Center
National Institutes of Biomedical
    Imaging and Bioengineering
Bethesda, Maryland

**Sanjiv S. Gambhir, MD, PhD**
Department of Radiology and
    Bioengineering
Stanford University School of
    Medicine
Stanford, California

**Glenn R. Gaudette, PhD**
Department of Biomedical
    Engineering
Worcester Polytechnic Institute
Worcester, Massachusetts

**Lisa M. Gazdzinski, PhD**
Toronto Centre for Phenogenomics
Hospital for Sick Children
Toronto, Ontario, Canada

**Margaret A. Goodell, PhD**
Department of Molecular and
    Human Genetics
Baylor College of Medicine
Houston, Texas

**Deepak M. Gupta, MD**
Department of Surgery
Stanford University School of
    Medicine
Stanford, California

**Raphael Guzman, MD**
Department of Neurosurgery and
Stanford University School of
    Medicine
Stanford, California

**T. Kevin Hitchens, PhD**
Department of Biological Sciences
Carnegie Mellon University
Pittsburgh, Pennsylvania

**Chien Ho, PhD**
Department of Biological Sciences
Carnegie Mellon University
Pittsburgh, Pennsylvania

**Yasuhiro Ikeda, DVM, PhD**
Department of Molecular Medicine
Mayo Clinic
Rochester, Minnesota

**Lynne L. Johnson, MD**
Departments of Medicine and
    Surgery
Columbia University
New York, New York

**Suzanne Kadereit, PhD**
Doerenkamp-Zbinden Lab for *In
Vitro* Toxicology and Biomedicine
University of Konstanz
Konstanz, Germany

**Amy R. Kontorovich, MD, PhD**
Department of Physiology and
Biophysics
Stony Brook University
Stony Brook, New York

**Dara L. Kraitchman, VMD, PhD,
FACC**
Russell H. Morgan Department
of Radiology and Radiological
Science
Johns Hopkins University School of
Medicine
Baltimore, Maryland

**Eddy S. M. Lee, PhD**
Radiology Department
Stanford University
Stanford, California

**Howard Leong-Poi, MD, FRCPC,
FASE**
St. Michael's Hospital
University of Toronto
Toronto, Ontario, Canada

**Maarten A. Lijkwan, MD**
Departments of Cardiothoracic
Surgery and Radiology
Stanford University School of
Medicine
Stanford, California

and

Department of Surgery
Leiden University Medical Center
Leiden, the Netherlands

**Michael T. Longaker, MD, MBA**
Department of Surgery
Stanford University School of
Medicine
Stanford, California

**Antônio J. Machado S., MD**
Russell H. Morgan Department
of Radiology and Radiological
Science
Johns Hopkins University School of
Medicine
Baltimore, Maryland

**Raj Makkar, MD**
Cedars-Sinai Heart Institute
Los Angeles, California

**Eduardo Marbán, MD, PhD,
FAHA, FACC**
Cedars-Sinai Heart Institute
Los Angeles, California

**Almudena Martinez-Fernandez,
PhD**
Departments of Medicine,
Molecular Pharmacology and
Experimental Therapeutics, and
Medical Genetics
Mayo Clinic
Rochester, Minnesota

**Robert J. McDonald, MD, PhD**
Department of Radiology
Mayo Clinic
Rochester, Minnesota

**Christine L. Mummery, PhD**
Department of Anatomy and
Embryology
Leiden University Medical Centre
Leiden, the Netherlands

**Timothy J. Nelson, MD, PhD**
Departments of Medicine,
    Molecular Pharmacology and
    Experimental Therapeutics, and
    Medical Genetics
Mayo Clinic
Rochester, Minnesota

**Adrian D. Nunn, PhD, FRSC**
Head Discovery Biology
Bracco Research USA Inc.
Princeton, New Jersey

**Nicholas J. Panetta, MD**
Department of Surgery
Stanford University School of
    Medicine
Stanford, California

**Carmen Perez-Terzic, MD**
Departments of Medicine,
    Molecular Pharmacology and
    Experimental Therapeutics, and
    Medical Genetics
Mayo Clinic
Rochester, Minnesota

**Mark F. Pittenger, PhD**
Department of Cardiac Surgery
University of Maryland School of
    Medicine
Baltimore, Maryland

and

Pearl Lifescience Partners, LLC
Baltimore, Maryland

**Hui Qiao, PhD**
Department of Radiology
University of Pennsylvania
Philadelphia, Pennsylvania

**Michael R. Rosen, MD**
Center for Molecular Therapeutics
Columbia University
New York, New York

**Brian K. Rutt, PhD**
Radiology Department
Stanford University
Stanford, California

**Lew C. Schon, MD**
Johns Hopkins University School of
    Medicine
Baltimore, Maryland

and

Georgetown University Medical
    Center
Washington, DC

and

Department of Orthopaedic Surgery
Union Memorial Hospital
Baltimore, Maryland

**Fiona See, PhD**
Departments of Medicine and
    Surgery
Columbia University
New York, New York

**Albert J. Sinusas, MD**
Departments of Medicine and
    Diagnostic Radiology
Yale University School of Medicine
New Haven, Connecticut

**Rachel Ruckdeschel Smith, PhD**
Cedars-Sinai Heart Institute
Los Angeles, California

**David E. Sosnovik, MD, FACC**
Harvard Medical School
Harvard-MIT Division of Health
  Sciences and Technology
Massachusetts General Hospital
Charlestown, Massachusetts

**Edouard G. Stanley**
Monash Immunology and Stem Cell
  Laboratories
Monash University
Clayton, Victoria, Australia

**Gary K. Steinberg, MD, PhD**
Department of Neurosurgery
Stanford University School of
  Medicine
Stanford, California

**Katsuyuki Taguchi, PhD**
Russell H. Morgan Department
  of Radiology and Radiological
  Science
Johns Hopkins University School of
  Medicine
Baltimore, Maryland

**Andre Terzic, MD, PhD**
Departments of Medicine,
  Molecular Pharmacology and
  Experimental Therapeutics, and
  Medical Genetics, and Physical
  Medicine and Rehabilitation
Mayo Clinic
Rochester, Minnesota

**Flordeliza S. Villanueva, MD**
Center for Ultrasound Molecular
  Imaging and Therapeutics
University of Pittsburgh
Pittsburgh, Pennsylvania

**Emmanuel J. Volanakis, MD**
Department of Oncology
St. Jude Children's Research
  Hospital
Memphis, Tennessee

**David C. Weksberg, MD, PhD**
Department of Molecular and
  Human Genetics and Stem Cells
  and Regenerative Medicine
  Center
Baylor College of Medicine
Houston, Texas

**Anthony J. White, MBBS, FRACP,
PhD**
Monash Cardiovascular Research
  Centre
Monash University
Clayton, Victoria, Australia

**Joseph C. Wu, MD, PhD, FACE**
Departments of Radiology and
  Medicine
Stanford University School of
  Medicine
Stanford, California

**Yijen Lin Wu, PhD**
Department of Biological Sciences
Carnegie Mellon University
Pittsburgh, Pennsylvania

**Shahriar S. Yaghoubi, PhD**
CellSight Technologies, Inc.
San Francisco, California

and

Department of Molecular and
  Medical Pharmacology
UCLA School of Medicine
Los Angeles, California

**Satsuki Yamada, MD, PhD**
Departments of Medicine,
  Molecular Pharmacology and
  Experimental Therapeutics, and
  Medical Genetics
Mayo Clinic
Rochester, Minnesota

**Qing Ye, MD**
Department of Biological Sciences
Carnegie Mellon University
Pittsburgh, Pennsylvania

**Haosen Zhang, PhD**
Department of Biological Sciences
Carnegie Mellon University
Pittsburgh, Pennsylvania

**Rong Zhou, PhD**
Department of Radiology
University of Pennsylvania
Philadelphia, Pennsylvania

# Introduction: Stem Cell Types Overview and Rationale for Labeling for Imaging

Dara L. Kraitchman and Joseph C. Wu

Cell labeling and tracking lend their roots to the development of methods to stain specific cellular components for microscopic evaluation. One of the best-known histological stains, hematoxylin, was originally used as a fabric dye and adapted by an amateur microscopist in the mid-1800s to stain the nuclear components in cells (Titford 2005). Subsequently, specific proteins (e.g., antigens or enzymes) were used to localize specific cells or components within the cell by microscopy. In 1939, Gomori developed a method to localize phosphatase activity as a brown-black precipitate for light microscopy (Gomori 1952). One of the earliest examples of translation of enzymatic histological techniques to cell tracking was the injection of horseradish peroxidase (HRP) into early-stage embryos. Transfer of the HRP into daughter cells enabled tracking of cells into lineage-specific cells of mesoderm, ectoderm, and endoderm (Balakier and Pedersen 1982; Cruz et al. 1991).

Antigen–antibody reactions for immunohistochemical staining were introduced by Albert H. Coons, who adopted procedures developed for identifying immune responses to tissue staining (Coons and Kaplan 1950). Coons' original work involved binding a fluorescent dye to an antibody *in vitro* that could form a complex with a specific antigen (i.e., an antigen–antibody complex) on specific cells or structures that could be viewed visually with ultraviolet light. This technique is still used for immunofluorescent staining of cells for pathological microscopic evaluation. Fluorescein, a green dye under ultraviolet light, in the form of fluorescein isothiocyanate (FITC) is one of the most commonly used antibody fluorescent markers for cell identification based on cell surface antigens (Figure I.1). Similar to enzymatic labeling for cell tracking, fluorescent probes (e.g., CellTracker™; Molecular Probes, Invitrogen) are widely available for cell tracking and can now be used in combination with histopathological fluorescent labeling (Figure I.1) One advantage of these cell tracker fluorescent dyes is that they are less cytotoxic than FITC.

The expansion of these techniques from histopathological examination to cell tracking *in vivo* was precipitated by a number of different pathways and clinical needs. The most widespread methods of cell tracking have

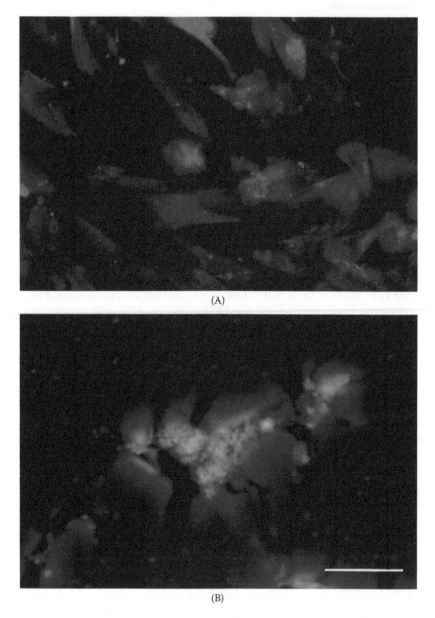

(A)

(B)

**FIGURE I.1 (See color insert.)**

(A) Mesenchymal stem cells (MSCs) exogenously labeled with cell tracker orange (CTO) appear bright orange using an epifluorescent microscope. (B) Photomicrograph showing orange fluorescence is present in both skeletal muscle and MSCs. However, only the MSCs were exogenously labeled with CTO. When implanted *in vivo*, the CTO label was released from dying MSCs and taken up by bystander cells. Thus, with direct labeling schemes, the label may no longer reflect the distribution of the labeled cells. Cell nuclei appear blue due to (DAPI) 4′,6-diamidino-2-phenylindole staining.

relied on the adaptation of radiotracer imaging to cell tracking. A long-lived radiotracer, 111-indium oxine, was approved by the Food and Drug Administration (FDA) in 1986 after extensive research in the late 1970s. The tracer was able to diffuse across membranes, and once intracellular, the oxine portion was cleaved, thereby trapping the radiotracer inside the cell. Thus, radiolabeling of cells was achieved similar to fluorescent marker labeling using direct incubation of the cells with the label. Now, it is standard clinical practice to label autologous leukocytes exogenously with 111-indium oxine and track their biodistribution with scintigraphy or single-photon emission computed tomography (SPECT) to identify regions of inflammation and infection (Segal et al. 1976). By contrast, enzymatic and antigen–antibody labeling of cells took longer to be clinically adopted, although these techniques offer several advantages. For instance, enzymatic and antigen–antibody labeling techniques can be used to target particular cells *in vitro* or *in vivo*. Proper choice of an enzyme or antigen can create labeling that is specific for a certain cell lineage. However, this specificity may work well in certain species but may not be universal for all species. Hence, clinical translation from preclinical animal studies may require modifications of the targeted probe. Nonetheless, a wide array of labels for noninvasive cell tracking has been implemented in recent years, with virtually all imaging modalities represented. The particulars of each imaging modality are discussed in more detail in the subsequent chapters of this book.

The other major technique adopted from postmortem labeling methodology is reporter gene labeling. Reporter gene techniques for histopathology consist of transfecting a cell so that it produces a specific enzyme, protein, or receptor that can be detected under the microscope. Perhaps the most well known in this class is green fluorescent protein (GFP). Mammalian cells that are induced to express GFP will then produce a bright green fluorescence when exposed to blue light. Only live cells will produce the reporter gene, whereas with direct cell labeling, the label may still be present in tissue long after the cell has died (Figure I.2).

While stem cell labeling was initiated originally for postmortem validation, there are many reasons to use it for tracking and labeling cells *in vivo*. The simplest rationale is the need to determine the location and distribution of cells anatomically immediately after delivery. One of the first clinical studies with dendritic vaccines radiolabeled and magnetic resonance imaging (MRI) labeled showed a dismal 50% success rate of delivering the cells to the targeted lymph node despite ultrasound image guidance (de Vries et al. 2005). Similarly, the ability to determine where cells go after delivery can have numerous critical implications. If the cells fail to remain in an organ of interest, will this more likely result in treatment failure? On the other hand, what is the consequence of delivery of cells to nontarget organs? Alternatively, are there certain time windows during which cells thrive and engraft better than other therapeutic intervals? In a small cardiovascular clinical cellular therapy trial, radiolabeled progenitor cells were tracked with positron

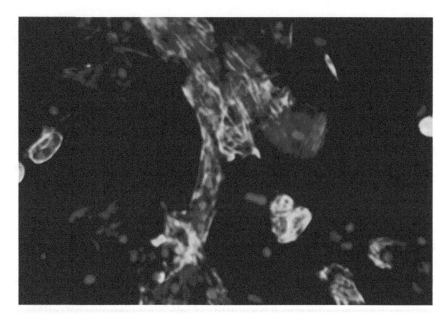

**FIGURE I.2**
A cell expressing the reporter gene for green fluorescent protein (GFP) appears bright green microscopically. One theoretical advantage of reporter gene imaging is that only viable cells will produce the reporter gene product, thereby limiting the bystander effect of direct labeling. (Cell nuclei appear blue due to DAPI staining.)

emission tomography (PET) to demonstrate that enrichment with a certain cell type led to a higher retention of cells within the heart after intracoronary delivery (Hofmann et al. 2005). While certain labeling methods may never become feasible for clinical applications, cell labeling can offer a method to study cell therapies serially, lending better insight into the dynamic distribution of cell therapies. Many lessons have already been learned from preclinical studies using labeled cells and noninvasive imaging, ranging from information about the best route of delivery (e.g., intravenous, intramuscular, intra-arterial) to the optimal dosing scheme to the survival of different types of stem cells after administration.

The development of stem cell labeling and imaging techniques bridges a variety of disciplines. The first chapters of the book provide background information and insight about the major classes of stem and progenitor cells. The subsequent chapters introduce the chief imaging techniques that are currently employed for stem cell tracking. The state-of-the-art techniques for stem cell tracking using these chief imaging modalities are covered in detail in the ensuing chapters. Finally, clinical perspectives of current and potential uses of stem cells and the impact of image-guided delivery and tracking in major organ systems are presented by scholars in the field to supply possible previews into the future of stem cell imaging. New developments in the field of stem cell imaging have been occurring at a frenetic pace, but these techniques are deeply rooted in the technological advances that have occurred to date. This text is intended to provide the fundamentals to understanding and employing the evolving concepts of stem cell labeling and tracking as the field continues to move forward.

Auxiliary material to accompany this book available at the CRC website http://www.crcpress.com/product/isbn/9781439807519.

## References

Balakier, H. and R.A. Pedersen. 1982. Allocation of cells to inner cell mass and tro-phectoderm lineages in preimplantation mouse embryos. *Dev Biol* 90, no. 2: 352–62.

Coons, A.H. and M.H. Kaplan. 1950. Localization of antigen in tissue cells; improvements in a method for the detection of antigen by means of fluorescent antibody. *J Exp Med* 91, no. 1: 1–13.

Cruz, Y.P., S.A. Sutherland, and J.D. Sutherland. 1991. Evaluation of anionic histological dyes as co-injectable cell markers in pre-implantation mouse embryos. *Int J Dev Biol* 35, no. 1: 57–62.

De Vries, I.J., W.J. Lesterhuis, J.O. Barentsz, P. Verdijk, J.H. Van Krieken, O.C. Boerman, W.J. Oyen, J.J. Bonenkamp, J.B. Boezeman, G.J. Adema, J.W. Bulte, T.W. Scheenen, C.J. Punt, A. Heerschap, and C.G. Figdor. 2005. Magnetic resonance tracking of dendritic cells in melanoma patients for monitoring of cellular therapy. *Nat Biotechnol* 23, no. 11: 1407–13.

Gomori, G. 1952. *Microscopic Histochemistry*. Chicago: University of Chicago Press.

Hofmann, M., K.C. Wollert, G.P. Meyer, A. Menke, L. Arseniev, B. Hertenstein, A. Ganser, W.H. Knapp, and H. Drexler. 2005. Monitoring of bone marrow cell homing into the infarcted human myocardium. *Circulation* 111, no. 17: 2198–202.

Segal, A.W., R.N. Arnot, M.L. Thakur, and J.P. Lavender. 1976. Indium-111-labelled leucocytes for localisation of abscesses. *Lancet* 2, no. 7994: 1056–58.

Titford, M. 2005. The long history of hematoxylin. *Biotech Histochem* 80, no. 2: 73–78.

# 1

## Human Embryonic Stem Cells

Richard P. Davis, Edouard G. Stanley, and Christine L. Mummery

### CONTENTS

## 1.1 Introduction

The first derivation of embryonic stem cells (ESCs) from human embryos in 1998 ushered in a new era in cell biology and biomedical research. Human ESCs (hESCs) are a unique cell type, defined by their unlimited self-renewal capacity and potential to differentiate to any cell type in the human body. Therefore, these cells not only provide the opportunity to investigate the developmental events that occur during human embryogenesis but also offer a new source of cells for regenerative medicine, drug discovery, and toxicity screening (Figure 1.1). In addition, these cells can be used to study cell commitment and (re)programming of fate, as well as the molecular mechanisms of complex diseases. However, to maximize the value of hESCs, we need to be able to control their differentiation potential and to be able to promote the development of specific cell types.

Because mouse ESCs (mESCs) were derived more than 15 years before the first hESC line, our understanding of ESC biology is largely based on studying mESCs. However, since the beginning of the twenty-first century have seen substantial advances in hESC research on the molecular and cellular mechanisms of self-renewal and differentiation. Indeed, this research assisted in the significant discovery that adult somatic cells can be repro-

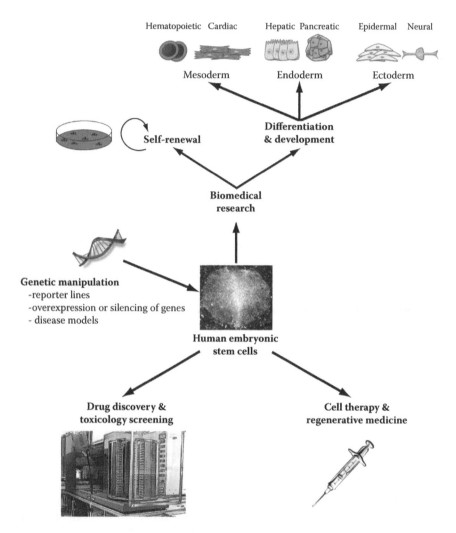

Hematopoietic   Cardiac     Hepatic Pancreatic     Epidermal   Neural

Mesoderm       Endoderm       Ectoderm

**Differentiation & development**

**Self-renewal**

**Biomedical research**

**Genetic manipulation**
-reporter lines
-overexpression or silencing of genes
- disease models

**Human embryonic stem cells**

**Drug discovery & toxicology screening**

**Cell therapy & regenerative medicine**

**FIGURE 1.1**

Applications for human embryonic stem cells (hESCs) and their derivatives in research and medicine. hESCs are amenable to genetic manipulation, allowing for the generation of cell lines that will be useful tools for assisting in deriving lineages and cell types of interest for developmental biology, drug discovery, and regenerative medicine. For example, the introduction of reporter genes under the control of a lineage-specific promoter enables screening for factors involved in differentiation and self-renewal. In addition, ectopic expression or silencing of key developmental genes can assist in developing directed differentiation protocols. Genetic modification can also be used to generate *in vitro* models of monogenetic diseases and potentially to correct disorders in transplantable cells for therapy. As optimized differentiation protocols are developed, both genetically modified and unmodified hESC-derived cardiac or hepatic cells are likely to be used in high-throughput systems to test the efficacy and toxicity of different drugs/chemicals. Some hESC-derived cells (e.g., some neural cell types) could also be used for transplantations; however, for most tissues, cell therapy is a long-term prospect.

grammed to an embryonic-like state by the overexpression of a few key pluripotency factors[1,2] (see Chapter 6, "Induced Pluripotent Stem Cells: ...").

## 1.2 Derivation, Maintenance, and Characterization of hESCs

It is generally thought that hESCs are an immortal epiblast-like cell type that exhibits two key characteristics. First, they can indefinitely self-renew as undifferentiated cells *in vitro* while preserving their original karyotype.[3] Second, they are pluripotent and therefore have the capacity to form all three primary embryonic germ layers (ectoderm, mesoderm, and endoderm) both *in vivo* and *in vitro* (reviewed in Reference 4).

hESCs are derived from the inner cell mass (ICM) of *in vitro* fertilized oocytes cultured to the blastocyst stage (approximately embryonic day [E] 5–7) (Figure 1.2A).[5,6] The ICMs are isolated mechanically or by immunosurgery[5–8] and form rounded cell colonies of small, tightly packed cells with a large nucleus-to-cytoplasmic ratio.[9] Following 1½ to 2 weeks in culture, the proliferating cells are dissociated into clumps and replated. This is repeated multiple times until a stable cell line is established (usually after 3–4 weeks). hESCs have also been derived from earlier morula-stage human embryos[10] and late blastocysts,[11] although it is unclear whether stem cells derived from these sources have the same developmental potential as those derived from early blastocysts.

While mESCs can be maintained in a pluripotent state as single cells in the presence of leukemia inhibitory factor (LIF),[12] the regulation of hESC growth is less well understood and differs from mESCs in that LIF has no role in the self-renewal of these cells.[13] It appears that the self-renewal of hESCs involves the Wnt signaling pathway as well as the basic fibroblast growth factor (FGF2) and transforming growth factor-β (TGF-β) signaling pathways.[14–20] Indeed, hESCs are often maintained as colonies on a mitotically inactivated feeder layer in FGF2-supplemented media.[3] These co-culture cell layers can include mouse embryonic fibroblasts (MEFs), human foreskin fibroblasts, or human epithelial or mesenchyme cells.[5,21–24] Alternatively, hESCs can also be grown on an extracellular matrix (ECM) such as Matrigel or laminin in media supplemented with MEF-conditioned medium (MEF-CM),[25] or in a serum-free, feeder-independent medium known as mTeSR1.[26,27] However, it should be noted that even the last culture system is still not completely defined as Matrigel is derived from a mouse tumor cell line and contains several growth factors, including TGF-β, which might contribute to hESC self-renewal.[15] Recombinant vitronectin was identified as a suitable and functional alternative to Matrigel for the culturing of undifferentiated hESCs in serum-free, nonconditioned media.[28]

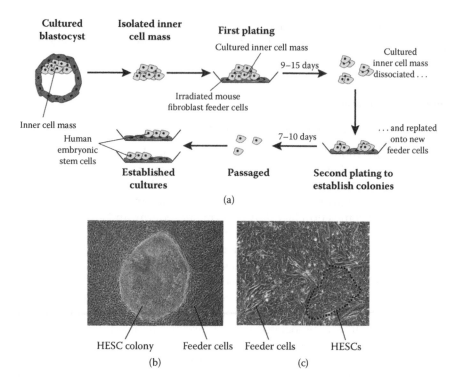

**FIGURE 1.2**

Derivation of human embryonic stem cells (hESCs). (a) The most common source of hESCs is the ICM of the blastocyst. To initiate new hESC lines, the cells of the ICM are plated on a feeder layer. As the ICM cells attach, they form colonies that can be isolated and passaged to derive new hESC lines. (b) An example of an undifferentiated hESC colony. (c) An example of enzymatically passaged hESCs grown on irradiated mouse fibroblast feeder cells.

Typically, undifferentiated hESCs are maintained in culture as colonies that are generally passaged once a week either mechanically or enzymatically as small clusters of cells (Figure 1.2B).[29,30] This method is preferred for long-term maintenance of hESCs as it reduces the level and frequency of stress associated with passaging and appears to minimize the occurrence of karyotypic abnormalities.[31] Alternatively, hESCs can also be expanded enzymatically as a single-cell suspension for short periods without the appearance of cells with an abnormal karyotype or changes detected by morphological and flow cytometric analysis (Figure 1.2C).[29,32] It is possible that the incidence of karyotypic changes occurs at the same frequency with both propagation methods, although the absolute number of cell divisions that occur during enzymatic passage may be higher than during mechanical passage, resulting in the genetically abnormal cells being more readily detected.[33] In both culture systems, the hESCs maintain their compact morphology with a high nuclear-to-cytoplasmic ratio and prominent nucleoli.

Due to the unavailability of a single definitive assay to measure hESC pluripotency, surrogate assays that assess markers or properties correlated with hESC developmental potential are utilized.[34] These include determining the expression of the transcription factors OCT 3/4, NANOG, and SOX2[35,36] and expression of the cell surface stage-specific antigens (SSEA-3 and -4) and keratin sulfate-related antigens (TRA-1-60 and TRA-1-81).[37] However, it is important to recognize that these markers are not specific for undifferentiated hESCs. For example, both SSEA-3 and -4 are expressed on a population of red blood cells,[38] and SSEA-4 is expressed in the human kidney.[39] The expression of OCT 3/4 can also be misleading with a less-than-twofold increase in expression, resulting in mESCs differentiating into primitive endoderm and mesoderm.[40] In addition, several pseudogenes for *OCT 3/4* are present in the genome, so care needs to be taken in interpreting real-time polymerase chain reaction (RT-PCR) expression data.[41] Therefore, it is critical that combinations of these markers are detected at a protein level when determining pluripotency. Currently, the most rigorous method for testing hESC pluripotency is by injecting them into immunodeficient mice and analyzing the resulting teratomas for the presence of cell types from all three embryonic germ layers.[42]

Data suggest that hESCs might actually represent cells from a late epiblast stage. Pluripotent epiblast cells have been derived from mice (mEpiSCs, mouse epiblast stem cells) that are not supported by the presence of LIF but rather require the addition of activin A and FGF2 to the media.[43,44] These cells display similar morphological characteristics as hESCs and can generate cells from all three germ layers but are unable to contribute to mouse chimeras. The gene expression profile of these cells is also closer to hESCs than to mESCs.[44] These similarities suggest that hESCs might be equivalent to cells of the early postimplantation epiblast, rather than the ICM progenitor, even though they are derived from the ICM.

## 1.3 Differentiation Strategies for hESCs

It has been observed that, *in vitro*, mESCs can recapitulate aspects of mammalian development as they occur *in vivo* (for reviews, see [4] and [45]). Under appropriate culture conditions, mESCs will differentiate in aggregates in suspension to embryoid bodies (EBs) that contain cells from all three germ layers. The gene expression pattern of the EBs, as they progress through differentiation, has been shown to mirror that of early postimplantation mouse embryos.[46] A similar temporal pattern of gene expression has been observed in differentiating hESCs to that seen in cultures of mouse EBs (reviewed in [45]). These results support the notion that *in vitro* differentiation of hESCs will model, at least to some extent, early human embryogenesis. This may provide new insights into the stages of normal human embryo development that are otherwise inaccessible for research.

Differentiation of hESCs *in vitro* can also be partially directed by addition of specific growth factors to the culture media (Figure 1.3).[47] Studies using mouse and human ESC differentiation as a model have highlighted that the signaling pathways involved in cell lineage specification are conserved between the two species, as well as with the developing mouse and *Xenopus* embryos.[48] Similar to mESCs, hESC differentiation strategies that recapitulate the processes of normal development have provided the best results. For example, as in vertebrate models, nodal signaling via ALK2 (which can be mimicked by activin A) promotes the formation of endodermal cell types from differentiating hESCs.[49,50] Similarly, bone morphogenetic protein 4 (BMP4) can induce the formation of ventral/posterior mesoderm (hematopoietic cells) from hESCs,[51–54] simulating the induction of blood cells promoted by overexpression of BMP4 in *Xenopus* embryos,[55] while the addition of both BMP4 and activin A leads to the induction of cardiac mesoderm.[56,57]

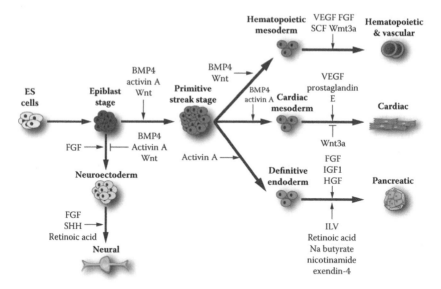

**FIGURE 1.3 (See color insert.)**
A schematic representation of *in vitro* ESC differentiation and factors involved in the generation of specific cell types. BMP, activin A, or Wnt signals induce an epiblast-like cell population to commit to a primitive streak-like population and subsequently to the mesoderm or definitive endoderm cell lineages. High concentrations of BMP are known to promote the formation of posterior-ventral hematopoietic mesoderm, while high amounts of activin A induce definitive endoderm. A combination of BMP and activin A induces anterior-dorsal cardiac mesoderm. If these signals are absent, the epiblast-like cells will differentiate to a neuroectodermal fate. Some of the growth factors and small molecules known either to promote or to inhibit the formation of more differentiated cell types from the three germ layers are also depicted in the figure and referred to in the text.

Two important criteria to consider when differentiating hESCs toward specific cell lineages are the efficiency and reproducibility of the protocol, with similar results ideally obtained using multiple hESC lines. Differentiation of hESCs can be induced in a variety of ways, each with its own set of advantages and disadvantages (Figure 1.4). A common technique involves the formation of cell aggregates referred to as embryoid bodies (EBs).[58–61] Alternative methods involve co-culturing the hESCs on stromal cells[62–67] or differentiating the hESCs on ECM proteins.[68–71] With all of these methods, ESCs can differentiate into a number of cell types.

However, each of these differentiation methods includes variables that can affect the reproducibility and precision of the differentiation. These variables include the quality of the starting hESC population, heterogeneity in EB size

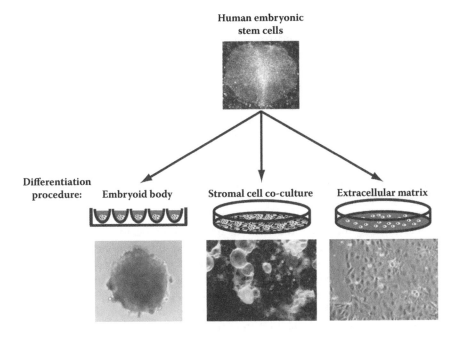

**FIGURE 1.4**
Three protocols for differentiating human embryonic stem cells (hESCs). Embryoid bodies (EBs) are initiated by culturing aggregates of hESCs in suspension. EBs are most commonly generated by suspension culture in bacterial-grade dishes, as hanging drops, or by forced aggregation in round- or conical-bottom multiwell plates ("spin EBs"). An embryoid body generated by the spin EB method is displayed in the left image. The embryoid bodies develop a three-dimensional structure with the outer layer of cells thought to be endodermal. Stromal cell co-culture differentiation involves plating the hESCs on a monolayer of cells that promotes differentiation of the hESCs to the desired cell type. The center image shows hESCs cultured on END2 cells to promote cardiac differentiation. The third method involves culturing hESCs on extracellular matrix proteins. hESCs differentiated on Matrigel in the presence of BMP4 are pictured in the image on the right. (Center photo is courtesy of D. Ward; photo on the right is courtesy of S. R. Braam.)

if clumps of hESCs are used to initiate EB formation, the influence of feeder cells or ECMs in monolayer cultures, and the presence of undefined stromal-derived factors in co-culture conditions. In addition, many of the media used in differentiation protocols include fetal calf serum or other serum derivatives that are undefined and are subject to batch variability. This has been detrimental to the reproducibility of these protocols and can confound the ability to meaningfully interpret the efficacy of growth factors or protocols on the differentiation of hESCs. Chemically defined, serum-free differentiation culture media, originally developed for mESCs, are now beginning to be adapted for use with hESC differentiation systems.[72–74] These are supplemented with specific inducers at key stages in the differentiation procedure to direct the differentiation to specific cell lineages, recapitulating the processes that occur in early embryonic development.

Methods for purifying specific differentiated cell types from a heterogeneous pool of cells are also required for many applications. The simplest purification procedure is based on visible morphological phenotypes, such as spontaneous beating that occurs when cardiomyocytes form. In this instance, clusters of beating cells can be dissected from the rest of the culture; however, the cardiomyocytes will still generally only comprise a small fraction of the total number of cells in the harvested clusters.[75] Alternatively, particular cell populations can be selected or excluded based on the expression of specific cell surface markers. These cell populations can be isolated by either fluorescence- or magnetic-activated cell sorting (FACS or MACS), with FACS providing the ability to purify the desired cell type quickly based on the expression of multiple surface markers.

For many progenitor cell types, antibodies are currently unavailable for suitable surface markers. Rather, the lineage-specific markers, which are often transcription factors, are only expressed intracellularly, preventing viable sorting and subsequent culture of these cells following identification by immunological analysis. Genetic modification of hESCs to introduce an antibiotic resistance gene under the control of a cell-specific promoter offers an alternative method for purifying desired populations, but this application is limited to already established differentiation protocols.[76,77] The insertion of fluorescent reporter proteins instead offers several advantages and, in combination with FACS, has been used successfully to isolate and enrich viable cell populations from differentiating hESCs that would otherwise have been unobtainable.[54,78,79] In addition, reporter knock-in lines will aid in the development of robust differentiation protocols.

hESC lines containing the fluorescent proteins EGFP (enhanced green fluorescent protein), hrGFP (humanized recombinant GFP), or tdRFP (tandem-dimer red fluorescent protein) have also been generated that stably express these markers in undifferentiated cells and their differentiated progeny, both *in vitro* and *in vivo*.[80–82] These lines are enormously valuable to hESC research as the fluorescent proteins allow tracking of the differentiated and undifferentiated hESCs in mixed cell cultures as well as when transplanted into animal models. These marked cells can be analyzed not only in tissues or organs *ex vivo* but also possibly in live animals by noninvasive whole-body imaging.[83]

In the following sections, we concentrate on the directed differentiation of hESCs to the hematopoietic, cardiac, neural, and pancreatic lineages as these are the areas in which most progress has been achieved. We also highlight key findings following the transplantation of these hESC derivatives into animal models.

## 1.4 Differentiation of hESCs to the Hematopoietic Lineage

The ability to direct the differentiation of hESCs *in vitro* to hematopoietic lineages is aiding our understanding of blood cell development. Many studies have examined the development of hematopoietic and endothelial progenitors in hESC cultures; however, the ability to develop transplantable hematopoietic stem cells (HSCs) is still elusive. The role of particular cytokines and growth factors on hematopoietic development from hESCs has begun to be investigated, but many of these early studies were conducted in the presence of fetal calf serum.[84] BMP4, a ventral mesoderm inducer, has been shown to strongly promote hematopoietic differentiation from hESCs being cultured as EBs, when combined with other hematopoietic cytokines.[85] The addition of vascular endothelial growth factor A (VEGF-A) further increased the efficiency of erythroid colony formation.[86] A modification of the EB system allowing for the formation of uniform-size "spin EBs" in a completely defined media free of animal product further defined the role of BMP4 for the development of hematopoietic mesoderm but less frequently supported the generation of hematopoietic progenitors.[51,52] However, the combination of BMP4 with VEGF, stem cell factor (SCF), and FGF2 enhanced the total yield of hematopoietic precursors.[52]

Wnt signaling pathways also appear to play roles in hematopoiesis. Woll et al. demonstrated that the inhibition of canonical Wnt signals in serum-induced cultures resulted in a decrease in the proportion of cells with hematopoietic potential.[87] This was further supported by the demonstration that a noncanonical Wnt (Wnt11) can direct hESCs toward mesoderm, whereas canonical Wnt3a causes cells already committed to the hematopoietic lineage to proliferate.[88]

The kinetics of hematopoietic lineage development within hESC differentiation parallels that observed with mESCs. Differentiating hESCs pass through a primitive streak stage defined by the expression of either BRACHYURY or MIXL1, markers of the primitive streak, to a KDR$^+$ or PDGFR$\alpha^+$ mesodermal progenitor, and subsequently to the yolk sac hematopoietic program.[51,53,54,89]

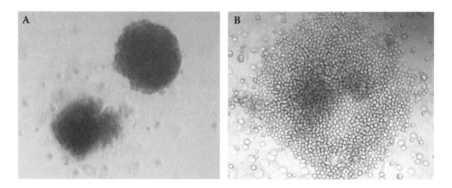

**FIGURE 1.5 (See color insert.)**
Hematopoietic cell types generated from hESCs. Differentiated hESCs were seeded as single cells in methylcellulose cultures containing hematopoietic-promoting growth factors. Erythroid (A) and myeloid (B) colonies can develop from these single cells. Original magnification ×100.

This is characterized by the formation of primitive nucleated erythroid progenitors and the absence of cells from the lymphoid lineage (Figure 1.5). The formation of hematopoietic progenitors from differentiating hESCs occurs within the first 7 days of differentiation.[53,54,65,89] Similar to mESCs and the mouse embryo, there is evidence to suggest that hematopoietic commitment in hESCs is marked by the development of the hemangioblast or blast colony-forming cell, a common progenitor of hematopoietic and endothelial cells. The hemangioblast is a transient population that develops following BMP4 stimulation between day 2 and day 4 of differentiation, prior to the development of the primitive erythroid lineage.[53,54,90]

Over a prolonged period of culture, more mature hematopoietic populations develop, characterized by the expression of CD45, a marker of fetal and adult hematopoietic cells, and the appearance of enucleated erythrocytes.[53,54,65] A maturation of hemoglobin expression is also observed with a switch from an embryonic yolk sac-like to a fetal liver-like pattern of erythropoiesis,[91] although it is unclear whether this is a maturation of a yolk sac-like primitive hematopoietic population or the onset of the definitive hematopoietic program.

The hematopoietic potential of hESCs has been extended to include lymphoid lineages, with both B and natural killer cells generated from co-culture of hESCs on either the stromal cell line OP9 or S17.[65,92] Using a similar differentiation strategy, T lymphoid progenitors have also been generated that can then be further matured through *in vivo* transplantation into immunodeficient mice containing human fetal liver and thymus.[93] On reisolation from the graft, the target cells exhibited *in vitro* responsiveness to T-cell antigen receptor-mimicking stimuli. While these cells could potentially be used to replenish T cells in diseases such as HIV-AIDS, a more defined *in vitro* culture system is first required so that the population that gave rise to these progenitors can be studied in more detail.

While it has been demonstrated that most of the blood lineages can be derived from hESCs *in vitro*, these populations display limited *in vivo* hematopoietic repopulating potential, and the generation of an HSC that is able to mediate long-term multilineage hematopoietic engraftment remains a challenge. Wang et al. provided the first report of a purified hESC-derived hematopoietic cell population that underwent further maturation and produced progeny, but only when directly engrafted into the bone marrow of the femur.[94] While these cells displayed some differentiation potential, unlike HSCs, they were unable to migrate from the femur to other sites in the recipient mice. In addition, the majority of injected mice died within 8 weeks following the transplantation due to clumping of the injected cells and the formation of pulmonary emboli. Another *in vivo* study, using a more heterogeneous mixture of hematopoietic and nonhematopoietic cells, was able to find the donor cells in the recipient mice more than 3 months later.[95] While longer-term engraftment was achieved, this was at very low levels and appeared to be restricted to the myeloid lineage, suggesting that the original donor cells might have been a primitive hematopoietic cell population.

A human/fetal sheep xenograft model has also been utilized to study the *in vivo* potential and function of human hematopoietic progenitors.[96] CD34+ cells, derived by co-culture of hESCs on S17 stromal cells and transplanted by intraperitoneal injections into fetal sheep, were still detected in the bone marrow and peripheral blood 5 to 17 months later, although again at very low levels (less than 1%).

## 1.5 Differentiation of hESCs to the Cardiac Lineage

For the differentiation of hESCs to cardiomyocytes, two protocols have been predominantly used. The first is an EB-based differentiation of hESCs stimulated with either serum or growth factors to generate cardiomyocytes,[57,97,98] while the second method involves co-culturing the hESCs with the mouse visceral endoderm cell line, END2.[63] As in the vertebrate embryo, where signals from the endoderm trigger the formation of cardiac progenitors from the adjacent mesoderm,[99] END2 cells induced the differentiating hESCs to form beating areas within 12 days.[63] By removing serum and insulin from the differentiation media, the efficiency of this co-culture system was further improved with approximately 60% of differentiating hESC colonies containing beating areas, and the beating areas consisting of approximately 25% cardiomyocytes.[75,100] While undefined stromal cell-derived factors make it difficult to precisely define the specific requirements for the differentiation of hESCs to cardiomyocytes, some of the proteins secreted by END2 cells have been determined.[101] Similarly, in an EB differentiation system, several groups have demonstrated that hESCs can be induced to form cardiomyocytes in an

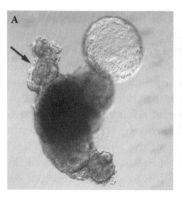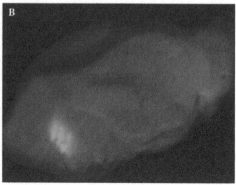

**FIGURE 1.6 (See color insert.)**
Differentiation of hESCs to cardiomyocytes and transplantation of hESC-derived cardiomyocytes. (A) Spontaneously contracting areas (black arrow) developed in hESCs differentiated
as spin EBs following culture in chemically defined media containing BMP4, VEGF, SCF, and
activin A. (B) hESC-CMs derived from an hESC line that expresses GFP were transplanted into
a mouse infarcted heart model. The grafted cells, detected by GFP epifluorescence, were still
present 3 weeks after transplantation. (Panel B is courtesy of L. W. van Laake.)

activin A and BMP4-suppplemented serum-free medium (Figure 1.6A).[56,57,74]
This has subsequently been further refined, with exogenous BMP4 and activin
A signals required only in the initial stages of differentiation, followed by
the addition of VEGF and the inhibition of canonical Wnt signaling.[57]

In an alternative approach, inhibitors of signaling pathways have been
included in differentiation media to generate cardiomyocytes. In an initial
study, inhibiting the p38 MAPK (mitogen-activated protein kinase) pathway
enhanced the cardiogenic activity of END2-conditioned media.[102] This was
subsequently further refined by the addition of prostaglandin E to a chemically defined media, resulting in a similar proportion of cardiomyocytes as
obtained with END2-conditioned media.[103] As a result of all of these studies,
the signaling pathways involved in cardiomyocyte differentiation are beginning to be identified.

Reports suggested that KDR, like its mouse ortholog Flk1, can also label
early cardiac progenitors.[57,104] In differentiating mESCs, a late-developing
Flk1+ population contains cardiovascular progenitors capable of generating
cardiac, endothelial, and vascular smooth muscle cells.[104] In differentiating
hESCs, a KDR$^{low}$ population also displays a similar cardiovascular potential.[57] However, KDR also labels hematopoietic progenitors from differentiating hESCs,[53] suggesting that it is likely to be marking a population of nascent
mesoderm and needs to be used in conjunction with other markers to isolate cardiovascular progenitors. Similarly, Bu et al. demonstrated that ISL1+
cells included cardiovascular progenitors that could give rise to a similar
range of cell lineages, although interestingly these cells were initially KDR-.[78]
However, only between 20% and 30% of the beating areas were derived from
an ISL1+ progenitor population, possibly reflecting the restricted expression

pattern of ISL1 in the human fetal heart, where ISL1[+] cells are only detected in the right atrium, left atrial wall, and outflow tract.[78]

When placed in adherent cultures, spontaneously rhythmically contracting areas within the differentiating colonies can be readily observed. RNA and protein analyses indicated that these regions express the cardiac transcription factors GATA4, MEF 2 and NKX2–5.[97,105] The ability to form beating areas also demonstrates, as confirmed by further expression studies, that the ion channels and signaling and contractile proteins necessary for functional excitation and contraction coupling are present.[63,97,105,106] Patch clamp analysis confirmed that cells with pacemaker, nodal, atrial, or ventricular phenotypes can be detected,[63,106,107] with the END2 co-culture system generating a predominance of ventricular-like cardiomyocytes.[63] Given that cardiomyocytes develop within 2 weeks of differentiation, it is not surprising that these cardiomyocytes display an immature phenotype resembling those of the fetal heart. The electrophysiological profile generally consists of low upstroke velocities and relatively positive resting potential, similar to that observed in cultured fetal cardiomyocytes.[63,106] The hESC-derived cardiomyocytes (hESC-CMs) also express proteins only detected in the embryonic heart and lack the rod-shape morphology typical of mature cells.[63,105] However, despite their immature phenotype, hESC-CMs display quantifiable appropriate responses to many pharmacological drugs (reviewed in Reference 108), suggesting these cells will potentially be useful to the pharmaceutical industry for drug screening.

Initial transplantation experiments of hESC-CMs into uninjured murine hearts demonstrated that they could form grafts of human myocardium, which continued to grow and mature over time.[98,109,110] Similarly, transplantation of cardiomyocytes into an immunosupressed pig or guinea pig model, in which conductance between the atrium and ventricle was blocked, demonstrated their ability to function as biological pacemakers.[111,112] Confirmation that the cells had integrated was further demonstrated by the presence of gap junction connections between the grafted and host cardiomyocytes.

Several studies have also transplanted hESC-CMs into murine myocardial infarct models, but while improvement of cardiac function was observed 4 weeks after transplantation (Figure 1.6B), [56,110,113,114] this recovery appears to be only transient.[110,115] One study, utilizing an hESC line that constitutively expresses GFP ("Envy")[80] to track the transplanted cardiomyocytes *in vivo*, extended the follow-up period up to 12 weeks.[110,115] The grafts at 10 weeks consisted almost entirely of mature cardiomyocytes that had formed intracellular contacts, but despite continued detection of the hESC-CMs in the heart, functional improvement did not last longer than 12 weeks. Indeed, the benefit observed at 4 weeks could be a transient paracrine effect due to the excretion of growth factors, triggering neovascularization of the host myocardium.[116] Interestingly, while the transplanted cardiomyocytes can form intracellular contacts between themselves, they are usually separated from the rodent myocardium by a layer of ECM proteins, which may prevent electrical coupling with the host.[110]

The predominant animal models currently used in these transplantation studies are immunocompromised mouse models with myocardial infarctions. The substantial difference in beating frequency between rodents and humans (500 beats per minute compared to 80 beats per minute, respectively) suggests that the rodent is unlikely to be the best model for these studies and could be a contributing factor in the failure to observe long-term cardiac function improvement. Therefore, larger animals such as pigs or primates that have slower heart rates could serve as better models for future transplantation studies. In addition, aside from the experimentally induced infarct, the animals used in these studies are otherwise healthy without confounding factors (i.e., high blood pressure, lifestyle, diabetes) also present in many humans with cardiac disease. Development of more appropriate animal models, including those with heart failure, therefore remains high on the research agenda for translational studies.

## 1.6 Differentiation of hESCs to Pancreatic Cells

Progress in the directed differentiation of hESCs to endoderm cells has been hampered by the lack of suitable markers to distinguish early definitive endoderm progenitors from extraembryonic, or visceral, endoderm. Indeed, with many *in vitro* differentiation strategies, it is likely that extraembryonic endoderm is the predominant cell lineage that is generated.[117] However, in the last couple of years, advancements have been made in generating definitive endoderm and pancreatic β-like cells from hESCs. D'Amour et al., recapitulating the developmental stages within the embryo, used a combination of low serum and activin A to induce hESC transition through the primitive streak.[49,69] These cells were then specified to a pancreatic fate by treating the cultures with FGF10 and retinoic acid and by inhibiting sonic hedgehog (SHH) signaling. Finally to promote β-cell maturation, exendin 4, insulin-like growth factor 1 (IGF1), and hepatocyte growth factor (HGF) were added to the media, and Notch signaling was inhibited. Some of the resulting cells expressed insulin and released C peptide, indicating that the insulin was produced by the cells and not absorbed from the culture media. However, the hESC-derived β cells were not responsive to glucose stimulation, suggesting that the cells might represent an immature pancreatic cell type.[69] Other groups have also described similar protocols for the generation of insulin-producing cells, all containing adaptations of embryological findings to ESC culture.[118,119] Jiang et al. used a combination of activin A and sodium butyrate to generate PDX1+ cells that were then matured to insulin-positive cells.[118] While the differentiation procedure took longer than that described by D'Amour et al., small clusters of more mature pancreatic cells, consisting not only of C-peptide-positive cells but also glucagon- and somatostatin-positive cells, were generated. These cells were also responsive to glucose stimulation.

The use of chemical compounds in directed differentiation protocols offers several benefits over the use of recombinant growth factors. Many can be manufactured synthetically and are not as structurally complex as the recombinant proteins purified from either bacteria or animal cells. The consequence of this is that they are generally cheaper to produce and provide a higher degree of reproducibility. Perhaps one of the most commonly used small molecules in differentiation protocols is retinoic acid. Like its role in the developing embryo, retinoic acid can promote the generation of Pdx1[+] cells from both differentiating mouse and human ESCs.[69,120] The molecule nicotinamide and the peptide exendin 4 are also often included in pancreatic differentiation protocols.[69,121] Again, the rationale for this is based on studies of human fetal pancreas cultures and rodent models in which they are inducers of endocrine differentiation.[122,123] It was demonstrated that mesendoderm could be efficiently formed by the combination of activin A and the phosphatidylinositol 3-kinase inhibitor, wortmannin.[124] When combined with the other chemical compounds, up to 25% of the differentiating cells express insulin. As more reproducible, robust, and defined differentiation procedures are developed for hESCs, it will become increasingly possible to perform high-content screening of chemical libraries to discover small molecules that efficiently generate desired cell types. An example of this was a high-content screen that identified the small molecule (–)-indolactam V (ILV) that, when combined with growth factors, can increase the production of PDX1-expressing pancreatic progenitors from hESCs that have already been committed to the endoderm lineage.[125]

The successful transplantation of cadaveric-derived islets as a treatment for type 1 diabetes[126] highlights the potential for hESC-based cell therapies if sufficient quantities of the appropriate cell types can be generated and the cells are glucose responsive *in vivo*. Three groups demonstrated pancreatic β-cells derived from hESCs could produce insulin and regulate blood glucose in streptozotocin-induced diabetic mice.[119,121,127] The initial study by Jiang demonstrated that the transplanted cells expressed human C peptide and PDX1 and were able to reverse hyperglycemia in 30% of the recipient mice.[121] Significantly no teratomas were detected up to 3 months after transplantation. In the other two studies, a more immature pancreatic progenitor population was transplanted into the mice and allowed to further mature *in vivo*. Functional β-like cells subsequently developed within the grafts that were able to rescue the mice following streptozotocin-induced hyperglycemia. While a higher proportion of mice remained euglycemic compared to the Jiang report,[121] teratomas were also detected in some of the mice,[127] perhaps indicative of the risk of injecting a more immature population of hESC-derived cells. It is worth noting that, similar to clinical islet transplantations, these experiments included the transplantation of non-β-cell pancreatic cell types, suggesting that highly purified populations of insulin-producing cells may not be the most appropriate source of cells to transplant.

## 1.7  Differentiation of hESCs to Neural Cell Types

The formation of ectodermal derivatives occurs frequently in spontaneously differentiating hESCs (Figure 1.7)[6] and is often considered a default developmental pathway. As with mESCs, trilineage neural progenitors can be readily generated from hESCs, which under appropriate culture conditions will form neurons, astrocytes, and oligodendrocytes.[128,129] For example, the directed differentiation of hESCs into primitive neuroectoderm can be achieved by culturing hESCs with the BMP antagonist, Noggin.[117,130–132] These Noggin-induced cells were capable of renewal in culture and generated neurospheres when cultured in serum-free conditions that formed neurons when plated. Alternatively, culturing the cells in serum-containing

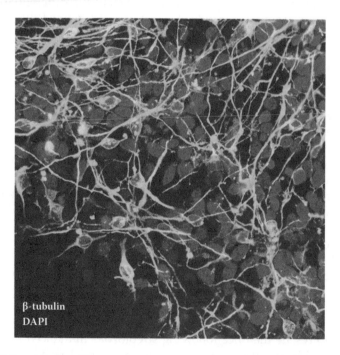

**FIGURE 1.7 (See color insert.)**
hESCs differentiated to neural cell types stained for the presence of β-tubulin, a marker expressed on neural cells. DAPI, 4′,6-diamidino-2-phenylindole. (Image is courtesy of S. R. Braam.)

medium resulted in the formation of the glial progenitors of astrocytes and oligodendrocytes. Culturing the cells in chemically defined media containing the growth factors insulin, EGF, and FGF2 also generated neural progenitors that were Nestin and PAX6 positive and could be expanded by more than 250,000-fold.[133] However it should be noted that while hESC-derived neural progenitors resemble adult and fetal neural progenitors, the gene expression profile and methylation status indicate significant differences between these progenitor populations.[134]

Again, the same signaling pathways that regulate neural cell fate in the embryo are also involved in generating the different types of neurons from hESCs (reviewed in Reference 135). The Notch signaling pathway has been demonstrated to be particularly important for establishing the neural progenitor cells from hESCs.[136] FGF and SHH signaling also play critical roles in neural specification (reviewed in Reference 137). FGF8 together with SHH will induce midbrain GABAergic- (γ-aminobutyric acid releasing), dopamine-, and serotonin-type neurons in a concentration- and temporal-dependent manner.[138] Not only is retinoic acid (RA) involved in endoderm specification, but also it is a strong differentiation inducer of caudally fated neuroepithelial cells.[139] Together with SHH, RA can induce spinal motor neurons and other neuron subtypes, depending on the time of exposure (reviewed in Reference 135). Importantly, it has also been demonstrated that hESC-derived neurons respond to neurotransmitters, generate action potentials, and form functional synapses.[139,140]

Several studies have reported the transplantation of hESC-derived neural cells into animal models of human disease or injury.[66,130,141–144] Oligodendrocyte precursors, generated by the addition of EGF, FGF, and RA to hESCs, are able to mature and remyelinate host axons when transplanted into the *shiverer* mouse model of dysmyelination.[143] These cells are of importance not only for the repair of traumatic injuries to the central nervous system, but also for regenerative repair for demyelinating diseases such as multiple sclerosis. Indeed, spinal cord injury is the setting for the first clinical trials of an hESC-derived product, with the Geron Corporation planning to initiate phase 1 trials of oligodendrocyte progenitor cells.[145]

Dopaminergic neurons are also of particular interest because of their central role in Parkinson's disease. There are now several different *in vitro* methods for producing these cells.[146,147] Importantly, the hESC-derived neurons express tyrosine hydroxylase, necessary for dopamine synthesis, and release dopamine on depolarization. A study by Ben Hur et al. involved the transplantation of hESC-derived neurospheres into the striatum of Parkinsonian rats.[130] These neurospheres differentiated further *in vivo* into dopaminergic neurons but at low efficiency. The transplanted neurons could be still identified 12 weeks after grafting and reduced drug-induced turning behavior; however, these neurons were not proliferating. Interestingly, more extensive predifferentiation *in vitro* to dopamine-producing cells has been shown to be detrimental to successful engraftment into the rat model.[66] This suggests that, like with transplantation of β-cells, the most efficacious transplanted material might contain a mixture of cell types.

## 1.8 Conclusions

Human ESCs have the capacity to differentiate into all somatic cell types in the body. Lineage-specific directed differentiation of hESCs is also possible based on the recapitulation of the processes that occur in vertebrate development. This demonstrates the potency of hESCs as an *in vitro* model system for studying the basic molecular and cellular mechanisms underlying the differentiation steps that occur during development, at stages that would otherwise not be accessible for analysis. The ability now to reprogram somatic cells to an ESC-like cell provides an additional tool with which to study development and particularly genetic diseases (see Chapter 6). A decade of research on developing differentiation protocols for hESCs is now proving directly applicable to human iPSCs. This field therefore benefits greatly from discoveries that have already been made with hESCs. In addition, with sets of "diseased" and normal pluripotent stem cells available, the mechanisms of complex genetic diseases that cannot be mimicked in other model systems now can be elucidated. This also presents possibilities for developing drug-targeting strategies as well as identifying factors that precipitate symptom onset at the cellular level.

Since 2000, there has also been considerable refinement of the culture and differentiation systems, leading to more reproducible procedures and the ability to generate highly enriched cell populations. As our ability to genetically manipulate hESCs improves, this will also assist in determining the combination of factors required to produce the desired cell type. The generation of lines in which reporter genes allow specific cell types to be identified and purified will be helpful in achieving this. These subpopulations can be further differentiated or developed in culture or transplanted and tracked in animal models. These advancements will lead to better characterization of cell populations that could eventually be used in a clinical setting. However, as illustrated in this review, there are only a few fields in which this treatment option might be available in the short term, and in most instances, hESC derivatives for use in regenerative medicine remains a long-term perspective. In the foreseeable future, there is greater potential for these cells in the field of drug discovery and safety pharmacology. Regardless, it is highly likely that hESC research will continue to make substantial contributions to human medicine and to our understanding of human development.

## Acknowledgments

We thank Stefan Braam, Linda van Laake, and Dorien Ward for the provision of figures. This work was supported by the EU FP7 ("InduStem" PIAP-GA-2008-230675) and the Australian Stem Cell Center.

## References

1. Takahashi, K., Tanabe, K., Ohnuki, M., Narita, M., Ichisaka, T., Tomoda, K., and Yamanaka, S., Induction of pluripotent stem cells from adult human fibroblasts by defined factors, *Cell* 131 (5), 861–72, 2007.
2. Yu, J., Vodyanik, M. A., Smuga-Otto, K., Antosiewicz-Bourget, J., Frane, J. L., Tian, S., Nie, J., Jonsdottir, G. A., Ruotti, V., Stewart, R., Slukvin, II, and Thomson, J. A., Induced pluripotent stem cell lines derived from human somatic cells, *Science* 318 (5858), 1917–20, 2007.
3. Amit, M., Carpenter, M. K., Inokuma, M. S., Chiu, C. P., Harris, C. P., Waknitz, M. A., Itskovitz-Eldor, J., and Thomson, J. A., Clonally derived human embryonic stem cell lines maintain pluripotency and proliferative potential for prolonged periods of culture, *Dev Biol* 227 (2), 271–78, 2000.
4. Keller, G., Embryonic stem cell differentiation: emergence of a new era in biology and medicine, *Genes Dev* 19 (10), 1129–55, 2005.
5. Thomson, J. A., Itskovitz-Eldor, J., Shapiro, S. S., Waknitz, M. A., Swiergiel, J. J., Marshall, V. S., and Jones, J. M., Embryonic stem cell lines derived from human blastocysts, *Science* 282 (5391), 1145–47, 1998.
6. Reubinoff, B. E., Pera, M. F., Fong, C. Y., Trounson, A., and Bongso, A., Embryonic stem cell lines from human blastocysts: Somatic differentiation *in vitro*, *Nat Biotechnol* 18 (4), 399–404, 2000.
7. Ellerstrom, C., Strehl, R., Moya, K., Andersson, K., Bergh, C., Lundin, K., Hyllner, J., and Semb, H., Derivation of a xeno-free human embryonic stem cell line, *Stem Cells* 24 (10), 2170–76, 2006.
8. Strom, S., Inzunza, J., Grinnemo, K. H., Holmberg, K., Matilainen, E., Stromberg, A. M., Blennow, E., and Hovatta, O., Mechanical isolation of the inner cell mass is effective in derivation of new human embryonic stem cell lines, *Hum Reprod* 22 (12), 3051–58, 2007.
9. Sathananthan, H., Pera, M., and Trounson, A., The fine structure of human embryonic stem cells, *Reprod Biomed Online* 4 (1), 56–61, 2002.
10. Strelchenko, N., Verlinsky, O., Kukharenko, V., and Verlinsky, Y., Morula-derived human embryonic stem cells, *Reprod Biomed Online* 9 (6), 623–29, 2004.
11. Stojkovic, M., Lako, M., Stojkovic, P., Stewart, R., Przyborski, S., Armstrong, L., Evans, J., Herbert, M., Hyslop, L., Ahmad, S., Murdoch, A., and Strachan, T., Derivation of human embryonic stem cells from day-8 blastocysts recovered after three-step in vitro culture, *Stem Cells* 22 (5), 790–97, 2004.

12. Williams, R. L., Hilton, D. J., Pease, S., Willson, T. A., Stewart, C. L., Gearing, D. P., Wagner, E. F., Metcalf, D., Nicola, N. A., and Gough, N. M., Myeloid leukaemia inhibitory factor maintains the developmental potential of embryonic stem cells, *Nature* 336 (6200), 684–87, 1988.
13. Daheron, L., Opitz, S. L., Zaehres, H., Lensch, W. M., Andrews, P. W., Itskovitz-Eldor, J., and Daley, G. Q., LIF/STAT3 signaling fails to maintain self-renewal of human embryonic stem cells, *Stem Cells* 22 (5), 770–78, 2004.
14. Xu, R. H., Peck, R. M., Li, D. S., Feng, X., Ludwig, T., and Thomson, J. A., Basic FGF and suppression of BMP signaling sustain undifferentiated proliferation of human ES cells, *Nat Methods* 2 (3), 185–90, 2005.
15. James, D., Levine, A. J., Besser, D., and Hemmati-Brivanlou, A., TGFbeta/activin/nodal signaling is necessary for the maintenance of pluripotency in human embryonic stem cells, *Development* 132 (6), 1273–82, 2005.
16. Xiao, L., Yuan, X., and Sharkis, S. J., Activin A maintains self-renewal and regulates fibroblast growth factor, Wnt, and bone morphogenic protein pathways in human embryonic stem cells, *Stem Cells* 24 (6), 1476–86, 2006.
17. Xu, C., Rosler, E., Jiang, J., Lebkowski, J. S., Gold, J. D., O'Sullivan, C., Delavan-Boorsma, K., Mok, M., Bronstein, A., and Carpenter, M. K., Basic fibroblast growth factor supports undifferentiated human embryonic stem cell growth without conditioned medium, *Stem Cells* 23 (3), 315–23, 2005.
18. Beattie, G. M., Lopez, A. D., Bucay, N., Hinton, A., Firpo, M. T., King, C. C., and Hayek, A., Activin A maintains pluripotency of human embryonic stem cells in the absence of feeder layers, *Stem Cells* 23 (4), 489–95, 2005.
19. Vallier, L., Alexander, M., and Pedersen, R. A., Activin/Nodal and FGF pathways cooperate to maintain pluripotency of human embryonic stem cells, *J Cell Sci* 118 (19), 4495–509, 2005.
20. Sato, N., Meijer, L., Skaltsounis, L., Greengard, P., and Brivanlou, A. H., Maintenance of pluripotency in human and mouse embryonic stem cells through activation of Wnt signaling by a pharmacological GSK-3-specific inhibitor, *Nat Med* 10 (1), 55–63, 2004.
21. Cheng, L., Hammond, H., Ye, Z., Zhan, X., and Dravid, G., Human adult marrow cells support prolonged expansion of human embryonic stem cells in culture, *Stem Cells* 21 (2), 131–42, 2003.
22. Hovatta, O., Mikkola, M., Gertow, K., Stromberg, A. M., Inzunza, J., Hreinsson, J., Rozell, B., Blennow, E., Andang, M., and Ahrlund-Richter, L., A culture system using human foreskin fibroblasts as feeder cells allows production of human embryonic stem cells, *Hum Reprod* 18 (7), 1404–9, 2003.
23. Lee, J. B., Lee, J. E., Park, J. H., Kim, S. J., Kim, M. K., Roh, S. I., and Yoon, H. S., Establishment and maintenance of human embryonic stem cell lines on human feeder cells derived from uterine endometrium under serum-free condition, *Biol Reprod* 72 (1), 42–49, 2005.
24. Richards, M., Fong, C. Y., Chan, W. K., Wong, P. C., and Bongso, A., Human feeders support prolonged undifferentiated growth of human inner cell masses and embryonic stem cells, *Nat Biotechnol* 20 (9), 933–36, 2002.
25. Xu, C., Inokuma, M. S., Denham, J., Golds, K., Kundu, P., Gold, J. D., and Carpenter, M. K., Feeder-free growth of undifferentiated human embryonic stem cells, *Nat Biotechnol* 19 (10), 971–74, 2001.

26. Ludwig, T. E., Bergendahl, V., Levenstein, M. E., Yu, J., Probasco, M. D., and Thomson, J. A., Feeder-independent culture of human embryonic stem cells, *Nat Methods* 3 (8), 637–46, 2006.

27. Ludwig, T. and Thomson, J. A., Defined culture media for human embryonic stem cells, in *Embyonic Stem Cells*, Masters, J. R., Palsson, B. O., and Thomson, J. A., eds., Springer, New York, 2007, pp. 1–16.

28. Braam, S. R., Zeinstra, L., Litjens, S., Ward-van Oostwaard, D., van den Brink, S., van Laake, L., Lebrin, F., Kats, P., Hochstenbach, R., Passier, R., Sonnenberg, A., and Mummery, C. L., Recombinant vitronectin is a functionally defined substrate that supports human embryonic stem cell self-renewal via αVβ5 integrin, *Stem Cells* 26 (9), 2257–65, 2008.

29. Costa, M., Sourris, K., Hatzistavrou, T., Elefanty, A. G., and Stanley, E. G., Expansion of human embryonic stem cells in vitro, *Curr Protoc Stem Cell Biol* 5, 1C.1.1–1C.1.7, 2008.

30. Ludwig, T. and Thomson, J. A., Defined, feeder-independent medium for human embryonic stem cell culture, *Curr Protoc Stem Cell Biol* 2, 1C.2.1–1C.2.16, 2007.

31. Mitalipova, M. M., Rao, R. R., Hoyer, D. M., Johnson, J. A., Meisner, L. F., Jones, K. L., Dalton, S., and Stice, S. L., Preserving the genetic integrity of human embryonic stem cells, *Nat Biotechnol* 23 (1), 19–20, 2005.

32. Cowan, C. A., Klimanskaya, I., McMahon, J., Atienza, J., Witmyer, J., Zucker, J. P., Wang, S., Morton, C. C., McMahon, A. P., Powers, D., and Melton, D. A., Derivation of embryonic stem-cell lines from human blastocysts, *N Engl J Med* 350 (13), 1353–56, 2004.

33. Baker, D. E., Harrison, N. J., Maltby, E., Smith, K., Moore, H. D., Shaw, P. J., Heath, P. R., Holden, H., and Andrews, P. W., Adaptation to culture of human embryonic stem cells and oncogenesis in vivo, *Nat Biotechnol* 25 (2), 207–15, 2007.

34. Ungrin, M., O'Connor, M., Eaves, C., and Zandstra, P. W., Phenotypic analysis of human embryonic stem cells, *Curr Protoc Stem Cell Biol* 2, 1B.3.1–1B.3.25, 2007.

35. Boiani, M. and Scholer, H. R., Regulatory networks in embryo-derived pluripotent stem cells, *Nat Rev Mol Cell Biol* 6 (11), 872–84, 2005.

36. Boyer, L. A., Lee, T. I., Cole, M. F., Johnstone, S. E., Levine, S. S., Zucker, J. P., Guenther, M. G., Kumar, R. M., Murray, H. L., Jenner, R. G., Gifford, D. K., Melton, D. A., Jaenisch, R., and Young, R. A., Core transcriptional regulatory circuitry in human embryonic stem cells, *Cell* 122 (6), 947–56, 2005.

37. Henderson, J. K., Draper, J. S., Baillie, H. S., Fishel, S., Thomson, J. A., Moore, H., and Andrews, P. W., Preimplantation human embryos and embryonic stem cells show comparable expression of stage-specific embryonic antigens, *Stem Cells* 20 (4), 329–37, 2002.

38. Tippett, P., Andrews, P. W., Knowles, B. B., Solter, D., and Goodfellow, P. N., Red cell antigens P (globoside) and Luke: identification by monoclonal antibodies defining the murine stage-specific embryonic antigens -3 and -4 (SSEA-3 and SSEA-4), *Vox Sang* 51 (1), 53–56, 1986.

39. Karr, J. F., Nowicki, B. J., Truong, L. D., Hull, R. A., Moulds, J. J., and Hull, S. I., pap-2-encoded fimbriae adhere to the P blood group-related glycosphingolipid stage-specific embryonic antigen 4 in the human kidney, *Infect Immun* 58 (12), 4055–62, 1990.

40. Niwa, H., Miyazaki, J., and Smith, A. G., Quantitative expression of Oct 3/4 defines differentiation, dedifferentiation or self-renewal of ES cells, *Nat Genet* 24 (4), 372–76, 2000.
41. Liedtke, S., Enczmann, J., Waclawczyk, S., Wernet, P., and Kogler, G., Oct4 and its pseudogenes confuse stem cell research, *Cell Stem Cell* 1 (4), 364–66, 2007.
42. Gertow, K., Przyborski, S., Loring, J. F., Auerbach, J. M., Epifano, O., Otonkoski, T., Damjanov, I., and Ahrlund-Richter, L., Isolation of human embryonic stem cell-derived teratomas for the assessment of pluripotency, *Curr Protoc Stem Cell Biol* 3, 1B.4.1–1B.4.29, 2007.
43. Brons, I. G., Smithers, L. E., Trotter, M. W., Rugg-Gunn, P., Sun, B., Chuva de Sousa Lopes, S. M., Howlett, S. K., Clarkson, A., Ahrlund-Richter, L., Pedersen, R. A., and Vallier, L., Derivation of pluripotent epiblast stem cells from mammalian embryos, *Nature* 448 (7150), 191–95, 2007.
44. Tesar, P. J., Chenoweth, J. G., Brook, F. A., Davies, T. J., Evans, E. P., Mack, D. L., Gardner, R. L., and McKay, R. D., New cell lines from mouse epiblast share defining features with human embryonic stem cells, *Nature* 448 (7150), 196–99, 2007.
45. Murry, C. E. and Keller, G., Differentiation of embryonic stem cells to clinically relevant populations: Lessons from embryonic development, *Cell* 132 (4), 661–80, 2008.
46. Hirst, C. E., Ng, E. S., Azzola, L., Voss, A. K., Thomas, T., Stanley, E. G., and Elefanty, A. G., Transcriptional profiling of mouse and human ES cells identifies SLAIN1, a novel stem cell gene, *Dev Biol* 293 (1), 90–103, 2006.
47. Schuldiner, M., Yanuka, O., Itskovitz-Eldor, J., Melton, D. A., and Benvenisty, N., Effects of eight growth factors on the differentiation of cells derived from human embryonic stem cells, *Proc Natl Acad Sci USA* 97 (21), 11307–12, 2000.
48. Smith, J., Wardle, F., Loose, M., Stanley, E., and Patient, R., Germ layer induction in ESC—following the vertebrate roadmap, *Curr Protoc Stem Cell Biol* 1, 1D.1.1–1D.1.22, 2007.
49. D'Amour, K. A., Agulnick, A. D., Eliazer, S., Kelly, O. G., Kroon, E., and Baetge, E. E., Efficient differentiation of human embryonic stem cells to definitive endoderm, *Nat Biotechnol* 23 (12), 1534–41, 2005.
50. McLean, A. B., D'Amour, K. A., Jones, K. L., Krishnamoorthy, M., Kulik, M. J., Reynolds, D. M., Sheppard, A. M., Liu, H., Xu, Y., Baetge, E. E., and Dalton, S., Activin a efficiently specifies definitive endoderm from human embryonic stem cells only when phosphatidylinositol 3-kinase signaling is suppressed, *Stem Cells* 25 (1), 29–38, 2007.
51. Ng, E. S., Davis, R. P., Azzola, L., Stanley, E. G., and Elefanty, A. G., Forced aggregation of defined numbers of human embryonic stem cells into embryoid bodies fosters robust, reproducible hematopoietic differentiation, *Blood* 106 (5), 1601–3, 2005.
52. Pick, M., Azzola, L., Mossman, A., Stanley, E. G., and Elefanty, A. G., Differentiation of human embryonic stem cells in serum free medium reveals distinct roles for BMP4, VEGF, SCF and FGF2 in hematopoiesis, *Stem Cells* 25 (9), 2206–14, 2007.
53. Kennedy, M., D'Souza, S. L., Lynch-Kattman, M., Schwantz, S., and Keller, G., Development of the hemangioblast defines the onset of hematopoiesis in human ES cell differentiation cultures, *Blood* 109 (7), 2679–87, 2007.

54. Davis, R. P., Ng, E. S., Costa, M., Mossman, A. K., Sourris, K., Elefanty, A. G., and Stanley, E. G., Targeting a GFP reporter gene to the MIXL1 locus of human embryonic stem cells identifies human primitive streak-like cells and enables isolation of primitive hematopoietic precursors, *Blood* 111 (4), 1876–84, 2008.

55. Dosch, R., Gawantka, V., Delius, H., Blumenstock, C., and Niehrs, C., Bmp-4 acts as a morphogen in dorsoventral mesoderm patterning in *Xenopus*, *Development* 124 (12), 2325–34, 1997.

56. Laflamme, M. A., Chen, K. Y., Naumova, A. V., Muskheli, V., Fugate, J. A., Dupras, S. K., Reinecke, H., Xu, C., Hassanipour, M., Police, S., O'Sullivan, C., Collins, L., Chen, Y., Minami, E., Gill, E. A., Ueno, S., Yuan, C., Gold, J., and Murry, C. E., Cardiomyocytes derived from human embryonic stem cells in pro-survival factors enhance function of infarcted rat hearts, *Nat Biotechnol* 25 (9), 1015–24, 2007.

57. Yang, L., Soonpaa, M. H., Adler, E. D., Roepke, T. K., Kattman, S. J., Kennedy, M., Henckaerts, E., Bonham, K., Abbott, G. W., Linden, R. M., Field, L. J., and Keller, G. M., Human cardiovascular progenitor cells develop from a KDR+ embryonic-stem-cell-derived population, *Nature* 453 (7194), 524–28, 2008.

58. Itskovitz-Eldor, J., Schuldiner, M., Karsenti, D., Eden, A., Yanuka, O., Amit, M., Soreq, H., and Benvenisty, N., Differentiation of human embryonic stem cells into embryoid bodies compromising the three embryonic germ layers, *Mol Med* 6 (2), 88–95, 2000.

59. Ng, E. S., Davis, R. P., Hatzistavrou, T., Stanley, E. G., and Elefanty, A. G., Directed differentiation of human embryonic stem cells as spin embryoid bodies and a description of the hematopoietic blast colony forming assay, *Curr Protoc Stem Cell Biol* 4, 1D.3.1–1D.3.23, 2008.

60. Burridge, P. W., Anderson, D., Priddle, H., Barbadillo Munoz, M. D., Chamberlain, S., Allegrucci, C., Young, L. E., and Denning, C., Improved human embryonic stem cell embryoid body homogeneity and cardiomyocyte differentiation from a novel V-96 plate aggregation system highlights interline variability, *Stem Cells* 25 (4), 929–38, 2007.

61. Ungrin, M. D., Joshi, C., Nica, A., Bauwens, C., and Zandstra, P. W., Reproducible, ultra high-throughput formation of multicellular organization from single cell suspension-derived human embryonic stem cell aggregates, *PLoS One* 3 (2), e1565, 2008.

62. Kaufman, D. S., Hanson, E. T., Lewis, R. L., Auerbach, R., and Thomson, J. A., Hematopoietic colony-forming cells derived from human embryonic stem cells, *Proc Natl Acad Sci USA* 98 (19), 10716–21, 2001.

63. Mummery, C., Ward-van Oostwaard, D., Doevendans, P., Spijker, R., van den Brink, S., Hassink, R., van der Heyden, M., Opthof, T., Pera, M., de la Riviere, A. B., Passier, R., and Tertoolen, L., Differentiation of human embryonic stem cells to cardiomyocytes: role of coculture with visceral endoderm-like cells, *Circulation* 107 (21), 2733–40, 2003.

64. Levenberg, S., Golub, J. S., Amit, M., Itskovitz-Eldor, J., and Langer, R., Endothelial cells derived from human embryonic stem cells, *Proc Natl Acad Sci USA* 99 (7), 4391–96, 2002.

65. Vodyanik, M. A., Bork, J. A., Thomson, J. A., and Slukvin, II, Human embryonic stem cell-derived CD34+ cells: efficient production in the coculture with OP9 stromal cells and analysis of lymphohematopoietic potential, *Blood* 105 (2), 617–26, 2005.

66. Park, C. H., Minn, Y. K., Lee, J. Y., Choi, D. H., Chang, M. Y., Shim, J. W., Ko, J. Y., Koh, H. C., Kang, M. J., Kang, J. S., Rhie, D. J., Lee, Y. S., Son, H., Moon, S. Y., Kim, K. S., and Lee, S. H., In vitro and in vivo analyses of human embryonic stem cell-derived dopamine neurons, *J Neurochem* 92 (5), 1265–76, 2005.

67. Roy, N. S., Cleren, C., Singh, S. K., Yang, L., Beal, M. F., and Goldman, S. A., Functional engraftment of human ES cell-derived dopaminergic neurons enriched by coculture with telomerase-immortalized midbrain astrocytes, *Nat Med* 12 (11), 1259–68, 2006.

68. Rambhatla, L., Chiu, C. P., Kundu, P., Peng, Y., and Carpenter, M. K., Generation of hepatocyte-like cells from human embryonic stem cells, *Cell Transplant* 12 (1), 1–11, 2003.

69. D'Amour, K. A., Bang, A. G., Eliazer, S., Kelly, O. G., Agulnick, A. D., Smart, N. G., Moorman, M. A., Kroon, E., Carpenter, M. K., and Baetge, E. E., Production of pancreatic hormone-expressing endocrine cells from human embryonic stem cells, *Nat Biotechnol* 24 (11), 1392–401, 2006.

70. Baharvand, H., Hashemi, S. M., and Shahsavani, M., Differentiation of human embryonic stem cells into functional hepatocyte-like cells in a serum-free adherent culture condition, *Differentiation* 76 (5), 465–77, 2008.

71. Van Hoof, D., Munoz, J., Braam, S. R., Pinkse, M. W., Linding, R., Heck, A. J., Mummery, C. L., and Krijgsveld, J., Phosphorylation dynamics during early differentiation of human embryonic stem cells, *Cell Stem Cell* 5 (2), 214–26, 2009.

72. Wiles, M. V. and Johansson, B. M., Embryonic stem cell development in a chemically defined medium, *Exp Cell Res* 247 (1), 241–48, 1999.

73. Ng, E. S., Davis, R. P., Stanley, E. G., and Elefanty, A. G., A protocol describing the use of a recombinant protein-based, animal product-free medium (APEL) for human embryonic stem cell differentiation as spin embryoid bodies, *Nat Protoc* 3 (5), 768–76, 2008.

74. Yao, S., Chen, S., Clark, J., Hao, E., Beattie, G. M., Hayek, A., and Ding, S., Long-term self-renewal and directed differentiation of human embryonic stem cells in chemically defined conditions, *Proc Natl Acad Sci USA* 103 (18), 6907–12, 2006.

75. Freund, C., Ward-van Oostwaard, D., Monshouwer-Kloots, J., van den Brink, S., van Rooijen, M., Xu, X., Zweigerdt, R., Mummery, C., and Passier, R., Insulin redirects differentiation from cardiogenic mesoderm and endoderm to neuroectoderm in differentiating human embryonic stem cells, *Stem Cells* 26 (3), 724–33, 2008.

76. Anderson, D., Self, T., Mellor, I. R., Goh, G., Hill, S. J., and Denning, C., Transgenic enrichment of cardiomyocytes from human embryonic stem cells, *Mol Ther* 15 (11), 2027–36, 2007.

77. Xu, X. Q., Zweigerdt, R., Soo, S. Y., Ngoh, Z. X., Tham, S. C., Wang, S. T., Graichen, R., Davidson, B., Colman, A., and Sun, W., Highly enriched cardiomyocytes from human embryonic stem cells, *Cytotherapy* 10 (4), 376–89, 2008.

78. Bu, L., Jiang, X., Martin-Puig, S., Caron, L., Zhu, S., Shao, Y., Roberts, D. J., Huang, P. L., Domian, I. J., and Chien, K. R., Human ISL1 heart progenitors generate diverse multipotent cardiovascular cell lineages, *Nature* 460 (7251), 113–17, 2009.

79. Xue, H., Wu, S., Papadeas, S. T., Spusta, S., Swistowska, A. M., Macarthur, C. C., Mattson, M. P., Maragakis, N. J., Capecchi, M. R., Rao, M. S., Zeng, X., and Liu, Y., A Targeted neuroglial reporter line generated by homologous recombination in human embryonic stem cells, *Stem Cells* 27 (8), 1836–46, 2009.

80. Costa, M., Dottori, M., Ng, E., Hawes, S. M., Sourris, K., Jamshidi, P., Pera, M. F., Elefanty, A. G., and Stanley, E. G., The hESC line Envy expresses high levels of GFP in all differentiated progeny, *Nat Methods* 2 (4), 259–60, 2005.

81. Irion, S., Luche, H., Gadue, P., Fehling, H. J., Kennedy, M., and Keller, G., Identification and targeting of the ROSA26 locus in human embryonic stem cells, *Nat Biotechnol* 25 (12), 1477–82, 2007.

82. Du, Z. W., Hu, B. Y., Ayala, M., Sauer, B., and Zhang, S. C., Cre recombination-mediated cassette exchange for building versatile transgenic human embryonic stem cells lines, *Stem Cells* 27 (5), 1032–41, 2009.

83. Hoffman, R., Imaging in mice with fluorescent proteins: From macro to subcellular, *Sensors* 8, 1157–73, 2008.

84. Chadwick, K., Wang, L., Li, L., Menendez, P., Murdoch, B., Rouleau, A., and Bhatia, M., Cytokines and BMP-4 promote hematopoietic differentiation of human embryonic stem cells, *Blood* 102 (3), 906–15, 2003.

85. Tian, X., Morris, J. K., Linehan, J. L., and Kaufman, D. S., Cytokine requirements differ for stroma and embryoid body-mediated hematopoiesis from human embryonic stem cells, *Exp Hematol* 32 (10), 1000–9, 2004.

86. Cerdan, C., Rouleau, A., and Bhatia, M., VEGF-A165 augments erythropoietic development from human embryonic stem cells, *Blood* 103 (7), 2504–12, 2004.

87. Woll, P. S., Morris, J. K., Painschab, M. S., Marcus, R. K., Kohn, A. D., Biechele, T. L., Moon, R. T., and Kaufman, D. S., Wnt signaling promotes hematoendothelial cell development from human embryonic stem cells, *Blood* 111 (1), 122–31, 2008.

88. Vijayaragavan, K., Szabo, E., Bosse, M., Ramos-Mejia, V., Moon, R. T., and Bhatia, M., Noncanonical Wnt signaling orchestrates early developmental events toward hematopoietic cell fate from human embryonic stem cells, *Cell Stem Cell* 4 (3), 248–62, 2009.

89. Zambidis, E. T., Peault, B., Park, T. S., Bunz, F., and Civin, C. I., Hematopoietic differentiation of human embryonic stem cells progresses through sequential hematoendothelial, primitive, and definitive stages resembling human yolk sac development, *Blood* 106 (3), 860–70, 2005.

90. Lu, S. J., Feng, Q., Caballero, S., Chen, Y., Moore, M. A., Grant, M. B., and Lanza, R., Generation of functional hemangioblasts from human embryonic stem cells, *Nat Methods* 4 (6), 501–9, 2007.

91. Qiu, C., Olivier, E. N., Velho, M., and Bouhassira, E. E., Globin switches in yolk sac-like primitive and fetal-like definitive red blood cells produced from human embryonic stem cells, *Blood* 111 (4), 2400–8, 2008.

92. Woll, P. S., Martin, C. H., Miller, J. S., and Kaufman, D. S., Human embryonic stem cell-derived NK cells acquire functional receptors and cytolytic activity, *J Immunol* 175 (8), 5095–103, 2005.

93. Galic, Z., Kitchen, S. G., Kacena, A., Subramanian, A., Burke, B., Cortado, R., and Zack, J. A., T lineage differentiation from human embryonic stem cells, *Proc Natl Acad Sci USA* 103 (31), 11742–47, 2006.

94. Wang, L., Menendez, P., Shojaei, F., Li, L., Mazurier, F., Dick, J. E., Cerdan, C., Levac, K., and Bhatia, M., Generation of hematopoietic repopulating cells from human embryonic stem cells independent of ectopic HOXB4 expression, *J Exp Med* 201 (10), 1603–14, 2005.

95. Tian, X., Woll, P. S., Morris, J. K., Linehan, J. L., and Kaufman, D. S., Hematopoietic engraftment of human embryonic stem cell-derived cells is regulated by recipient innate immunity, *Stem Cells* 24 (5), 1370–80, 2006.

96. Narayan, A. D., Chase, J. L., Lewis, R. L., Tian, X., Kaufman, D. S., Thomson, J. A., and Zanjani, E. D., Human embryonic stem cell-derived hematopoietic cells are capable of engrafting primary as well as secondary fetal sheep recipients, *Blood* 107 (5), 2180–83, 2006.

97. Kehat, I., Kenyagin-Karsenti, D., Snir, M., Segev, H., Amit, M., Gepstein, A., Livne, E., Binah, O., Itskovitz-Eldor, J., and Gepstein, L., Human embryonic stem cells can differentiate into myocytes with structural and functional properties of cardiomyocytes, *J Clin Invest* 108 (3), 407–14, 2001.

98. Laflamme, M. A., Gold, J., Xu, C., Hassanipour, M., Rosler, E., Police, S., Muskheli, V., and Murry, C. E., Formation of human myocardium in the rat heart from human embryonic stem cells, *Am J Pathol* 167 (3), 663–71, 2005.

99. Nascone, N. and Mercola, M., An inductive role for the endoderm in *Xenopus* cardiogenesis, *Development* 121 (2), 515–23, 1995.

100. Passier, R., Oostwaard, D. W., Snapper, J., Kloots, J., Hassink, R. J., Kuijk, E., Roelen, B., de la Riviere, A. B., and Mummery, C., Increased cardiomyocyte differentiation from human embryonic stem cells in serum-free cultures, *Stem Cells* 23 (6), 772–80, 2005.

101. Kang, Y., Nagy, J. M., Polak, J. M., and Mantalaris, A., Proteomic characterisation of the conditioned media produced by the visceral endoderm-like cell lines hepG2 and end2: towards a defined medium for the osteogenic/chondrogenic differentiation of embryonic stem cells, *Stem Cells Dev* 18 (1), 77–91, 2009.

102. Graichen, R., Xu, X., Braam, S. R., Balakrishnan, T., Norfiza, S., Sieh, S., Soo, S. Y., Tham, S. C., Mummery, C., Colman, A., Zweigerdt, R., and Davidson, B. P., Enhanced cardiomyogenesis of human embryonic stem cells by a small molecular inhibitor of p38 MAPK, *Differentiation* 76 (4), 357–70, 2008.

103. Xu, X. Q., Graichen, R., Soo, S. Y., Balakrishnan, T., Rahmat, S. N., Sieh, S., Tham, S. C., Freund, C., Moore, J., Mummery, C., Colman, A., Zweigerdt, R., and Davidson, B. P., Chemically defined medium supporting cardiomyocyte differentiation of human embryonic stem cells, *Differentiation* 76 (9), 958–70, 2008.

104. Kattman, S. J., Huber, T. L., and Keller, G. M., Multipotent flk-1+ cardiovascular progenitor cells give rise to the cardiomyocyte, endothelial, and vascular smooth muscle lineages, *Dev Cell* 11 (5), 723–32, 2006.

105. Xu, C., Police, S., Rao, N., and Carpenter, M. K., Characterization and enrichment of cardiomyocytes derived from human embryonic stem cells, *Circ Res* 91 (6), 501–8, 2002.

106. He, J. Q., Ma, Y., Lee, Y., Thomson, J. A., and Kamp, T. J., Human embryonic stem cells develop into multiple types of cardiac myocytes: action potential characterization, *Circ Res* 93 (1), 32–39, 2003.

107. Satin, J., Kehat, I., Caspi, O., Huber, I., Arbel, G., Itzhaki, I., Magyar, J., Schroder, E. A., Perlman, I., and Gepstein, L., Mechanism of spontaneous excitability in human embryonic stem cell derived cardiomyocytes, *J Physiol* 559 (2), 479–96, 2004.

108. Denning, C. and Anderson, D., Cardiomyocytes from human embryonic stem cells as predictors of cardiotoxicity, *Drug Discov Today: Ther Strategies*, 5, 223–32, 2008.

109. Dai, W., Field, L. J., Rubart, M., Reuter, S., Hale, S. L., Zweigerdt, R., Graichen, R. E., Kay, G. L., Jyrala, A. J., Colman, A., Davidson, B. P., Pera, M., and Kloner, R. A., Survival and maturation of human embryonic stem cell-derived cardiomyocytes in rat hearts, *J Mol Cell Cardiol* 43 (4), 504–16, 2007.

110. van Laake, L. W., Passier, R., Monshouwer-Kloots, J., Verkleij, A. J., Lips, D. J., Freund, C., den Ouden, K., Ward-van Oostwaard, D., Korving, J., Tertoolen, L. G., van Echteld, C. J., Doevendans, P. A., and Mummery, C. L., Human embryonic stem cell-derived cardiomyocytes survive and mature in the mouse heart and transiently improve function after myocardial infarction, *Stem Cell Res* 1 (1), 9–24, 2007.

111. Kehat, I., Khimovich, L., Caspi, O., Gepstein, A., Shofti, R., Arbel, G., Huber, I., Satin, J., Itskovitz-Eldor, J., and Gepstein, L., Electromechanical integration of cardiomyocytes derived from human embryonic stem cells, *Nat Biotechnol* 22 (10), 1282–89, 2004.

112. Xue, T., Cho, H. C., Akar, F. G., Tsang, S. Y., Jones, S. P., Marban, E., Tomaselli, G. F., and Li, R. A., Functional integration of electrically active cardiac derivatives from genetically engineered human embryonic stem cells with quiescent recipient ventricular cardiomyocytes: Insights into the development of cell-based pacemakers, *Circulation* 111 (1), 11–20, 2005.

113. Caspi, O., Huber, I., Kehat, I., Habib, M., Arbel, G., Gepstein, A., Yankelson, L., Aronson, D., Beyar, R., and Gepstein, L., Transplantation of human embryonic stem cell-derived cardiomyocytes improves myocardial performance in infarcted rat hearts, *J Am Coll Cardiol* 50 (19), 1884–93, 2007.

114. Leor, J., Gerecht, S., Cohen, S., Miller, L., Holbova, R., Ziskind, A., Shachar, M., Feinberg, M. S., Guetta, E., and Itskovitz-Eldor, J., Human embryonic stem cell transplantation to repair the infarcted myocardium, *Heart* 93 (10), 1278–84, 2007.

115. van Laake, L. W., Passier, R., Doevendans, P. A., and Mummery, C. L., Human embryonic stem cell-derived cardiomyocytes and cardiac repair in rodents, *Circ Res* 102 (9), 1008–10, 2008.

116. van Laake, L. W., Passier, R., den Ouden, K., Schreurs, C., Monshouwer-Kloots, J., Ward-van Oostwaard, D., van Echteld, C. J., Doevendans, P. A., and Mummery, C. L., Improvement of mouse cardiac function by hESC-derived cardiomyocytes correlates with vascularity but not graft size, *Stem Cell Res*, 3 (2–3), 106–12, 2009.

117. Pera, M. F., Andrade, J., Houssami, S., Reubinoff, B., Trounson, A., Stanley, E. G., Ward-van Oostwaard, D., and Mummery, C., Regulation of human embryonic stem cell differentiation by BMP-2 and its antagonist noggin, *J Cell Sci* 117 (Pt 7), 1269–80, 2004.

118. Jiang, J., Au, M., Lu, K., Eshpeter, A., Korbutt, G., Fisk, G., and Majumdar, A. S., Generation of insulin-producing islet-like clusters from human embryonic stem cells, *Stem Cells* 25 (8), 1940–53, 2007.

119. Shim, J. H., Kim, S. E., Woo, D. H., Kim, S. K., Oh, C. H., McKay, R., and Kim, J. H., Directed differentiation of human embryonic stem cells towards a pancreatic cell fate, *Diabetologia* 50 (6), 1228–38, 2007.

120. Micallef, S. J., Janes, M. E., Knezevic, K., Davis, R. P., Elefanty, A. G., and Stanley, E. G., Retinoic acid induces Pdx1-positive endoderm in differentiating mouse embryonic stem cells, *Diabetes* 54 (2), 301–5, 2005.

121. Jiang, W., Shi, Y., Zhao, D., Chen, S., Yong, J., Zhang, J., Qing, T., Sun, X., Zhang, P., Ding, M., Li, D., and Deng, H., In vitro derivation of functional insulin-producing cells from human embryonic stem cells, *Cell Res* 17 (4), 333–44, 2007.

122. Otonkoski, T., Beattie, G. M., Mally, M. I., Ricordi, C., and Hayek, A., Nicotinamide is a potent inducer of endocrine differentiation in cultured human fetal pancreatic cells, *J Clin Invest* 92 (3), 1459–66, 1993.

123. Xu, G., Stoffers, D. A., Habener, J. F., and Bonner-Weir, S., Exendin-4 stimulates both beta-cell replication and neogenesis, resulting in increased beta-cell mass and improved glucose tolerance in diabetic rats, *Diabetes* 48 (12), 2270–76, 1999.
124. Zhang, D., Jiang, W., Liu, M., Sui, X., Yin, X., Chen, S., Shi, Y., and Deng, H., Highly efficient differentiation of human ES cells and iPS cells into mature pancreatic insulin-producing cells, *Cell Res* 19 (4), 429–38, 2009.
125. Chen, S., Borowiak, M., Fox, J. L., Maehr, R., Osafune, K., Davidow, L., Lam, K., Peng, L. F., Schreiber, S. L., Rubin, L. L., and Melton, D., A small molecule that directs differentiation of human ESCs into the pancreatic lineage, *Nat Chem Biol* 5 (4), 258–65, 2009.
126. Shapiro, A. M., Lakey, J. R., Ryan, E. A., Korbutt, G. S., Toth, E., Warnock, G. L., Kneteman, N. M., and Rajotte, R. V., Islet transplantation in seven patients with type 1 diabetes mellitus using a glucocorticoid-free immunosuppressive regimen, *N Engl J Med* 343 (4), 230–38, 2000.
127. Kroon, E., Martinson, L. A., Kadoya, K., Bang, A. G., Kelly, O. G., Eliazer, S., Young, H., Richardson, M., Smart, N. G., Cunningham, J., Agulnick, A. D., D'Amour, K. A., Carpenter, M. K., and Baetge, E. E., Pancreatic endoderm derived from human embryonic stem cells generates glucose-responsive insulin-secreting cells in vivo, *Nat Biotechnol* 26 (4), 443–52, 2008.
128. Reubinoff, B. E., Itsykson, P., Turetsky, T., Pera, M. F., Reinhartz, E., Itzik, A., and Ben-Hur, T., Neural progenitors from human embryonic stem cells, *Nat Biotechnol* 19 (12), 1134–40, 2001.
129. Zhang, S. C., Wernig, M., Duncan, I. D., Brustle, O., and Thomson, J. A., In vitro differentiation of transplantable neural precursors from human embryonic stem cells, *Nat Biotechnol* 19 (12), 1129–33, 2001.
130. Ben-Hur, T., Idelson, M., Khaner, H., Pera, M., Reinhartz, E., Itzik, A., and Reubinoff, B. E., Transplantation of human embryonic stem cell-derived neural progenitors improves behavioral deficit in Parkinsonian rats, *Stem Cells* 22 (7), 1246–55, 2004.
131. Gerrard, L., Rodgers, L., and Cui, W., Differentiation of human embryonic stem cells to neural lineages in adherent culture by blocking bone morphogenetic protein signaling, *Stem Cells* 23 (9), 1234–41, 2005.
132. Itsykson, P., Ilouz, N., Turetsky, T., Goldstein, R. S., Pera, M. F., Fishbein, I., Segal, M., and Reubinoff, B. E., Derivation of neural precursors from human embryonic stem cells in the presence of noggin, *Mol Cell Neurosci* 30 (1), 24–36, 2005.
133. Joannides, A. J., Fiore-Heriche, C., Battersby, A. A., Athauda-Arachchi, P., Bouhon, I. A., Williams, L., Westmore, K., Kemp, P. J., Compston, A., Allen, N. D., and Chandran, S., A scaleable and defined system for generating neural stem cells from human embryonic stem cells, *Stem Cells* 25 (3), 731–37, 2007.
134. Shin, S., Sun, Y., Liu, Y., Khaner, H., Svant, S., Cai, J., Xu, Q. X., Davidson, B. P., Stice, S. L., Smith, A. K., Goldman, S. A., Reubinoff, B. E., Zhan, M., Rao, M. S., and Chesnut, J. D., Whole genome analysis of human neural stem cells derived from embryonic stem cells and stem and progenitor cells isolated from fetal tissue, *Stem Cells* 25 (5), 1298–306, 2007.
135. Zhang, S. C., Li, X. J., Johnson, M. A., and Pankratz, M. T., Human embryonic stem cells for brain repair?, *Philos Trans R Soc Lond B Biol Sci* 363 (1489), 87–99, 2008.

136. Lowell, S., Benchoua, A., Heavey, B., and Smith, A. G., Notch promotes neural lineage entry by pluripotent embryonic stem cells, *PLoS Biol* 4 (5), e121, 2006.
137. Rao, B. M. and Zandstra, P. W., Culture development for human embryonic stem cell propagation: molecular aspects and challenges, *Curr Opin Biotechnol* 16 (5), 568–76, 2005.
138. Yan, Y., Yang, D., Zarnowska, E. D., Du, Z., Werbel, B., Valliere, C., Pearce, R. A., Thomson, J. A., and Zhang, S. C., Directed differentiation of dopaminergic neuronal subtypes from human embryonic stem cells, *Stem Cells* 23 (6), 781–90, 2005.
139. Li, X. J., Du, Z. W., Zarnowska, E. D., Pankratz, M., Hansen, L. O., Pearce, R. A., and Zhang, S. C., Specification of motoneurons from human embryonic stem cells, *Nat Biotechnol* 23 (2), 215–21, 2005.
140. Carpenter, M. K., Inokuma, M. S., Denham, J., Mujtaba, T., Chiu, C. P., and Rao, M. S., Enrichment of neurons and neural precursors from human embryonic stem cells, *Exp Neurol* 172 (2), 383–97, 2001.
141. Zeng, X., Cai, J., Chen, J., Luo, Y., You, Z. B., Fotter, E., Wang, Y., Harvey, B., Miura, T., Backman, C., Chen, G. J., Rao, M. S., and Freed, W. J., Dopaminergic differentiation of human embryonic stem cells, *Stem Cells* 22 (6), 925–40, 2004.
142. Tabar, V., Panagiotakos, G., Greenberg, E. D., Chan, B. K., Sadelain, M., Gutin, P. H., and Studer, L., Migration and differentiation of neural precursors derived from human embryonic stem cells in the rat brain, *Nat Biotechnol* 23 (5), 601–6, 2005.
143. Nistor, G. I., Totoiu, M. O., Haque, N., Carpenter, M. K., and Keirstead, H. S., Human embryonic stem cells differentiate into oligodendrocytes in high purity and myelinate after spinal cord transplantation, *Glia* 49 (3), 385–96, 2005.
144. Keirstead, H. S., Nistor, G., Bernal, G., Totoiu, M., Cloutier, F., Sharp, K., and Steward, O., Human embryonic stem cell-derived oligodendrocyte progenitor cell transplants remyelinate and restore locomotion after spinal cord injury, *J Neurosci* 25 (19), 4694–705, 2005.
145. Alper, J., Geron gets green light for human trial of ES cell-derived product, *Nat Biotechnol* 27 (3), 213–4, 2009.
146. Perrier, A. L., Tabar, V., Barberi, T., Rubio, M. E., Bruses, J., Topf, N., Harrison, N. L., and Studer, L., Derivation of midbrain dopamine neurons from human embryonic stem cells, *Proc Natl Acad Sci USA* 101 (34), 12543–48, 2004.
147. Schulz, T. C., Noggle, S. A., Palmarini, G. M., Weiler, D. A., Lyons, I. G., Pensa, K. A., Meedeniya, A. C., Davidson, B. P., Lambert, N. A., and Condie, B. G., Differentiation of human embryonic stem cells to dopaminergic neurons in serum-free suspension culture, *Stem Cells* 22 (7), 1218–38, 2004.

# 2

# Mesenchymal Stem Cells from Bone Marrow

Mark F. Pittenger

## CONTENTS

## 2.1 Introduction

Bone marrow contains a number of stem cells and progenitor cells, and because bone marrow is a tissue that is thought to regenerate itself, investigators have seen it as a renewable source of stem and progenitor cells. Bone marrow serves as a reliable source for acquiring hematopoietic stem cells (HSCs) and mesenchymal stem cells (MSCs) for research or therapeutic applications. The MSCs can be isolated from other tissues, but bone marrow is the most reliable. MSCs have been shown to differentiate with lineage fidelity in a robust manner to form several different cell types in defined conditions *in vitro*. They appear to form additional cell types *in vivo*. Beyond providing undifferentiated cells for tissue regeneration, the MSCs interact with the repairing tissue to enhance repair and regeneration through the release of growth factors and cytokines. The MSCs have interesting immune-modulating capabilities, which allow their use in allogeneic recipients, although questions arise regarding whether the autologous MSCs may provide a longer-lived therapeutic cell population. Current clinical trials are taking advantage of the remarkable abilities of the MSCs, and at least 20 trials from phase 1 to phase 3 are under way in the United States, with another 85 ongoing in 2009 worldwide (http://www.clinicaltrials.gov/ct2/resu lts?term=mesenchymal+stem+cells). For research purposes, MSCs have been labeled by virtually every technique available to further understand this important progenitor cell population. This chapter discusses the origins of MSC research, MSC characteristics, their immunological properties, and their ability to migrate to a site of injury, such as cardiac infarcts.

## 2.2 Background

Mesenchymal stem cells, also known as multipotential stromal cells, are progen-itor cells for mesenchymal lineages and may be isolated from several sources. MSCs have been isolated from many species, from mouse through humans, including rat, guinea pig, rabbit, dog, goat, pig, horse, macaque, and baboon; all have similar properties. Most of the information in this chapter applies to MSCs from rat through humans. Unfortunately, the mouse and all its attrac-tive mutant strains have proven to be the most difficult species to work with as mouse MSC proliferation is always accompanied by hematopoietic cells, and final purification by immunoselection is needed. The human MSCs have been routinely isolated from bone marrow as it is thought this tissue is a renew-able source that can be harvested with limited tissue trauma (unlike bone, skin, muscle, etc.). MSCs are rarer than HSCs in bone marrow; for comparison, HSCs constitute about 0.1% of the nucleated cells in bone marrow, while MSCs are present at 0.01–0.001%. However, MSCs expand readily in proper culture condi-tions, while HSCs are more difficult to expand by *in vitro* culture.

MSCs or similar cells closely related to MSCs appear to occur in all tissues, and there is evidence that microvascular pericytes (mural cells) have proper-ties similar to MSCs. This means that while bone marrow is a good reservoir of MSCs, there are similar cell populations along blood vessels throughout the body that may perform similar roles described for MSCs in responding to tissue inflammation, trauma, and subsequent tissue regeneration. Different studies have used several similar names, including mesenchymal progenitor cells, multipotential stromal cells, and others, as well as MSCs to describe the cells under study. With some limitations on our understanding of its mean-ing and limitations, the name *MSC* continues to be used until a better, more refined, and more complete MSC candidate is isolated or created by improv-ing the culture methods or altering gene expression.

There is an ongoing debate whether MSCs should be considered true stem cells as they cannot be cultured indefinitely, and they cannot be shown to dif-ferentiate to *all* mesenchymal lineages. However, the MSCs can be expanded a millionfold or more *in vitro* and retain most characteristics and abilities that they have at the early passages. The cultured MSCs differentiate readily to several mesenchymal lineages, but they have not been shown to proceed down all mesenchymal lineages (Figure 2.1). For example, the *in vitro* conditions that promote MSC differentiation to the osteogenic, chondrogenic, and adipogenic lineages work well, but we do not understand how to encourage MSCs similarly to differentiate to a muscle lineage *in vitro*, and limited evidence of MSCs becoming muscle cells *in vivo* has been reported (Table 2.1). Similarly, hematopoietic cells are *mesenchymal* in origin; however, the elusive hemangioblast, capable of differentiating to hematopoietic and all other mesenchymal lineages, has not been isolated (although some reports for a mouse

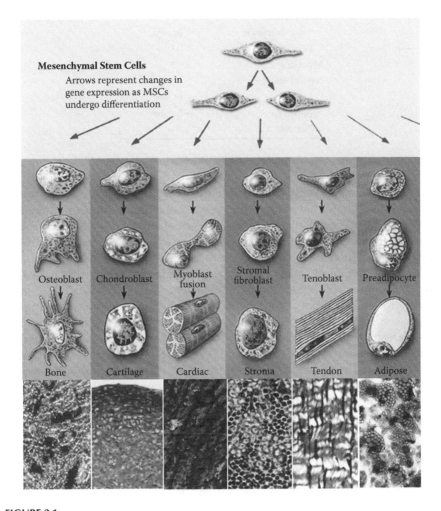

**FIGURE 2.1**
Mesenchymal stem cell differentiation to multiple lineages. Nucleated cells from a human bone marrow aspirate are cultured under appropriate conditions and produce a homogeneous population of cells as analyzed by flow cytometry. The cells can be cultured in different conditions to cause differentiation to particular lineages, such as (left to right) osteoblasts, chondrocytes, cardiac muscle (*in vivo*), hematopoietic supporting stroma, tenocytes, and adipocytes.

**TABLE 2.1**

Differentiation to Lineages

| Cell Lineage | Ease of Differentiation |
|---|---|
| Osteoblast | +++ |
| Chondrocyte | +++ |
| Adipocyte | +++ |
| Stromal | +++ |
| Tenocytes | ++ *In vivo* |
| Endothelial | |
|    Smooth muscle | + *In vivo* |
|    Pericytes | ? |
| Myoblast | ? |
| Neural lineages | |
|    Glial | + *In vivo* |
|    Astrocyte | + *In vivo* |
|    Neuron | — |

hemangioblast appear in the literature). So, here again the MSCs do not satisfy this criterion as an MSC. With an appreciation of such limitations, MSCs still provide a remarkable research tool for understanding stem cells and a powerful therapeutic agent for tissue regeneration and immune modulation.

An important concept for studying stem cells of any kind is whether we analyze the qualities of stem cells with *in vitro* assays or *in vivo* assays. What is needed is to demonstrate the properties of stem cells with high fidelity, in controlled, reproducible conditions, while recognizing that proper differentiation and regeneration will need to occur in the context of healing tissue. *In vitro* studies remove the complexity of the *in vivo* setting but have their limits to impart understanding about the roles of stem cells *in vivo*. While the *in vivo* studies have much greater experimental complexity, this limits our ability to extract understanding about the input of stem cells. Therefore, by combining results from *in vitro* and *in vivo* experiments, the most insight into understanding MSCs can be gained. In this regard, the labeling and imaging techniques in this book provide the most comprehensive view of assembled data available for MSCs.

## 2.3 Early Studies on Bone-Forming Cells

Early studies by Marshall Urist showed that when pieces of bone were implanted, surrounding cells formed new bone; this work eventually led to the identification of bone morphogenic proteins (BMPs) (Urist 2009; Urist and McLean 1952). Following this lead, experiments to identify the bone-forming cells present in bone marrow were conducted by Alexander Friedenstein starting in the 1960s.

These studies showed that it was possible to culture fibroblastic cells from guinea pig bone marrow, and that when the cells were placed in chambers and reimplanted, many of the chambers showed histological evidence of cartilage and bone (Friedenstein 1976; Friedenstein et al. 1968; Luria et al. 1987). Several additional studies from Maureen Owen and colleagues extended this work to show both osteogenic and adipogenic differentiation of rat bone marrow cells (Ashton et al. 1980; Beresford et al. 1992; Owen and Friedenstein 1988; Owen et al. 1987). Owen developed the concept in the 1980s of mesenchymal lineage development based on the lineage diagrams drawn for hematopoiesis at that time. There were studies of cultured human bone marrow stroma cells as feeder-layer support cells for hematopoietic progenitor cells, but they were never tested for differentiation to other lineages (Castro-Malaspina et al. 1980, 1982).

A number of studies of bone-forming cells from rat bone marrow were performed by Arnold Caplan and colleagues in the 1980s at Case Western Reserve University. These studies then led Steve Haynesworth and Caplan to the first efforts to isolate human bone marrow cells with the characteristics of MSCs with the planned goal of tissue regeneration in humans (Caplan 1991; Haynesworth et al. 1992). The use of an implantable open-pore ceramic matrix filled with potential candidate MSCs was an important step because it (a) provided a reservoir of implanted cells that was also a recoverable implant open to interaction with the host environment, (b) provided an osteoconductive setting for favorable differentiation, and (c) importantly, allowed the selection of fetal bovine serum (FBS), which promoted the isolation of multipotential human cells, thereby allowing consistent results necessary for further developments. The intended use for these cells was for autologous therapeutic purposes in orthopedic indications (bone and cartilage), but oncology colleagues Stan Gerson and Hillard Lazarus were interested in using the cells to assist in the engraftment of HSCs following bone marrow transplantation or, as is most often used, peripheral blood mononuclear cells (PBMCs). The early safety analysis for these infused MSCs in the early 1990s led to enthusiasm for their use in bone marrow transplantation to aid engraftment of PBMCs in cancer patients (Koc et al. 2000; Lazarus et al. 1995; Reese et al. 1999). Several studies soon followed to evaluate the therapeutic benefit of allogeneic MSCs in children with the genetic disease osteogenesis imperfecta (Horwitz et al. 1999, 2001, 2002). Early clinical studies of MSC transplantation were also conducted for the glycogen storage diseases Hurler syndrome and metachromatic leukodystrophy, with promising early results (Koç et al., 2002).

## 2.4 Characteristics of Human MSCs

The characterization of any cell type relies on the identification of surface markers, isoenzymes, growth characteristics, replicative stability, chromosome analysis, gene expression analysis, DNA methylation patterns, and more. For cells capable of differentiating to several lineages, analysis of the differentiated phenotype is also warranted, including many of these criteria. The first demonstration of the isolation of a *human* cell population capable of differentiation to several mesenchymal lineages was performed by Haynesworth and Caplan at Case Western University, and when the bone marrow was placed in culture, this cell population proliferated and became quite homogeneous when analyzed by flow cytometry for surface markers. These investigators developed monoclonal antibodies (SH-2 and SH-3) reactive with surface molecules on this cell population, and these antibodies stained rare cells in the bone marrow. When the cultured cells were placed on ceramic cubes and implanted in athymic mice, and then processed for histology and analyzed, the cells had formed bone and cartilage in the pores of the cube. This *in vivo* assay for bone and cartilage formation has been instrumental in selecting individual lots of FBS that maintain the multipotential nature of the isolated cells. As few as 30% of tested FBS batches pass this test, so it is worth the effort to select serum before embarking on MSC studies. Commercial sources of FBS should provide the results of their analysis to investigators, and caution should be used when interpreting results with MSCs grown in different serum or growth factor supplements. A number of individual growth factors have been demonstrated to promote MSC growth, including epidermal growth factor (EGF), platelet-derived growth factor (PDGF), insulin-like growth factor (IGF), and fibroblast growth factor (FGF), particularly FGF2. These factors can work as well as FBS but do not provide the attachment factors present in serum. A useful compromise is to add FGF2 at 10 μg/mL to FBS. A number of investigators are examining the commercial human blood products platelet-rich plasma and AB serum to provide all-human-derived factors for human MSC expansion, thereby avoiding the use of bovine or other xenogeneic reagents. A variety of cell culture basal media is used to culture MSCs. These include low-glucose Dulbecco's modified Eagle medium (DMEM), α-MEM (modified Eagle medium), and 50:50 DMEM/MEM, among others. These media provide similar amino acids and other nutrients for anabolic activity of the cells, and few consistent differences have been demonstrated among the basal media for MSC expansion.

The development of *in vitro* assays for mesenchymal lineages provides additional evidence of the multilineage potential of MSCs. The differentiation to osteogenic, chondrogenic, and adipogenic lineages occurs readily *in vitro* (Bruder et al. 1997; Johnstone et al. 1998; Mackay et al. 1998; Pittenger et al. 1999). Under carefully controlled conditions, the differentiation can be quite complete to each of these lineages, with the overwhelming majority of the MSCs responding and without evidence of other lineages by mRNA (messenger RNA) analysis (Figure 2.2) or histology (Figure 2.1). These assays indicate that MSCs require specific signaling to progress along particular lineage pathways, and conditions for one lineage inhibit the other lineages. That is, MSC differentiation is influenced by growth factors, cytokines, basal

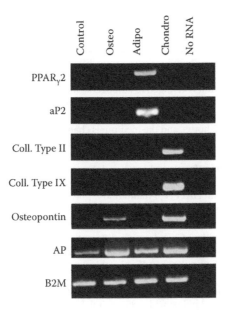

**FIGURE 2.2**

*In vitro* differentiation of MSCs does not result in differentiation to multiple lineages. Messenger RNA isolated from control and *in vitro* differentiated human MSCs was probed with cDNAs (complementary DNAs) for adipogenic markers (PPARγ2 and aP2), chondrogenic markers (collagens II and IX), and an osteogenic marker (osteopontin). Under the conditions, human MSCs differentiated with fidelity to the desired lineage, and RNA analysis did not detect differentiation to the other lineages. (Reprinted from Pittenger, M.F., A.M. Mackay, S.C. Beck, R.K. Jaiswal, R. Douglas, J.D. Mosca, M.A. Moorman, D.W. Simonetti, S. Craig, and D.R. Marshak. 1999. *Science* 284, no. 5411: 143–47.)

nutrients, cell density, spatial organization, and mechanical forces to perpetuate or inhibit lineage progression. Table 2.2 provides the *in vitro* differentiation methods for the adipogenic, chondrogenic, and osteogenic lineages.

When analyzed by flow cytometry, these carefully grown multipotential cells exhibit a regular and reproducible phenotype with the reliable presence of many surface molecules and the absence of others (Table 2.3). Clonal expansion of individual prospective MSCs demonstrated that the cells could be expanded a millionfold, approximately 22 population doublings, and still retained their multilineage differentiation ability (Pittenger et al. 1999). While no single marker, or even a collection of markers, can be claimed to denote a stem cell without evidence of differentiation, the addition of evidence from the *in vitro* multilineage assays combines to create a strong case for the stem cell nature of these MSCs (Dominici et al. 2006; Jaiswal et al. 2000; Pittenger et al. 1999). The surface molecules also indicate which direct cell-cell interactions can occur between MSCs and other cells. The undifferentiated MSCs can serve as a stromal cell feeder layer for the propagation of HSCs and embryonic stem cells, providing an all-human system that avoids the concerns of xenogeneic feeder layers. The addition of other assays

**TABLE 2.2**

Differentiation Methods for MSCs *In Vitro*

**Adipogenic Differentiation.** To induce adipogenic differentiation, MSCs are cultured as monolayers in DMEM and allowed to become confluent. The cells are cultured for 3–7 days more, and then the medium is changed to adipogenic induction medium (MDI + I medium) containing 0.5 mM methyl-isobutylxanthine, 1 µM dexamethasone and 10 µg/mL insulin, 100 µM indomethacin, and 10% FBS in low glucose (1 g/L) DMEM. The MSCs are incubated in this medium for 48–72 hours, and the medium is changed to adipogenic maintenance (AM) medium containing 10 µg/mL insulin and 10% FBS in DMEM for 24 hours. The cells are then re-treated with (MDI + I) for second and third treatment rounds. The cultures are maintained in AM medium for about 1 week and then assayed. Nile red is a fluorescent vital dye, and staining is used to quantify lipid vacuoles using a UV plate reader and counterstaining DNA with DAPI (4′,6-diamidino-2-phenylindole) to label DNA content as described (Pittenger et al. 1999). The adipogenic MSCs can then be fixed and stained with oil red O for nonquantitative histological evaluation.

**Chondrogenic Differentiation.** Chondrogenic differentiation is induced by placing $2.5 \times 10^5$ human MSCs into defined chondrogenic medium and subjecting them to gentle centrifugation (800$g$ for 5 minutes) in a 15-mL conical polypropylene tube, where they consolidate into a pellet within 24 hours. Chondrogenic media consists of high-glucose (4.5 g/L) DMEM supplemented with 6.25 µg/mL insulin, 6.25 µg/mL transferrin, 6.25 µg/mL selenous acid, 5.33 µg/mL linoleic acid, 1.25 mg/mL bovine serum albumin (ITS+, Collaborative Research, Cambridge, MA), 0.1 µM dexamethasone, 10 ng/mL TGF-β3, 50 µg/mL ascorbate 2-phosphate, 2 mM pyruvate, and antibiotics. [Note: (1) The TGF-β3 is stored in small aliquots at −80°C to avoid freeze-thawing damage. (2) For rat MSCs, 10 ng/mL BMP2 is also included for efficient chondrogenic differentiation.] The chondrogenic differentiation medium is changed every day due to the labile TGF. After 2–3 weeks, the pellets are fixed in 4% formaldehyde in phosphate-buffered saline (PBS), paraffin embedded for histology, sectioned and analyzed by immunostaining for collagen II expression, collagen I, and so on. Sections are also stained with safranin O for detection of proteoglycans and toluidine blue for metachromasia. For rat MSCs, chondrogenic differentiation is enhanced by the addition of 1 µg/mL BMP2.

**Osteogenic Differentiation.** To promote osteogenic differentiation of MSCs, approximately $3 \times 10^4$ cells are seeded onto 35-mm dishes in low-glucose DMEM with 10% FBS. After 24 hours, this medium is replaced with osteogenic differentiation medium composed of the same medium supplemented with 50 µM ascorbate 2-phosphate, 10 mM β-glycerol phosphate, and 100 nM dexamethasone. The medium is changed about every 3 days, and after 10–14 days the cells are stained with alizarin red and compared to MSCs maintained in normal culture medium. In separate cultures, mineralization is quantified by measuring calcium deposition (Kit 587-M, Sigma Chemical Co.).

will strengthen the evaluation of MSC differentiation, and data suggest that MSCs may differentiate to endothelial cells as they form vessel-like structures *in vivo* that express smooth muscle actin, and increased blood flow is seen in regions where MSCs have been implanted (Pittenger and Martin 2004, Shake et al. 2000, Quevedo et al. 2000). However, these vessel-like structures formed from MSCs have not been demonstrated to connect with the existing circulation and to convey blood cells.

MSCs produce a number of growth factors and cytokines that influence other cell types (Majumdar et al. 1998; Pittenger et al. 1999; Reese et al. 1999). As a stromal cell, they support the expansion of hematopoietic progenitor

**TABLE 2.3**

Surface Molecules Expressed by Cultured MSCs

| Present | Absent |
|---------|--------|
| CD13+ | CD11a or CD11b– |
| CD29+ | CD14– |
| CD44+ | CD18– |
| CD49b+ | CD31– |
| CD49e+ | CD34– |
| CD54+ | CD45–* |
| CD71+ | CD49d– |
| CD73+ | CD50– |
| CD90+ | CD62E– |
| CD105+ | CD117– |
| CD106+ | CD133– |
| CD166+ | |
| CD271+ | |

*Other Markers Expressed by MSCs*
Galectin 1+
Nestin+
TRK A, B, C+
HLA ABC+
HLA DR– (+ when interferon added)
SSEA3+
SSEA4+
Oct 4+

* Positive at isolation, lost in culture.

cells by the release of granulocyte colony-stimulating factor (G-CSF), macrophage colony-stimulating factor (M-CSF), and granulocyte-macrophage colony-stimulating factor (GM-CSF). The MSCs produce a large number of cytokines that influence not only nearby cells but also cells at a distance; MSCs produce the interleukins IL-1a, IL-1b, IL-6, IL-7, IL-8, IL-10, IL-11, IL-12, IL-14, and IL-15. MSCs have been transduced with expression vectors for IL-3 to further support hematopoietic cell expansion (Allay et al. 1997). The human MSCs can also support human embryonic stem cell expansion, providing an all-human system necessary for clinical studies of embryonic stem cells (Cheng et al. 2003).

## 2.5 Labeling of MSCs

A variety of methods has been used to identify MSCs in co-culture with other cells or to find them after *in vivo* implantation. Without going into details about labeling MSCs by various methods for different objectives (as this is discussed in other chapters), it is important to acknowledge that labeling methods should not disrupt the known properties of the cells under study. Therefore, *in vitro* differentiation assays are useful to examine at what level the labeling procedures impair the differentiation of MSCs. Similarly, the labeling should not interfere with MSC attachment, proliferation, contact inhibition, or interaction with other cell types. The labeled MSCs should also not differ in their expression of growth factors and cytokine or immunologic properties from their unlabeled counterparts. Table 2.4 lists 13 ways in which MSCs have been labeled for later identification for either *in vitro* or *in vivo* studies. While the MSCs labeled with these methods have been examined for influences of the labeling methods, not all known MSC characteristics are examined, and some caution should be taken in the interpretation of results. Nevertheless, these labeling methods have all been used successfully to learn more about MSCs *in vitro* and *in vivo*. These methods include ways to identify the unmodified cells or identify a label that has been attached to the cells, or the presence of the viable MSCs can be inferred from the continued expression of a gene product they produce. Many methods have been used to label MSCs and include membrane dyes, DNA intercalating dyes, antibody stains, gene tags, radiolabels, and magnetic resonance imaging (MRI) labels. Other methods do not require exogenous labeling of the MSCs but detect inherent marker labels, such as the Y chromosome or natural mutations. No one method should be relied on exclusively, and multiple approaches should be used whenever possible.

**TABLE 2.4**

Methods Used to Label and Detect MSCs *In Vitro* and *In Vivo*

1. DNA labeling by DAPI
2. Membrane labels such as carboxymethyl-1,1′-dioctadecyl-3,3,3′,3′-tetramethylindocarbocyanine perchlorate (DiI) or 3,3′-dioctadecyloxacarbocyanine perchlorate (DiO)
3. Species-specific antibody labeling, such as human MSCs in athymic mice
4. Gene tagging and polymerase chain reaction (PCR) detection
5. Gene tagging and antibody isolation or detection by fluorescence microscopy
6. Detection with antibodies by fluorescence microscopy of fluorescence-activated cell sorting (FACS)
7. In situ Y chromosome labeling for fluorescence microscopy
8. PCR for Y chromosome
9. Tritiated thymidine labeling of MSCs, detection by scintillation counting
10. Iridium labeling of MSCs and detection by neutron activation analysis
11. Radioactive indium labeling and detection by single-photon emission computed tomography (SPECT) imaging
12. Ferumoxide labeling and detection by magnetic resonance imaging (MRI)
13. Luciferase transduction and bioluminesence imaging (BLI) followed by biopsy quantification

## 2.6 *In Vitro* Cultured MSCs versus the *In Vivo* MSC

The reader should understand that the majority of our knowledge of MSCs comes from *in vitro* cultured cells. That is, MSCs are rare in bone marrow, constituting about 0.01–0.001% of the nucleated cells from bone marrow. This number comes from MSC plating experiments wherein the bone marrow nucleated cells are counted, diluted with culture medium, and incubated until visible colonies form, each colony believed to be derived from a single primary MSC (rather than two cells or a cluster of cells from a niche). This assumes that all MSCs survive the traumatic transfer from the natural bone marrow setting to tissue culture conditions, and the plating efficiency is 100%. This is not clearly correct because the subculturing of culture-adapted MSCs is not 100%; rather, subsequent passages have a plating efficiency of about 75–85%. Thus, the primary colony formation, referred to as colony-forming unit (CFU), is the most useful starting point. The passaged MSCs are the cells most often discussed, and with care in culturing, the MSCs can be expanded manyfold and still retain their multilineage potential *in vitro*. Most researchers use MSCs that have been expanded from bone marrow for 3–4 weeks and have undergone 3–5 passages and 6–15 population doublings.

Some researchers believe that MSCs can only be evaluated *in vivo*, but several parameters associated with *in vivo* studies, such as delivery location, implantation, viability, engraftment, site perfusion, as well as animal-to-animal variability, are difficult to control. Thus, *in vivo* studies always have more inherent "noise" than *in vitro* studies. Therefore, *in vivo* studies need support from *in vitro* counterparts and vice versa. The *in vitro* assays are important and convenient during the expansion and evaluation of MSCs. However, the ultimate goal of most MSC studies is to return them to the body to repair tissue, alleviate inflammation, or learn more about their healing potential as well as any dangers that may come from cultured multipotential cells. Subsequent chapters contain extensive information on *in vivo* implantation and evaluation.

## 2.7 Immunology of MSCs

The early clinical studies to coinfuse MSCs as stromal support cells with hematopoietic progenitors in cancer patients undergoing bone marrow transplantation coincided with *in vitro* studies examining the interaction of MSCs with immune cells. Particularly, research was under way to test the ability of MSCs to act as antigen-presenting cells to stimulate T lymphocytes. The MSCs have surface major histocompatibility class I (MHC class 1 or human leukocyte antigen [HLA] A, B, C); intracellular adhesion molecule 1 (ICAM1);

vascular cell adhesion molecule (VCAM); and variable leukocyte antigen 3 (VLA3), and these can interact with surface receptors on T lymphocytes to stimulate their proliferation. The *in vitro* studies first gave the surprising result that MSCs did *not* stimulate the allogeneic T cells to proliferate, even when the MSCs were gene transduced to express co-stimulatory molecules B7-1 or B7-2 as shown in Figure 2.3 (Klyushnenkova et al. 1998, 2005). That is, T cells did not see allogeneic MSCs as foreign. It was also demonstrated that MSCs could suppress the T-cell proliferation caused by a third unrelated donor cell as stimulator or an ongoing proliferation. The effect was not a toxic effect on T cells, as T cells, which were removed from the presence of MSCs, could then respond to another stimulator population of allogeneic cells. A series of peer-reviewed articles confirmed and extended these results to show that soluble factors were involved as well as cell-cell contact (Bartholomew et al. 2002; Di Nicola et al. 2002; Krampera et al. 2003; Le Blanc et al. 2003; Potian et al. 2003; Tse et al. 2003). Investigations into the agents involved in suppressing the T-lymphocyte proliferation identified as many as 11 soluble factors, including transforming growth factor β (TGFβ), hepatocyte growth factor (HGF), indoleamine 2,3-dioxygenase (IDO), prostaglandin E2 (PGE2), IL-6,

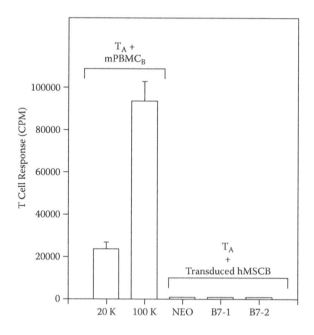

**FIGURE 2.3**
Allogeneic MSCs do not stimulate T-lymphocyte proliferation. T lymphocytes from donor A proliferated as shown by the incorporation of $^3$H-thymidine when incubated with peripheral blood mononuclear cells (PBMC$_B$) from donor B but did not proliferate when incubated with MSCs expressing B7-1 or B7-2 co-stimulatory molecules or the vector alone (Neo). (Reprinted from McIntosh, Mosca, and Klyushnenkova, Figure 1, U.S. Patent 6,368,636, at http://www.USPTO.gov.)

IL-10, IL-1 receptor antagonist (IL-1Rag), inducible nitric oxide synthetase (iNOS), galectin 1 (Gal-1), tumor necrosis factor α- (TNFα-) induced protein 6 (TSG-6), and HLA-G (Aggarwal and Pittenger 2004; Beyth et al. 2005; Kadri et al. 2005; Meisel et al. 2004; Ortiz et al. 2007; Sato et al. 2007). This large number of factors allows MSCs to interact with different immune cells rather than simply subduing all immune responses. When individual immune cells were isolated and tested for their interaction with MSCs, specific responses could be identified (Figure 2.4) (Aggarwal and Pittenger 2004). Similar studies with isolated natural killer (NK) cells and isolated B cells demonstrated that MSCs could modulate their activities as well (Corcione et al. 2006; Nasef et al. 2007; Rasmusson et al. 2007; Sotiropoulou et al. 2006).

During this time, there were several clinical studies that took advantage of the immune-modulating activity of MSCs along with their ability to support HSCs and used allogeneic MSCs co-transplanted with donor PBMCs (Ball et al. 2007; Fouillard et al. 2003; Lazarus et al. 2005; Le Blanc et al. 2004; Ringden et al. 2006). Cancer patients receiving allogeneic MSCs for their stromal support function of allogeneic PBMCs were noticed to have less graft-versus-host disease (GvHD) than expected. GvHD is a frequent complication of the transplanted donor immune cells present in PBMCs attacking the patient host and, if not controlled by steroids, becomes life threatening with the involvement of skin, gut, and liver in these immunosuppressed patients. The affect of MSC treatment can be seen in the study by LeBlanc et al. (2004) in which a 9-year-old boy, who had life-threatening grade IV GvHD when he received $2 \times 10^6$ MSCs per kilogram from a partially matched donor (his mother), made a full recovery. When steroids were removed to enhance the graft-versus-leukemia effect, the boy relapsed with grade IV GvHD. This time, he was treated with the remaining donor MSCs at a dose of $1 \times 10^6$ per kilogram, and he again recovered. The MSCs appear to have a more dramatic positive effect in patients with the worst GvHD. Understanding MSC interactions with immune cells is an important step in using MSCs to treat not only GvHD but also for every other clinical application.

## 2.8 Delivery of MSCs and Their Migration to Sites of Injury

A number of methods have been examined to deliver MSCs to damaged tissue. MSCs have been directly injected, delivered on a biocompatible matrix, and delivered systemically. Investigators have applied each of these methods to deliver MSCs to a myocardial infarction in hopes of having a meaningful impact on heart recovery following an acute event or chronic heart failure. Numerous preclinical cardiac studies have been performed in mice, rats, dogs, and pigs. Clinical trials involving MSCs for treatment of cardiac infarct are underway, and advancement to phase 2 has been made (http://

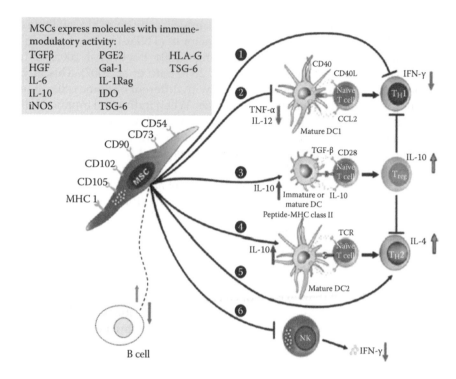

**FIGURE 2.4**

MSCs interact with immune cells in a complex manner. MSCs interact with T lymphocytes to reduce the inflammatory response by decreasing interferon-γ (IFNγ) expression from TH1 cells and increasing the expression of anti-inflammatory molecules IL-10 from TREG and IL-4 from TH2 cells (pathways 1, 3, and 5). MSCs can also decrease the expression of TNFα and IL-12 from DC1 cells (pathway 2). MSCs can interact with immature DCs (pathway 3) to cause release of IL-10 and promote TREG induction from immature T cells and release of additional IL-10. MSCs interaction with NK cells decreased IFNγ expression and inhibited NK proliferation due to HLA-G (pathway 6). MSCs inhibit B-cell proliferation and maturation, resulting in less secretion, but can also increase B-cell maturation responses in other conditions. (Updated from Aggarwal, S. and M.F. Pittenger. 2005. Blood 105(4): 1815–22.)

www.clinicaltrials.gov/ct2/results?term=mesenchymal+stem+cells). While direct injection provides a large number of MSCs at the injury site, thought to be necessary for robust tissue repair, systemic delivery provides the cells in a natural way at the level of the capillaries. Systemic delivery in the mouse first revealed the migration or homing of MSCs to sites of tissue damage as further detailed next.

Systemic delivery through a central vein is routinely used to deliver PBMCs to patients receiving a bone marrow transplant, and this route of administration has been used in many clinical trials to deliver MSCs. Early studies assumed this route would deliver PBMCs and MSCs to the greatest number of niches for engraftment, and this method proved to be safe and effective (Ball et al. 2007; Fouillard et al. 2003; Lazarus et al. 2005; Le Blanc et al. 2004;

Ringden et al. 2006). Systemic delivery spreads MSCs to the widest array of tissues and tests the ability of MSCs to engraft and differentiate when surrounded by mature, healthy, functioning cells in different tissues. It is known that cultured MSCs are relatively large and will become trapped in blood capillaries when injected intravenously, but over time many of the MSCs pass through the capillaries. Caplan and colleagues examined the biodistribution of rat MSCs (rMSCs) labeled with [111]Indium-oxime and delivered intravenously and intra-arterially (Gao et al. 2001). They found that over 48 hours, rMSCs were found first in the lungs and later in the liver and spleen, followed by other tissues but to a much lower level. Injecting a vasodilator such as nitroprusside prior to rMSC delivery accelerated the redistribution of the MSCs from the lungs and allowed more cells to reach the bone marrow. To avoid the first-pass trapping of venous-delivered MSCs in the lungs, Toma et al. used injection of human MSCs into the left ventricle of the heart of the SCID/*beige* immune-deficient mouse to achieve systemic delivery and analyzed the rare MSCs engrafted in the heart. The results showed that while the MSCs could be found in the heart at early time points, no differentiation was evident until 3–6 weeks, when expression of cardiac troponin I, connexin 43, and myosin heavy chain were evident and structural organization resembled the neighboring host cardiomyocytes (Toma et al. 2002).

Delivery of large numbers of MSCs to localized injuries such as tissue trauma, orthopedic injuries, or myocardial infarcts was thought to require delivery methods other than intravenous infusion. Many orthopedic studies performed for the repair of bone cartilage and tendons utilized compatible biodegradable carriers. In cardiac studies, direct injection into the damaged ventricle wall in an open-chest pig model demonstrated that the autologous porcine MSCs could be delivered, were retained in the heart wall, and improved the wall motion and decreased the ventricle wall thinning that normally occurs (Shake et al. 2002). In a different pig study, the first closed-chest delivery of allogeneic porcine MSCs, the cells were labeled for MRI and delivered via catheter to the infarcted region of the heart to demonstrate that cell delivery could be followed by fluoroscopy and the MSCs tracked by MRI (Kraitchman et al. 2003). Catheter delivery continues to be an effective method for delivery of MSCs to infarcted tissue and provides a clinically acceptable method to enhance local delivery of MSCs. Work by Hare and colleagues using catheter delivery of MSCs continued to show engraftment and indicated some differentiation of MSCs to cardiomyocytes, vascular muscle, and endothelial cells (Quevedo et al. 2009).

While direct injection via needles or catheters will certainly continue to be a delivery method of choice, there is good evidence that MSCs delivered intravenously also migrate to sites of tissue injury. This migration of bone-marrow-derived cells to sites of repairing muscle was first seen in the heart and skeletal muscle of mdx mice (Bittner et al. 1999). Soon thereafter, Chiu and colleagues demonstrated that, in the rat infarct model, rMSCs could be delivered by tail vein injection, and analysis revealed the MSCs were present at the site

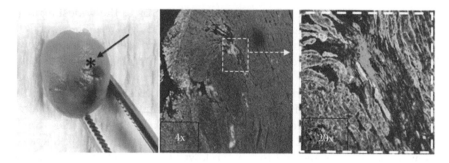

**FIGURE 2.5**

Migration (homing) of MSCs to sites of tissue injury. Human MSCs were transduced with the lacZ gene and infused via the tail vein into athymic rats. Three days later, the rats received an experimental infarct by tying off the left anterior descending artery (at * in the left panel) for 30 minutes. Animals were allowed to recover and after 2 weeks were sacrificed; the hearts were stained overnight to reveal the location of lacZ-containing cells (left panel). The hearts were then sectioned and stained with antibody to lacZ (red staining in center [×4 objective] and right panel [×20 objective]). Note that the MSCs reside almost exclusively in the remodeling infarcted tissue.

of infarction (Bittira et al. 2003). This study was the first of many to show that MSCs migrate to sites of wounding and tissue repair. Moreover, tissue damage increases MSC engraftment—that is, more MSCs are retained and more MSCs engraft in different damaged tissues. As shown in Figure 2.5 in the rat model of infarction, systemically delivered MSCs engrafted and were retained within the infarcted tissue (Pittenger and Martin 2004). This does not appear to be merely MSCs leaving the blood vessel into the injury site and being unable to exit the site. MSCs seem to be mobilized locally and at distant sites at the time of tissue injury. In the majority of the *in vivo* experimental studies, it is important to note that only the labeled cells are followed, and that the host MSCs are presumed to respond in a similar fashion, but this is not shown or proven. In other words, the *in vitro* cultured MSCs may or may not be representative of the endogenous *in vivo* MSCs in their natural environment.

The imaging of the migration of MSCs to cardiac infarcts was performed by labeling the MSCs with the radionuclide [111]Indium for single-photon emission computed tomography (SPECT) with the sensitivity to detect small numbers of cells in the heart as early as 24 hours (Kraitchman et al. 2005). In this canine model, the allogeneic MSCs were first seen in the lungs, but redistributed to the infarcted heart, as well as liver, and persisted in the heart for at least 7 days. Studies using an isolated heart model demonstrated that MSCs undergo morphological changes and begin to express some cardiac markers only hours after reaching the infarcted heart (Penna et al. 2008). This type of MSC migration to injury is not confined only to the heart or muscle. For example, radiation injury in a primate model can stimulate the migration and engraftment of MSCs to sites of damage, even when damage is not immediate and traumatic (Francois et al. 2006).

The mechanism for MSC migration and homing is partially understood. MSC migration studies have shown a number of factors can stimulate directed motility *in vitro* and likely operate *in vivo*. The Stromal cell-Derived Factor 1 (SDF1) and its receptor CXCR4 are certainly involved, but other factors may play a role, such as epac and integrin receptors (Carmona et al. 2008; Haider et al. 2008; Karp and Leng Teo 2009; Otsuru et al. 2008). For MSCs to migrate efficiently to sites of injury via the blood, they must readily navigate the capillary bed and a better understanding of the deformability and transit time in the blood is needed. Some in situ observations of this process have been made, but more work is needed (Toma et al. 2009).

Multiple studies in animal models of infarction have shown that MSCs engraft, express myogenic proteins, and improve the function of the infarcted heart, as measured by ejection fraction, wall thinning, compliance, diastolic pressure, and new vessel formation. While studies to date have demonstrated improvement in heart function, there is not any good evidence for the improved systolic function that would require differentiation to a mature cardiomyocyte with proper adherens junctions, gap junctions, and integration with surrounding cardiomyocytes along with well-developed innervations and vascularity. Within the remodeling infarcted tissue in the adult, this is not easy to achieve.

## 2.9 Summary

Much has been learned about the interactions of MSCs with other cells and in tissue from *in vitro* studies and preclinical studies using MSCs labeled for histological analysis as well as noninvasive imaging. Initial clinical trials using MSCs have been performed and are ongoing, but rarely have the MSCs been labeled for cell tracking. However, insights from these clinical trials are guiding ongoing animal studies to further understand the complex interactions in tissue.

## References

Aggarwal, S. and M.F. Pittenger. 2004. Human mesenchymal stem cells modulate allogeneic immune cell responses. *Blood* 105(4): 1815–22.

Allay, J.A., J.E. Dennis, S.E. Haynesworth, M.K. Majumdar, D.W. Clapp, L.D. Shultz, A.I. Caplan, and S.L. Gerson. 1997. Lacz and interleukin-3 expression in vivo after retroviral transduction of marrow-derived human osteogenic mesenchymal progenitors. *Hum Gene Ther* 8, no. 12: 1417–27.

Ashton, B.A., T.D. Allen, C.R. Howlett, C.C. Eaglesom, A. Hattori, and M. Owen. 1980. Formation of bone and cartilage by marrow stromal cells in diffusion chambers in vivo. *Clin Orthop Relat Res*, 151: 294–307.

Ball, L.M., M.E. Bernardo, H. Roelofs, A. Lankester, A. Cometa, R.M. Egeler, F. Locatelli, and W.E. Fibbe. 2007. Cotransplantation of ex vivo expanded mesenchymal stem cells accelerates lymphocyte recovery and may reduce the risk of graft failure in haploidentical hematopoietic stem-cell transplantation. *Blood* 110, no. 7: 2764–67.

Bartholomew, A., C. Sturgeon, M. Siatskas, K. Ferrer, K. Mcintosh, S. Patil, W. Hardy, S. Devine, D. Ucker, R. Deans, A. Moseley, and R. Hoffman. 2002. Mesenchymal stem cells suppress lymphocyte proliferation in vitro and prolong skin graft survival in vivo. *Exp Hematol* 30, no. 1: 42–48.

Beresford, J.N., J.H. Bennett, C. Devlin, P.S. Leboy, and M.E. Owen. 1992. Evidence for an inverse relationship between the differentiation of adipocytic and osteogenic cells in rat marrow stromal cell cultures. *J Cell Sci* 102 (Pt 2): 341–51.

Beyth, S., Z. Borovsky, D. Mevorach, M. Liebergall, Z. Gazit, H. Aslan, E. Galun, and J. Rachmilewitz. 2005. Human mesenchymal stem cells alter antigen-presenting cell maturation and induce T-cell unresponsiveness. *Blood* 105, no. 5: 2214–19.

Bittira, B., D. Shum-Tim, A. Al-Khaldi, and R.C. Chiu. 2003. Mobilization and homing of bone marrow stromal cells in myocardial infarction. *Eur J Cardiothorac Surg* 24, no. 3: 393–98.

Bittner, R.E., C. Schofer, K. Weipoltshammer, S. Ivanova, B. Streubel, E. Hauser, M. Freilinger, H. Hoger, A. Elbe-Burger, and F. Wachtler. 1999. Recruitment of bone-marrow-derived cells by skeletal and cardiac muscle in adult dystrophic mdx mice. *Anat Embryol (Berl)* 199, no. 5: 391–96.

Bruder, S.P., N. Jaiswal, and S.E. Haynesworth. 1997. Growth kinetics, self-renewal, and the osteogenic potential of purified human mesenchymal stem cells during extensive subcultivation and following cryopreservation. *J Cell Biochem* 64, no. 2: 278–94.

Caplan, A.I. 1991. Mesenchymal stem cells. *J Orthop Res* 9, no. 5: 641–50.

Carmona, G., E. Chavakis, U. Koehl, A.M. Zeiher, and S. Dimmeler. 2008. Activation of epac stimulates integrin-dependent homing of progenitor cells. *Blood* 111, no. 5: 2640–46.

Castro-Malaspina, H., R.E. Gay, S.C. Jhanwar, J.A. Hamilton, D.R. Chiarieri, P.A. Meyers, S. Gay, and M.A. Moore. 1982. Characteristics of bone marrow fibroblast colony-forming cells (cfu-f) and their progeny in patients with myeloproliferative disorders. *Blood* 59, no. 5: 1046–54.

Castro-Malaspina, H., R.E. Gay, G. Resnick, N. Kapoor, P. Meyers, D. Chiarieri, S. Mckenzie, H.E. Broxmeyer, and M.A. Moore. 1980. Characterization of human bone marrow fibroblast colony-forming cells (cfu-f) and their progeny. *Blood* 56, no. 2: 289–301.

Cheng, L., H. Hammond, Z. Ye, X. Zhan, and G. Dravid. 2003. Human adult marrow cells support prolonged expansion of human embryonic stem cells in culture. *Stem Cells* 21, no. 2: 131–42.

Corcione, A., F. Benvenuto, E. Ferretti, D. Giunti, V. Cappiello, F. Cazzanti, M. Risso, F. Gualandi, G.L. Mancardi, V. Pistoia, and A. Uccelli. 2006. Human mesenchymal stem cells modulate B-cell functions. *Blood* 107, no. 1: 367–72.

Di Nicola, M., C. Carlo-Stella, M. Magni, M. Milanesi, P.D. Longoni, P. Matteucci, S. Grisanti, and A.M. Gianni. 2002. Human bone marrow stromal cells suppress T-lymphocyte proliferation induced by cellular or nonspecific mitogenic stimuli. *Blood* 99, no. 10: 3838–43.

Dominici, M., K. Le Blanc, I. Mueller, I. Slaper-Cortenbach, F. Marini, D. Krause, R. Deans, A. Keating, D. Prockop, and E. Horwitz. 2006. Minimal criteria for defining multipotent mesenchymal stromal cells. The International Society for Cellular Therapy position statement. *Cytotherapy* 8, no. 4: 315–17.

Fouillard, L., M. Bensidhoum, D. Bories, H. Bonte, M. Lopez, A.M. Moseley, A. Smith, S. Lesage, F. Beaujean, D. Thierry, P. Gourmelon, A. Najman, and N.C. Gorin. 2003. Engraftment of allogeneic mesenchymal stem cells in the bone marrow of a patient with severe idiopathic aplastic anemia improves stroma. *Leukemia* 17, no. 2: 474–76.

Francois, S., M. Bensidhoum, M. Mouiseddine, C. Mazurier, B. Allenet, A. Semont, J. Frick, A. Sache, S. Bouchet, D. Thierry, P. Gourmelon, N.C. Gorin, and A. Chapel. 2006. Local irradiation not only induces homing of human mesenchymal stem cells at exposed sites but promotes their widespread engraftment to multiple organs: A study of their quantitative distribution after irradiation damage. *Stem Cells* 24, no. 4: 1020–29.

Friedenstein, A.J. 1976. Precursor cells of mechanocytes. *Int Rev Cytol.* 47: 327–55.

Friedenstein, A.J., K.V. Petrakova, A.I. Kurolesova, and G.P. Frolova. 1968. Heterotopic transplants of bone marrow: analysis of precursor cells for osteogenic and haematopoietic tissues. *Transplantation* 6: 230–47.

Gao, J., J.E. Dennis, R.F. Muzic, M. Lundberg, and A.I. Caplan. 2001. The dynamic in vivo distribution of bone marrow-derived mesenchymal stem cells after infusion. *Cell Tissues Organs* 169: 12–20.

Haider, H., S. Jiang, N.M. Idris, and M. Ashraf. 2008. IGF-1-overexpressing mesenchymal stem cells accelerate bone marrow stem cell mobilization via paracrine activation of SDF-1alpha/CXCR4 signaling to promote myocardial repair. *Circ Res* 103, no. 11: 1300–8.

Haynesworth, S.E., M.A. Baber, and A.I. Caplan. 1992. Cell surface antigens on human marrow-derived mesenchymal cells are detected by monoclonal antibodies. *Bone* 13, no. 1: 69–80.

Horwitz, E.M., P.L. Gordon, W.K. Koo, J.C. Marx, M.D. Neel, R.Y. Mcnall, L. Muul, and T. Hofmann. 2002. Isolated allogeneic bone marrow-derived mesenchymal cells engraft and stimulate growth in children with osteogenesis imperfecta: implications for cell therapy of bone. *Proc Natl Acad Sci USA* 99, no. 13: 8932–7.

Horwitz, E.M., D.J. Prockop, L.A. Fitzpatrick, W.W. Koo, P.L. Gordon, M. Neel, M. Sussman, P. Orchard, J.C. Marx, R.E. Pyeritz, and M.K. Brenner. 1999. Transplantability and therapeutic effects of bone marrow-derived mesenchymal cells in children with osteogenesis imperfecta. *Nat Med* 5, no. 3: 309–13.

Horwitz, E.M., D.J. Prockop, P.L. Gordon, W.W. Koo, L.A. Fitzpatrick, M.D. Neel, M.E. Mccarville, P.J. Orchard, R.E. Pyeritz, and M.K. Brenner. 2001. Clinical responses to bone marrow transplantation in children with severe osteogenesis imperfecta. *Blood* 97, no. 5: 1227–31.

Jaiswal, R.K., N. Jaiswal, S.P. Bruder, G. Mbalaviele, D.R. Marshak, and M.F. Pittenger. 2000. Adult human mesenchymal stem cell differentiation to the osteogenic or adipogenic lineage is regulated by mitogen-activated protein kinase. *J Biol Chem* 275, no. 13: 9645–52.

Johnstone, B., T.M. Hering, A.I. Caplan, V.M. Goldberg, and J.U. Yoo. 1998. In vitro chondrogenesis of bone marrow-derived mesenchymal progenitor cells. *Exp Cell Res* 238, no. 1: 265–72.

Kadri, T., J.J. Lataillade, C. Doucet, A. Marie, I. Ernou, P. Bourin, R. Joubert-Caron, M. Caron, and D. Lutomski. 2005. Proteomic study of galectin-1 expression in human mesenchymal stem cells. *Stem Cells Dev* 14, no. 2: 204–12.

Karp, J.M. and G.S. Leng Teo. 2009. Mesenchymal stem cell homing: the devil is in the details. *Cell Stem Cell* 4, no. 3: 206–16.

Klyushnenkova, E., J.D. Mosca, and K.R. McIntosh. 1998. Human mesenchymal stem cells suppress allogeneic T cell responses in vitro: implications for allogeneic transplantation *Blood* 92: 642a.

Klyushnenkova, E., J.D. Mosca, V. Zernetkina, M.K. Majumdar, K.J. Beggs, D.W. Simonetti, R.J. Deans, and K.R. McIntosh. 2005. T cell responses to allogeneic human mesenchymal stem cells: Immunogenicity, tolerance, and suppression. *J Biomed Sci* 12, no. 1: 47–57.

Koç, ON., S.L. Gerson, B.W. Cooper, S.M. Dyhouse, S.E. Haynesworth, A.I. Caplan, and H.M. Lazarus. 2000. Rapid hematopoietic recovery after coinfusion of autologous-blood stem cells and culture-expanded marrow mesenchymal stem cells in advanced breast cancer patients receiving high-dose chemotherapy. *J Clin Oncol* 18, no. 2: 307–16.

Koç, ON., J. Day, M. Nieder, S.L. Gerson, H.M. Lazarus, W. Kruivit. 2002. Allogeneic mesenchymal stem cell infusion for treatment of metachromatic leukodystrophy (MLD) and Hurler syndrome (MPS-IH). *Bone Marrow Transplant*. August, 30(4): 215–22.

Kraitchman, D.L., A.W. Heldman, E. Atalar, L.C. Amado, B.J. Martin, M.F. Pittenger, J.M. Hare, and J.W. Bulte. 2003. In vivo magnetic resonance imaging of mesenchymal stem cells in myocardial infarction. *Circulation* 107, no. 18: 2290–3.

Kraitchman, D.L., M. Tatsumi, W.D. Gilson, T. Ishimori, D. Kedziorek, P. Walczak, W.P. Segars, H.H. Chen, D. Fritzges, I. Izbudak, R.G. Young, M. Marcelino, M.F. Pittenger, M. Solaiyappan, R.C. Boston, B.M. Tsui, R.L. Wahl, and J.W. Bulte. 2005. Dynamic imaging of allogeneic mesenchymal stem cells trafficking to myocardial infarction. *Circulation* 112, no. 10: 1451–61.

Krampera, M., S. Glennie, J. Dyson, D. Scott, R. Laylor, E. Simpson, and F. Dazzi. 2003. Bone marrow mesenchymal stem cells inhibit the response of naive and memory antigen-specific T cells to their cognate peptide. *Blood* 101, no. 9: 3722–9.

Lazarus, H.M., S.E. Haynesworth, S.L. Gerson, N.S. Rosenthal, and A.I. Caplan. 1995. Ex vivo expansion and subsequent infusion of human bone marrow-derived stromal progenitor cells (mesenchymal progenitor cells): implications for therapeutic use. *Bone Marrow Transplant* 16, no. 4: 557–64.

Lazarus, H.M., O.N. Koc, S.M. Devine, P. Curtin, R.T. Maziarz, H.K. Holland, E.J. Shpall, P. Mccarthy, K. Atkinson, B.W. Cooper, S.L. Gerson, M.J. Laughlin, F.R. Loberiza, Jr., A.B. Moseley, and A. Bacigalupo. 2005. Cotransplantation of HLA-identical sibling culture-expanded mesenchymal stem cells and hematopoietic stem cells in hematologic malignancy patients. *Biol Blood Marrow Transplant* 11, no. 5: 389–98.

Le Blanc, K., I. Rasmusson, B. Sundberg, C. Gotherstrom, M. Hassan, M. Uzunel, and O. Ringden. 2004. Treatment of severe acute graft-versus-host disease with third party haploidentical mesenchymal stem cells. *Lancet* 363, no. 9419: 1439–41.

Le Blanc, K., L. Tammik, B. Sundberg, S.E. Haynesworth, and O. Ringden. 2003. Mesenchymal stem cells inhibit and stimulate mixed lymphocyte cultures and mitogenic responses independently of the major histocompatibility complex. *Scand J Immunol* 57, no. 1: 11–20.

Luria, E.A., A.J. Friedenstein, J.F. Morris, and S.A. Kuznetzow. 1987. Bone formation in organ cultures of bone marrow. *Cell Tissue Res* 248: 449–54.

Mackay, A.M., S.C. Beck, J.M. Murphy, F.P. Barry, C.O. Chichester, and M.F. Pittenger. 1998. Chondrogenic differentiation of cultured human mesenchymal stem cells from marrow. *Tissue Eng* 4, no. 4: 415–28.

Majumdar, M.K., M.A. Thiede, J.D. Mosca, M. Moorman, and S.L. Gerson. 1998. Phenotypic and functional comparison of cultures of marrow-derived mesenchymal stem cells (MSCs) and stromal cells. *J Cell Physiol* 176, no. 1: 57–66.

Meisel, R., A. Zibert, M. Laryea, U. Gobel, W. Daubener, and D. Dilloo. 2004. Human bone marrow stromal cells inhibit allogeneic T-cell responses by indoleamine 2,3-dioxygenase-mediated tryptophan degradation. *Blood* 103, no. 12: 4619–21.

Nasef, A., N. Mathieu, A. Chapel, J. Frick, S. Francois, C. Mazurier, A. Boutarfa, S. Bouchet, N.C. Gorin, D. Thierry, and L. Fouillard. 2007. Immunosuppressive effects of mesenchymal stem cells: Involvement of HLA-G. *Transplantation* 84, no. 2: 231–7.

Ortiz, L.A., M. Dutreil, C. Fattman, A.C. Pandey, G. Torres, K. Go, and D.G. Phinney. 2007. Interleukin 1 receptor antagonist mediates the antiinflammatory and anti-fibrotic effect of mesenchymal stem cells during lung injury. *Proc Natl Acad Sci USA* 104, no. 26: 11002–7.

Otsuru, S., K. Tamai, T. Yamazaki, H. Yoshikawa, and Y. Kaneda. 2008. Circulating bone marrow-derived osteoblast progenitor cells are recruited to the bone-forming site by the CXCR4/stromal cell-derived factor-1 pathway. *Stem Cells* 26, no. 1: 223–34.

Owen, M. and A.J. Friedenstein. 1988. Stromal stem cells: marrow-derived osteogenic precursors. *CIBA Found Symp* 136: 42–60.

Owen, M.E., J. Cave, and C.J. Joyner. 1987. Clonal analysis in vitro of osteogenic differentiation of marrow cfu-f. *J Cell Sci* 87 (Pt 5): 731–8.

Penna, C., S. Raimondo, G. Ronchi, R. Rastaldo, D. Mancardi, S. Cappello, G. Losano, S. Geuna, and P. Pagliaro. 2008. Early homing of adult mesenchymal stem cells in normal and infarcted isolated beating hearts. *J Cell Mol Med* 12, no. 2: 507–21.

Pittenger, M.F., A.M. Mackay, S.C. Beck, R.K. Jaiswal, R. Douglas, J.D. Mosca, M.A. Moorman, D.W. Simonetti, S. Craig, and D.R. Marshak. 1999. Multilineage potential of adult human mesenchymal stem cells. *Science* 284, no. 5411: 143–7.

Pittenger, M.F. and B.J. Martin. 2004. Mesenchymal stem cells and their potential as cardiac therapeutics. *Circ Res* 95, no. 1: 9–20.

Potian, J.A., H. Aviv, N.M. Ponzio, J.S. Harrison, and P. Rameshwar. 2003. Veto-like activity of mesenchymal stem cells: functional discrimination between cellular responses to alloantigens and recall antigens. *J Immunol* 171, no. 7: 3426–34.

Quevedo, H.C., K.E. Hatzistergos, B.N. Oskouei, G.S. Feigenbaum, J.E. Rodriguez, D. Valdes, P.M. Pattany, J.P. Zambrano, Q. Hu, I. McNiece, A.W. Heldman, and J.M. Hare. 2009. Allogeneic mesenchymal stem cells restore cardiac function in chronic ischemic cardiomyopathy via trilineage differentiating capacity. *Proc Natl Acad Sci USA* 106, no. 33: 14022–7.

Rasmusson, I., K. Le Blanc, B. Sundberg, and O. Ringden. 2007. Mesenchymal stem cells stimulate antibody secretion in human B cells. *Scand J Immunol* 65, no. 4: 336–43.

Reese, J.S., O.N. Koç, and S.L. Gerson. 1999. Human mesenchymal stem cells provide stromal support for efficient CD34+ transduction. *J Hematother Stem Cell Res* 8, no. 5: 515–23.

Ringden, O., M. Uzunel, I. Rasmusson, M. Remberger, B. Sundberg, H. Lonnies, H.U. Marschall, A. Dlugosz, A. Szakos, Z. Hassan, B. Omazic, J. Aschan, L. Barkholt, and K. Le Blanc. 2006. Mesenchymal stem cells for treatment of therapy-resistant graft-versus-host disease. *Transplantation* 81, no. 10: 1390–7.

Sato, K., K. Ozaki, I. Oh, A. Meguro, K. Hatanaka, T. Nagai, K. Muroi, and K. Ozawa. 2007. Nitric oxide plays a critical role in suppression of T-cell proliferation by mesenchymal stem cells. *Blood* 109, no. 1: 228–34.

Shake, J.G., P.J. Gruber, W.A. Baumgartner, G. Senechal, J. Meyers, J.M. Redmond, M.F. Pittenger, and B.J. Martin. 2002. Mesenchymal stem cell implantation in a swine myocardial infarct model: Engraftment and functional effects. *Ann Thorac Surg* 73, no. 6: 1919–25; discussion 26.

Sotiropoulou, P.A., S.A. Perez, A.D. Gritzapis, C.N. Baxevanis, and M. Papamichail. 2006. Interactions between human mesenchymal stem cells and natural killer cells. *Stem Cells* 24, no. 1: 74–85.

Toma, C., M.F. Pittenger, K.S. Cahill, B.J. Byrne, and P.D. Kessler. 2002. Human mesenchymal stem cells differentiate to a cardiomyocyte phenotype in the adult murine heart. *Circulation* 105, no. 1: 93–8.

Toma, C., W.R. Wagner, S. Bowry, A. Schwartz, and F. Villanueva. 2009. Fate of culture-expanded mesenchymal stem cells in the microvasculature: *In vivo* observations of cell kinetics. *Circ Res* 104, no. 3: 398–402.

Tse, W.T., J.D. Pendleton, W.M. Beyer, M.C. Egalka, and E.C. Guinan. 2003. Suppression of allogeneic T-cell proliferation by human marrow stromal cells: Implications in transplantation. *Transplantation* 75, no. 3: 389–97.

Urist, M.R. 2009. The classic: A morphogenetic matrix for differentiation of bone tissue. *Clin Orthop Relat Res* 467, no. 12: 3068–70.

Urist, M.R. and F.C. Mclean. 1952. Osteogenetic potency and new-bone formation by induction in transplants to the anterior chamber of the eye. *J Bone Joint Surg Am* 34, no. A(2): 443–76.

# 3

## *Hematopoietic Stem Cells*

David C. Weksberg, Megan T. Baldridge,
Emmanuel J. Volanakis, and Margaret A. Goodell

### CONTENTS

## 3.1 Hematopoietic Stem Cells: An Overview

The hematopoietic system faces a lifelong need for replenishment as its cellular elements are lost to physiologic turnover or in response to insult from infection or trauma. The engine that drives this cycle of loss and renewal is the hematopoietic stem cell (HSC)—a rare, self-renewing, multipotent cell type that resides in the adult bone marrow in mammals. The HSC sits at the base of a well-defined hematopoietic hierarchy, giving rise to progressively more differentiated progenitor and, ultimately, mature cell populations.

While hypothesized to exist as early as 1917 [1], the first direct evidence for the HSC came in landmark experiments by Till and McCulloch in 1961, who observed that transplanted bone marrow cells were capable of forming clonal macroscopic hematopoietic colonies in the spleens of recipient mice (colony-forming-unit-spleen; CFU-S), and, furthermore, that these colonies could subsequently form new spleen colonies after retransplantation into secondary recipients [2,3]. In the decades since, experiments have conclusively demonstrated that this repopulation activity resides in a single cell type and have shown that HSCs are incredibly potent: a single HSC is capable of providing long-term reconstitution of a lethally irradiated mouse [4,5]. This potency, coupled with ease of isolation and the development of robust functional assays, has made the HSC a powerful system for the study of stem cell biology and has made the HSC an especially attractive target for developing cell-based therapies. The study of HSC biology thus has broad implications, ranging from a basic understanding of hematopoietic development, to common features of other adult and embryonic stem cell populations, to therapeutic application for a host of clinical problems.

## 3.2 Clinical and Basic Relevance of the HSC

### 3.2.1 Therapeutic Use

As a therapeutic tool, the HSC already has a firmly established place in medical practice. Every year, thousands of patients receive bone marrow transplantations (BMTs) as treatment for neoplastic, immunologic, hematologic, or metabolic disorders—a procedure aimed, in effect, at transplanting HSCs to provide long-term reconstitution of the hematopoietic system. Of all the types of stem cells that have been identified, HSCs have the longest history, and the broadest current scope, of therapeutic application. The primary clinical use of the HSC, both historically and in current practice, is the reconstitution of the hematopoietic compartment after iatrogenic ablation or suppression. Historically, this treatment was first accomplished using bone marrow as a source of HSCs for transplantation; however, current practices draw on additional sources (i.e., peripheral blood and umbilical cord blood). Such treatment is indicated for some patients undergoing aggressive chemotherapy for cancer or for patients whose genetically defective marrow must be replaced.

Furthermore, the well-developed clinical protocols for autologous BMT make the HSC an attractive therapeutic target for gene therapy approaches wherein a patient's HSCs can be isolated, gene corrected, and reintroduced to the patient [6]. Thus, as gene therapy technologies continue to improve in efficacy and safety, the power and scope of the traditional therapeutic use of HSCs will likewise expand.

Transplantation of HSCs is most often performed as part of the treatment for cancer, including forms of leukemia and lymphoma, as well as for nonhematologic solid tumors, such as breast cancer, testicular cancer, small cell lung cancer, and colon cancer. In these clinical settings, the goal of transplantation is to allow recovery from aggressive chemotherapeutic treatment that ablates or transiently suppresses the patient's hematopoietic compartment even as it destroys cancer cells. HSC transplantation is also curative in nonmalignant hematologic diseases such as aplastic anemia, sickle cell anemia, and β-thalassemia. Other nonmalignant, nonhematologic diseases that are treated with HSC transplantation include immune deficiency diseases (e.g., severe combined immunodeficiency disease [SCID]), and metabolic diseases (e.g. Gaucher's disease) [7]. There has been interest in the potential use of HSC transplantation to treat autoimmune disease, such as refractory rheumatoid arthritis, and clinical trials are under way to investigate this possibility [8]. With these growing clinical applications for HSC therapeutics, BMT therapy has grown from its first use in the late 1950s by E. Donall Thomas to over 25,000 autologous and 15,000 allogenic transplants worldwide by 2003 [9].

### 3.2.2 HSCs as a Paradigm for Stem Cell Biology

Owing to the relatively long history of study of hematopoiesis, the HSC developmental hierarchy has been well characterized, and robust methods are available for the prospective isolation and study of these cells. Thus, the HSC serves as a useful model for stem cell systems in general and has provided insight into the biology both of other adult tissue-specific stem cells and of various pluripotential stem cell types; for example, the transcription factor Zfx has been shown to be required for self-renewal (through the inhibition of apoptosis) in both mouse HSC and embryonic stem cell contexts [10].

Insights gleaned from the study of HSCs also have a growing role in cancer research [11]. On a most basic level, HSCs represent a potential substrate for leukemogenesis, as these self-renewing cells persist for the lifetime of an animal and are thus subject to the accrual of mutagenic and epigenetic changes that may promote transformation of normal HSCs (or their downstream hematopoietic progeny). However, an emerging hypothesis suggests that cancers themselves may be heterogeneous with respect to tumor-initiating capacity—and that tumor recurrence after therapy may be driven by a resistant "cancer stem cell" population that behaves analogously to tissue-specific stem cells [12]. Thus, the study of HSC self-renewal, in addition to its intimate association with governance of the cell cycle, may ultimately bear on mechanisms of cancer self-renewal as well.

## 3.3 Fundamentals of HSC biology

### 3.3.1 Ontogeny: Primitive and Definitive Hematopoiesis

Hematopoiesis begins during early embryonic development, in which two waves of hematopoiesis contribute to blood in mammals. In mice, the

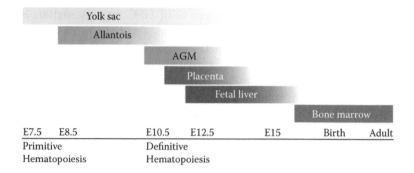

**FIGURE 3.1**

Timeline of murine hematopoietic development. Extraembryonic HSCs begin to give rise to blood around E7.5 in a wave known as primitive hematopoiesis. Both the yolk sac and allantois show early hematopoietic potential. Definitive hematopoiesis begins at E10.5 in the aorta-gonadal-mesonephros (AGM) region, later shifting to the fetal liver and finally to the bone marrow, where HSCs reside throughout adulthood.

"primitive" wave of hematopoiesis appears at about embryonic day (E) 7.5 and is driven by extraembryonic HSCs localized to the yolk sac (Figure 3.1). These cells derive from putative hemangioblasts (a common mesodermal precursor of endothelial and hematopoietic lineages) and give rise to primitive erythrocytes only, providing a rudimentary circulatory system [13]. The allantois also shows early hematopoietic potential [14]. The second, "definitive," wave begins with the appearance, at E10.5, of HSCs in mesenchymal tissue known as the aorta-gonadal-mesonephros (AGM) region and in the placenta [15,16]. The wave proceeds with HSC colonization of the fetal liver, after which HSCs undergo a final migration to the bone marrow, where adult hematopoiesis occurs [17,18].

The genetic controls on primitive and definitive hematopoiesis are complex and overlap to some extent. A detailed discussion is outside the scope of this text, but it is important to note that the molecular mechanisms that specify fetal HSCs are in part separable from those that regulate adult HSC homeostasis. An example comes through study of the genes *Scl* and *Bmi*; *Scl* has been demonstrated as essential for primitive, but not definitive, hematopoiesis; by contrast, *Bmi* is dispensable for both phases of embryonic hematopoiesis yet is required to maintain normal adult HSC function [19,20].

### 3.3.2 HSC Self-Renewal and Differentiation

Classically, stem cells are defined as having two essential characteristics: the ability to give rise to multiple adult cell types (multilineage differentiation) and the ability to self-renew (i.e., at some point, the progeny of a stem cell division must include another stem cell). HSCs satisfy both of these criteria; transplantation studies have demonstrated that HSCs give rise to all the

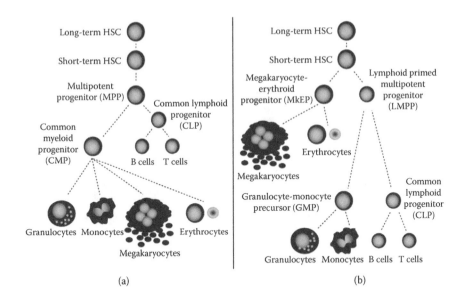

**FIGURE 3.2**
HSCs give rise to all of the cellular elements of blood. Hematopoietic stem cells have both the ability to self-renew and the ability to give rise to all the cell lineages of blood. HSCs start in a multipotential state and differentiate into a series of progenitor cell types, with an accompanying restriction of potential. Classically, this pathway was thought to begin with strict split into lymphoid and myeloid lineages (a); however, new evidence suggests that loss of megakaryocyte and erythroid potential may be the first step (b). Importantly, in either model, HSCs are the only cell types that possess extensive self-renewal capacity.

cellular elements of blood and can expand *in vivo*, such that they can reconstitute through multiple rounds of serial transplantation.

HSCs can be divided into two pools: long-term HSCs (LT-HSCs) with indefinite potential for self-renewal, and short-term HSCs (ST-HSCs), which have a limited self-renewal capacity (ST-HSCs exhaust their self-renewal potential after approximately 8 weeks in mice) [21]. The classic model (Figure 3.2a) of HSC self-renewal and differentiation begins with the indefinitely renewing LT-HSC population, which gives rise to ST-HSCs, which in turn yield multipotent progenitor (MPP) cells. The prevailing model holds that the path then bifurcates, with MPP cells giving rise to common lymphoid and common myeloid progenitors (CLPs and CMPs, respectively). However, findings have led to a proposed modification to this model (Figure 3.2b), suggesting that the first step of lineage restriction is not a split into the lymphoid and myeloid pathways, but rather a split into one population of progenitors that give rise to megakaryocyte and erythroid precursors and another population that gives rise to all other cell types [22]. Yet, while MPP cells ultimately give rise to multiple differentiated cell types, they have a limited capacity for self-renewal—a fact that distinguishes progenitor cells from stem cells [23].

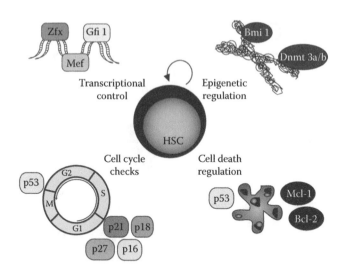

**FIGURE 3.3**

Control of HSC self-renewal is a heavily regulated process involving a variety of factors. HSC activity is controlled on many levels, including transcriptional control by multiple factors that activate or repress proliferation and differentiation. Epigenetic regulation also plays an important role in determining the chromatin methylation state, thereby also affecting gene expression. Direct regulators of the cell cycle also critically help control the HSC, a mostly quiescent cell that only occasionally enters the cell cycle. In addition, cell death regulators maintain healthy HSCs by preventing premature apoptosis, thereby allowing self-renewal to occur.

For self-renewal to occur, an HSC must undergo division, avoid apoptosis, and yield at least one daughter HSC. In addition, over multiple rounds of division, HSCs must maintain telomere length and protect genomic integrity. Control of self-renewal operates at several levels, including, but not limited to, cell cycle regulation and regulated cell death (Figure 3.3). Recent work indicated that signaling from the bone marrow microenvironment is also important in maintaining self-renewal, with Wnt, transforming growth factor β (TGFβ), and Notch signaling pathways implicated.

Self-renewal is fundamentally an aspect of cell division, and this fact is reflected in our growing understanding of regulators of HSC function that are components of the cell cycle machinery. Mice deficient for p21$^{cip1/waf1}$, a cyclin-dependent kinase inhibitor that regulates the G1 cell cycle checkpoint, were found to have increases in absolute HSC number as well as HSC proliferation—plus premature exhaustion in serial transplantation experiments [24]. Conversely, work on the p27$^{kip1}$ kinase inhibitor demonstrated that p27 does not govern pool size but does restrict the proliferation of downstream progenitors, leading to enhanced competition in transplant assays [25]. Deletion of a cyclin regulator, p18$^{INK4c}$, leads to increased self-renewing divisions [26], and the absence of p16$^{INK4a}$ rescues some age-related functional deficits in HSCs, including increased apoptosis [27].

Polycomb and Trithorax group genes exert control of HSC quiescence and cell cycle entry through epigenetic mechanisms. The gene products of the Polycomb and Trithorax groups participate in chromatin-modifying complexes that maintain silencing or transcriptional activation, respectively, of specific genes crucial to preserving HSC functional integrity. For example, ablation of *Bmi1*, a member of the Polycomb repressive complex PRC1, results in early catastrophic bone marrow failure, and its overexpression enhances HSC self-renewal. The failure of *Bmi1*[-/-] HSCs to self-renew is partially attributable to derepression of the the *Ink4a/Arf* locus, which encodes two functionally distinct tumor suppressors, p16[Ink4a] and p19[Arf]. Co-deletion of *Ink4a/Arf* partially rescues the *Bmi1*[-/-] hematopoietic phenotype by forestalling HSC senescence and apoptosis [20,28–31]. Our understanding of epigenetic regulators of HSC function is still growing, as underscored by recent findings that DNA methyltransferases (*Dnmt3a* and *Dnmt3b)* are essential to HSC self-renewal [32].

In addition to control of HSC proliferation by direct regulators of cell cycle progression, several transcription factors have been implicated in hematopoietic homeostasis. The Mef/Elf4 transcription factor promotes proliferation as its deletion leads to overly quiescent HSCs resistant to proliferative stimulation [33]. Conversely, HSCs deficient for the zinc-finger repressor Gfi1 display elevated proliferation yet are functionally compromised in competitive transplant assays [34]. Thus, strict control of HSC proliferation is required for normal function, in agreement with the experimental observation that, on serial HSC transplantation, hematopoietic reconstitution typically fails by the fifth consecutive recipient [35]. Finally, the FoxO transcription factors integrate environmental cues, including metabolic and oxidative stress, to preserve HSC function in the face of changing physiologic conditions [36].

As HSC self-renewal fundamentally involves cell division, it is also intricately related to the regulation of programmed cell death [37]. Apoptotic regulators were first implicated in self-renewal through the creation of a transgenic mouse that overexpresses the antiapoptotic gene *Bcl-2*. These mice have an increased LT-HSC compartment, without increased HSC proliferation, suggesting that this effect is achieved through the inhibition of cell death [38,39]. Bcl-2 is a prosurvival protein, the hallmark of a family that also includes Bcl-$x_L$ and Mcl-1. These proteins oppose apoptosis by interacting with Bac and Bak proteins, which trigger downstream mechanisms of cell death; and while Bcl-2 and Bcl-$x_L$ appear to be dispensable for relatively normal HSC function [40], *Mcl-1* null animals exhibit bone marrow failure from loss of early hematopoietic progenitors, and the protein is essential for HSC function [41].

Beyond the immediate regulation of the pathways that trigger apoptosis, work has led to a growing understanding of more complex regulatory networks that control HSC apoptosis in the face of cellular stress. Following DNA damage, the p53 tumor suppressor (itself a regulator of HSC pool size and engraftment capacity) [42] drives cells to apoptosis through the transcription of *Puma*, among other genes. However, in HSCs and progenitor cells, stress also induces the p53-mediated transcription of *Slug*, which encodes a transcriptional repressor that protects against apoptosis [43,44].

### 3.3.3 HSC Cycling: From the Quiescence Niche to Activation

In addition to their ability to self-renew, another essential aspect of HSC biology is the balance between HSC proliferation and quiescence. Under homeostatic conditions in the adult animal, HSCs exist as a largely quiescent population—that is, they divide at low rates and maintain themselves in a pool of relatively constant size. Yet, HSCs can be induced to cycle and expand in several settings, including BMT and the administration of chemotherapeutic agents such as 5-fluorouracil (5-FU). Along with understanding the regulation of self-renewal, elucidating the controls on HSC quiescence versus activation have direct relevance to potential therapeutic applications, bearing on the ability to expand and manipulate HSCs *ex vivo*.

In recent years, the interaction of HSCs with the complex microenvironment in which they reside has garnered increasing attention. First elucidated in 1978 [45,46], the concept of a "stem cell niche" has grown to describe HSC interactions with a host of cells in the marrow environment, including stromal cells, osteoblasts, and less-primitive hematopoietic progenitors.

Given the complexity of the HSC microenvironment, it is unsurprising that the study of how HSCs are localized within and interact with their niche has been a controversial area of study. Although our understanding is certain to evolve continuously, currently there are two chief schools of thought with regard to the cellular microenvironment: the osteoblastic niche hypothesis and the endothelial niche hypothesis.

The osteoblastic niche hypothesis derives from evidence showing that HSCs can be localized in situ along the endosteal surface of the bone marrow cavity [45]. There, cell-cell adhesion and cell-cell signaling maintain a quiescence niche (Figure 3.4). This model suggests that spindle-shaped N-cadherin+ CD45- osteoblasts (SNO cells) engage in reciprocal signaling to help regulate the HSC. *In vitro*, osteoblasts have been found to support HSCs in co-culture [47], and recent work has provided compelling *in vivo* evidence that the osteoblast is a key cellular component of the niche. It has been found that ablation of developing osteoblasts via induction of thymidine kinase leads to HSC depletion [48]. Conversely, an effective increase in osteoblast number, such as that which occurs in transgenic mice with an inducible deletion of BMP receptor type 1A, correlates with a subsequent increase in the number of observed HSCs, as well as SNO cells attached to HSCs *in vivo* [49]. In addition, experimental activation of the parathyroid hormone axis, which drives bone hypertrophy and increases Notch pathway signaling from osteoblasts to HSCs, leads to an increase in HSC number [50]. Finally, niche osteoblasts also express angipoietin 1 (Ang1), which is the ligand for

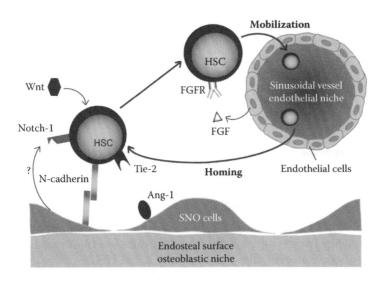

**FIGURE 3.4**

HSCs reside in a specialized bone marrow microenvironment. In the bone marrow, HSCs reside within a supportive niche composed of a variety of different cell types. HSCs are preferentially located against the endosteal surface, in contact with supporting osteoblasts (SNO cells). The osteoblastic niche provides a variety of signals that help to maintain HSCs in their baseline quiescent state, including signaling based on cell-cell contact (e.g., N-cadherin, Tie2-Ang1) as well as soluble factors (Wnt). HSCs may also localize to the endothelial niche along sinusoidal vessels; FGF signaling may contribute to HSC recruitment to the endothelial surface. Transitioning toward the sinusoidal vessels may represent a step toward HSC mobilization, while HSC homing occurs when HSCs leave the vasculature for the osteoblastic niche.

the HSC-expressed receptor tyrosine kinase Tie2. Tie2-Ang1 interactions enhance HSC quiescence as well as adhesion to the stroma [51].

In addition to regulatory signals mediated by cell-cell contact between HSCs and the niche, soluble factors in the niche milieu, including Kit ligand, thrombopoietin, various cytokines, and Wnt proteins, play critical roles in HSC maintenance and expansion [52].

The endothelial niche hypothesis proposes an alternative HSC niche located along the sinusoidal endothelial cells of the bone marrow and spleen vasculature, where HSCs have been found to localize [53]. Like osteoblasts, endothelial cells can help maintain HSCs in culture, and *in vivo* endothelial cell ablation, performed using antibody against anti-vascular endothelial cadherin, leads to hematopoietic failure [54,55]. The molecular regulation of this recently postulated niche remains mostly uncharacterized. However, fibroblast growth factor (FGF) signaling may help attract HSCs to the vascular niche, as HSCs express FGF receptors (FGFRs) and respond to FGF signaling *in vitro* [56,57]. Moreover, FGF recruits hematopoietic progenitors to the vasculature *in vivo* [55].

It is likely that both niches play a role in HSC maintenance; it has been proposed that the osteoblastic niche may represent a microenvironment for more quiescent HSCs, while the endothelial niche promotes HSC proliferation and differentiation [56,58]. In addition, the transition from osteoblastic niche to endothelial niche may offer an intermediate step in HSC mobilization to the periphery. However, additional studies are necessary to further elucidate the roles of the osteoblastic and endothelial niches in HSC biology.

## 3.4 Techniques for the Isolation and Study of HSCs

### 3.4.1 Prospective Identification of HSCs

A prerequisite for the successful study of any tissue or cell population is the ability to isolate pure samples of the tissue for experimentation, and the HSC is no exception. However, due to the scarcity of the HSC (~0.05% of the bone marrow) and the marked heterogeneity of the source material (bone marrow contains a variety of progenitor, stromal, vascular, and other cellular components), obtaining highly purified populations is not trivial, and several techniques have been developed to this end, each providing various levels of enrichment for HSCs.

Originally, techniques for isolation of stem cells were based on the fractionation of whole bone marrow by physical properties, such as gradient separations (Ficoll, Percoll) and centrifugal elutriation (reviewed in Reference 59). However, the coupling of flow cytometry techniques with monoclonal antibodies to cell surface markers has allowed for superior enrichment strategies. As a current first-pass purification scheme, differentiated cell types express a variety of lineage markers that are not found on HSCs (e.g., CD4 and CD8 are expressed by T-cell types and are absent on HSCs). Thus, an enriched HSC population can be obtained using a cell sorter or magnetic enrichment to exclude all cells that express these markers; this population is referred to as *lineage negative* (Lin−) [60]. Furthermore, murine HSCs have been extensively characterized based on their expression of the cell surface proteins Sca1, c-kit, and Thy1.1. Under this scheme, HSCs from appropriate mouse strains can be identified by sorting cells that are Sca1+, c-kit+, Thy1.1$^{lo}$, and Lin$^{lo}$/−. This population, abbreviated KTLS, is enriched about 1,000-fold for HSCs and represents roughly 0.05% of whole bone marrow. KTLS cells can be further subdivided into LT-HSCs and ST-HSCs based on Lin− (LT-HSCs) versus Lin$^{lo}$ (ST-HSCs) as well as Flk2 expression [61]. For cell surface-based purification of human HSCs, Lin− or CD34+, CD38− cells can be sorted [62].

Another technique for HSC purification is based on the efflux of Hoechst 33342, a vital DNA-binding dye that is preferentially effluxed by HSCs (Figure 3.5a). Staining whole bone marrow with Hoechst dye allows the resolution of a distinct verapamil-sensitive side population (SP) of cells, which

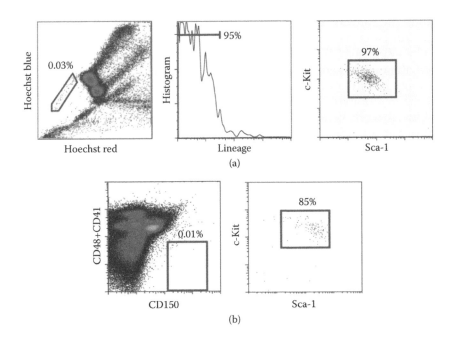

**FIGURE 3.5**

HSCs can be identified and isolated using several different flow cytometry marking systems. (a) HSCs preferentially efflux the fluorescent dye Hoechst 33342, allowing resolution of the "side population (SP)" of cells (gate) with low Hoechst fluorescence. SP cells represent a highly enriched stem cell population and are largely negative for lineage markers while displaying the positive stem cell markers Sca-1 and c-kit. SP cells are also predominantly CD34−/low and Flk2−. (b) HSCs can also be identified using the SLAM surface marker system. CD150+, CD48−, CD41− cells are strongly enriched for the positive HSC markers Sca-1 and c-kit (normally this population is only approximately 0.3% of whole bone marrow). Like the SP, SLAM cells are also CD34−/low and Flk2−.

shows less Hoechst fluorescence and has been shown to be highly enriched (~3,000-fold) for HSC activity [63,64]. SP cells have been demonstrated for murine and human HSCs and exist in several other tissue types as well [65,66].

Finally, work has led to the identification of the SLAM family of cell surface receptors as an alternative method for identifying HSCs and other progenitor populations. In this scheme, HSCs are identified as CD150+, CD48−, CD41−, CD244− cells—and the relative simplicity of the marker identification scheme allows this technique to be used for identifying HSCs in tissue sections as well as by flow cytometry (Figure 3.5b) [53,67,68]. However, as has been previously reported [69,70], stringent gating criteria must be used to identify CD150+ cells as the HSC fraction is contained in the brightest 1% of CD150+ cells. This finding is of particular relevance for imaging applications as technologies seeking to exploit the CD150 phenotype must be able to distinguish the rare, bright HSCs within the CD150 population.

Thus, the advent of improved flow cytometry and cell-sorting techniques has allowed investigators to prospectively identify stem cells based on phenotype (e.g., SP, KTLS, SLAM) and then interrogate the isolated cell population with functional assays. However, while the techniques for HSC isolation out of bone marrow are well developed, there remains a substantial gap in our ability to image these cells in the microenvironment *in vivo*, and future work along these lines would open up major avenues of investigation. Yet, an essential component of any scheme for HSC isolation is the use of functional assays to provide rigorous verification that the population in question does in fact represent the true HSC compartment; these assays are reviewed next.

### 3.4.2 *In Vivo* Assays: Transplantation as the Gold Standard

The gold standard for assaying HSC function is transplantation—a technique that allows the cells under study to demonstrate both multilineage potential and self-renewal. For mammalian studies, particularly with mice, BMT comprises the bulk of *in vivo* experimentation. In this experimental system, a test population of donor cells is intravenously injected into a lethally irradiated recipient (if not rescued with supporting marrow, the animal will die within 2 weeks) and assayed for the ability to sustain long-term (12–16 weeks) engraftment with multiple adult blood lineages (e.g., granulocyte, erythroid, lymphoid). An essential component of any BMT experiment is the ability to conclusively identify donor-derived cells, and there are several methods by which to achieve this. Most commonly, BMT experiments take advantage of an allelic variation in CD45—a cell surface tyrosine phosphatase that marks nucleated hematopoietic cells and whose functionally equivalent variants (CD45.1 and CD45.2) can be discriminated by specific monoclonal antibodies [60]. Thus, CD45.2 test cells can be transplanted into irradiated CD45.1 hosts, and the peripheral blood (or bone marrow, spleen, etc.) can be assayed for 45.1/45.2 chimerism by flow cytometry (Figure 3.6). Other options for marking donor cells include use of retroviral marking (which allows for clonal analysis), reporter transgenes (e.g., green fluorescent protein [GFP], *lacZ*), and gender (e.g., transplanting male into female and detecting the Y chromosome with fluorescence in situ hybridization [FISH]).

Generally, BMT experiments can be divided into two categories: competitive and noncompetitive. In a competitive transplant [71], the test population is transplanted along with wild-type marrow to assess whether the test population has a competitive advantage or disadvantage relative to the wild-type standard. A noncompetitive transplant, as the name implies, involves the administration of the test population only. These conditions place great selective pressure on the test marrow; thus, only those stem cell populations with a severe defect fail to engraft, leading in extreme cases to the death of the transplant recipient.

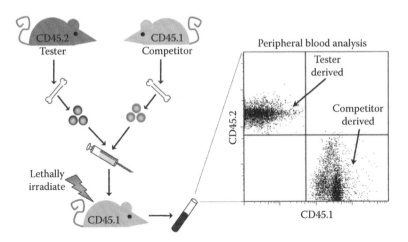

**FIGURE 3.6**

A competitive bone marrow transplantation using the CD45.1/2 system. Bone marrow transplantation experiments may take advantage of an allelic variation in CD45 that provides functionally equivalent variants (CD45.1 and CD45.2). In this example, HSCs isolated from the bone marrow of a mouse with the CD45.2 variant competed against HSCs from a mouse with the CD45.1 variant. These HSCs were mixed and injected into the circulatory system of a lethally irradiated CD45.1 recipient. The HSCs home to the recipient bone marrow and begin to contribute to the production of peripheral blood, which can be isolated from the recipient at various time points posttransplantation. CD45.1 and CD45.2 can be discriminated by specific monoclonal antibodies; thus, their respective contribution to the peripheral blood can be assessed by flow cytometry. In this example, the tester HSCs could be derived from wild-type and knockout mice; these models could then be compared in this functional assay.

The test population used in a BMT can vary depending on the specific question addressed. Whole bone marrow or purified stem cells can be transplanted; the test population may consist of marrow derived from a knockout mouse or stem cells that have been manipulated *ex vivo* to express a reporter or transgene. Single cells can be transplanted with distinguishable bone marrow as a carrier—either unmanipulated or depleted of Lin⁻ cells or of LT-HSCs. As few as 30–40 HSCs can be transplanted without carrier and rescue a lethally irradiated mouse [60].

It is important to note that BMT, while the gold standard for defining a functional HSC, is by no means a simple assay. Following intravenous administration of a test population of cells, the HSCs contained therein must undergo several steps between injection and readout as successfully engrafted. They must (a) circulate and relocate, or "home," to the bone marrow; (b) find the appropriate microenvironment within the bone marrow; (c) expand to establish the HSC compartment (self-renew); and (d) differentiate to reconstitute the peripheral blood of the irradiated host. Defects in any of these steps can result in the failure to sustain hematopoiesis following transplantation; several modifications can be made to the basic BMT rubric to specifically assay homing or self-renewal properties of a given cell population.

### 3.4.3 *In Vitro* Assays

Over the years, a multiplicity of *in vitro* assays has been developed for the characterization of HSCs (reviewed in Reference 59). In essence, these assays seek to replicate the functional demands that BMT would place on the HSCs or, alternatively, expose HSCs to a variety of progrowth stimuli (sera, feeder cells, specific cytokines, etc.) to induce and test their differentiation potential. The most general assay of this type is to expose HSCs to a specific cytokine cocktail (stem cell factor [SCF], interleukin 3 [IL-3], EPO, and granulocyte-macrophage colony-stimulating factor [GM-CSF]) in methylcellulose (a semi-solid culture media) and assay for the number of myeloerythroid colonies formed after 12–14 days.

In addition to semisolid media assays, another family of assays has been developed based on the ability of a cultured HSC population to reinitiate hematopoietic colonies over an extended period of time. These assays are known as long-term colony-initiating cells (LTC-ICs) and extended long-term colony-initiating cells (ELTC-ICs), and under these protocols, test cells are grown on a stromal feeder layer for a period of 35–60 or 60–100 days, respectively [72–74]. At successive time points, aliquots of culture are transferred to cytokine-containing media, which will induce colony formation if stem cells are still present. As most other assays detect progenitor populations, rather than true stem cells, LTC-IC and ELTC-IC assays are the best assays for detecting primitive cells *in vitro*. However, these time-consuming and labor-intensive procedures still fall short compared to *in vivo* studies in measuring HSC activity.

### 3.4.4 Xenograft Assays

For obvious ethical reasons, experimental BMT is limited in human subjects, meaning that the investigation of human HSC function must rely partially on the *in vitro* options detailed in this chapter. However, animal models have been developed to allow the testing of human HSC function in a xenograft setting—including the use of immunodeficient mice [75] and fetal sheep [76]. Nonobese diabetic (NOD)/SCID mice currently serve as the most widely used xenograft model for human HSC function. Essentially, because the immune-deficient NOD/SCID mice fail to recognize the human HSCs as foreign, the transplanted HSCs achieve multilineage engraftment in the murine hosts.

---

## 3.5 Summary

In summary, hematopoiesis is a lifelong process in which the cellular elements of blood are derived from HSCs—a rare, multipotent cell type found in bone marrow and capable of indefinite self-renewal. Under normal circumstances, LT-HSCs are maintained in a quiescent state (with low numbers of cells in cycle); however, when subjected to appropriate stimuli, HSCs enter a cycle of activation that generally involves a hyperquiescent stage, followed by proliferative stages, and finally a return to quiescence [77]. In quiescence, HSCs are thought to reside in an osteoblastic niche, which may regulate the size of the stem cell compartment as well as the maintenance of quiescence. Finally, while some inroads have been made, overall, the genetic and molecular mechanisms that regulate HSC self-renewal and cycling are not well understood, and much work remains to be done. Importantly, our current ability to exploit HSCs for therapeutic applications is limited by our inability to expand HSCs *ex vivo*—unlike embryonic stem cells, which can be cultured indefinitely, HSCs can be at best maintained after removal from their microenvironment. Thus, an improved understanding of the regulatory cross-talk between HSCs and their microenvironment may prove essential to realizing the full potential of HSC therapeutics.

## References

1. Pappenheim, A. 1917. Prinzipien der neueren morphologischen Haematozytologie nach zytogenetischer Grundlage. *Folia Haematol* 21:91.
2. Till, J.E. and E. McCulloch. 1961. A direct measurement of the radiation sensitivity of normal mouse bone marrow cells. *Radiat Res* 14:213–222.
3. Till, J.E., E.A. McCulloch, and L. Siminovitch. 1964. A stochastic model of stem cell proliferation, based on the growth of spleen colony-forming cells. *Proc Natl Acad Sci USA* 51:29–36.
4. Osawa, M., K. Hanada, H. Hamada, and H. Nakauchi. 1996. Long-term lymphohematopoietic reconstitution by a single CD34-low/negative hematopoietic stem cell. *Science* 273:242–245.
5. Smith, L.G., I.L. Weissman, and S. Heimfeld. 1991. Clonal analysis of hematopoietic stem-cell differentiation in vivo. *Proc Natl Acad Sci USA* 88:2788–2792.
6. Cavazzana-Calvo, M. and A. Fischer. 2007. Gene therapy for severe combined immunodeficiency: are we there yet? *J Clin Invest* 117:1456–1465.
7. Armitage, J.O. 1994. Bone marrow transplantation. *N Engl J Med* 330:827–838.
8. Sullivan, K.M., R. Parkman, and M.C. Walters. 2000. Bone marrow transplantation for non-malignant disease. *Hematology (Am Soc Hematol Educ Program)* 319–338.
9. Loberiza, F. 2003. Report on state of the art in blood and marrow transplantation. *IBMTR/ABMTR Newsletter* 10(1):7–10.
10. Galan-Caridad, J.M., S. Harel, T.L. Arenzana, Z.E. Hou, F.K. Doetsch, L.A. Mirny, and B. Reizis. 2007. Zfx controls the self-renewal of embryonic and hematopoietic stem cells. *Cell* 129:345–357.
11. Reya, T., S.J. Morrison, M.F. Clarke, and I.L. Weissman. 2001. Stem cells, cancer, and cancer stem cells. *Nature* 414:105–111.

12. Diehn, M., R.W. Cho, and M.F. Clarke. 2009. Therapeutic implications of the cancer stem cell hypothesis. *Semin Radiat Oncol* 19:78–86.

13. Choi, K., M. Kennedy, A. Kazarov, J.C. Papadimitriou, and G. Keller. 1998. A common precursor for hematopoietic and endothelial cells. *Development* 125:725–732.

14. Corbel, C., J. Salaun, P. Belo-Diabangouaya, and F. Dieterlen-Lievre. 2007. Hematopoietic potential of the pre-fusion allantois. *Dev Biol* 301:478–488.

15. Ottersbach, K. and E. Dzierzak. 2005. The murine placenta contains hematopoietic stem cells within the vascular labyrinth region. *Dev Cell* 8:377–387.

16. Medvinsky, A. and E. Dzierzak. 1996. Definitive hematopoiesis is autonomously initiated by the AGM region. *Cell* 86:897–906.

17. Dzierzak, E. 2003. Ontogenic emergence of definitive hematopoietic stem cells. *Curr Opin Hematol* 10:229–234.

18. Zon, L.I. 1995. Developmental biology of hematopoiesis. *Blood* 86:2876–2891.

19. Mikkola, H.K., J. Klintman, H. Yang, H. Hock, T.M. Schlaeger, Y. Fujiwara, and S.H. Orkin. 2003. Haematopoietic stem cells retain long-term repopulating activity and multipotency in the absence of stem-cell leukaemia SCL/tal-1 gene. *Nature* 421:547–551.

20. Park, I.K., D. Qian, M. Kiel, M.W. Becker, M. Pihalja, I.L. Weissman, S.J. Morrison, and M.F. Clarke. 2003. Bmi-1 is required for maintenance of adult self-renewing haematopoietic stem cells. *Nature* 423:302–305.

21. Morrison, S.J. and I.L. Weissman. 1994. The long-term repopulating subset of hematopoietic stem cells is deterministic and isolatable by phenotype. *Immunity* 1:661–673.

22. Adolfsson, J., R. Mansson, N. Buza-Vidas, A. Hultquist, K. Liuba, C.T. Jensen, D. Bryder, L. Yang, O.J. Borge, L.A. Thoren, K. Anderson, E. Sitnicka, Y. Sasaki, M. Sigvardsson, and S.E. Jacobsen. 2005. Identification of Flt3+ lympho-myeloid stem cells lacking erythro-megakaryocytic potential: A revised road map for adult blood lineage commitment. *Cell* 121:295–306.

23. Weissman, I.L., D.J. Anderson, and F. Gage. 2001. Stem and progenitor cells: origins, phenotypes, lineage commitments, and transdifferentiations. *Annu Rev Cell Dev Biol* 17:387–403.

24. Cheng, T., N. Rodrigues, H. Shen, Y. Yang, D. Dombkowski, M. Sykes, and D.T. Scadden. 2000b. Hematopoietic stem cell quiescence maintained by p21cip1/waf1. *Science* 287:1804–1808.

25. Cheng, T., N. Rodrigues, D. Dombkowski, S. Stier, and D.T. Scadden. 2000a. Stem cell repopulation efficiency but not pool size is governed by p27(kip1). *Nat Med* 6:1235–1240.

26. Yuan, Y., H. Shen, D.S. Franklin, D.T. Scadden, and T. Cheng. 2004. In vivo self-renewing divisions of haematopoietic stem cells are increased in the absence of the early G1-phase inhibitor, p18INK4C. *Nat Cell Biol* 6:436–442.

27. Janzen, V., R. Forkert, H.E. Fleming, Y. Saito, M.T. Waring, D.M. Dombkowski, T. Cheng, R.A. DePinho, N.E. Sharpless, and D.T. Scadden. 2006. Stem-cell ageing modified by the cyclin-dependent kinase inhibitor p16INK4a. *Nature* 443:421–426.

28. Lessard, J. and G. Sauvageau. 2003. Bmi-1 determines the proliferative capacity of normal and leukaemic stem cells. *Nature* 423:255–260.

29. Iwama, A., H. Oguro, M. Negishi, Y. Kato, Y. Morita, H. Tsukui, H. Ema, T. Kamijo, Y. Katoh-Fukui, H. Koseki, M. van Lohuizen, and H. Nakauchi. 2004. Enhanced self-renewal of hematopoietic stem cells mediated by the polycomb gene product Bmi-1. *Immunity* 21:843–851.

30. Jacobs, J.J., K. Kieboom, S. Marino, R.A. DePinho, and M. van Lohuizen. 1999. The oncogene and Polycomb-group gene bmi-1 regulates cell proliferation and senescence through the ink4a locus. *Nature* 397:164–168.

31. Oguro, H., A. Iwama, Y. Morita, T. Kamijo, M. van Lohuizen, and H. Nakauchi. 2006. Differential impact of Ink4a and Arf on hematopoietic stem cells and their bone marrow microenvironment in Bmi1-deficient mice. *J Exp Med* 203:2247–2253.

32. Tadokoro, Y., H. Ema, M. Okano, E. Li, and H. Nakauchi. 2007. De novo DNA methyltransferase is essential for self-renewal, but not for differentiation, in hematopoietic stem cells. *J Exp Med* 204:715–722.

33. Lacorazza, H.D., T. Yamada, Y. Liu, Y. Miyata, M. Sivina, J. Nunes, and S.D. Nimer. 2006. The transcription factor MEF/ELF4 regulates the quiescence of primitive hematopoietic cells. *Cancer Cell* 9:175–187.

34. Hock, H., M.J. Hamblen, H.M. Rooke, J.W. Schindler, S. Saleque, Y. Fujiwara, and S.H. Orkin. 2004. Gfi-1 restricts proliferation and preserves functional integrity of haematopoietic stem cells. *Nature* 431:1002–1007.

35. Siminovitch, L., J.E. Till, and E.A. McCulloch. 1964. Decline in colony-forming ability of marrow cells subjected to serial transplantation into irradiated mice. *J Cell Physiol* 64:23–31.

36. Tothova, Z. and D.G. Gilliland. 2007. FoxO transcription factors and stem cell homeostasis: Insights from the hematopoietic system. *Cell Stem Cell* 1:140–152.

37. Oguro, H., and A. Iwama. 2007. Life and death in hematopoietic stem cells. *Curr Opin Immunol* 19:503–509.

38. Domen, J., K.L. Gandy, and I.L. Weissman. 1998. Systemic overexpression of BCL-2 in the hematopoietic system protects transgenic mice from the consequences of lethal irradiation. *Blood* 91:2272–2282.

39. Domen, J., S.H. Cheshier, and I.L. Weissman. 2000. The role of apoptosis in the regulation of hematopoietic stem cells: overexpression of Bcl-2 increases both their number and repopulation potential. *J Exp Med* 191:253–264.

40. Motoyama, N., F. Wang, K.A. Roth, H. Sawa, K. Nakayama, K. Nakayama, I. Negishi, S. Senju, Q. Zhang, S. Fujii, et al. 1995. Massive cell death of immature hematopoietic cells and neurons in Bcl-x-deficient mice. *Science* 267:1506–1510.

41. Opferman, J.T., H. Iwasaki, C.C. Ong, H. Suh, S. Mizuno, K. Akashi, and S.J. Korsmeyer. 2005. Obligate role of anti-apoptotic MCL-1 in the survival of hematopoietic stem cells. *Science* 307:1101–1104.

42. TeKippe, M., D.E. Harrison, and J. Chen. 2003. Expansion of hematopoietic stem cell phenotype and activity in Trp53-null mice. *Exp Hematol* 31:521–527.

43. Inoue, A., M.G. Seidel, W. Wu, S. Kamizono, A.A. Ferrando, R.T. Bronson, H. Iwasaki, K. Akashi, A. Morimoto, J.K. Hitzler, T.I. Pestina, C.W. Jackson, R. Tanaka, M.J. Chong, P.J. McKinnon, T. Inukai, G.C. Grosveld, and A.T. Look. 2002. Slug, a highly conserved zinc finger transcriptional repressor, protects hematopoietic progenitor cells from radiation-induced apoptosis in vivo. *Cancer Cell* 2:279–288.

44. Wu, W.S., S. Heinrichs, D. Xu, S.P. Garrison, G.P. Zambetti, J.M. Adams, and A.T. Look. 2005. Slug antagonizes p53-mediated apoptosis of hematopoietic progenitors by repressing puma. *Cell* 123:641–653.

45. Gong, J.K. 1978. Endosteal marrow: A rich source of hematopoietic stem cells. *Science* 199:1443–1445.

46. Schofield, R. 1978. The relationship between the spleen colony-forming cell and the haemopoietic stem cell. *Blood Cells* 4:7–25.
47. Taichman, R.S. and S.G. Emerson. 1994. Human osteoblasts support hematopoiesis through the production of granulocyte colony-stimulating factor. *J Exp Med* 179:1677–1682.
48. Visnjic, D., Z. Kalajzic, D.W. Rowe, V. Katavic, J. Lorenzo, and H.L. Aguila. 2004. Hematopoiesis is severely altered in mice with an induced osteoblast deficiency. *Blood* 103:3258–3264.
49. Zhang, J., C. Niu, L. Ye, H. Huang, X. He, W.G. Tong, J. Ross, J. Haug, T. Johnson, J.Q. Feng, S. Harris, L.M. Wiedemann, Y. Mishina, and L. Li. 2003. Identification of the haematopoietic stem cell niche and control of the niche size. *Nature* 425:836–841.
50. Calvi, L.M., G.B. Adams, K.W. Weibrecht, J.M. Weber, D.P. Olson, M.C. Knight, R.P. Martin, E. Schipani, P. Divieti, F.R. Bringhurst, L.A. Milner, H.M. Kronenberg, and D.T. Scadden. 2003. Osteoblastic cells regulate the haematopoietic stem cell niche. *Nature* 425:841–846.
51. Arai, F., A. Hirao, M. Ohmura, H. Sato, S. Matsuoka, K. Takubo, K. Ito, G.Y. Koh, and T. Suda. 2004. Tie2/angiopoietin-1 signaling regulates hematopoietic stem cell quiescence in the bone marrow niche. *Cell* 118:149–161.
52. Reya, T., A.W. Duncan, L. Ailles, J. Domen, D.C. Scherer, K. Willert, L. Hintz, R. Nusse, and I.L. Weissman. 2003. A role for Wnt signalling in self-renewal of haematopoietic stem cells. *Nature* 423:409–414.
53. Kiel, M.J., O.H. Yilmaz, T. Iwashita, O.H. Yilmaz, C. Terhorst, and S.J. Morrison. 2005. SLAM family receptors distinguish hematopoietic stem and progenitor cells and reveal endothelial niches for stem cells. *Cell* 121:1109–1121.
54. Li, W., S.A. Johnson, W.C. Shelley, and M.C. Yoder. 2004. Hematopoietic stem cell repopulating ability can be maintained in vitro by some primary endothelial cells. *Exp Hematol* 32:1226–1237.
55. Avecilla, S.T., K. Hattori, B. Heissig, R. Tejada, F. Liao, K. Shido, D.K. Jin, S. Dias, F. Zhang, T.E. Hartman, N.R. Hackett, R.G. Crystal, L. Witte, D.J. Hicklin, P. Bohlen, D. Eaton, D. Lyden, F. de Sauvage, and S. Rafii. 2004. Chemokine-mediated interaction of hematopoietic progenitors with the bone marrow vascular niche is required for thrombopoiesis. *Nat Med* 10:64–71.
56. Kopp, H.G., S.T. Avecilla, A.T. Hooper, and S. Rafii. 2005. The bone marrow vascular niche: home of HSC differentiation and mobilization. *Physiology (Bethesda)* 20:349–356.
57. de Haan, G., E. Weersing, B. Dontje, R. van Os, L.V. Bystrykh, E. Vellenga, and G. Miller. 2003. In vitro generation of long-term repopulating hematopoietic stem cells by fibroblast growth factor-1. *Dev Cell* 4:241–251.
58. Yin, T. and L. Li. 2006. The stem cell niches in bone. *J Clin Invest* 116:1195–1201.
59. Ramos, C.A., T.A. Venezia, F.A. Camargo, and M.A. Goodell. 2003. Techniques for the study of adult stem cells: be fruitful and multiply. *Biotechniques* 34:572–578, 580–574, 586–591.
60. Spangrude, G.J., S. Heimfeld, and I.L. Weissman. 1988. Purification and characterization of mouse hematopoietic stem cells. *Science* 241:58–62.
61. Morrison, S.J., E. Lagasse, and I.L. Weissman. 1994. Demonstration that Thy(lo) subsets of mouse bone marrow that express high levels of lineage markers are not significant hematopoietic progenitors. *Blood* 83:3480–3490.

62. Wognum, A.W., A.C. Eaves, and T.E. Thomas. 2003. Identification and isolation of hematopoietic stem cells. *Arch Med Res* 34(6):461–475.
63. Goodell, M.A., K. Brose, G. Paradis, A.S. Conner, and R.C. Mulligan. 1996. Isolation and functional properties of murine hematopoietic stem cells that are replicating in vivo. *J Exp Med* 183:1797–1806.
64. Goodell, M.A., M. Rosenzweig, H. Kim, D.F. Marks, M. DeMaria, G. Paradis, S.A. Grupp, C.A. Sieff, R.C. Mulligan, and R.P. Johnson. 1997. Dye efflux studies suggest that hematopoietic stem cells expressing low or undetectable levels of CD34 antigen exist in multiple species. *Nat Med* 3:1337–1345.
65. Asakura, A. and M.A. Rudnicki. 2002. Side population cells from diverse adult tissues are capable of in vitro hematopoietic differentiation. *Exp Hematol* 30:1339–1345.
66. Welm, B.E., S.B. Tepera, T. Venezia, T.A. Graubert, J.M. Rosen, and M.A. Goodell. 2002. Sca-1(pos) cells in the mouse mammary gland represent an enriched progenitor cell population. *Dev Biol* 245:42–56.
67. Kim, I., S. He, O.H. Yilmaz, M.J. Kiel, and S.J. Morrison. 2006. Enhanced purification of fetal liver hematopoietic stem cells using SLAM family receptors. *Blood* 108:737–744.
68. Yilmaz, O.H., M.J. Kiel, and S.J. Morrison. 2006. SLAM family markers are conserved among hematopoietic stem cells from old and reconstituted mice and markedly increase their purity. *Blood* 107:924–930.
69. Weksberg, D.C., S.M. Chambers, N.C. Boles, and M.A. Goodell. 2008a. CD150⁻ side population cells represent a functionally distinct population of long-term hematopoietic stem cells. *Blood* 111:2444–2451.
70. Weksberg, D.C., S.M. Chambers, and M.A. Goodell. 2008b. The CD150high compartment is not the exclusive reservoir of LT-HSCs within the bone marrow. *Blood* 111:4414–4415.
71. Harrison, D.E. 1980. Competitive repopulation: a new assay for long-term stem cell functional capacity. *Blood* 55:77–81.
72. Hao, Q.L., F.T. Thiemann, D. Petersen, E.M. Smogorzewska, and G.M. Crooks. 1996. Extended long-term culture reveals a highly quiescent and primitive human hematopoietic progenitor population. *Blood* 88:3306–3313.
73. Sutherland, H.J., C.J. Eaves, A.C. Eaves, W. Dragowska, and P.M. Lansdorp. 1989. Characterization and partial purification of human marrow cells capable of initiating long-term hematopoiesis in vitro. *Blood* 74:1563–1570.
74. Sutherland, H.J., P.M. Lansdorp, D.H. Henkelman, A.C. Eaves, and C.J. Eaves. 1990. Functional characterization of individual human hematopoietic stem cells cultured at limiting dilution on supportive marrow stromal layers. *Proc Natl Acad Sci USA* 87:3584–3588.
75. Vormoor, J., T. Lapidot, F. Pflumio, G. Risdon, B. Patterson, H.E. Broxmeyer, and J.E. Dick. 1994. SCID mice as an in vivo model of human cord blood hematopoiesis. *Blood Cells* 20:316–320; discussion 320–322.
76. Zanjani, E.D., G. Almeida-Porada, and A.W. Flake. 1996. The human/sheep xenograft model: a large animal model of human hematopoiesis. *Int J Hematol* 63:179–192.
77. Venezia, T.A., A.A. Merchant, C.A. Ramos, N.L. Whitehouse, A.S. Young, C.A. Shaw, and M.A. Goodell. 2004. Molecular signatures of proliferation and quiescence in hematopoietic stem cells. *PLoS Biol* 2:e301.

# 4

## Adipose-Derived Adult Stem Cells

Deepak M. Gupta, Nicholas J. Panetta, and Michael T. Longaker

## CONTENTS

## 4.1 Introduction

Stem cells have increasingly been the subject of considerable excitement. The swift pace with which new data are being generated is astonishing and has significant implications for modern medicine. The term *stem cell biology* has largely been equated with *embryonic* stem cell biology; today, however, the term *stem cell biology* also applies to adult stem cells, tissue-specific stem cells, cancer stem cells, and induced pluripotent stem cells (iPSCs). With the exponential growth of our understanding of stem cells, it is an exciting time to realize the clinical potential of stem cell biology that was not conceivable a mere decade ago.

In contemporary medicine, dead and diseased tissue is discarded from the human body without an ideal alternative to replace it. In most cases, nonfunctional and nonanatomic scar tissue replaces native tissue that has been devitalized.[1,2] In some cases, autologous tissue can be used to surgically restore a tissue defect; however, this solution is not ideal and often necessitates further surgery and creates donor site morbidity. Alternatively, there are several alloplastic biomaterials and bioengineered constructs with which tissue replacement may be performed.[3-6] Among them, hydroxyapatite, metal alloys, glass, polymethylmethacrylate, plaster of Paris, and various polymers have also been employed for reconstructive purposes.[7-9] These resources are associated with several inherent risks and shortcomings, including infection, contour irregularities, and structural failure. The myriad different techniques and supplies available to the surgeon for reconstruction reflect both the inadequacies of the individual materials and the necessity for more effective tactics for bone regeneration.

As an alternative, it may be possible to *regenerate* instead of replace or reconstruct the tissue. Several studies have suggested such a paradigm. In this model, building blocks (progenitor cells) are delivered in such a configuration (biomaterial scaffold) that they may respond temporospatially to appropriate signals (i.e., growth factors, gene therapy, etc.) and differentiate toward de novo functional tissue units in situ. This strategy, which has come to be known as *regenerative medicine*, promises novel and improved therapies when the native tissue degenerates or dies.

As part of this three-prong approach to tissue regeneration, progenitor cells play perhaps the most significant role. While a variety of biocompatible/biomimetic scaffolds and methodologies to "encourage" cells down particular pathways exist, the choice of progenitor cell with which to design a novel therapy poses unique challenges. First, the progenitor cell type must ideally be autologous. For a novel therapy to be attractive and widely available, most, if not all, individuals must be able to provide these progenitor cells. As an alternative, the progenitor cell type must at least be allogeneic and nonimmunogenic; that is, when transplanted into an individual, the cells must not elicit an immune reaction that may mediate rejection or require immunosuppressants, which are associated with their own adverse side effects. The cells must also be available in sufficient quantities to realistically be able to seed a construct that will mediate tissue regeneration. The harvest of progenitor cells from an individual must also ideally be convenient. The procedure should not introduce significant morbidity to an affected individual and should be able to be performed expeditiously and easily by a clinical practitioner. The progenitor cells should also be quickly and easily identifiable and isolatable from surrounding cells so that they may be applied to the affected area in a single procedure, without the need for long-term *in vitro* or *ex vivo* culturing, sorting, or other laboratory processing.

## 4.2 Adipose-Derived Stem Cells

A number of cell populations arguably meet these criteria, but perhaps the best candidate to fill this role is the adipose-derived stem cell (ASC). ASCs are present in subcutaneous adipose (fat) tissue in abdominal, inner thigh, outer thigh, buttock, flank, and upper arm areas. These cells also have several inherent advantages compared to other progenitor cell populations.[10,11] First, adipose tissue is present in relatively abundant quantities, even in thin individuals. Moreover, ASCs automatically have the advantage of being autologous. In addition, ASCs are easily harvested by surgical procedure, which can be performed under local anesthesia. Lipoaspiration procedures can safely be performed with little-to-no donor site morbidity in most cases. Last, ASCs are easily isolated by enzymatic digestion. As investigated previously, ASCs have similar or enhanced proliferative capacity and multipotential to differentiate when compared to other progenitor cell types.[12,13]

The multipotent nature of ASCs was first reported in 2001 by Zuk et al. in a report concerning elective and surgical isolation of these cells from lipoaspirate, which would normally be discarded.[13] In this study, ASCs were isolated by enzymatic digestion of lipoaspirate and differentiated *in vitro* toward adipogenic, myogenic, chondrogenic, and osteogenic lineages using chemically defined media. This report was released shortly after a report by Pittenger et al. in 1999, in which multipotent bone marrow stromal cells (MSCs) were isolated and differentiated toward these mesodermal lineages.[14] Given the age-related replacement of bone marrow with fat, it was perhaps thought that adipose tissue contained a cognate population of stromal cells that were capable of differentiating toward these same mesodermal lineages. Since these seminal reports, the multipotent nature of ASCs has been investigated in many studies around the globe and across a number of disciplines, from developmental biology to materials sciences and from stem cell biology to cardiology to bioengineering.

ASCs have been increasingly studied in the last decade. Most of these studies have been performed in the animal model, specifically in the mouse model. Mouse ASCs have been used *in vitro* and *in vivo* to study commitment to a number of mesodermal lineages, as described, including chondrogenic, myogenic, and adipogenic. However, studies regarding ASC commitments to these lineages form only the minority of the literature on ASCs. Instead, the vast majority of the literature addresses the potential of ASCs to undergo osteogenesis for bone formation.

Mouse ASCs have been shown to deposit mineralized extracellular matrix like their lineage-committed osteoblastic controls, as well as upregulate key genes that participate in or control the osteogenic pathway.[13] Specifically, mouse ASCs significantly upregulate *Cbfa1/Runx2*, a "master" transcription factor for osteogenic commitment during early osteogenesis. At the same time, alkaline phosphatase has also been shown to be upregulated. While this last marker is not specific for osteoblastic activity, its activity is present only in low levels in fresh ASCs; thus, its presence during early differentiation in chemically defined osteogenic differentiation media has been correlated to osteoblastic commitment. Osteocalcin is also significantly upregulated during osteogenic differentiation of mouse ASCs, indicating terminal differentiation toward the osteogenic lineage. These findings in mouse ASCs are significant and have provided a standard for subsequent studies of ASCs, as recent studies still cite these markers as controls for osteogenic differentiation.

As the literature underlying ASCs has matured, several distinct themes have emerged. First, bone morphogenetic proteins (BMPs) and fibroblast growth factors (FGFs) have been increasingly studied in this model of cell-based tissue engineering of bone. These growth factors are well described, especially in their roles during embryonic development and bone homeostasis, and undoubtedly have important roles in mediating osteogenesis and bone formation by ASCs. Other well-characterized growth factors, such as Wnt and Hedgehog, have also been studied in ASCs and may also play a role in tissue engineering applications in the future; however, further studies are still necessary. Second, the exact role that ASCs assume in mediating cell proliferation and differentiation has been under increasing scrutiny. Several theories have been proposed regarding how ASCs affect tissue biology. Some have suggested that ASCs do not directly contribute to de novo tissue cellular makeup but only provide growth factors or paracrine signals that affect endogenous host cells. Others have suggested that ASCs affect tissue growth and differentiation by fusing with host cells. Yet others have suggested that ASCs differentiate to functional lineage-committed cells to contribute directly to de novo tissue growth and function. Third, the need to study *human* ASCs further has been recognized if these cells are to move toward clinical applications. Human cells have been increasingly applied to *in vitro* studies as well as immunocompromised animal models, with promising results. For example, Lendeckel and colleagues have published a case report of ASCs applied clinically to reconstruct bone.[15] In this report, a 7-year-old girl with significant calvarial bone loss, after several failed reconstructive attempts using traditional surgical methods, received autologous ASCs mixed with bone chips in a biocompatible matrix. Within 3 months, the authors noted an uneventful postoperative course and near-complete calvarial continuity by traditional radiological examination.

These three areas described have been confirmed by Zuk's initial description of the multipotent nature of ASCs, as well as an important 2004 report by Cowan and colleagues.[16] In this report, mouse ASCs were rigorously tested against MSCs and osteoblasts (control) in a mouse model of critical-size calvarial defects. In this model, critical-size calvarial defects have been shown to lack any signs of healing if left untreated or treated with an empty apatite-coated poly-lactic-co-glycolic acid (PLGA) scaffold that does not contain cells.[17] In this study, empty scaffold did not induce any healing; however, when these scaffolds were seeded with osteoblasts, MSCs, or ASCs, near-complete continuity of the calvarium was noted by 12 weeks.[16] Furthermore, greater than 92% of the cells within the regenerate were derived from donor cells, not host/wound. This seminal report was among the first showing that ASCs had a similar potential to undergo osteogenesis and bone formation *in vivo* compared to MSCs. This study was also important as it specifically tracked (histomorphometrically) the cells within the regenerated tissue, supporting the notion that donor, and not host, cells were responsible for mediating the observed phenotype.

## 4.3 Potential Therapeutic Targets

After demonstrating the osteogenic potential of ASCs in an *in vivo* mouse model, our laboratory began to focus on accelerating bone formation with the *ex vivo* addition of various growth factors. Using the same critical-size mouse defect model, we added BMP2 and retinoic acid (RA) to ASCs prior to implantation.[18] ASCs seeded on PLGA scaffolds were cultured for 4 weeks in osteogenic media containing BMP2 and RA and then implanted into critical-size calvarial defects. Bone bridging of the defects was observed as early as 2 weeks with implantation of the treated ASCs. However, although bone formation was initially accelerated, osteoclasts were noted by 4 weeks, which led to bone resorption; one hypothesis was that RA that was carried over by the scaffold led to the observed osteoclastogenesis.

To investigate the mechanisms underlying RA and BMP2-mediated osteogenic differentiation, Wan et al. investigated the BMP receptor expression profile in ASCs exposed to these factors.[19] They found a dramatic upregulation of the expression of the BMP receptor IB (BMPR-IB) isoform during osteogenesis. To demonstrate the critical role of BMPR-IB during ASC osteogenic differentiation, the receptor isoform was suppressed using RNA interference (RNAi). RNAi-mediated suppression impaired the ability of ASCs to form bone *in vitro*, as evidenced by lack of alizarin red staining. From these data, it appears that ASC osteogenesis in mice requires RA, which enhances expression of the essential BMPR-IB isoform.

Given the finding of osteoclastogenesis with RA and BMP2 treatment, our attention turned to refining and optimizing the approach to priming ASCs for osteogenic differentiation. To do this, ASCs in a monolayer were exposed to a shorter, 15-day course of RA before implantation into critical-size calvarial defects.[20] With this abbreviated course of RA pretreatment in monolayer, which was washed off before implantation, significantly accelerated bone formation was observed with the "primed" ASCs as compared to ASCs alone without concomitant osteoclastic resorption. Therefore, *in vivo* priming of ASCs in monolayer with RA or other osteogenic agonists may be developed to allow for earlier patient mobilization and return to activity after skeletal reconstruction with the novel strategies.

A number of other cytokines and signaling factors have also been shown to be critical mediators of osteogenic differentiation of ASCs. In addition to BMPs, our laboratory has concentrated on the effect of FGFs on ASC biology during osteogenesis.[21] After exposing ASCs to FGF2, we found increased proliferation of these cells. Furthermore, exposure of ASCs to FGF2, by either retroviral transduction or exogenous supplementation, sustained the proliferative and osteogenic potential of the cells. Thus, FGF2 may assist in the proliferation and enrichment of osteoprogenitors within ASCs for skeletal tissue engineering. In addition, other factors, such as Wnts, transforming growth factor $\beta$, and Hedgehog, have been implicated in osteogenesis and regulation of bone development and therefore may be exploited for regeneration applications.

## 4.4 Toward Clinical Translation

Given these exciting findings in the mouse model and the climate of stem cell biology, we have turned our attention to studies in *human* ASCs for tissue engineering applications. One fundamental difference between mouse and human ASCs is the process by which these cells are procured. In the human setting, we have recognized an increasing clinical trend toward the use of lipoaspiration that includes various forms of ultrasound, which affects the removal of the lipoaspirate from the subcutaneous tissue. Ultrasound has been shown in many unrelated studies to affect cell biology significantly; as such, we sought to investigate whether ultrasound application to ASCs would affect their biology for potential tissue engineering applications.[11] To answer such a question, we collected ASCs from several patients by both standard lipoaspiration and ultrasound-assisted devices (third generation) and documented their ability to proliferate and undergo osteogenic differentiation *in vitro*. We found that ASCs collected from ultrasound-assisted samples were equivalent in numbers to those from standard lipoaspirate, and they proliferated and underwent osteogenic differentiation *in vitro* similarly to ASCs collected by lipoaspiration alone. These data were supported by *in vitro* cell counting, osteogenic differentiation in chemically defined media, subsequent matrix staining and quantification with alizarin red, and qRT-PCR analyses of underlying gene expression.

Early reports of human ASCs are encouraging for future stem cell therapies, but to realize this potential, rigorous *in vivo* studies are necessary. In one *in vivo* study, we implanted human ASCs into critical-size calvarial defects in nude, immunocompromised mice. Our preliminary data are encouraging for significant calvarial bone healing noted with implantation of cells compared to scaffold and empty controls. Several other groups have documented bone regeneration mediated by human ASCs in immunocompromised mice; however, we are also interested in elucidating the precise contribution by human ASCs in the observed phenotype. In a related report, we have also most recently documented the relationship between BMP signaling and human ASCs.[22] Briefly, human ASCs were found to respond to rhBMP2 (recombinant human BMP2), a growth factor approved by the Food and Drug Administration (FDA), in a dose-dependent fashion and significantly upregulate extracellular matrix mineralization and osteogenic gene expression *in vitro*. Furthermore, we documented that human ASCs actually elaborate their own BMP signaling elements, which are not only *sufficient* to allow these cells to undergo osteogenesis and bone formation but also *necessary*.

Given these exciting data, our laboratory is looking toward the future for potential tissue engineering strategies. We have initiated several studies to identify new paradigms with which novel biomaterials may effect bone formation by the controlled release of one or more growth factors; for example, perhaps it is possible to expand the osteoprogenitor population within ASCs with early release of FGF2 and then drive them toward bone formation with delayed release of rhBMP2. This novel strategy, among others, promises improved clinical outcomes.

## 4.5 Role of ASCs in Myocardial Infarction

While our laboratory has maintained a focus on potential applications of tissue engineering for craniofacial bone, other groups have looked to ASCs with a slightly different hope. Increasing numbers of reports have begun to populate the literature regarding the potential use of ASCs to mediate improved clinical outcomes of cardiac remodeling after myocardial infarction. These studies likely come from preceding reports on the potential use of MSCs to mediate improved clinical outcomes of cardiac remodeling after myocardial infarction, which was perhaps first shown in 2001 by Orlic et al.[23]

Since this time, the use of ASCs after myocardial infarction has been the subject of considerable debate and attention. Several reports have documented improved cardiac histopathology after myocardial infarction when ASCs were injected, either systemically or locally, into the affected myocardial area.[24–26] In these reports, myocardial fibrosis was reduced, angiogenesis was increased, and infarct size and subsequent ventricular dilation were limited. These postmortem findings translated to higher cardiac output, increased left ventricular ejection fractions, and improved left ventricular end diastolic pressures and volumes, despite suggesting a mechanism by which these phenotypes were affected. However, other reports have disagreed with these findings. For example, one report suggested that neither bone marrow MSCs nor ASCs engraft within injected myocardium beyond 4 to 5 weeks due to "drastic donor cell death."[27] In this study, "transplantation of either cell type was not capable of preserving ventricular function and dimensions, as confirmed by pressure-volume-loops and histology ... resulting in acute donor cell death and a subsequent loss of cardiac function similar to control groups." This study was important in that it employed novel noninvasive *in vivo* imaging to track cell location and count, which may avoid sampling bias that can occur with the use of multiple animals that need to be sacrificed at multiple time points to perform conventional postmortem histological analyses. As such, this study has significant clinical relevance. This study was also in agreement with several high-powered studies that showed poor cell engraftment of MSCs following myocardial injection after myocardial infarction.[28–30] In one exceptional study, however, MSCs that were modified by viral transduction with Akt1 were able to withstand the harsh, hypoxic environment of the infarcted myocardium, "reducing intramyocardial inflammation, collagen deposition and cardiac myocyte hypertrophy, regenerated 80–90% of lost myocardial volume, and completely normalized systolic and diastolic cardiac function."[31] Furthermore, "mesenchymal stem cells transduced with Akt1 restored fourfold greater myocardial volume than equal numbers of cells transduced with the reporter gene lacZ." This study was consistent with another report, in which MSCs that were transduced with Bcl-2 underwent significantly decreased apoptosis in the hypoxic myocardial tissue and significantly enhanced vascular endothelial growth factor secretion by more than 60%.[32] This last report highlights another important area of investigation, which is the role that ASCs (or MSCs) may play in hypoxic myocardial injuries.

Briefly, there are numerous reports that suggest ASCs may undergo cardiomyocyte differentiation in chemically defined media *in vitro*.[33,34] *In vivo*, however, only a small number of studies have suggested that the actual donor-derived cells expressed cardiac gene markers after a period of *in vivo* proliferation and differentiation.[26] These data also stand in contrast to studies that refuted transdifferentiation of MSCs and instead purported possible paracrine effects of these cells that may increase angiogenesis, reduce fibrosis, and so on.[27,35] Due to the inconsistent results, there is a great need for further research to understand if and how ASCs mediate improved clinical outcomes after myocardial infarction.

Despite these unsettled mechanistic issues, improvements in functional outcome have been shown in several animal studies. Perhaps as a result, researchers have looked toward clinical applications, most recently in 2007 with the initiation of two clinical trials. The two trials, PRECISE (A Randomized Clinical Trial of adiPose-deRived stEm & Regenerative Cells in the Treatment of Patients With NonrevaScularizable ischEmic Myocardium) and APOLLO (A Randomized Clinical Trial of AdiPOse-Derived Stem ceLLs in the Treatment of Patients With ST-Elevation myOcardial Infarction) are registered at http://www.clinicaltrials.gov. Both are prospective, double-blind, randomized, placebo-controlled trials, currently in phase 1, focusing on safety as the primary outcome measure as determined by major adverse cardiac and cerebral events (MACCE). Key inclusion criteria for the PRECISE trial are (a) ability to provide written informed consent; (b) males or females 20 to 75 years of age, inclusive; (c) coronary artery disease not amenable to any type of revascularization procedure (percutaneous or surgical) in the target area; (d) hemodynamic stability; (e) ability to undergo liposuction; (f) ability to walk on a treadmill; and (g) negative urine pregnancy test (females only). Key inclusion criteria for the APOLLO trial are (a) acute myocardial infarction (AMI); (b) clinical symptoms consistent with AMI for a minimum of 2 and a maximum of 12 hours from onset of symptoms to percutaneous coronary intervention (PCI) and unresponsive to nitroglycerin; (c) successful revascularization of the culprit lesion in the major epicardial vessel; (d) area of hypo- or akinesia corresponding to the culprit lesion, as determined by left ventriculogram at the time of primary PCI; (e) left ventricular ejection fraction (LVEF) of 30% or greater and less than or equal to 50% by left ventricular angiography at the time of successful revascularization; and (f) ability to undergo liposuction.

## 4.6 Novel Discovery and Implications for Stem Cell Biology

This review has highlighted much of the research of ASCs over the last decade. On an exciting and more recent note, our laboratory has released a report by Sun et al., in which ASCs were able to be induced to pluripotency by viral transduction with Oct4, Sox2, Klf4, and c-MYC, otherwise known as the "Yamanaka four" factors (Figure 4.1).[36] Induced pluripotency has previously been shown in fibroblasts; however, our studies suggest that ASCs undergo reprogramming with greater efficiency, with greater expediency, and without xenobiotic feeder reagents, all advantages that have significant implications for translational therapies and stem cell applications.[36] The advantages of using ASCs instead of fibroblasts perhaps stem from the already multipotent state of ASCs, instead of the lineage-committed state of fibroblasts. ASCs are not as far along in their lineage commitments, and perhaps for this reason, they undergo reprogramming to stem-like states with greater efficiency. In our study, iPSCs were derived from human ASCs *in vitro*. Multiple gene expression analyses showed embryonic stem cell-like phenotypes in these newly derived iPSCs. Subsequently, these iPSCs (derived from human ASCs) were injected into immunocompromised mice. The result was teratoma formation containing all three germ layers, suggesting that any cell in the human body could potentially be derived from iPSCs derived from adult, human ASCs. This discovery has significant implications for tissue regeneration not limited to bone and the heart, but may be extended to other organs of the body.

## 4.7 Noninvasive Imaging of Adipose-Derived Stem Cells

As part of the aforementioned investigations, adipose-derived stem cells have been transplanted, systemically or intravenously delivered, topically applied, seeded on scaffolds, and injected directly into tissue. The wide variety of delivery methods brings versatility to the use of ASCs; however, it has also introduced a confounding variable into their study.

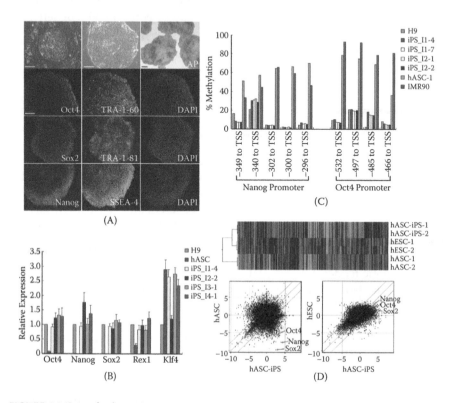

**FIGURE 4.1 (See color insert.)**
Characterization of human adipose stromal cell-derived induced pluripotent stem (hASC-iPS) cells. (A) Immunostaining of hASC-iPS cell colonies with common hES (human embryonic stem) cell markers. The two phase contrast microscopies show a typical hASC-iPS cell colony growing on MEF (mouse embryonic fibroblast) feeder cells and feeder-free Matrigel surface, respectively (bars, 100 μm). (B) Quantitative PCR (polymerase chain reaction) analyzing pluripotency gene expression level within hASCs (human adipose-derived stromal cells) and hASC-iPS cells relative to those in H9 hES cells. iPS_I1-4 denotes iPS cell line 4 derived from individual 1. (C) Bisulfite pyrosequencing measuring methylation status within the promoter region of Oct4 and Nanog genes in H9 hES cells, hASC-iPS cells, hASCs, and IMR90 cells. TSS, transcription start site. (D) Microarray data comparing global gene expression profiles of hASCs, hASC-iPS cells, and hES cells. Upper panel, heat map and hierarchical clustering analysis by Pearson correlation showing hASC-iPS cells are similar to hES cells and distinct from hASCs. Lower panel, scatter plots comparing global gene expression patterns between hASCs, hASC-iPS cells, and hES cells. Highlighted are the pluripotency genes Oct4, Sox2, and Nanog (red arrows). The green diagonal lines indicate linear equivalent and fivefold changes in gene expression levels between paired samples.

The modalities by which ASCs are delivered to tissue have not tradition-ally been accompanied by rigorous methodology to track these cells once they have been applied. As such, there are conflicting reports in the litera-ture regarding the presence and direct contribution of cells within tissue that may mediate observed phenotypes, migration of cells to wounds that mediate healing, and paracrine action versus transdifferentiation of cells within injury models. The ability to track ASCs promises higher-quality data, applicability to the clinical setting, and elimination of biases and tech-nical variation across experiments.

Several studies have reported the use of novel noninvasive imaging to track ASCs. As described previously, van der Bogt et al. tracked ASCs for 6 weeks after injection into infarcted myocardium.[27] This strategy allowed for tracking of cell number as well as location over the experimental time frame. It also eliminated sampling bias that may be introduced by interval sacrific-ing of animals for postmortem histomorphometric analyses. This study used a luciferase transgene reporter that was introduced into ASCs and a highly sensitive CCD (charge-coupled device) camera to detect their luminescence *in vivo*, as have others.[37-39] In several other reports, ASCs were labeled with superparamagnetic iron oxide and tracked by magnetic resonance imag-ing (MRI).[40-42] The shortcoming of this technique, however, is that it does not allow for tracking of cell counts as iron particles are divided between daughter cells during cell proliferation and ingested by macrophages on cell death.[43] In yet another study, MRI was again used, but this time to track Gadofluorine M uptake as it compared to Gadolinium-Diethylene-triamine-penta-acetic Acid (Gd-DTPA) and Gadomer uptake in human ASCs.[44] These three contrast agents were effective in tracking ASCs and correlated with postmortem histological analysis and staining. Finally, another study reported the use of micro-PET (positron emission tomography) and micro-CT (computed tomography) to track ASCs.[45]

In addition to the aforementioned studies that have made valuable basic discoveries, FDA requirements must be met to realize potential future stem cell therapies. Such hurdles include the use of xenobiotic reagents, viral-mediated genetic engineering, transmission of infectious diseases, and concerns of tumorigenicity. Furthermore, periodic noninvasive imaging of stem cells is likely to become a cornerstone of novel therapies given (a) the potential multiple areas of applications, particularly with consideration for recent excitement regarding reprogramming ASCs to pluripotency; (b) the potential risks of migration and tumorigenicity; and (c) a lack of under-standing regarding how, when, where, and why ASCs affect cell biology. As mechanistic questions are answered, noninvasive imaging of ASCs will play an increasingly important role in accelerating the transition from bench to bedside.

## References

1. Gurtner, G.C., Callaghan, M.J., and Longaker, M.T. Progress and potential for regenerative medicine. *Annual Review of Medicine* 58, 299–312 (2007).
2. Gurtner, G.C., Werner, S., Barrandon, Y., and Longaker, M.T. Wound repair and regeneration. *Nature* 453, 314–21 (2008).
3. Langer, R. and Vacanti, J.P. Tissue engineering. *Science* 260, 920–26 (1993).
4. Langer, R. and Vacanti, J.P. Artificial organs. *Sci Am* 273, 130–33 (1995).
5. Vacanti, J.P. and Langer, R. Tissue engineering: The design and fabrication of living replacement devices for surgical reconstruction and transplantation. *Lancet* 354 Suppl 1, SI32–SI34 (1999).
6. Khademhosseini, A., Vacanti, J.P., and Langer, R. Progress in tissue engineering. *Sci Am* 300, 64–71 (2009).
7. Eppley, B.L., Pietrzak, W.S., and Blanton, M.W. Allograft and alloplastic bone substitutes: a review of science and technology for the craniomaxillofacial surgeon. *J Craniofac Surg* 16, 981–89 (2005).
8. Parikh, S.N. Bone graft substitutes: Past, present, future. *J Postgrad Med* 48, 142–48 (2002).
9. Wan, D.C., Nacamuli, R.P., and Longaker, M.T. Craniofacial bone tissue engineering. *Dent Clin North Am* 50, 175–90, vii (2006).
10. De Ugarte, D.A., Morizono, K., Elbarbary, A., et al. Comparison of multi-lineage cells from human adipose tissue and bone marrow. *Cells Tissues Organs* 174, 101–9 (2003).
11. Panetta, N.J., Gupta, D.M., Kwan, M.D., et al. Tissue harvest by means of suction-assisted or third-generation ultrasound-assisted lipoaspiration has no effect on osteogenic potential of human adipose-derived stromal cells. *Plast Reconstr Surg* 124, 65–73 (2009).
12. Zuk, P.A. Stem cell research has only just begun. *Science* 293, 211–12 (2001).
13. Zuk, P.A., Zhu, M., Mizuno, H., et al. Multilineage cells from human adipose tissue: implications for cell-based therapies. *Tissue Eng* 7, 211–28 (2001).
14. Pittenger, M.F., Mackay, A.M., Beck, S.C., et al. Multilineage potential of adult human mesenchymal stem cells. *Science* 284, 143–47 (1999).
15. Lendeckel, S., Jodicke, A., Christophis, P., et al. Autologous stem cells (adipose) and fibrin glue used to treat widespread traumatic calvarial defects: Case report. *J Craniomaxillofac Surg* 32, 370–73 (2004).
16. Cowan, C.M., Shi, Y.Y., Aalami, O.O., et al. Adipose-derived adult stromal cells heal critical-size mouse calvarial defects. *Nat Biotechnol* 22, 560–67 (2004).
17. Aalami, O.O., Nacamuli, R.P., Lenton, K.A., et al. Applications of a mouse model of calvarial healing: differences in regenerative abilities of juveniles and adults. *Plast Reconstr Surg* 114, 713–20 (2004).
18. Cowan, C.M., Aalami, O.O., Shi, Y.Y., et al. Bone morphogenetic protein 2 and retinoic acid accelerate in vivo bone formation, osteoclast recruitment, and bone turnover. *Tissue Eng* 11, 645–58 (2005).
19. Wan, D.C., Shi, Y.Y., Nacamuli, R.P., et al. Osteogenic differentiation of mouse adipose-derived adult stromal cells requires retinoic acid and bone morphogenetic protein receptor type IB signaling. *Proc Natl Acad Sci USA* 103, 12335–40 (2006).

20. Wan, D.C., Siedhoff, M.T., Kwan, M.D., et al. Refining retinoic acid stimulation for osteogenic differentiation of murine adipose-derived adult stromal cells. *Tissue Eng* 13, 1623–31 (2007).

21. Quarto, N. and Longaker, M.T. FGF-2 inhibits osteogenesis in mouse adipose tissue-derived stromal cells and sustains their proliferative and osteogenic potential state. *Tissue Eng* 12, 1405–18 (2006).

22. Panetta, N.J., Gupta, D.M., Lee, J.K., et al. Human adipose-derived stromal cells respond to and elaborate bone morphogenetic protein-2 during in vitro osteogenic differentiation. *Plast Reconstr Surg* 125(2), 483–93 (2010).

23. Orlic, D., Kajstura, J., Chimenti, S., et al. Bone marrow cells regenerate infarcted myocardium. *Nature* 410, 701–5 (2001).

24. Chachques, J.C. Cellular cardiac regenerative therapy in which patients? *Expert Rev Cardiovasc Ther* 7, 911–19 (2009).

25. Hoke, N.N., Salloum, F.N., Loesser-Casey, K.E., and Kukreja, R.C. Cardiac regenerative potential of adipose tissue-derived stem cells. *Acta Physiol Hung* 96, 251–65 (2009).

26. Madonna, R., Geng, Y.J., and De Caterina, R. Adipose tissue-derived stem cells. Characterization and potential for cardiovascular repair. *Arterioscler Thromb Vasc Biol* 29(11):1723–29 (2009).

27. van der Bogt, K.E.A., Schrepfer, S., Yu, J., et al. Comparison of transplantation of adipose tissue- and bone marrow-derived mesenchymal stem cells in the infarcted heart. *Transplantation* 87, 642–52 (2009).

28. Amsalem, Y., Mardor, Y., Feinberg, M.S., et al. Iron-oxide labeling and outcome of transplanted mesenchymal stem cells in the infarcted myocardium. *Circulation* 116, I38–45 (2007).

29. Muller-Ehmsen, J., Krausgrill, B., Burst, V., et al. Effective engraftment but poor mid-term persistence of mononuclear and mesenchymal bone marrow cells in acute and chronic rat myocardial infarction. *J Mol Cell Cardiol* 41, 876–84 (2006).

30. Nakamura, Y., Wang, X., Xu, C., et al. Xenotransplantation of long-term-cultured swine bone marrow-derived mesenchymal stem cells. *Stem Cells* 25, 612–20 (2007).

31. Mangi, A.A., Noiseux, N., Kong, D., et al. Mesenchymal stem cells modified with Akt prevent remodeling and restore performance of infarcted hearts. *Nat Med* 9, 1195–1201 (2003).

32. Li, W., Ma, N., Ong, L.L., et al. Bcl-2 engineered MSCs inhibited apoptosis and improved heart function. *Stem Cells* 25, 2118–27 (2007).

33. Rangappa, S., Fen, C., Lee, E.H., Bongso, A., and Sim, E.K. Transformation of adult mesenchymal stem cells isolated from the fatty tissue into cardiomyocytes. *Ann Thorac Surg* 75, 775–79 (2003).

34. Miyahara, Y., Nagaya, N., Kataoka, M., et al. Monolayered mesenchymal stem cells repair scarred myocardium after myocardial infarction. *Nat Med* 12, 459–65 (2006).

35. Balsam, L.B., Wagers, A.J., Christensen, J.L., et al. Haematopoietic stem cells adopt mature haematopoietic fates in ischaemic myocardium. *Nature* 428, 668–73 (2004).

36. Sun, N., Panetta, N.J., Gupta, D.M., et al. Feeder-free derivation of induced pluripotent stem cells from adult human adipose stem cells. *Proc Natl Acad Sci USA* 106, 15720–25 (2009).

37. Wolbank, S., Peterbauer, A., Wassermann, E., et al. Labelling of human adipose-derived stem cells for non-invasive in vivo cell tracking. *Cell Tissue Bank* 8, 163–77 (2007).

38. Degano, I.R., Vilalta, M., Bago, J.R., et al. Bioluminescence imaging of calvarial bone repair using bone marrow and adipose tissue-derived mesenchymal stem cells. *Biomaterials* 29, 427–37 (2008).
39. Vilalta, M., Degano, I.R., Bago, J., et al. Biodistribution, long-term survival, and safety of human adipose tissue-derived mesenchymal stem cells transplanted in nude mice by high sensitivity non-invasive bioluminescence imaging. *Stem Cells Dev* 17, 993–1003 (2008).
40. Rice, H.E., Hsu, E.W., Sheng, H., et al. Superparamagnetic iron oxide labeling and transplantation of adipose-derived stem cells in middle cerebral artery occlusion-injured mice. *Am J Roentgenol* 188, 1101–8 (2007).
41. Liu, Z.Y., Wang, Y., Wang, G.Y., et al. [In vivo magnetic resonance imaging tracking of transplanted adipose-derived stem cells labeled with superparamagnetic iron oxide in rat hearts]. *Zhongguo Yi Xue Ke Xue Yuan Xue Bao* 31, 187–91 (2009).
42. Wang, L., Deng, J., Wang, J., et al. Superparamagnetic iron oxide does not affect the viability and function of adipose-derived stem cells, and superparamagnetic iron oxide-enhanced magnetic resonance imaging identifies viable cells. *Magn Reson Imaging* 27, 108–19 (2009).
43. Li, Z., Suzuki, Y., Huang, M., et al. Comparison of reporter gene and iron particle labeling for tracking fate of human embryonic stem cells and differentiated endothelial cells in living subjects. *Stem Cells* 26, 864–73 (2008).
44. Giesel, F.L., Stroick, M., Griebe, M., et al. Gadofluorine m uptake in stem cells as a new magnetic resonance imaging tracking method: An *in vitro* and in vivo study. *Invest Radiol* 41, 868–73 (2006).
45. Lee, S.W., Padmanabhan, P., Ray, P., et al. Stem cell-mediated accelerated bone healing observed with in vivo molecular and small animal imaging technologies in a model of skeletal injury. *J Orthop Res* 27, 295–302 (2009).

# 5

# *Umbilical Cord Stem Cells*

Suzanne Kadereit

## CONTENTS

## 5.1 Introduction

Umbilical cord blood and its surrounding tissue have become increasingly prominent as a source rich in different types of stem cells. In 1974, it was shown that cord blood contains functional hematopoietic progenitors (Knudtzon 1974). Additional groups have demonstrated that multilineage colony-forming units were present in cord blood and that cryopreservation was possible, thus setting the stage for cord blood as a suitable source for clinical applications (Koike 1983; Nakahata and Ogawa 1982). In 1989, cord blood from an HLA- (human leukocyte antigen-) identical sibling was used for the hematopoietic reconstitution of a young patient with Fanconi's anemia (Gluckman et al. 1989). Since then, numerous studies using cord blood have demonstrated that it contains hematopoietic stem cells (HSCs) capable of engrafting patients with hematologic diseases. It has emerged that cord blood grafts have multiple unique clinical advantages compared to bone marrow (BM) and mobilize peripheral blood stem cell sources. These include a greater tolerance for HLA mismatches, thus increasing significantly the donor pool for patients lacking HLA-matched donors; reduced graft-versus-host disease (GvHD) severity, with maintained graft-versus-leukemia (GVL) effect; collection without danger to the donor; and ease of banking (Laughlin et al. 2004; Rubinstein 2006). Particularly, the advantage of easy collection and banking capability make umbilical cord blood an attractive source for cellular therapies. More than 8,000 cord blood transplants had been performed by

2007, in large majority for hematologic diseases, with only rare occasions of donor-derived tumor development (Harris 2008; Matsunaga et al. 2005).

Data demonstrated that certain adult stem cells lose their potential with increasing age or with disease. It has been shown that HSCs in older adults are less regenerative than HSCs from younger adults, although they are more proliferative (Lansdorp et al. 1993). The numbers of circulating endothelial progenitor cells (EPCs) and their functionality are reduced in smokers and patients with certain pathological conditions, such as diabetes and coronary artery disease (CAD) (Tepper et al. 2002; Vasa et al. 2001). In addition, the numbers and functionality of mesenchymal stem cells (MSCs) is also reduced in the BM of the aging population (Stenderup et al. 2003). As such, performing autologous transplants for older patients, that is using their own BM stem cells, may likely not be beneficial. Stem cells from the umbilical cord are thought to be more primitive and would thus provide a more suitable source for transplantation for degenerative diseases in older adults.

Stem cells present in the umbilical cord have been used to derive tissue of mesodermal, endodermal, and ectodermal lineages. Stem and progenitor cells in cord blood include EPCs, MSCs, and the primitive unrestricted somatic stem cells (USSCs) (Alejandro Erices 2000; Asahara et al. 1997; Kogler et al. 2004). Accordingly, the umbilical cord is now under investigation as a potential cell source for nonhematologic reconstitution, such as regeneration of spinal cord injury or neurodegenerative diseases (e.g., Alzheimer's and Parkinson diseases), and as a potential source for immunotherapy and gene therapy (Garbuzova-Davis et al. 2006; Kohn et al. 1995; Newman et al. 2004; van de Ven et al. 2007).

Particularly interesting for clinical regeneration is that stem and progenitor cells in the umbilical cord appear to be more primitive, with higher proliferative capacity than their counterparts from adult BM or mobilized peripheral blood. The umbilical cord is also an abundant source for procurement of stem cells, as daily hundreds of thousands of births take place around the world. Importantly, however, as umbilical cord blood collections represent the general population giving birth, the HLA type of the collected stem cells is more representative of the general population. It is thus a good source of transplantable cells for ethnic minorities who are usually underrepresented in transplant registries and have difficulties in finding a compatible donor (Johansen et al. 2008). One could envision future therapies derived from the umbilical cord for a large population of patients with nonhematologic diseases such as cardiovascular disease, neurodegenerative diseases, regeneration of muscular dystrophies, and more. A word of caution, however, has to be added. If umbilical cord stem cells are not used in the original donor, the product is then an allogeneic transplant presumably entailing graft rejection problems in patients with an intact immune system. Although graft survival has improved significantly over the years with improved immune-suppressive therapy, both acute and chronic rejection still are major problems hampering the transplantation field (Lechler et al. 2005).

## 5.2 Umbilical Cord Blood-Derived
## Hematopoietic Stem Cells (HSCs)

Umbilical cord blood is rich in HSCs and much has been learned in the last years from this abundant source of human HSCs. While in the mouse the HSC is defined as c-kit+/Thy-1.1lo/Lin−/Sca-1+ (KTLS cell) and can be transplanted as a single cell and regenerate an entire mouse, identification of the equivalent cell in the human system has been challenging (Wagers et al. 2002). To identify and quantify the engraftment capacity of the human HSCs experimentally, the cells were injected into a tail vein of sublethally irradiated nonobese diabetic/severe combined immunodeficient (NOD/SCID) mice. Despite a xenogeneic environment, the human HSCs engrafted in this context, and the repopulating human cells are named SCID-repopulating cells (SRCs). Cord blood contains approximately 0.1–0.2% of such SRCs, compared to only 0.01–0.1% of SRCs in the adult BM and even less in mobilized peripheral blood (Coulombel 2004; Larochelle et al. 1996; Wang et al. 1997).

Cord blood also contains HSCs that appear to be more primitive than the HSCs in the BM or mobilized adult peripheral blood, with longer telomeres and a higher proliferation potential (Lansdorp et al. 1993, Vaziri et al. 1994). In standard *in vitro* assays, the cord blood HSCs also generate significantly more cells per colony in methylcellulose clonogenic assays colony forming cells (CFCs). Studies on the single-cell level showed that cord blood HSCs generate significantly more cells and more often give rise to two instead of one daughter long-term culture-initiating cell (LTC-IC) (Theunissen and Verfaillie 2005).

It is generally accepted that expression of CD34 correlates with a stem cell phenotype, and the number of transplanted CD34+ cells has been shown to correlate directly with human engraftment (Wagner et al. 2002). Another hallmark of the human HSC is the absence of expression of CD38, a marker gained on HSC differentiation. Within the CD34+/CD38− cell population in cord blood, 1 cell in 617 is capable of repopulating in the SRC assay, while no repopulating activity is contained within the CD38+ cell population (Bhatia et al. 1997).

Cord blood also contains primitive HSCs, lacking the expression of both CD34 and CD38 but capable of repopulating the NOD/SCID mouse. These so-called CD34− SRCs are present at two orders of magnitude lower than the CD34+ SRCs in cord blood, and contrary to the CD34+, SRCs give rise to CD4+ and CD8+ lymphocytes. Furthermore, the repopulating capacity increased from 1 SRC in 125,000 in CD34− cells to 1 SRC in 38,000 when CD34− cells were cultured for 4 days prior to injection into the NOD/SCID mice (Bhatia et al. 1998). The frequency of engraftment of CD34− cells could be further increased when the cells were injected directly into the BM. These CD34− cells engrafted not only in the injected BM but also in other bones, giving rise to

CD34+ HSCs and extensive lymphoid and myeloid progeny. The CD34+ progeny were, moreover, capable of engrafting in secondary recipients, indicative of the presence of long-term repopulating HSCs (Wang et al. 2003). The relevance of the CD34− cells in the human transplantation context is, however, not yet clear. Recipients of CD34+ BM grafts in which the donor-derived multilineage hematopoiesis has been followed over more than 7 years, are lacking donor-derived CD34−/Lin− cells in their BM (Verfaillie 2002).

Additional markers have been identified in the human HSC. One such marker is the detoxifying enzyme aldehyde dehydrogenase (ALDH). In cord blood, all *in vivo* SRC activity seems to be contained within the ALDH$^{hi}$ population, although the ALDH$^{lo}$ population also included CD34+/CD38− cells (Hess et al. 2004). Moreover, the ALDH$^{hi}$ population also contained CD34− cells with long-term SRC activity (Pearce and Bonnet 2007). Another marker, CD133, is already used in the clinical context (Lang et al. 2004). It is expressed mostly on CD34+ cells, and CD133+/CD34+ cells have a higher regeneration potential than CD133−/CD34+ cells. Analyses in NOD/SCID mice showed that purification of cells based on high ALDH activity and coexpression of CD133 increased the repopulating proportions of the cells, enriching for cells capable of repopulating secondary recipients (Hess et al. 2006). Interestingly, CD133 is also expressed on the hemangioblast, the common precursor of the HSC and the EPC (Gehling 2006).

## 5.3 Umbilical Cord Blood-Derived Endothelial Progenitor Cells (EPCs)

Heart failure is the leading cause of death worldwide (Lloyd-Jones et al. 2010). Millions of cardiac patients could benefit from cellular therapies aimed at improving or restoring blood flow. It is thus not surprising that since their initial isolation from peripheral adult blood and successful use for therapeutic angiogenesis over 10 years ago, EPCs have been the focus of intensive investigation (Asahara et al. 1997; Zhang et al. 2006). Many groups now have shown that these cells have angiogenic potential and are capable of contributing to vascular regeneration. In fact, it is now known that endogenous, circulating EPCs contribute to neoangiogenesis during ischemia, wound healing, and a variety of other pathological conditions in the adult.

EPCs reside in the BM in very small numbers and can proliferate, migrate, home to sites of neovascularization, and differentiate into endothelial cells (Asahara 2005; Asahara et al. 1999). On growth factor and cytokine release in the context of pathological conditions, EPCs can be mobilized into the circulation, where they usually account for about 0.01% of all circulating cells (Khoo et al. 2008; Walter and Dimmeler 2002). For example, it was shown that vascular trauma and acute myocardial infarction resulted in a rapid but transient mobilization of EPCs into the peripheral blood (Gill et al. 2001; Shintani et al. 2001; Takahashi et al. 1999). Conversely, it has become clear that the aging process is associated with endothelial dysfunction, as well as with decreased EPC function and mobilization and reduced numbers of circulating EPCs (Hill et al. 2003). These defects result in a decreased capacity of neovascularization in the aged and an increased risk for cardiovascular events (Hoenig et al. 2008). Moreover, the number of circulating EPCs and their functionality are reduced in patients with certain pathological conditions, such as diabetes and CAD (Kawamoto and Asahara 2007; Naruse et al. 2005).

In addition to BM and peripheral blood, EPCs can be isolated from umbilical vein and cord blood (Ingram et al. 2005; Murohara et al. 2000). EPCs isolated from cord blood have a similar marker expression profile; they express CD34, CD133, and vascular endothelial growth factor receptor 2 (VEGFR2) (Zhang et al. 2006). When comparing directly EPCs derived from cord blood with those from BM, cord blood EPCs expressed differing levels of stromal markers. CXCR4 and Stro-1 were expressed in a higher proportion in cord blood-derived EPCs, while CD73 and CD105 were expressed at lower levels compared to BM-derived EPCs. Expression of endothelial markers (CD146, CD31, and VE-cadherin) was similar (Finney et al. 2006).

EPCs from cord blood have a higher clonogenic and proliferative potential compared to peripheral blood EPCs (Ingram et al. 2004; Ott et al. 2005). In culture, cord blood EPC-derived colonies emerged 1 week earlier and were larger than colonies from adult blood-derived EPCs and formed capillary-like structures faster. When plating single cells, cord blood EPCs averaged a 100-fold greater number of cell progeny than the number of cells derived from an individual adult cell. In addition, cord blood contains a small subpopulation of EPCs that retains high levels of telomerase activity, has high population doubling (at least 100-fold), and could be replated into secondary and tertiary colonies (Ingram et al. 2004).

EPCs from cord blood have been used successfully in preclinical models of therapeutic neovascularization for hind limb ischemia, cardiac regeneration, and diabetic neuropathy (Finney et al. 2006; Murohara et al. 2000; Naruse et al. 2005; Ott et al. 2005). When comparing durability and function of EPC-derived vessels *in vivo*, it was found that EPCs derived from adult peripheral blood formed blood vessels that were not stable and regressed within 3 weeks. In contrast, umbilical cord blood-derived EPCs formed stable blood vessels with normal blood flow lasting for more than 4 months (Au et al. 2008). Overall, these studies suggest that cord blood EPCs could be used for clinical organ vascularization and regeneration and could restore long-lasting organ function. Use of cord blood EPCs could be of particular interest for patients with impaired autologous EPC function, such as diabetics and patients with CAD. A word of caution, however, has to be added. Animal models used to test the regenerative capacity of EPCs are usually immunocompromised mice, which (depending on the mouse strain) only poorly, if at all, reject the xenogeneic human cells. In the human patient receiving cord blood-derived EPCs, the transplanted cells would presumably be rejected as foreign by the immune system of the patient. Rejection could be attenuated by immuno-suppressive therapy and HLA matching, which would be facilitated by large cord blood banks containing cord blood representative of the general popu-lation giving blood and donating their cord blood.

## 5.4 Mesenchymal Stem Cells

It was long thought that cord blood did not contain MSCs, but recently it was shown that clonally expanded MSCs could be grown from cord blood, and that similar to MSCs from the BM-derived MSCs, these cells were of mul-tilineage differentiation potential. Their immunophenotype was consistent with that of BM MSCs and, similar to BM MSCs, these cells were shown to differentiate into bone, cartilage, and fat. In addition, these MSC-like cells from cord blood were able to differentiate into neuroglial- and hepatocyte-like cells (Lee et al. 2004). Many laboratories were able to isolate MSCs from cord blood and have shown that cord blood-derived MSCs are very simi-lar to BM MSCs. However, cord blood is a poor source of MSCs, likely due to generally low numbers of MSCs found in free-flowing blood. A much richer source of MSCs is the surrounding connective tissue of the umbili-cal cord, the Wharton's jelly. From there, cells can be isolated easily and are called Wharton's jelly cells (WJCs). These cells can be expanded to 80 pop-ulation doublings and display characteristics of BM MSCs, such as plastic adherence, spindle-shaped morphology, similar surface marker expression, extensive self-renewal capacity, and differentiation potential to bone, carti-lage, and adipose cells (Mitchell et al. 2003). Similar to BM MSCs, WJCs can support growth, expansion, and engraftment of HSCs and have immuno-modulatory capacities (Friedman et al. 2007; Troyer and Weiss 2008). Several points, however, distinguish WJCs from BM MSCs. Similar to other stem cell populations in cord blood, WJCs have a greater proliferation and expansion potential compared to BM MSCs (Baksh et al. 2007). Moreover, compared to BM MSCs, WJCs express higher levels of stem cell markers, such as FLK-1, ABCG2, CXCR4, and TERT, and WJCs can also be distinguished from BM MSCs based on their cytokine profile (Weiss et al. 2006).

WJCs have been differentiated into neural cells and used successfully to regenerate rodent neurodegenerative models for Parkinson's disease, retinal degeneration, and stroke (Koh et al. 2008; Lund et al. 2007; Mitchell et al. 2003; Weiss et al. 2006). WJCs have also been used to derive tissue-engineered heart valves, pulmonary conduits, and cartilage and bone constructs (Bailey et al. 2007; Hoerstrup et al. 2002; Schmidt et al. 2006).

The ease of procurement of the umbilical cord combined with the ease of isolation, expansion, *in vitro* differentiation, and successful application in regenerative animal models make WJCs an interesting cell source for future tissue engineering work and regenerative medicine. Moreover, as these cells are easily transfected and produce a large array of cytokines, WJCs should have broad applicability in human cell-based therapies.

## 5.5 The Unrestricted Somatic Stem Cell

Accumulating data suggested that cord blood also contains pluripotent stem cells capable of generating tissue of endodermal, mesodermal, and ecto-dermal origin. Generation of mesodermal and neural cells (neuronal and glial) was reported from cord blood, as well as generation of hepatocytes, after transplantation into NOD/SCID mice (Bicknese et al. 2002; Buzanska et al. 2002; Goodwin et al. 2001; Kakinuma et al. 2003; Newsome et al. 2003). Accordingly, stem cells could be isolated that are capable of differentiating into cells of all three germ layers. These pluripotent cells have been named unrestricted somatic stem cells (USSCs), display robust *in vitro* prolifera-tive capacity without spontaneous differentiation, and can be culture iso-lated from approximately 43% of fresh cord blood units (Kögler et al. 2004). Isolation from cryopreserved cord blood is, however, much poorer, with a success rate of only 19% (Kögler et al. 2010).

USSCs grow adherently, have a spindle-shaped morphology, maintain a normal karyotype over 40 population doublings, and maintain longer telomeres than MSCs from adult BM. USSCs share overlapping features with MSCs of fetal origin, such as immunophenotype and differentiation potential (O'Donoghue and Fisk 2004). *In vitro*, the USSCs can be differenti-ated into homogeneous cultures of osteoblasts, chondroblasts, and adipo-cytes, as well as into hematopoietic and neural cells, including astrocytes and neurons.

When transplanted into nude rats, the USSCs were able to initiate bone formation and repair critical size bone defects. When transplanted into the intact brain, USCCs differentiated into cells of typical neuronal morphol-ogy and persisted for up to 3 months. In the developmental model of the preimmune fetal sheep, USSCs contributed to cardiac tissue and differen-tiated into hematopoietic cells and hepatocytes (Kögler et al. 2004). Unlike fetal and adult MSCs, which predominantly engraft in the Purkinje fiber sys-tem and poorly generate cardiomyocytes in the fetal sheep model, the USSCs also generated cardiomyocytes (Airey et al. 2004; Kögler et al. 2004). Cardiac regeneration with USSCs could also be shown in an infarct model, demon-strating that these stem cells can contribute to developing tissue, as well as to injured and regenerating tissue (Kim et al. 2005).

## 5.6 Conclusions

Data accumulated over the last 10–20 years have clearly demonstrated that the umbilical cord and its surrounding tissue is rich in stem cells, which can readily be used for basic and clinical research. Its ample availability has, for example, enabled an in-depth characterization of human HSCs. Classically, stem cells (e.g., HSCs, MSCs) were isolated from human BM, which is far less abundant source than cord blood. Collection of BM is, moreover, not trivial and may exhibit impaired hematopoietic function in the donor decades later. Furthermore, stem cells in BM appear to be "older" and to have lower potential than their counterparts in umbilical cord.

Cord blood is increasingly banked around the world, both publicly and privately. With the insight gained into the types of stem cells contained in the umbilical cord, it may become possible to use this readily available source of human stem cells for regeneration of solid tissues, such as heart, bone, vessels, brain, and more. Its application for hematological regeneration for years has shown that the risk of tumorigenesis from transplanted cells appears to be extremely low (Matsunaga et al. 2005). Whether this will hold true in the context of transplantation of culture-expanded or culture-differentiated cord EPCs, MSCs, or USSCs remains to be seen.

## References

Airey, J.A., G. Almeida-Porada, E.J. Colletti, C.D. Porada, J. Chamberlain, M. Movsesian, J.L. Sutko, and E.D. Zanjani. 2004. Human mesenchymal stem cells form purkinje fibers in fetal sheep heart. *Circulation 109*, no. 11: 1401–7.

Alejandro Erices, P.C.J.J.M. 2000. Mesenchymal progenitor cells in human umbilical cord blood. *Br J Haematol* 109, no. 1: 235–42.

Asahara, T. 2005. Stem cell biology for vascular regeneration. *Ernst Schering Res Found Workshop* no. 54: 111–29.

Asahara, T., H. Masuda, T. Takahashi, C. Kalka, C. Pastore, M. Silver, M. Kearne, M. Magner, and J.M. Isner. 1999. Bone marrow origin of endothelial progenitor cells responsible for postnatal vasculogenesis in physiological and pathological neovascularization. *Circ Res* 85, no. 3: 221–8.

Asahara, T., T. Murohara, A. Sullivan, M. Silver, R. Van Der Zee, T. Li, B. Witzenbichler, G. Schatteman, and J. Isner. 1997. Isolation of putative progenitor endothelial cells for angiogenesis. *Science* 275, no. 5302: 964–7.

Au, P., L.M. Daheron, D.G. Duda, K.S. Cohen, J.A. Tyrrell, R.M. Lanning, D. Fukumura, D.T. Scadden, and R.K. Jain. 2008. Differential in vivo potential of endothelial progenitor cells from human umbilical cord blood and adult peripheral blood to form functional long-lasting vessels. *Blood* 111, no. 3: 1302–5.

Bailey, M.M., L. Wang, C.J. Bode, K.E. Mitchell, and M.S. Detamore. 2007. A comparison of human umbilical cord matrix stem cells and temporomandibular joint condylar chondrocytes for tissue engineering temporomandibular joint condylar cartilage. *Tissue Eng* 13, no. 8: 2003–10.

Baksh, D., R. Yao, and R.S. Tuan. 2007. Comparison of proliferative and multilineage differentiation potential of human mesenchymal stem cells derived from umbilical cord and bone marrow. *Stem Cells* 25, no. 6: 1384–92.

Bhatia, M., D. Bonnet, B. Murdoch, O.I. Gan, and J.E. Dick. 1998. A newly discovered class of human hematopoietic cells with SCID-repopulating activity. *Nat Med* 4, no. 9: 1038–45.

Bhatia, M., J.C. Wang, U. Kapp, D. Bonnet, and J.E. Dick. 1997. Purification of primitive human hematopoietic cells capable of repopulating immune-deficient mice. *Proc Natl Acad Sci USA* 94, no. 10: 5320–5.

Bicknese, A.R., H.S. Goodwin, C.O. Quinn, V.C. Henderson, S.N. Chien, and D.A. Wall. 2002. Human umbilical cord blood cells can be induced to express markers for neurons and glia. *Cell Transplant* 11, no. 3: 261–4.

Buzanska, L., E.K. Machaj, B. Zablocka, Z. Pojda, and K. Domanska-Janik. 2002. Human cord blood-derived cells attain neuronal and glial features in vitro. *J Cell Sci* 115, Pt 10: 2131–8.

Coulombel, L. 2004. Identification of hematopoietic stem/progenitor cells: Strength and drawbacks of functional assays. *Oncogene* 23, no. 43: 7210–22.

Finney, M.R., J. Greco, S. Haynesworth, J. Martin, D. Hedrick, J. Swan, D. Winter, S. Kadereit, M. Joseph, P. Fu, V. Pompili, and M. Laughlin. 2006. Direct comparison of umbilical cord blood versus bone marrow-derived endothelial precursor cells in mediating neovascularization in response to vascular ischemia. *Biol Blood Marrow Transplant* 12, no. 5: 585–93.

Friedman, R., M. Betancur, L. Boissel, H. Tuncer, C. Cetrulo, and H. Klingemann. 2007. Umbilical cord mesenchymal stem cells: Adjuvants for human cell transplantation. *Biol Blood Marrow Transplant* 13, no. 12: 1477–86.

Garbuzova-Davis, S., A.E. Willing, S. Saporta, P.C. Bickford, C. Gemma, N. Chen, C.D. Sanberg, S.K. Klasko, C.V. Borlongan, and P.R. Sanberg. 2006. Novel cell therapy approaches for brain repair. *Prog Brain Res* 157: 207–22.

Gehling, U.M. 2006. Hemangioblasts and their progeny. *Methods Enzymol* 419: 179–93.

Gill, M., S. Dias, K. Hattori, M.L. Rivera, D. Hicklin, L. Witte, L. Girardi, R. Yurt, H. Himel, and S. Rafii. 2001. Vascular trauma induces rapid but transient mobilization of VEGFR2(+)AC133(+) endothelial precursor cells. *Circ Res* 88, no. 2: 167–74.

Gluckman, E., H.A. Broxmeyer, A.D. Auerbach, H.S. Friedman, G.W. Douglas, A. Devergie, H. Esperou, D. Thierry, G. Socie, P. Lehn, et al. 1989. Hematopoietic reconstitution in a patient with Fanconi's anemia by means of umbilical-cord blood from an HLA-identical sibling. *N Engl J Med* 321, no. 17: 1174–8.

Goodwin, H.S., A.R. Bicknese, S.N. Chien, B.D. Bogucki, C.O. Quinn, and D.A. Wall. 2001. Multilineage differentiation activity by cells isolated from umbilical cord blood: expression of bone, fat, and neural markers. *Biol Blood Marrow Transplant* 7, no. 11: 581–8.

Harris, D. 2008. Cord blood stem cells: A review of potential neurological applications. *Stem Cell Rev Rep* 4, no. 4: 269–74.

Hess, D.A., T.E. Meyerrose, L. Wirthlin, T.P. Craft, P.E. Herrbrich, M.H. Creer, and J.A. Nolta. 2004. Functional characterization of highly purified human hematopoietic repopulating cells isolated according to aldehyde dehydrogenase activity. *Blood* 104, no. 6: 1648–55.

Hess, D.A., L. Wirthlin, T.P. Craft, P.E. Herrbrich, S.A. Hohm, R. Lahey, W.C. Eades, M.H. Creer, and J.A. Nolta. 2006. Selection based on CD133 and high aldehyde dehydrogenase activity isolates long-term reconstituting human hematopoietic stem cells. *Blood* 107, no. 5: 2162–9.

Hill, J.M., G. Zalos, J.P. Halcox, W.H. Schenke, M.A. Waclawiw, A.A. Quyyumi, and T. Finkel. 2003. Circulating endothelial progenitor cells, vascular function, and cardiovascular risk. *N Engl J Med* 348, no. 7: 593–600.

Hoenig, M.R., C. Bianchi, A. Rosenzweig, and F.W. Sellke. 2008. Decreased vascular repair and neovascularization with ageing: Mechanisms and clinical relevance with an emphasis on hypoxia-inducible factor-1. *Curr Mol Med* 8, no. 8: 754–67.

Hoerstrup, S.P., A. Kadner, C. Breymann, C.F. Maurus, C.I. Guenter, R. Sodian, J.F. Visjager, G. Zund, and M.I. Turina. 2002. Living, autologous pulmonary artery conduits tissue engineered from human umbilical cord cells. *Ann Thorac Surg* 74, no. 1: 46–52.

Ingram, D.A., L.E. Mead, D.B. Moore, W. Woodard, A. Fenoglio, and M.C. Yoder. 2005. Vessel wall-derived endothelial cells rapidly proliferate because they contain a complete hierarchy of endothelial progenitor cells. *Blood* 105, no. 7: 2783–6.

Ingram, D.A., L.E. Mead, H. Tanaka, V. Meade, A. Fenoglio, K. Mortell, K. Pollok, M.J. Ferkowicz, D. Gilley, and M.C. Yoder. 2004. Identification of a novel hierarchy of endothelial progenitor cells using human peripheral and umbilical cord blood. *Blood* 104, no. 9: 2752–60.

Johansen, K., J. Schneider, M. Mccaffree, and G. Woods. 2008. Efforts of the United States' national marrow donor program and registry to improve utilization and representation of minority donors. *Transfus Med* 18, no. 4: 250–9.

Kakinuma, S., Y. Tanaka, R. Chinzei, M. Watanabe, K. Shimizu-Saito, Y. Hara, K. Teramoto, S. Arii, C. Sato, K. Takase, T. Yasumizu, and H. Teraoka. 2003. Human umbilical cord blood as a source of transplantable hepatic progenitor cells. *Stem Cells* 21, no. 2: 217–27.

Kawamoto, A. and T. Asahara. 2007. Role of progenitor endothelial cells in cardiovascular disease and upcoming therapies. *Catheter Cardiovasc Interv* 70, no. 4: 477–84.

Khoo, C.P., P. Pozzilli, and M.R. Alison. 2008. Endothelial progenitor cells and their potential therapeutic applications. *Regen Med* 3, no. 6: 863–76.

Kim, B.O., H. Tian, K. Prasongsukarn, J. Wu, D. Angoulvant, S. Wnendt, A. Muhs, D. Spitkovsky, and R.K. Li. 2005. Cell transplantation improves ventricular function after a myocardial infarction: a preclinical study of human unrestricted somatic stem cells in a porcine model. *Circulation* 112, no. 9 Suppl: I96–104.

Knudtzon, S. 1974. In vitro growth of granulocytic colonies from circulating cells in human cord blood. *Blood* 43, no. 3: 357–61.

Kögler, G., T. Radke, and R. Sorg. 2010. The unrestricted somatic stem cell (USSC). In *Umbilical cord stem cells: A future for regenerative medicine?* Eds. Kadereit, S. and Udolph, G. Singapore: World Scientific.

Kögler, G., S. Sensken, J.A. Airey, T. Trapp, M. Muschen, N. Feldhahn, S. Liedtke, R.V. Sorg, J. Fischer, C. Rosenbaum, S. Greschat, A. Knipper, J. Bender, O. Degistirici, J. Gao, A.I. Caplan, E.J. Colletti, G. Almeida-Porada, H.W. Muller, E. Zanjani, and P. Wernet. 2004. A new human somatic stem cell from placental cord blood with intrinsic pluripotent differentiation potential. *J Exp Med*. 200, no. 2: 123–35.

Koh, S.H., K.S. Kim, M.R. Choi, K.H. Jung, K.S. Park, Y.G. Chai, W. Roh, S.J. Hwang, H.J. Ko, Y.M. Huh, H.T. Kim, and S.H. Kim. 2008. Implantation of human umbilical cord-derived mesenchymal stem cells as a neuroprotective therapy for ischemic stroke in rats. *Brain Res* 1229: 233–48.

Kohn, D.B., K.I. Weinberg, J.A. Nolta, L.N. Heiss, C. Lenarsky, G.M. Crooks, M.E. Hanley, G. Annett, J.S. Brooks, A. El-Khoureiy, et al. 1995. Engraftment of gene-modified umbilical cord blood cells in neonates with adenosine deaminase deficiency. *Nat Med* 1, no. 10: 1017–23.

Koike, K. 1983. Cryopreservation of pluripotent and committed hemopoietic progenitor cells from human bone marrow and cord blood. *Pediatr Int* 25, no. 3: 275–83.

Lang, P., P. Bader, M. Schumm, T. Feuchtinger, H. Einsele, M. Fuhrer, C. Weinstock, R. Handgretinger, S. Kuci, D. Martin, D. Niethammer, and J. Greil. 2004. Transplantation of a combination of CD133+ and CD34+ selected progenitor cells from alternative donors. *Br J Haematol* 124, no. 1: 72–9.

Lansdorp, P.M., W. Dragowska, and H. Mayani. 1993. Ontogeny-related changes in proliferative potential of human hematopoietic cells. *J Exp Med* 178, no. 3: 787–91.

Larochelle, A., J. Vormoor, H. Hanenberg, J.C. Wang, M. Bhatia, T. Lapidot, T. Moritz, B. Murdoch, X.L. Xiao, I. Kato, D.A. Williams, and J.E. Dick. 1996. Identification of primitive human hematopoietic cells capable of repopulating NOD/SCID mouse bone marrow: Implications for gene therapy. *Nat Med* 2, no. 12: 1329–37.

Laughlin, M.J., M. Eapen, P. Rubinstein, J.E. Wagner, M.J. Zhang, R.E. Champlin, C. Stevens, J.N. Barker, R.P. Gale, H.M. Lazarus, D.I. Marks, J.J. Van Rood, A. Scaradavou, and M.M. Horowitz. 2004. Outcomes after transplantation of cord blood or bone marrow from unrelated donors in adults with leukemia. *N Engl J Med* 351, no. 22: 2265–75.

Lechler, R.I., M. Sykes, A.W. Thomson, and L.A. Turka. 2005. Organ transplantation—how much of the promise has been realized? *Nat Med* 11, no. 6: 605–13.

Lee, O.K., T.K. Kuo, W.M. Chen, K.D. Lee, S.L. Hsieh, and T.H. Chen. 2004. Isolation of multipotent mesenchymal stem cells from umbilical cord blood. *Blood* 103, no. 5: 1669–75.

Lloyd-Jones, D., R.J. Adams, T.M. Brown, M. Carnethon, S. Dai, G. De Simone, T.B. Ferguson, E. Ford, K. Furie, C. Gillespie, A. Go, K. Greenlund, N. Haase, S. Hailpern, P.M. Ho, V. Howard, B. Kissela, S. Kittner, D. Lackland, L. Lisabeth, A. Marelli, M.M. Mcdermott, J. Meigs, D. Mozaffarian, M. Mussolino, G. Nichol, V.L. Roger, W. Rosamond, R. Sacco, P. Sorlie, R. Stafford, T. Thom, S. Wasserthiel-Smoller, N.D. Wong, and J. Wylie-Rosett. 2010. Executive summary: heart disease and stroke statistics—2010 update: A report from the American Heart Association. *Circulation* 121, no. 7: 948–54.

Lund, R.D., S. Wang, B. Lu, S. Girman, T. Holmes, Y. Sauve, D.J. Messina, I.R. Harris, A.J. Kihm, A.M. Harmon, F.-Y. Chin, A. Gosiewska, and S.K. Mistry. 2007. Cells isolated from umbilical cord tissue rescue photoreceptors and visual functions in a rodent model of retinal disease. *Stem Cells* 25, no. 3: 602–11.

Matsunaga, T., K. Murase, M. Yoshida, A. Fujimi, S. Iyama, K. Kuribayashi, T. Sato, K. Kogawa, Y. Hirayama, S. Sakamaki, K. Kohda, and Y. Niitsu. 2005. Donor cell derived acute myeloid leukemia after allogeneic cord blood transplantation in a patient with adult t-cell lymphoma. *Am J Hematol* 79, no. 4: 294–98.

Mitchell, K.E., M.L. Weiss, B.M. Mitchell, P. Martin, D. Davis, L. Morales, B. Helwig, M. Beerenstrauch, K. Abou-Easa, T. Hildreth, and D. Troyer. 2003. Matrix cells from Wharton's jelly form neurons and glia. *Stem Cells* 21, no. 1: 50–60.

Murohara, T., H. Ikeda, J. Duan, S. Shintani, K. Sasaki, H. Eguchi, I. Onitsuka, K. Matsui, and T. Imaizumi. 2000. Transplanted cord blood-derived endothelial precursor cells augment postnatal neovascularization. *J Clin Invest* 105, no. 11: 1527–36.

Nakahata, T. and M. Ogawa. 1982. Hemopoietic colony-forming cells in umbilical cord blood with extensive capability to generate mono- and multipotential hemopoietic progenitors. *J Clin Invest* 70, no. 6: 1324–28.

Naruse, K., Y. Hamada, E. Nakashima, K. Kato, R. Mizubayashi, H. Kamiya, Y. Yuzawa, S. Matsuo, T. Murohara, T. Matsubara, Y. Oiso, and J. Nakamura. 2005. Therapeutic neovascularization using cord blood-derived endothelial progenitor cells for diabetic neuropathy. *Diabetes* 54, no. 6: 1823–28.

Newman, M.B., D.F. Emerich, C.V. Borlongan, C.D. Sanberg, and P.R. Sanberg. 2004. Use of human umbilical cord blood (HUCB) cells to repair the damaged brain. *Curr Neurovasc Res* 1, no. 3: 269–81.

Newsome, P.N., I. Johannessen, S. Boyle, E. Dalakas, K.A. Mcaulay, K. Samuel, F. Rae, L. Forrester, M.L. Turner, P.C. Hayes, D.J. Harrison, W.A. Bickmore, and J.N. Plevris. 2003. Human cord blood-derived cells can differentiate into hepatocytes in the mouse liver with no evidence of cellular fusion. *Gastroenterology* 124, no. 7: 1891–900.

O'Donoghue, K. and N.M. Fisk. 2004. Fetal stem cells. *Best Pract Res Clin Obstet Gynaecol* 18, no. 6: 853–75.

Ott, I., U. Keller, M. Knoedler, K. Gotze, K. Doss, P. Fischer, K. Urlbauer, G. Debus, N. Von Bubnoff, M. Rudelius, A. Schomig, C. Peschel, and R. Oostendorp. 2005. Endothelial-like cells expanded from CD34$^+$ blood cells improve left ventricular function after experimental myocardial infarction. *FASEB J* 19, no. 8: 992–4.

Pearce, D.J. and D. Bonnet. 2007. The combined use of hoechst efflux ability and aldehyde dehydrogenase activity to identify murine and human hematopoietic stem cells. *Exp Hematol* 35, no. 9: 1437–46.

Rubinstein, P. 2006. Why cord blood? *Hum Immunol* 67, no. 6: 398–404.

Schmidt, D.R., A. Mol, B. Odermatt, S. Neuenschwander, C. Breymann, M. Gössi, M. Genoni, G. Zund, and S.P. Hoerstrup. 2006. Engineering of biologically active living heart valve leaflets using human umbilical cord-derived progenitor cells. *Tissue Eng* 12, no. 11: 3223–32.

Shintani, S., T. Murohara, H. Ikeda, T. Ueno, T. Honma, A. Katoh, K. Sasaki, T. Shimada, Y. Oike, and T. Imaizumi. 2001. Mobilization of endothelial progenitor cells in patients with acute myocardial infarction. *Circulation* 103, no. 23: 2776–9.

Stenderup, K., J. Justesen, C. Clausen, and M. Kassem. 2003. Aging is associated with decreased maximal life span and accelerated senescence of bone marrow stromal cells. *Bone* 33, no. 6: 919–26.

Takahashi, T., C. Kalka, H. Masuda, D. Chen, M. Silver, M. Kearney, M. Magner, J.M. Isner, and T. Asahara. 1999. Ischemia- and cytokine-induced mobilization of bone marrow-derived endothelial progenitor cells for neovascularization. *Nat Med* 5, no. 4: 434–8.

Tepper, O.M., R.D. Galiano, J.M. Capla, C. Kalka, P.J. Gagne, G.R. Jacobowitz, J.P. Levine, and G.C. Gurtner. 2002. Human endothelial progenitor cells from type II diabetics exhibit impaired proliferation, adhesion, and incorporation into vascular structures. *Circulation* 106, no. 22: 2781–6.

Theunissen, K. and C.M. Verfaillie. 2005. A multifactorial analysis of umbilical cord blood, adult bone marrow and mobilized peripheral blood progenitors using the improved ML-IC assay. *Exp Hematol* 33, no. 2: 165–72.

Troyer, D.L. and M.L. Weiss. 2008. Concise review: Wharton's jelly-derived cells are a primitive stromal cell population. *Stem Cells* 26, no. 3: 591–9.

van de Ven, C., D. Collins, M.B. Bradley, E. Morris, and M.S. Cairo. 2007. The potential of umbilical cord blood multipotent stem cells for nonhematopoietic tissue and cell regeneration. *Exp Hematol* 35, no. 12: 1753–65.

Vasa, M., S. Fichtlscherer, A. Aicher, K. Adler, C. Urbich, H. Martin, A.M. Zeiher, and S. Dimmeler. 2001. Number and migratory activity of circulating endothelial progenitor cells inversely correlate with risk factors for coronary artery disease. *Circ Res* 89, no. 1: E1–E7.

Vaziri H, Dragowska W, Allsopp RC, Thomas TE, Harley CB, Lansdorp PM. 1994. Evidence for a mitotic clock in human hematopoietic stem cells: loss of telomeric DNA with age. *Proc Natl Acad Sci USA*, no. 91:9857-60.

Verfaillie, C.M. 2002. Hematopoietic stem cells for transplantation. *Nat Immunol* 3, no. 4: 314–7.

Wagers, A.J., R.I. Sherwood, J.L. Christensen, and I.L. Weissman. 2002. Little evidence for developmental plasticity of adult hematopoietic stem cells. *Science* 297, no. 5590: 2256–9.

Wagner, J.E., J.N. Barker, T.E. Defor, K.S. Baker, B.R. Blazar, C. Eide, A. Goldman, J. Kersey, W. Krivit, M.L. Macmillan, P.J. Orchard, C. Peters, D.J. Weisdorf, N.K. Ramsay, and S.M. Davies. 2002. Transplantation of unrelated donor umbilical cord blood in 102 patients with malignant and nonmalignant diseases: Influence of CD34 cell dose and HLA disparity on treatment-related mortality and survival. *Blood* 100, no. 5: 1611–8.

Walter, D.H. and S. Dimmeler. 2002. Endothelial progenitor cells: Regulation and contribution to adult neovascularization. *Herz* 27, no. 7: 579–88.

Wang, J., T. Kimura, R. Asada, S. Harada, S. Yokota, Y. Kawamoto, Y. Fujimura, T. Tsuji, S. Ikehara, and Y. Sonoda. 2003. SCID-repopulating cell activity of human cord blood-derived CD34⁻ cells assured by intra-bone marrow injection. *Blood* 101, no. 8: 2924–31.

Wang, J.C., M. Doedens, and J.E. Dick. 1997. Primitive human hematopoietic cells are enriched in cord blood compared with adult bone marrow or mobilized peripheral blood as measured by the quantitative in vivo SCID-repopulating cell assay. *Blood* 89, no. 11: 3919–24.

Weiss, M.L., S. Medicetty, A.R. Bledsoe, R.S. Rachakatla, M. Choi, S. Merchav, Y. Luo, M.S. Rao, G. Velagaleti, and D. Troyer. 2006. Human umbilical cord matrix stem cells: preliminary characterization and effect of transplantation in a rodent model of Parkinson's disease. *Stem Cells* 24, no. 3: 781–92.

Zhang, L., R. Yang, and Z. Han. 2006. Transplantation of umbilical cord blood-derived endothelial progenitor cells: A promising method of therapeutic revascularisation. *Eur J Haematol* 76, no. 1: 1–8.

# 6

## Induced Pluripotent Stem Cells

Timothy J. Nelson, Satsuki Yamada, Robert J. McDonald,
Almudena Martinez-Fernandez, Carmen Perez-Terzic,
Yasuhiro Ikeda, and Andre Terzic

## CONTENTS

## 6.1 Introduction

Stem cells are present throughout the lifespan. Naturally derived stem cells include embryonic stem cells, perinatal stem cells, and adult stem cells [1–8]. These endogenous pools of stem cells provide the foundation for cardiovascular development in utero and tissue renewal throughout adulthood [9–11]. Beyond natural sources of stem cells that are limited by availability, immune intolerance, and lineage specification, bioengineered stem cell platforms are rapidly being developed for regenerative medicine applications. This chapter provides an overview of the current state of the art for induced pluripotent stem (iPS) cells and highlights the emerging techniques available to image the transformation of ordinary cells into stem cell counterparts that have acquired the ability to functionally repair the damaged heart.

## 6.2 Bioengineered Stem Cells

The newest platform of bioengineered stem cells offers the ability to provide an unlimited supply of progenitor cells for virtually all cell types and tissues of the adult body, starting from ordinary self-derived tissues (Figure 6.1). By exploiting epigenetics and the microenvironment of somatic nuclei, the aim is to reverse the cell fate of common, readily available cell types to achieve conversion from mature back to the embryonic ground state [12]. Advancements have built on the pioneering work of somatic cell nuclear transfer (SCNT) technology that demonstrated the efficacy of transacting factors present within the mammalian oocytes, conserved across species, to reprogram somatic cell nuclei to an undifferentiated state [13,14]. Thus, therapeutic cloning refers to SCNT, in which the nuclear content of a somatic cell from an individual is transferred into an enucleated donor egg to derive blastocysts that contain pluripotent embryonic-like stem cells. In this way, SCNT has produced cloned embryonic stem cells from multiple mammalian somatic cell biopsies [15–16]. The pluripotency of derived cells

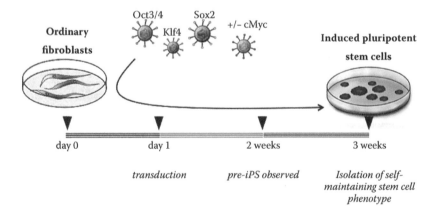

**FIGURE 6.1**
Nuclear reprogramming of somatic tissue into stem cell-like progenitors. Ordinary fibroblasts, derived from multiple sources, including adult patient-specific tissues, are transduced with ectopic stemness factors Oct3/4, Sox2, and Klf4 with or without c-Myc. The coerced expression of stemness-related factors induces global changes within the somatic cells over 2–4 weeks and results in the production of a self-maintaining stem cell-like population. Clonal expansion of these induced pluripotent stem (iPS) cells demonstrates the ability to propagate indefinitely and allows subsequent detailed analysis of full stem cell functionality.

has been confirmed through germline transmission and reproductive cloning. However, due to technological limitations, cloned human blastocytes have only recently been achieved, albeit in low efficiency [17], and successful isolation of embryonic stem cells from the inner cell mass has yet to be demonstrated with human protocols.

### 6.2.1 Principles of Nuclear Reprogramming

Nuclear reprogramming of adult somatic cells through ectopic introduction of a small number of pluripotency-associated transcription factors is a streamlined approach to induce an embryonic stem cell-like phenotype [18–21]. Transcription factor sets Oct4, Sox2, c-Myc and Klf4, or alternatively, Oct4, Sox2, Nanog, and Lin28 [22] are sufficient to reprogram human somatic cells by inducing a sequential reversal into a pluripotent phenotype (Figure 6.1). The process of nuclear reprogramming requires controlled expression of specific stemness factors in the proper stoichiometry for a defined period of time [23–29]. Multiple source tissues have been successfully reprogrammed, such as ordinary fibroblasts [30], keratinocytes [31], hematopoietic lineages [32], or adipose tissue [33]. The balanced exposure of ectopic factors is sufficient to induce telomere elongation [34], histone modifications [35], secondary gene expression profiles [36], and cellular metamorphosis that collectively reestablish a self-stabilizing phenotype of pluripotency [37]. Reprogramming typically occurs within weeks of coerced equilibrium of the trans-acting factors that can be delivered to the nucleus by plasmids, viruses, or bioengineered proteins (Figure 6.1). Thereby, ectopic transgene expression initiates a sequence of stochastic events that eventually transforms a small fraction of cells (<0.5%) to acquire this imposed pluripotent state characterized by a stable epigenetic environment indistinguishable from the blastocyst-derived natural stem cell milieu. The acquired pluripotent ground state culminates in the maintenance of the unique developmental potential to differentiate into all germ layers. In this way, iPS cells with the ability to derive patient-specific progenitor cells should largely eliminate the concern of stem cell shortage, immune rejection of nonautologous sources, and inadequate capacity for lineage specification [38–40]. Moreover, iPS cell-based technology will facilitate the production of cell line panels that closely reflect the genetic diversity of a population, enabling the discovery, development, and validation of diagnostic and therapeutics tailored for each individual [41].

Such platforms bypass the need for embryo extraction to generate true pluripotent stem cell phenotypes from autologous sources. In the mouse, bioengineering has yielded iPS cells sufficient for complete de novo embryogenesis as the highest evidence of pluripotent stringency [42,43]; in humans, by giving rise to all three germ layers, bioengineering has ensured comprehensive multilineage tissue differentiation. Self-derived iPS cells will be recognized within the transplanted hosts as native tissue due to their autologous status but will also require a new level of protection from dysregulated growth. The next generation of bioengineered stem cells will likely include specialized properties for improved stress tolerance, streamlined differentiation capacity, and increased engraftment/survival to improve regenerative potential.

### 6.2.1.1 First-Generation Technology

Retroviral and lentiviral approaches offered the initial methodology that launched the field and established the technological basis of nuclear reprogramming with rapid confirmation across integrating vector systems [18,22,23,31,44–50]. The risk of oncogenic genes and insertional mutagenesis inherent to stable genomic integration have been recognized as potential limitations from the outset of this technology. However, distinct advantages of the retroviral-based vector systems have generated critical insights to the mechanisms of reprogramming. Retroviral and lentiviral systems have built-in sequences that silence the process of transcription on pluripotent induction, thus temporarily restricting the persistent exposure to ectopic gene expression at the time of reinduction of pluripotency. This allows an essential observation to be made in that successful self-maintenance of the pluripotent state was possible without long-term transgene expression. This made it possible to envision systems for transient production of stemness-related genes without integration into the genome to improve the safety and efficacy of nuclear reprogramming. The first proof of principle was achieved by nonintegrating viral vector systems, such as adenovirus [51], and confirmed by repeated exposure to extrachromosomal plasmid-based transgenes [52]. Importantly, these reports demonstrated that expression of stemness-related factors was required for only a limited time frame until the progeny developed autonomous self-renewal, establishing nuclear reprogramming as a bioengineered process that resets a sustainable pluripotent cell fate independent of permanent genomic modifications. The inherent inefficiency of nonintegrated technologies has, however, hindered broader applicability and stimulated the search for more efficient methodology.

### 6.2.1.2 Second-Generation Technology

Nonviral approaches capable of high-efficiency production have advanced iPS cell technology toward clinical applications [53,54]. These approaches are dependent on short sequences of mobile genetic elements that can be used to integrate transgenes into host cell genomes and provide a genetic tag to "cut and paste" flanked genomic DNA sequences [55]. The piggyBac (PB) system couples enzymatic cleavage with sequence-specific recognition using a transposon/transposase interaction to ensure high-efficiency removal of flanked DNA without residual footprint. This technology achieves a traceless transgenic approach in which nonnative genomic sequences, transiently required for nuclear reprogramming, can be removed on induction of pluripotency. Specifically, using the PB transposition system with randomly integrated stemness-related transgenes, studies have demonstrated that disposal of ectopic genes could be efficiently regulated on induction of self-maintaining pluripotency, according to expression of the transposase enzyme without infringement on genomic stability [54]. This state-of-the-art system is qualified to allow safe integration and removal of ectopic transgenes, improving the efficiency of iPS cell production and facilitating a minimally invasive methodology without permanent modifications to the progeny. Alternatively, the security of unmodified genomic intervention can be achieved with non-integrating episomal vectors [56]. Collectively, these emerging strategies have accelerated the feasibility of regenerative medicine, bringing it a step closer to clinical applicability by allowing genetically unmodified progenitor cells to acquire the capacity of pluripotency.

Alternatively, high-stringency iPS cells have also been produced with proteins in the absence of genetic material [57,58]. The protein-only approach has successfully induced reprogramming either with whole-cell extract enriched in four stemness factors used in combination with pharmacological induction of cell permeability or with stemness factors modified by cell-permeating polyarginine tag [57]. Although the reprogramming efficiency compared to genetic methodology is reduced, strategies have been developed that complement the influence of stemness factors within somatic cells. Namely, small molecules targeting histone modifications have increased reprogramming efficiencies [59], along with the latest discovery that the tumor suppressor gene p53 is responsible for inhibiting the reprogramming process [60–65]. Thereby, transient knockdown of p53 using small interfering RNA (siRNA) strategies that target the breakdown of mRNA (messenger RNA) or overexpression of MDM-2 to increase p53 protein degradation has proven successful to increase overall efficiency by one or two orders of magnitude, with up to 20% of selected cells undergoing bona fide reprogramming [60–65]. Together, rapid advancements in nuclear reprogramming have brought bioengineered pluripotent stem cells closer to the milestones required for clinical translation.

### 6.2.1.3 Models of Nuclear Reprogramming

Nuclear reprogramming through ectopic gene expression offers a revolutionary framework to derive pluripotent stem cells from somatic biopsy. Two models have been used to explain the reprogramming process. The first is the "elite model," in which a small number of partially preprogrammed progenitor cells are able to respond to additional stemness factors. The "stochastic model," in contrast, postulates that virtually any ordinary cell type can be fully reprogrammed in the presence of the proper combination of conditions, depending on the nature and environment of the target cell [21]. Although both models are plausible and have received support from experimental data, the stochastic model has received additional credibility with the finding that mature tissue such as mature lymphoid cell types, validated according to V(D)J recombination, are capable of dedifferentiating into stable pluripotent stem cells [49,62]. This supports the stochastic model by showing that cell fate is fully reversible, even for mature tissue sources if they are exposed to proper intracellular and extracellular environments.

## 6.2.2 Imaging iPS Cells from Discovery to Application

Stem cells and their natural or engineered products, collectively known as *biologics*, provide the functional components of a regenerative therapeutic regimen [8]. Stem cells have a unique ability to differentiate into specialized cell types and form new tissue, thereby providing the active ingredient for regenerative therapy [66]. Cell-based medicinal products often involve cell samples of limited amount, mostly to be used in a patient-specific manner. This raises specific issues pertaining to quality control testing designed for each product under examination. Cell tracking and imaging are important prerequisites to ensure safe and effective therapies, starting with the manufacturing of cell-based medicinal products for product consistency and traceability [67]. Screening for purity, potency, infectious contamination, and karyotype stability has become necessary, making up release criteria that must be met for compliance with standard operating practices for production and banking of cell-based products. Accordingly, the U.S. Food and Drug Administration and the European Medicines Agency have imposed regulatory guidelines for risk assessment, quality of manufacturing, preclinical and clinical development, and postmarketing surveillance of stem cell biologics for translation from bench to bedside to populations [67,68]. Increasingly, the importance of "real-time" imaging of stem cells and their derived biologics is recognized as a valuable contributor to the search for new therapeutic applications.

### 6.2.2.1 Imaging iPS Cellular Structure

On ectopic expression of human pluripotent genes for OCT3/4, SOX2, c-MYC, and KLF4 into fibroblasts, cellular morphology dramatically changes with successful nuclear reprogramming (Figure 6.1). The real-time light microscopy is sufficient to identify the transformation of large, flat parental fibroblasts as they undergo conversion into small, rounded cells that condense into compact clusters, reminiscent of embryonic stem cells (Figure 6.2). Typically, within 10–14 days, nuclear reprogramming produces colonies that can be individually isolated by picking a stem cell-like colony to allow clonal expansion of rapidly dividing iPS cell lines. Although iPS cell lines were originally obtained according to drug-selectable markers under control of stem cell factors, the robust characteristics of nuclear reprogramming now allow routine isolation of iPS cells based on morphology alone [69].

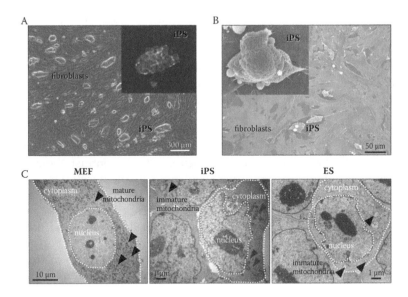

**FIGURE 6.2**
Microscopy to visualize induced pluripotent stem cell phenotype. (A) Bright field imaging demonstrates the characteristic growth pattern of induced pluripotent stem (iPS) cells as compact colonies on a monolayer of feeder cells. Confocal microscopy with immunofluorescence detection of pluripotent-associated antigens illustrates the established cell localization pattern of SSEA1 on iPS cells. (B) Scanning electron microscopy demonstrates at high resolution the densely compacted clusters of iPS cells as outgrowths from the parental fibroblast monolayer on nuclear reprogramming. (C) Field emission scanning microscopy of thin sections of parental fibroblasts illustrates a relatively small nuclear volume compared to the large volume of the densely packed cytoplasm with mature mitochondria (left). Embryonic stem (ES) cells demonstrate low cytoplasmic density of immature mitochondria with a relatively large nuclear-to-cytoplasmic ratio (right). iPS cells recapitulate the pattern of ES cells with similar nuclear-to-cytoplasmic ratio and low density of immature mitochondria (middle).

Typical morphology of iPS cell lines maintained in culture is in fact indistinguishable from embryonic stem cell counterparts (Figure 6.2A).

Cellular morphology can be further resolved in more detail by utilizing antibodies to detect and bind to antigens expressed specifically on stem cells. As nuclear reprogramming modulates the target cell, an early transformation is the acquisition of rapid cellular proliferation, which can be observed by the expression of proteins required for reentry into the cell cycle. One protein, Ki-67, is present during the active stages of the cell cycle (G1, S, G2, and mitosis) but is notably absent from resting cells (G0) and thus identifies the parental cell source as its progeny begin the process of self-renewal [69,70]. Immunostaining according to cell type-specific SSEA-1 expressed on stem cells or alkaline phosphatase staining enables rapid and robust quantification of iPS cells compared to parental fibroblasts that have not fully acquired the pluripotent ground state (Figure 6.2A, inset). Although not available for real-time measurements, immunostaining offers a reliable approach to validate cellular metamorphosis during the nuclear reprogramming process.

Further zooming into the cell at the time of reprogramming, electron microscopy provides a subcellular vantage point that highlights the ultra-structure of the metamorphosing cells. Parental fibroblasts are characterized by scanning electron microscopy as large, flat cytotypes compared to the distinctive three-dimensional appearance of reprogrammed cells that shrinks and becomes rounded (Figure 6.2B). Furthermore, the parental fibroblasts contain a large cytoplasm-to-nuclear ratio and mature mitochondria coupled with an extensive ribosomal network (Figure 6.2C). In contrast, the stem cell-like colonies contain cytotypes that display a lack of subcellular organization [70]. These stem cell-like progeny are characterized by a small amount of cytoplasm with an enlarged nucleus, reminiscent of embryonic stem cells. The scant cytoplasm is sparsely populated with organelles, and the mitochondria that are present have an immature phenotype without cristae [71]. Thus, subcellular imaging clearly demarcates parameters with distinct profiles in parental fibroblasts compared to reprogrammed progeny.

### 6.2.2.2 Imaging iPS Cell Differentiation

To determine the potential of lineage differentiation from iPS cells, multiple systems have been used from *in vitro* cell culture (embryoid bodies) to *in vivo* differentiation in immunocompromised host animals, to in utero differentiation during normal embryonic development (Figure 6.3). Using parental fibroblasts and embryonic stem cells for baseline comparisons of prototypical ordinary cells versus stem cells, reprogrammed progeny differentiate into multiple lineages similar to embryonic stem cells. On nuclear reprogramming, derived progenitors can be differentiated in three-dimensional cultures to allow spontaneous germ layer formation and subsequent tissue-specific differentiation *in vitro* [71]. This method of differentiation produces characteristic aggregation of progeny that demonstrate the ability to spontaneously beat within 7–8 days and confirms the acquired cardiogenic potential of iPS cells (Supplemental Movie 1). Real-time imaging of the excitation-contraction coupling can be observed with coordinated calcium transients in conjunction with mechanical contractions (Figure 6.3, top row).

Alternatively, the *in vivo* differentiation potential of iPS cells can be grossly visualized following subcutaneous injection into immunodeficient mice in whose bodies pluripotent cell clones are able to give rise to tumors known as teratomas [69]. In comparison to parental fibroblasts that have no ability to grow or differentiate further, iPS cell clones can give rise to tumors following an injection of 500,000 cells to about 1 cm in diameter within 4 weeks. These iPS cell-derived tumors were found to be encapsulated within the localized site of transplantation (Figure 6.3, middle row) and demonstrated a heterogeneous appearance on gross inspection. Subsequent tissue histology was sufficient to reveal the cellular architecture of mesoderm, ectoderm,

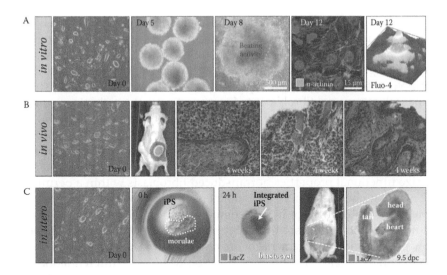

**FIGURE 6.3**

Imaging iPS cell differentiation from *in vitro* analysis to chimeric offspring. (A) Differentiating iPS cells *in vitro* produces day 5 embryoid bodies capable of multilineage differentiation, with bona fide cardiogenesis demonstrated by beating activity at day 8, expression of cardiac sarcomeric proteins (middle panels), and functional excitation-contraction coupling demonstrated by calcium transients (far right panel). (B) Injection of iPS cells into the subcutaneous tissue of immunodeficient host produces engraftment of labeled progeny detected by bioluminescence 4 weeks after transplantation. The stable engraftment enables teratoma formation to depict trilineage differentiation *in vivo* and to validate pluripotent differentiation capacity of iPS cells. (C) Transplantation of labeled iPS cells into host morula demonstrates the functional equivalency of iPS cells compared to native blastomeres as the chimeric tissue develops through gastrulation and gives rise to early-stage embryos. The growth of iPS cell-derived chimeric embryos can be monitored in real time with *in vivo* imaging according to luciferase expression within pregnant mothers throughout development and confirmed with *lacZ* staining (far right panels).

and endoderm lineages, as well as the persistence of poorly differentiated cytotypes. Together, these experimental systems documented the multiple tissues derived from *in vivo* differentiation and spontaneous formation of complex cytoarchitecture derived from iPS cell clones.

Ultimately, normal differentiation of stem cells is validated in a developmental model system that tests the ability of progenitor cells to function equivalently to native embryonic stem cells. This can be done through either blastocyst injection or aggregation techniques. The chimeric offspring can be directly analyzed for proper tissue-specific differentiation or used to determine germline transmission. The highest stringency for pluripotency from iPS cells has been documented by tetraploid aggregation in which the entire embryo proper is derived from transplanted stem cells within a host embryonic environment that has been compromised by electrofusion at the two-cell stage to prevent native blastomeres from developing into embryonic tissue [42,43]. Thus, the resulting embryo that develops and gives rise to adult offspring is solely derived from the transplanted pluripotent stem cells. An efficient model to screen clones for functional developmental potential has been demonstrated by diploid aggregation in which iPS cells are incorporated into eight-cell embryos to produce a chimeric morula [69,70]. Only true pluripotent stem cells will be accepted into the host embryo, as demonstrated by the lack of parental fibroblasts to incorporate into the early-stage embryo. If a chimeric morula is obtained within the first 24 hours of aggregation, then embryos can be transplanted into surrogate mothers to allow subsequent stages of development to be examined carefully (Figure 6.3, bottom row). Diploid aggregation thus provides an efficient ex utero methodology to characterize functional properties of transduced fibroblasts within a permissive embryonic niche. Beyond *ex vivo* characterization, chimeric embryos also allow the establishment of in situ competency of transduced progeny during natural embryogenesis through adulthood. Adult chimeric offspring allow physiologic measurements of not only cardiovascular performance but also integrated organ systems as a function of iPS cell differentiation [72]. Furthermore, stem cell differentiation can be imaged *in vivo* with state-of-the-art reporter systems in which tissue-specific gene expression is revealed according to real-time readouts from transgenic reporter constructions.

### 6.2.2.3 Imaging iPS Cell Host Engraftment

Cell-based interventions, such as iPS cells for heart disease, require a real-time readout of biodistribution unique to transplanted cells that have the potential for systemic circulation, self-renewal, and dysregulated growth. Tissue engraftment is a product of environmental exposure (i.e., systemic injection or intratissue transplantation) and the responsiveness of the cells according to receptors and signaling machinery functioning within the pool of cells. Furthermore, the initial site of engraftment may not be the final destination of differentiated progeny. Therefore, highly sensitive imaging modalities to track the migration and engraftment of individual cells along with survival and tissue homeostasis of the resulting progeny are essential to the discovery of application of iPS cell technology in disease model systems, such as the postischemic myocardium [70].

To advance iPS cell treatment for cardiac repair, the recurring questions of optimal cell dose and tumor formation must be answered. Due to the pluripotent nature of iPS cells, it was initially uncertain whether intramyocardial delivery would result in localized engraftment and de novo tissue formation in the context of postischemic injury. A standard methodology to begin to address this question was built on the experience of *in vivo* imaging with optical bioluminescence technology [72]. On transplantation into murine hearts, iPS cells were tracked following intraperitoneal injection of D-luciferin with the IVIS 200 Bioluminescence Imaging System (Xenogen). Signal intensities and tissue localization were analyzed with Living Image Software (Figure 6.4). On permanent occlusion of epicardial coronary vasculature and immediate

**FIGURE 6.4**

Imaging iPS cell engraftment with *in vivo* optical technology. Time course of intracardiac transplanted iPS cells labeled with luciferase enables high-sensitivity detection of stably engrafted progeny over a 4-week follow-up. The lack of detectable cells in noncardiac tissue suggests no metastatic dissemination within this model system of acute myocardial infarction. Stable expression of luciferase activity within the heart field suggests engraftment and persistent retention of iPS cell progeny.

microsurgical transfer of genetically labeled iPS cells, bioluminescence emitted from differentiated progeny was specifically localized within the ischemic hearts by 2 weeks posttransplantation [70]. There was no evidence of metastatic dissemination during a 3-month follow-up, suggesting a low metastatic potential of iPS cells transplanted into ischemic myocardial tissue [70].

Although the two-dimensional observations consistently localized to the left side of the chest within the cardiac silhouette, the technical limitations of bioluminescence prevent detailed tissue localization within the regions of the myocardium. Theoretically, *in vivo* noninvasive imaging would simultaneously address the degree of nondisruptive engraftment, allow detection at the single-cell level, allow linear quantification over a wide range of detection, provide stable expression of surrogate markers, and avoid any harm to the physiological model system required for longitudinal evaluations. Magnetic resonance imaging (MRI) and positron emission tomographic (PET) imaging methods offer attractive alternatives to bioluminescence/fluorescence techniques to monitor cellular engraftment as these imaging approaches permit three-dimensional reconstruction of cardiac anatomy and do not suffer from tissue depth-induced signal attenuation in larger animal and human studies.

PET imaging techniques benefit from the superior signal-to-noise characteristics intrinsic to radiotracers that permit use of very low substrate concentrations and the ability to co-register with anatomic data. Most PET methods compatible with stem cell imaging are reporter gene constructs that rely on intracellular transcriptional machinery to generate or uptake a radiotracer substrate. Two of the more promising technologies that have already seen early use in cardiac stem cell imaging are the versatile triple-fusion (TF) construct (*fluc-mrfp-ttk*) and the sodium-iodine symporter (NIS) reporter system. The TF reporter gene construct permits simultaneous preliminary cell sorting (monomeric red fluorescent protein [*mrfp*] using fluorescence-activated cell sorting [FACS]), and *in vivo* imaging using firefly luciferase (*fluc*) for high-throughput bioluminescence imaging or the herpes simplex virus type 1-truncated thymidine kinase reporter gene (*ttk*) using [$^{18}$F]-FHBG for PET imaging [73,74]. Due to the risk of possible teratoma formation, the *ttk* gene also acts as a ganciclovir-sensitive "suicide gene" in response to ganciclovir therapy that can be used to kill off all implanted cells [72]. The NIS expression system capitalizes on a highly tissue-specific membrane transporter capable of exchanging Iodine ($^{124}$I) or pertechnetate (technetium-99m [$^{99m}$Tc]) for intracellular sodium. Using this system in cardiac-derived stem cells, ectopic NIS expression has been visualized in the heart with minimal background signal originating from tissues (thyroid, stomach, salivary glands) that endogenously express this symport gene [74].

Cardiac MRI offers the highest anatomic resolution of any current noninvasive imaging modality. To date, MR-based imaging approaches have relied on cell labeling/tagging with MR-active contrast agents or, more recently, use of MR reporter genes. Both negative contrast-inducing agents (e.g., dextran-coated iron oxide nanoparticles) and positive contrast-inducing agents (e.g., gadolinium chelates) have been utilized in preclinical studies as a means to image and track stem cells [75–78]. Evolution of these contrast agents to multimodality MRI substrates such as dextran/fluorophore-coated cross-linked iron oxide (CLIO) nanoparticles or quantum dots conjugated with gadolinium chelates offers added versatility and flexibility beyond early MR contrast targets [79–81]. Antibody or antigen-conjugated MR contrast agents further enhance the capacity to image stem cells without the drawbacks of exogenous labeling of stem cells prior to delivery. In one particularly salient example, a hybrid quantum dot/gadolinium lipid micelle system was developed with a VEGF (vascular endothelial growth factor) homing peptide (cyclic RGD) that permits visualization of angiogenesis in a model using both fluorescence and MRI [82,83].

In addition to contrast agents, MR reporter genes have been developed to permit continuous production of a MR-compatible marker. These developments are perhaps the most compelling avenues in MRI as such report genes can reduce or eliminate the need for exogenous contrast reagents and their undesirable side effects. Examples of these efforts include expression of a novel lysine-rich peptide compatible with chemical exchange saturation transfer (CEST) imaging [84]; expression of *magA* from magnetotactic bacteria that allows for controllable postimplantation uptake of endogenous iron [85,86]; and β-gal-mediated activation of the MR contrast agent 1-(2-(β-galactopyranosyloxy)propyl)-4,7,10-tris(carboxymethyl)-1,4,7,10-tetraazacyclododecane)gadolinium(III) (EgadMe) via *LacZ* expression [87].

### 6.2.2.4 *Imaging iPS Cell-Based Therapeutic Repair*

Therapeutic application of iPS cell methodology requires optimization of tissue-specific regeneration in response to injury with *in vivo* remuscularization and vascularization, leading to restoration of damaged cardiac structure and function. Imaging for cardiac structure and function serves as a surrogate endpoint to monitor therapeutic benefit following cell-based interventions. Echocardiography has been established as a standard imaging modality to evaluate postischemic cardiac performance because it is noninvasive (Supplemental Movie 2), is efficiently performed with portable equipment, and involves no harmful radiation [88]. Previously limited by technology engineered for clinical studies, small-animal imaging with echocardiography required the development of specialized equipment that has now advanced to keep pace with other imaging modalities in the preclinical setting (Figure 6.5). By utilizing enhancement strategies such as vascular contrast, left ventricular volumes, mass, and ejection fraction can be more accurately measured by an experienced sonographer. In addition, echocardiography includes the ability to quantify regional wall thickening and endocardial wall motion in real time under physiologic conditions. Coupled with Doppler ultrasound measurements, pressure gradients can be routinely determined to assist with diastolic calculations. In randomized postischemic small-animal cohorts transplanted with either parental fibroblasts or iPS cell-derived stem cells [70], significant differences in regional wall motion and cardiac function as calculated by fractional shortening and ejection fraction can be quantified by two-dimensional echocardiography (Figure 6.5). In contrast to parental fibroblasts, iPS cell intervention in the acute stages of myocardial infarction can improve cardiac contractility, which is sustained throughout a 4-week follow-up period. Functional benefit in response to iPS cell therapy can be further verified by echocardiography by improvements in regional septal wall thickness during contraction that demonstrate coordinated concentric contractions visualized by long-axis and short-axis two-dimensional imaging. Beyond functional deterioration, maladaptive remodeling with detrimental structural changes prognosticates poor outcome following ischemic injury. As reported, iPS cell intervention can also attenuate global left ventricular diastolic diameter predictive of decompensated heart disease [70]. A consequence of pathologic structural remodeling was also evident by prolongation of the QT interval, which increases risk of life-threatening arrhythmias. Successful iPS cell treatment prevents structural remodeling and avoids the deleterious effects on electrical conductivity, as predicted by the lack of maladaptive structural remodeling determined by echocardiography [70].

As an alternative modality for small-animal cardiac performance, MRI provides a noninvasive means to acquire high-resolution structural images and can provide functional images of the heart throughout the entire cardiac systole-diastole cycle by the use of gated-imaging techniques [89]. Taking advantage of the natural differences in spectra obtained from rapidly circulating blood and more static tissue dynamics, MRI is able to obtain high-resolution images with the appearance of contrast studies (Figure 6.6A,B). Therefore, precise mapping of the three-dimensional geometry of the

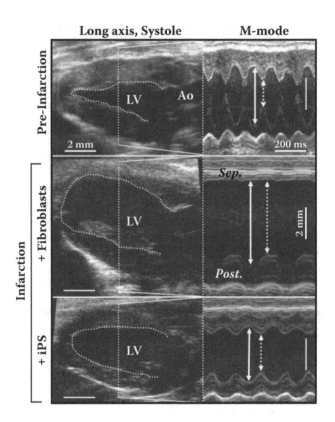

**FIGURE 6.5**

Echocardiography of cardiac structure and function following iPS cell-based treatment of acute myocardial infarction. Long-axis views of mouse hearts prior to infarction demonstrate normal systolic function, with representative M-mode imaging illustrating synchronized septal and posterior wall motion at the mid-left ventricle (top row). Hearts infarcted by surgical ligation of the anterior coronary arteries consistently demonstrate decreased regional septal wall motion that does not significantly improve following transplantation with fibroblasts (middle row). Transplantation of iPS progeny into the peri-infarcted tissue of the left ventricle leads to recovery of anterior wall and septal wall motion with restoration of systolic function (bottom row). Solid arrow, diastole; dotted arrow, systole; LV, left ventricle; Ao, aorta; Sep, septal wall; Post., posterior wall.

myocardium and calculating volumes used to extrapolate functional significance will continue to be employed as equipment becomes more accessible (Supplemental Movie 3). MRI offers additional advantages with emergence of cell-tracking and -labeling techniques. By coupling structure, function, and de novo tissue reconstruction, MRI will provide valuable insight to mechanism of cell-based repair utilizing noninvasive technology that will have direct applicability to preclinical and clinical studies.

As real-time evaluation of cardiac structure and function is essential to further validate the therapeutic value of pluripotent stem cells in ischemic heart disease, emerging technology will continue to refine the accuracy

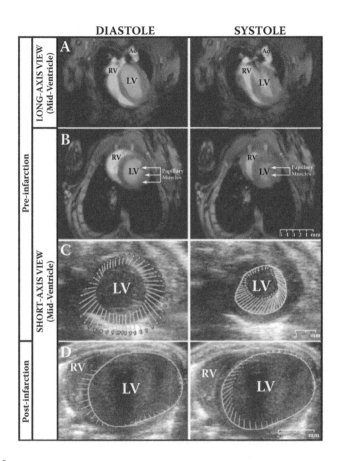

**FIGURE 6.6**

Advanced imaging techniques: cine MRI and echocardiography with strain data. Cine-derived gated MRI of the murine cardiac cycle of a noninfarcted heart using upright wide-bore 17-T nuclear magnetic resonance (NMR). Long- (A) and short-axis (B) sections are shown at systole (right) and diastole (left). Short-axis echocardiographic views of a pre- (C) and postinfarcted (D) murine heart demonstrate transition from normal (synchronized anterior and posterior wall motion at the mid-left ventricle) to abnormal (decreased regional anterior wall motion and intraventricular discordance) systolic function as depicted by green vector representations of regional myocardial dynamics. RV, right ventricle (red in A and B); LV, left ventricle (blue in A and B); Ao, aorta (yellow in A and B).

of therapeutic repair by mapping regional myocardium regeneration and vasculature density. Tissue Doppler imaging (TDI) avoids high-frequency signal dominated by blood flow and allows focused imaging of myocardial tissue velocities. In related technology, speckle tracking utilizes acoustic interference patterns within the native tissue to determine precise tissue movement. These parameters enable the calculation of strain within regions of the myocardium and provide a highly sensitive methodology to quantify early changes in postischemic myocardium (Figure 6.6C,D). In addition, three-dimensional ultrasound is available for small-animal protocols, which in principle should improve the accuracy of left ventricular function calculations compared to traditional two-dimensional ultrasound, which fails to account for changes in ventricular geometry of diseased hearts.

Electromechanical mapping provides an additional technique to augment the monitoring of disease progression and the benefit of cell-based therapy [91]. Nonfluoroscopic, electroanatomical navigation systems use ultralow magnetic field sources and catheter electrodes to track the trajectory of a catheter independently of contrast or direct visualization. These systems allow precise guidance of a diagnostic catheter or delivery catheter along with the acquisition of spatial, electrophysiological, and mechanical information to re-create a three-dimensional map of the viable myocardium. The diagnostic utility of electromechanical mapping has been validated in preclinical and clinical studies to quantify the location and degree of postischemic myocardium. This detailed functional map is dependent on the magnitude and timing of the infarction, thus enabling the system to be useful to guide cell-based interventions for direct delivery of cells to regions of particular interest [90].

## 6.3 Conclusion

Triggered nuclear reprogramming through ectopic expression of stemness factors has revolutionized a strategy for embryo-independent derivation of pluripotent stem cells. iPS cell technology offers breakthroughs in stem cell biology by promising to invalidate chronological age, reverse cell fate, and restore an atavistic embryonic-like potential of ordinary adult cells. To these ends, iPS cell clones have demonstrated functions previously demonstrated only by natural embryonic stem cells to produce all tissue types. Furthermore, the regenerative potential of iPS cells has been confirmed in head-to-head studies between parental fibroblasts and reprogrammed progeny in the setting of ischemic heart disease, providing the first direct evidence of the potential therapeutic value of bioengineered pluripotent stem cells in the setting of heart disease [71]. Specifically, transplantation of iPS cells in the acutely infarcted myocardium yielded structural and functional repair to secure performance recovery as qualified clones contributed to *in vivo* tissue reconstruction with "on-demand" cardiovasculogenesis. Although the full clinical impact of iPS cell-based technology has yet to be realized, proof-of-principle studies coupled with real-time *in vivo* imaging modalities are being conducted to establish the safety and efficacy profiles, enabling further optimization of clinical-grade production and matching the right cell type to the most appropriate target disease. Therefore, converting self-derived fibroblasts, ubiquitous in any patient, into reparative progenitors can now be considered as a feasible goal of regenerative medicine to individualize treatment algorithms for a diverse scope of degenerative diseases.

## Supplemental Movies

Movie 1. Beating activity of iPS derived cardiomyocytes

Movie 2. Echocardiography of postischemic myocardium

Movie 3. Axial-plane cine-MRI of complete cardiac cycle

## References

1. Kajstura, J., Hosoda, T., Bearzi, C., et al. 2008. The human heart: A self-renewing organ. *Clin Transl Sci* 1:80–86.
2. Quaini, F., Urbanek, K., Beltrami, A.P., et al. 2002. Chimerism of the transplanted heart. *N Engl J Med* 346:5–15.
3. Torella, D., Ellison, G.M., Méndez-Ferrer, S., Ibanez, B., Nadal-Ginard, B. 2006. Resident human cardiac stem cells: Role in cardiac cellular homeostasis and potential for myocardial regeneration. *Nat Clin Pract Cardiovasc Med* 3 Suppl 1:S8–S13.
4. Bergmann, O., Bhardwaj, R.D., Bernard, S., et al. 2009. Evidence for cardiomyocyte renewal in humans. *Science* 324:98–102.
5. Kubo, H., Jaleel, N., Kumarapeli, A., et al. 2008. Increased cardiac myocyte progenitors in failing human hearts. *Circulation* 118:649–657.
6. Rupp, S., Koyanagi, M., Iwasaki, M., et al. 2008. Characterization of long-term endogenous cardiac repair in children after heart transplantation. *Eur Heart J* 29:1867–1872.
7. Urbanek, K., Torella, D., Sheikh, F., et al. 2005. Myocardial regeneration by activation of multipotent cardiac stem cells in ischemic heart failure. *Proc Natl Acad Sci USA* 102:8692–8697.
8. Nelson, T.J., Behfar, A., Yamada, S., Martinez-Fernandez, A., Terzic, A. 2009. Stem cell platforms for regenerative medicine. *Clin Transl Sci* 2:222–227.
9. Klimanskaya, I., Rosenthal, N., Lanza, R. 2008. Derive and conquer: Sourcing and differentiating stem cells for therapeutic applications. *Nat Rev Drug Discov* 7:131–142.
10. Morrison, S.J., Spradling, A.C. 2008. Stem cells and niches: mechanisms that promote stem cell maiteneance throughout life. *Cell* 132:598–611.
11. Surani, M.A., McLaren, A. 2006. Stem cells: A new route to rejuvenation. *Nature* 443:284–285.
12. Jaenisch, R., Young, R. 2008. Stem cells, the molecular circuitry of pluripotency and nuclear reprogramming. *Cell* 132:567–582.

13. Yang, X., Smith, S.L., Tian, X.C., Lewin, H.A., Renard, J.P., Wakayama, T. 2007. Nuclear reprogramming of cloned embryos and its implications for therapeutic cloning. *Nat Genet* 39:295–302.

14. Beyhan, Z., Iager, A.E., Cibelli, J.B. 2007. Interspecies nuclear transfer: implications for embryonic stem cell biology. *Cell Stem Cell* 1:502–512.

15. Hall, V.J., Stojkovic, M. 2006. The status of human nuclear transfer. *Stem Cell Rev* 2:301–308.

16. Byrne, J.A., Pedersen, D.A., Clepper, L.L., et al. 2007. Producing primate embryonic stem cells by somatic cell nuclear transfer. *Nature* 450:497–502.

17. French, A.J., Adams, C.A., Anderson, L.S., Kitchen, J.R., Hughes, M.R., Wood, S.H. 2008. Development of human cloned blastocysts following somatic cell nuclear transfer with adult fibroblasts. *Stem Cells* 26:485–493.

18. Takahashi, K., Yamanaka, S. 2006. Induction of pluripotent stem cells from mouse embryonic and adult fibroblast cultures by defined factors. *Cell* 126:663–676.

19. Yamanaka, S. 2008. Pluripotency and nuclear reprogramming. *Philos Trans R Soc Lond B Biol Sci* 363:2079–2087.

20. Yamanaka, S. 2009. A fresh look at iPS cells. *Cell* 137:13–17.

21. Yamanaka, S. 2009. Elite and stochastic models for induced pluripotent stem cell generation. *Nature* 460:49–52.

22. Yu, J., Vodyanik, M.A., Smuga-Otto, K., et al. 2007. Induced pluripotent stem cell lines derived from human somatic cells. *Science* 318:1917–1920.

23. Meissner, A., Wernig, M., Jaenisch, R. 2007. Direct reprogramming of genetically unmodified fibroblasts into pluripotent stem cells. *Nat Biotechnol* 25:1177–1181.

24. Maherali, N., Sridharan, R., Xie, W., et al. 2007. Directly reprogrammed fibroblasts show global epigenetic remodeling and widespread tissue contribution. *Cell Stem Cell* 1:55–70.

25. Takahashi, K., Tanabe, K., Ohnuki, M., et al. 2007. Induction of pluripotent stem cells from adult human fibroblasts by defined factors. *Cell* 131:861–872.

26. Yamanaka, S. 2007. Strategies and new developments in the generation of patient-specific pluripotent stem cells. *Cell Stem Cell* 1:39–49.

27. Park, I.H., Lerou, P.H., Zhao, R., Huo, H., Daley, G.Q. 2008. Generation of human-induced pluripotent stem cells. *Nat Protoc* 3:1180–1186.

28. Park, I.H., Arora, N., Huo, H., et al. 2008. Disease-specific induced pluripotent stem cells. *Cell* 134:877–886.

29. Papapetrou, E.P., Tomishima, M.J., Chambers, S.M., et al. 2009. Stoichiometric and temporal requirements of Oct4, Sox2, Klf4, and c-Myc expression for efficient human iPSC induction and differentiation. *Proc Natl Acad Sci USA* 106:12759–12764.

30. Takahashi, K., Okita, K., Nakagawa, M., Yamanaka, S. 2007. Induction of pluripotent stem cells from fibroblast cultures. *Nat Protoc* 2:3081–3089.

31. Aasen, T., Raya, A., Barrero, M.J., et al. 2008. Efficient and rapid generation of induced pluripotent stem cells from human keratinocytes. *Nat Biotechnol* 26:1276–1284.

32. Loh, Y., Agarwal, S., Park, I., et al. 2009. Generation of induced pluripotent stem cells from human blood. *Blood* 113:5476–5479.

33. Sun, N., Panett, N.J., Gupta, D.M., et al. 2009. Feeder-free derivation of induced pluripotent stem cells from adult human adipose stem cells. *Proc Natl Acad Sci USA* 106:15720–15725.

34. Marion, R.M., Strati, K., Li, H., et al. 2009. Telomeres acquire embryonic stem cell characteristics in induced pluripotent stem cells. *Cell Stem Cell* 4:141–154.

35. Deng, J., Shoemaker, R., Xie, B., et al. 2009. Targeted bisulfite sequencing reveals changes in DNA methylation associated with nuclear reprogramming. *Nat Biotechnol* 27:353–360.

36. Mikkelsen, T.S., Hanna, J., Zhang, X., et al. 2008. Dissecting direct reprogramming through integrative genomic analysis. *Nature* 454:49–55.

37. Silva, J., Nichols, J., Theunissen, T.W., et al. 2009. Nanog is the gateway to the pluripotent ground state. *Cell* 138:722–737.

38. Nishikawa, S., Goldstein, R.A., Nierras, C.R. 2008. The promise of human induced pluripotent stem cells for research and therapy. *Nat Rev Mol Cell Biol* 9:725–729.

39. Nakagawa, M., Koyanagi, M., Tanabe, K., et al. 2008. Generation of induced pluripotent stem cells without Myc from mouse and human fibroblasts. *Nat Biotechnol* 26:101–106.

40. Park, I.H., Zhao, R., West, J.A., et al. 2008. Reprogramming of human somatic cells to pluripotency with defined factors. *Nature* 451:141–146.

41. Waldman, S.A., Terzic, A. 2008. Therapeutic targeting: A crucible for individualized medicine. *Clin Pharmacol Ther* 83:651–654.

42. Zhao, X.Y., Li, W., Lv, Z., et al. 2009. iPS cells produce viable mice through tetraploid complementation. *Nature* 461:86–90.

43. Boland, M.J., Hazen, J.L., Nazor, K.L., et al. 2009. Adult mice generated from induced pluripotent stem cells. *Nature* 461:91–94.

44. Okita, K., Ichisaka, T., Yamanaka, S. 2007. Generation of germline-competent induced pluripotent stem cells. *Nature* 448:313–317.

45. Aoi, T., Yae, K., Nakagawa, M., et al. 2008. Generation of pluripotent stem cells from adult mouse liver and stomach cells. *Science* 321:699–702.

46. Huangfu, D., Osafune, K., Maehr, R., et al. 2008. Induction of pluripotent stem cells from primary human fibroblasts with only Oct4 and Sox2. *Nat Biotechnol* 26:1269–1275.

47. Eminli, S., Utikal, J., Arnold, K., et al. 2008. Reprogramming of neural progenitor cells into induced pluripotent stem cells in the absence of exogenous Sox2 expression. *Stem Cells* 26:2467–2474.

48. Kim, J.B., Zaehres, H., Wu, G., et al. 2008. Pluripotent stem cells induced from adult neural stem cells by reprogramming with two factors. *Nature* 454:646–650.

49. Hanna, J., Markoulaki, S., Schorderet, P., et al. 2008. Direct reprogramming of terminally differentiated mature B lymphocytes to pluripotency. *Cell* 133:250–264.

50. Feng, B., Jiang, J., Kraus, P., et al. 2009. Reprogramming of fibroblasts into induced pluripotent stem cells with orphan nuclear receptor Esrrb. *Nat Cell Biol* 11:197–203.

51. Stadtfeld, M., Nagaya, M., Utikal, J., et al. 2008. Induced pluripotent stem cells generated without viral integration. *Science* 322:945–949.

52. Okita, K., Nakagawa, M., Hyenjong, H., et al. 2008. Generation of mouse induced pluripotent stem cells without viral vectors. *Science* 322:949–953.

53. Kaji, K., Norrby, K., Paca, A., et al. 2009. Virus-free induction of pluripotency and subsequent excision of reprogramming factors. *Nature* doi:10.1038/nature07864.

54. Woltjen, K., Michael, I.P., Mohseni, P., et al. 2009. piggyBac transposition reprograms fibroblasts to induced pluripotent stem cells. *Nature* doi:10.1038/nature07863.

55. Nelson, T.J., Terzic, A. 2009. Induced pluripotent stem cells: reprogrammed without a trace. *Regen Med* 4:333–355.

56. Yu, J., Hu, K., Smuga-Otto, K., et al. 2009. Human induced pluripotent stem cells free of vector and transgene sequences. *Science* 324:797–801.

57. Zhou, H., Wu, S., Joo, J.Y., et al. 2009. Generation of induced pluripotent stem cells using recombinant proteins. *Cell Stem Cell* 4:381–384.

58. Kim, D., Kim, C.H., Moon, J.I., et al. 2009. Generation of human induced pluripotent stem cells by direct delivery of reprogramming proteins. *Cell Stem Cell* 6:472–476.

59. Shi, Y., Desponts, C., Do, J.T., Hahm, H.S., Schöler, H.R., Ding, S. 2008. Induction of pluripotent stem cells from mouse embryonic fibroblasts by Oct4 and Klf4 with small-molecule compounds. *Cell Stem Cell* 3:568–574.

60. Banito, A., Rashid, S.T., Acosta, J.C., et al. 2009. Senescence impairs successful reprogramming to pluripotent stem cells. *Genes Dev* 23:2134–2139.

61. Hong, H., Takahashi, K., Ichisaka, T., et al. 2009. Suppression of induced pluripotent stem cell generation by the p53-p21 pathway. *Nature* 460:1132–1135.

62. Utikal, J., Polo, J.M., Stadtfeld, M., et al. 2009. Immortalization eliminates a roadblock during cellular reprogramming into iPS cells. *Nature* 460:1145–1148.

63. Marión, R.M., Strati, K., Li, H., et al. 2009. A p53-mediated DNA damage response limits reprogramming to ensure iPS cell genomic integrity. *Nature* 460:1149–1153.

64. Li, H., Collado, M., Villasante, A., et al. 2009. The Ink4/Arf locus is a barrier for iPS cell reprogramming. *Nature* 460:1136–1139.

65. Kawamura, T., Suzuki, J., Wang, Y.V., et al. 2009. Linking the p53 tumour suppressor pathway to somatic cell reprogramming. *Nature* 460:1140–1144.

66. Nelson, T.J., Behfar, A., Terzic, A. 2008. Stem cells: biologics for regeneration. *Clin Pharmacol Ther* 84:620–623.

67. European Medicines Agency. 2006. *Guideline on human cell-based medicinal products*. EMEA/CHMP/410869/1-24. European Medicines Agency, London.

68. Halme, D.G., Kessler, D.A. 2006. FDA regulation of stem-cell-based therapies. *N Engl J Med* 355:1730–1735.

69. Nelson, T.J., Martinez-Fernandez, A.J., Yamada, S., Mael, A.A., Terzic, A., Ikeda, Y. 2009. Induced pluripotent reprogramming from promiscuous human stemness-related factors. *Clin Translation Sci* 2:118–126.

70. Nelson, T.J., Martinez-Fernandez, A., Yamada, S., Perez-Terzic, C., Ikeda, Y., Terzic, A. 2009. Repair of acute myocardial infarction with human stemness factors induced pluirpotent stem cells. *Circulation* 120:408–416.

71. Martinez-Fernandez, A., Nelson, T.J., Yamada, S., et al. 2009. iPS programmed without c-MYC yield proficient cardiogenesis for functional heart chimerism. *Circ Res* 105:648–656.

72. Cao, F., Lin, S., Xie, X., et al. 2006. In vivo visualization of embryonic stem cell survival, proliferation, and migration after cardiac delivery. *Circulation* 113:1005–1014.

73. Love, Z., Wang, F., Denniset, J., et al. 2007. Imaging of mesenchymal stem cell transplant by bioluminescence and PET. *J Nucl Med* 48:2011–2020.

74. Terrovitis, J., Kwok, K.F., Lautamäki, R., et al. 2008. Ectopic expression of the sodium-iodide symporter enables imaging of transplanted cardiac stem cells in vivo by single-photon emission computed tomography or positron emission tomography. *J Am Coll Cardiol* 52:1652–1660.

75. Reimer, P., Weissleder, R., Lee, A.S., Wittenberg, J., Brady, T.J. 1990. Receptor imaging: application to MR imaging of liver cancer. *Radiology* 177:729–734.

76. Josephson, L., Groman, E.V., Menz, E., Lewis, J.M., Bengele, H. 1990. A functionalized superparamagnetic iron oxide colloid as a receptor directed MR contrast agent. *Magn Reson Imaging* 8:637–646.

77. Rudelius, M., Daldrup-Link, H.E., Heinzmann, U., et al. 2003. Highly efficient paramagnetic labelling of embryonic and neuronal stem cells. *Eur J Nucl Med Mol Imaging* 30:1038–1044.

78. Daldrup-Link, H.E., Rudelius, M., Oostendorp, R.A., et al. 2003. Targeting of hematopoietic progenitor cells with MR contrast agents. *Radiology* 228:760–767.

79. Mulder, W.J., Koole, R., Brandwijk, R.J., et al. 2006. Quantum dots with a paramagnetic coating as a bimodal molecular imaging probe. *Nano Lett* 6:1–6.

80. Schellenberger, E.A., Bogdanov, A. Jr., Högemann, D., Tait, J., Weissleder, R., Josephson, L. 2002. Annexin V-CLIO: A nanoparticle for detecting apoptosis by MRI. *Mol Imaging* 1:102–107.

81. Sosnovik, D.E., Nahrendorf, M., Deliolanis, N., et al. 2007. Fluorescence tomography and magnetic resonance imaging of myocardial macrophage infiltration in infarcted myocardium in vivo. *Circulation* 115:1384–1391.

82. Mulder, W.J., Castermans, K., van Beijnum, J.R., et al. 2009. Molecular imaging of tumor angiogenesis using alphavbeta3-integrin targeted multimodal quantum dots. *Angiogenesis* 12:17–24.

83. van Tilborg, G.A., Mulder, W.J., van der Schaft, D.W., et al. 2008. Improved magnetic resonance molecular imaging of tumor angiogenesis by avidin-induced clearance of nonbound bimodal liposomes. *Neoplasia* 10:1459–1469.

84. Gilad, A.A., McMahon, M.T., Walczak, P., et al. 2007. Artificial reporter gene providing MRI contrast based on proton exchange. *Nat Biotechnol* 25:217–219.

85. Zurkiya, O., Chan, A.W., Hu, X. 2008. MagA is sufficient for producing magnetic nanoparticles in mammalian cells, making it an MRI reporter. *Magn Reson Med* 59:1225–1231.

86. Goldhawk, D.E., Lemaire, C., McCreary, C.R., et al. 2009. Magnetic resonance imaging of cells overexpressing MagA, an endogenous contrast agent for live cell imaging. *Mol Imaging* 8:129–139.

87. Louie, A.Y., Hüber, M.H., Ahrens, E.T., et al. 2000. In vivo visualization of gene expression using magnetic resonance imaging. *Nat Biotechnol* 18:321–325.

88. Kahn, A., DeMaria, A.N. 2009. The role of cardiac ultrasound in stem cell therapy. *J Cardiovasc Transl Res* 2:2–8.

89. From, A.H.L., Ugurbil, K. 2009. The use of magnetic resonance methods in translational cardiovascular research. *J Cardiovasc Transl Res* 2:39–47.

90. Psaltis, P., Worthley, S.G. 2009. Endoventricular electromechanical mapping—the diagnostic and therapeutic utility of the NOGA® XP cardiac navigation system. *J Cardiovasc Transl Res* 2:48–62.

# 7

## Radionuclide Approaches to Imaging Stem Cells and Their Biological Effects on the Myocardium

Fiona See and Lynne L. Johnson

### CONTENTS

## 7.1 Introduction

Heart failure (HF) affects 5–7 million Americans, with 400,000 new cases every year, killing some 300,000 people annually (NHLBI, National Heart, Lung, and Blood Institute*). Coronary artery disease is among the greatest contributors to HF because it leads to the loss of myocardium from infarctions and chronic ischemia. Current medical and surgical approaches are limited in efficacy; therefore, novel therapeutic options such as cell-based therapies are in critical demand. To this end, the development of a safe and effective technology to monitor the outcome of new approaches to optimize engraftment and viability of transplanted cells is important.

A large body of animal data provides evidence for the efficacy of stem cell-based therapy for myocardial ischemia. The beneficial outcomes include improved left ventricular (LV) contractile function, augmentation of the angiogenic response, limitation of scar formation, protection of resident cells against apoptotic cell death, and even regeneration of myocytes [1–3]. However, the results of multicenter clinical trials have been mixed for endpoints such as treadmill time or global or regional LV function [4–7]. Factors

---

* www.heart.org

contributing to the success of cell therapy for myocardial ischemia are likely to include method and timing of cell delivery, type of cell, number of cells, and local tissue environment. To investigate the effects of altering each of these variables as well as other variables on improving cell therapy requires a tool for tracking stem cell homing, engraftment, and survival *in vivo*. Nuclear imaging has the potential to fulfill this role.

## 7.2  Types of Cells

Investigations into the therapeutic potential of human embryonic stem cells have been hampered by ethical issues surrounding tissue source. Experimental studies examining the effects of transplantation of human or murine embryonic stem cells into mice have shown a high incidence of teratoma formation in the graft [8,9]. Such findings have reduced the enthusiasm for this cellular approach and have shifted focus to other types of progenitor cells. Candidate cells currently in clinical trials include bone marrow mononuclear cells (MNCs); subpopulations of bone marrow MNCs, including endothelial progenitor cells (EPCs), mesenchymal stem cells (MSC), and mesenchymal progenitor cells (MPCs); as well as cells derived from other tissue sources, such as cardiac progenitor cells, adipose-derived stem cells, and skeletal myoblasts (http://www.clinicaltrials.gov). Irrespective of cell type, nuclear imaging may be applied to track cells following administration.

## 7.3  Direct Cell Labeling

Radiochemicals such as [111]indium oxine, [18]F-fluorodeoxyglucose (FDG), and [99m]technetium exametazime (HMPAO) have been used to label cells directly prior to transplantation into animal and human subjects. Imaging of labeled cells in real time with highly sensitive single-photon emission computed tomographic (SPECT) technology has yielded both spatiotemporal and quantitative data on the kinetics of cell distribution following administration. A summary of four studies reporting imaging of directly labeled cells with [111]In oxine or oxyquinoline or [99m]Tc in rat models of myocardial infarction (MI) is presented in Table 7.1 [10–13]. Findings from these studies provide some insights into effects of routes and timing of administration of cells on homing, engraftment, and biodistribution. In one study, [111]In-oxine-tagged EPCs were found to home in on the infarcted rat heart. Overall radioactivity in the heart was determined to be approximately 4.7% following intraventricular

**TABLE 7.1**

Examples of Some of the Limitations of Direct Labeling of Cells

| Radiotracer | Cells | Labeling Efficiency | Model and Injection Route | Viability 96 hr in Culture | Cell Function in Culture | % Cells in Heart | Comments |
|---|---|---|---|---|---|---|---|
| [111]In oxine [10] | CD34+ huHPCs | 3 2 ± 11% | Nude rats with MI, LV cavity | 70% | Reduced | 1.0% | Blood pool from binding to transferrin |
| [111]In oxine [11] | huEPCs | 67 ± 13% | Nude rats with MI, LV cavity | 9 6 ± 6% | No change | 3.0% | |
| [111]In oxyquinoline [12] | H9c2 rat cardiomyoblasts | 85% | rats with acute MI, intramyocardial | Not addressed | Not addressed | 20% 96 h | Free In-111 on autorads |
| [99m]Tc exametazime [13] | MSCs | | Rats with acute MI, LV cavity | 99% | Not addressed | 0.9 % | |

*Note:* huHPCs = human hematopoietic progenitor cells, huEPCs = human endothelial progenitor cells, MSCs = mesenchymal stem cells, MI = myocardial infarction.

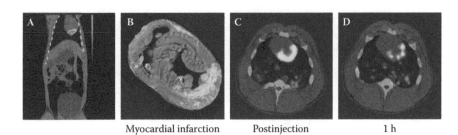

Myocardial infarction       Postinjection              1 h

**FIGURE 7.1**

(A) Tracer uptake in the inferolateral wall of the LV following injection of $^{18}$F-FDG-labeled progenitor cells into the left circumflex coronary artery of a normal swine using a balloon occlusion technique. (B)–(D) Swine underwent LCx occlusion for 90 minutes followed by injection of $^{18}$F-FDG-labeled progenitor cells. (B) Triphenyltetrazolium chloride-stained myocardial section showing inferolateral necrosis (white), (C) single-frame transverse hybrid PET/CT image taken immediately after intracoronary injection of labeled cells showing myocardial activity localized to infarct territory, and (D) similar image taken 1 hour after injection showing decrease in activity in infarct zone when compared with postinjection image. (From Doyle, B., et al., *J Nucl Med* 2007;48(10):1708–14. With permission.)

administration, whereas radioactivity following intravenous injection was only 2% [11]. Interestingly, injection of labeled human bone marrow cells into the LV cavity of MI rats resulted in low levels of cardiac engraftment, with a risk of pulmonary uptake, across three separate studies [10,11,13]. In contrast, intramyocardial injection of labeled rat H9C2 cells resulted in a considerably higher rate of engraftment in the ischemic myocardium [12]. These imaging studies suggest that a number of factors may be at play in determining the success of engraftment, including cell type, donor-host histocompatibility, route of injection, and timing of cell delivery relative to injury. The studies summarized in Table 7.1 also provide examples of some of the limitations of direct labeling of cells. Among the studies that employed $^{111}$In-labeled cells, there is some variability in the levels of cell viability, which may be attributed to radiation levels associated with the dose of $^{111}$In used for incubation [10–12]. Binding of $^{111}$In to transferin may increase blood pool activity [10]. The relatively short physical half-lives of these radiotracers limit follow-up imaging to 120 hours as the longest reported imaging time following injection of $^{111}$In [12].

$^{18}$F-FDG has an even shorter half-life but has been used to monitor homing and tissue distribution following intracoronary injection of labeled cells [14,15]. The method has been useful for assessing different catheter-based approaches to cell delivery into the coronary arteries (Figure 7.1). With optimal temperatures and incubation times, the labeling efficiency was reported as greater than 90%. An important finding from these studies is the greater cell retention in nonreperfused MIs, in contrast to the immediate washout of cells from the heart following reperfusion. This observation has consequence for clinical trials.

## 7.4 Genetic Labeling with Reporter Genes

Reporter gene technology is a more refined method to determine cell viability and subcellular events such as differentiation *in vivo*. In this approach, stem cells are transduced with a reporter gene encoding for intracellular enzymes or transmembrane proteins. The majority of studies have used herpes simplex virus type 1 thymidine kinase (HSV1-tk) as the reporter gene and a positron emission tomographic- (PET-) labeled nucleoside substrate as the probe [16–19]. The advantage of this approach is that the radiolabeled probe can be reinjected over time and can specifically identify viable engrafted cells by selectively binding to the reporter gene product as long as the cells express the enzyme or receptor. An alternative reporter gene under investigation is the human sodium iodine symporter (hNIS) gene [20–22]. The NIS is constitutively localized to the thyroid, salivary glands, gastric mucosa, and lactating mammary glands, where it facilitates cellular accumulation of iodide. The "probes" for this reporter gene include readily available radiotracers such as $^{125}I$, $^{123}I$, $^{131}I$, and $^{99m}Tc$. Terrovitis and colleagues reported SPECT and PET imaging of rat cardiac stem cells (CSCs), transduced with a lentiviral vector containing an hNIS plasmid, following syngeneic transplantation into a rodent model of acute myocardial ischemia [14] (Figure 7.2). Serial imaging of hNIS-positive CSC-treated rats demonstrated the gradual loss of cells from the site of injection from day 1 to day 6 following injection, consistent with data obtained by quantitative polymerase chain reaction (PCR) determination of cell persistence at these time points.

The use of viral vectors to introduce DNA into target cells raises the risk of integration into the host genome and associated safety and ethical concerns. The development of nonreplicating viral vectors is an approach to improve safety. Alternatively, transfection of hNIS mRNA (messenger RNA) represents a novel and efficient approach to achieving transient hNIS expression in stem cells. *In vitro*-synthesized mRNA may be delivered into

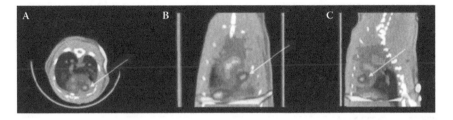

**FIGURE 7.2**
Injection of Tc-99m 1 day after engraftment of hNIS-expressing rat cardiac-derived stem cells into anterior wall following LAD ligation. Thallium-201 was injected to demarcate myocardial perfusion. Tc-99m activity is shown in red and thallium activity in green. The Tc-99m uptake corresponds to the stem cell graft that is localized within the infarct zone delineated by the thallium defect (yellow arrows). (From Terrovitis, J., *J Am Coll Cardiol* 2008;52:1652–60. With permission.)

the cytoplasm by cationic lipids or electroporation, where its translation results in a functional protein product in viable cells [23]. Unlike transfection of plasmid DNA, mRNA transfection does not require nuclear transfer; therefore, quiescent cells may be targeted. In addition, there is no risk of insertional mutagenesis or oncogenesis. Using this approach, mRNA-mediated gene delivery into human progenitor cells has been shown to promote highly efficient protein expression of nerve growth factor receptor, over and above plasmid-based nucleofection, without effect on multipotency or cellular viability [24]. Similarly, mRNA transfection of the cell surface receptor CXCR4-GFP (green fluorescent protein) fusion product into MSCs resulted in sustained plasma membrane CXCR4 (chemokine receptor type 4) protein expression, which demonstrated functional binding of its ligand, stromal cell-derived factor 1 (SDF1) [25].

We performed a pilot study to determine the feasibility of mRNA transfection as a method to induce functional hNIS reporter gene expression in adult human MPCs for nuclear imaging following transplantation. Polyadenylated hNIS mRNA was synthesized *in vitro* for transfection into culture-expanded MPCs by cationic lipid delivery. The level of hNIS mRNA remained steady for 3 days and declined thereafter, but remained detectable up to day 7. hNIS mRNA was not detected in nontransfected control cells. hNIS expressed by transfected MPCs was functional; the activity of known numbers of hNIS-positive MPCs pulsed with Tc-99m correlated with cell number and was inhibited by sodium perchlorate. To examine the feasibility of imaging hNIS mRNA-transfected hMPCs *in vivo*, hNIS-positive MPCs or nontransfected control cells were suspended in Matrigel for injection into subcutaneous pockets in the abdominal wall of nude rats to create a reservoir of cells. The Matrigel patch was allowed to vascularize over 3 days before rats were injected with Tc-99m for imaging on a Bioscan nanoSPECT/CT (computed tomographic) scanner. Percentage injected dose and *ex vivo* counts of excised Matrigel patches were consistent with the number of hNIS-positive hMPCs delivered (Figure 7.3). These pilot animal studies provide encouraging data on the feasibility of quantitative imaging of hNIS-positive MPCs *in vivo* and suggest that mRNA transfection of hNIS reporter gene into MPCs may represent a clinically applicable method to enable tracking of cellular distribution, engraftment, and fate by nuclear imaging following administration *in vivo*.

## 7.5 Importance of the Milieu of the Engraftment Site

A major limitation that is common to existing approaches to cell therapy for HF is the transient and low rates of engraftment of administered cells [26–28]. Successful engraftment of transplanted cells in host tissue may be

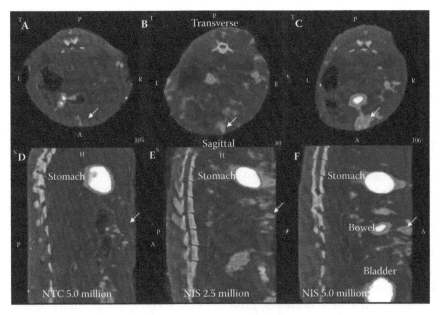

| | Non-Transfected Control hMPCs ($5 \times 10^6$) | hNIS + hMPCs ($2.5 \times 10^6$) | hNIS + hMPCs ($5 \times 10^6$) |
|---|---|---|---|
| % ID scan | $0.1 \times 10^{-3}$ | $1.28 \times 10^{-3}$ | $2.4 \times 10^{-3}$ |
| *Ex vivo* (counts/g) | $109,885$ | $236,341$ | $952,167$ |

**FIGURE 7.3**
*In vivo* SPECT/CT imaging of hNIS mRNA transfected hMPCs implanted with Matrigel into subcutaneous pocket in the abdomen of nude rats. Three days after transplantation, rats were injected with Tc-99m and imaged. Minimal signal was detected with nontransfected cells (NTC) (A, D), weak signal from $2.5 \times 10^6$ cells (B, E) and stronger signal from $5 \times 10^6$ cells (D, F). The table shows relationship between Tc-99m activity and cell number.

considered as a function of a series of cellular processes: migration, adhesion, and survival. Notably, transplanted stem cells do not engraft in the heart in the absence of ischemic injury [11]. These observations suggest that the ischemic myocardium may be both a source of signals that influence stem cell engraftment and the physical substrate within which cells must persist.

In the acute phase post-MI, a constellation of neurohumoral factors, cytokines, and growth factors of local and systemic origin are activated to support LV function and to promote repair of the injured myocardium. In this setting, injured myocytes may undergo necrotic or apoptotic cell death [29]. Injured myocytes and vascular cells also release chemotactic signals, such as monocyte chemoattractant protein 1 (MCP1) and tumor necrosis factor α (TNFα) [30–32]. An inflammatory response ensues in which dead cells are removed by phagocytosis, primarily by infiltrating macrophages [33]. Inflammatory cells migrating to the site of infarct also act to amplify the neurohumoral response, delivering various cytokines and growth factors such as interleukin 6 (IL-6), transforming growth factor β1 (TGFβ1), and vascular endothelial growth factor (VEGF) to the site of MI, which in turn act along paracrine pathways to stimulate their further synthesis and release by resident cells [34,35]. These factors stimulate neovascularization of the spared myocardium as well as reparative scar formation at the site of infarct to replace lost myocytes and thereby reinforce the structural integrity of the heart [35]. With respect to cell therapy, the biochemical environment of the ischemic myocardium described influences stem cell engraftment (e.g., migration, cell-cell interactions, and matrix attachment) and survival (e.g., reduced vascular supply and inflammation) and further offers targets for molecular imaging.

The observed therapeutic effects of adult stem or progenitor cells have been shown to occur, at least in part, through the release of paracrine factors at the site of engraftment, which augment neovascularization and improve survival of jeopardized myocardial cells [1,36–39]. Two important classes of these beneficial paracrine factors are proangiogenic and antiapoptotic. Bone marrow-derived cells express high levels of angiogenic ligands and cytokines, including SDF1 basic fibroblast growth factor (bFGF), VEGF, angiopoietin 1 (Ang1), IL-1β, and TNFα [37–39]. The expression of these important factors can be amplified by preconditioning the stem cells before delivery [40,41].

Angiogenesis offers a signal for radionuclide molecular imaging. Capillary sprouting requires an interaction between the endothelial cell growth and the surrounding supporting tissue or extracellular matrix. This interaction, which is necessary for new growth, is mediated by the integrins, a family of heterodimeric transmembrane glycoproteins consisting of an alpha and beta subunit [42]. Integrins recognize extracellular matrix proteins such as vitronectin, fibrinogen, and fibronectin, and they facilitate the binding of growth factor receptor and ligand [43]. The αvβ3 integrin receptor has been identified as a principal integrin expressed in neovascular growth and is associated with PDGF (platelet-derived growth factor) and VEGF receptors [44]. Extracellular matrix proteins interact with the integrins via the amino acid sequence arginine-glycine-aspartic acid (RGD) [45]. This observation forms the basis for development of radiolabeled ligands with RGD motifs for *in vivo* imaging of angiogenesis. Haubner et al. were the first to report the synthesis and biological evaluation of glycosylated RGD-containing peptides [46,47].

Myocardial hypoxia induced by coronary artery ligation upregulates hypoxia-inducible factor 1α (HIF-1α), which stimulates the local release of VEGF into areas of myocardial ischemia and infarction [48]. Meoli et al. reported an animal study using dual-isotope imaging of an [111]In-labeled quinolone ([111]In-RP748) targeting αvβ3 integrin and a radiolabeled perfusion agent. In a chronic infarction model in rats, they showed an inverse relationship between uptake of RP748 and [201]thallium (Tl) in myocardial tissue [49]. In a canine model of chronic infarction, they performed *in vivo* SPECT

imaging and showed uptake of the radiotracer in the zone of chronic infarction (perfusion defect), which corresponded on histopathology to areas of
lectin-positive sprouting capillaries [49] (Figure 7.4).

Separating the angiogenic effects of engrafted stem cells from the host
response to the ischemic insult in acute or chronic MI is potentially challenging. However, Dobrucki et al. demonstrated greater uptake of a $^{99m}$Tc-labeled
RGD peptide in rats injected at the time of LAD (left anterior descending
artery) ligation with an angiogenic growth factor than controls that correlated
with greater improvement in LV function [50]. Further complicating interpretation of αvβ3-targeted imaging in chronic infarction is the finding that RGD
peptides also bind to integrin moieties expressed on myofibroblasts associated with collagen formation, fibrosis, and ventricular remodeling [51,52].

Chronic ischemic dysfunctional myocardium presents a pathological substrate more favorable to imaging effects of stem cells to promote angiogenesis.

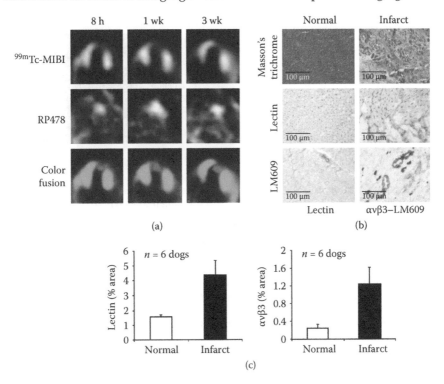

**FIGURE 7.4**
(a) Dual-isotope SPECT serial imaging of dog with 2-hour LAD occlusion at 8 hours, 1 week,
and 3 weeks after injection of $^{111}$In-RP786 (RGD peptomimetic quinolone) targeting αvβ3 integrins. The color fusion images show uptake of the $^{111}$In-RP748 tracer (red color) corresponding
to the thallium (green) perfusion defect in the LAD (apical) territory. (b) Immunohistochemical
staining of canine infarct. Mason's trichrome staining demonstrates increased vascular density in the central infarct region. Angiogenesis was confirmed by lectin staining and staining
for αvβ3 (LM609). (From Meoli, D.F., et al., *J Clin Invest* 2004;113:1684–1691. With permission.)

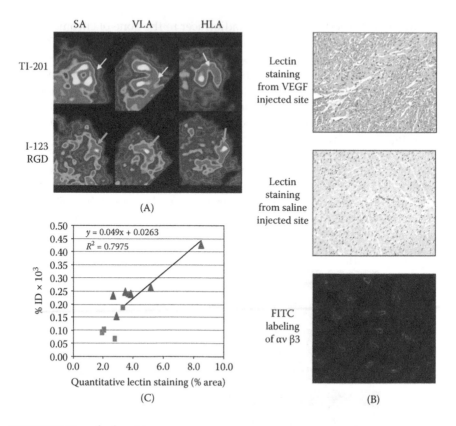

**FIGURE 7.5 (See color insert.)**
(A) Short-axis (SA), vertical long-axis (VLA), and horizontal long-axis (HLA) SPECT slices from a swine with chronic myocardial hibernation (ameroid constrictor on LCx) treated with phVEGF$_{165}$ delivered into the lateral wall and 20 days later injected with [$^{123}$I]Gluco-RGD targeting αvβ3 and thallium-201 for myocardial perfusion. Focal uptake of [$^{123}$I]Gluco-RGD corresponds to the apical lateral wall perfusion defect on the thallium scan. (B) Immunohistochemical staining of tissue from the lateral wall from animal treated with VEGF and one treated with saline (control) shows evidence for angiogenesis on lectin staining and on fluorescent staining for αvβ3. (C) Percentage injected dose (ID) per gram of tissue for [$^{123}$I]Gluco-RGD plotted versus lectin staining as percentage area for VEGF-injected animals (triangles) and saline-injected (squares) ones. (From Johnson, L.L., et al., *J Am Coll Cardiol Vasc Imag* 2008;1:501–510.)

Johnson et al. imaged angiogenesis response to VEGF therapy in pigs with hibernating myocardium produced by ameroid constrictors placed on the LCx (left circumflex) coronary arteries [53]. Three weeks after ameroid placement, using unipolar voltage-guided mapping, they injected either phVEGF$_{165}$ or saline into the ameroid territory and 20 ± 6 days later injected animals with [$^{123}$I]Gluco-RGD to perform SPECT imaging. Focal uptake of [$^{123}$I]Gluco-RGD corresponding to $^{201}$Tl defects was seen in VEGF-injected but not in saline-injected animals. A significant correlation was found between percentage injected dose and quantitative lectin staining (Figure 7.5). These studies suggest the potential for use of radiolabeled RGD peptides targeting $\alpha v \beta 3$ to noninvasively and quantitatively assess the paracrine actions of engrafted stem cells to salvage myocardium by stimulating angiogenesis. In the acute and post-MI setting, therapeutic effects will have to be separated from naturally occurring angiogenesis in response to hypoxia in the infarct zone, which may limit their usefulness.

Another mechanism for the beneficial effects of stem cell-mediated neo-vascularization to improve myocardial function is by reducing ischemia-induced myocyte apoptosis. Kocher et al. injected human CD34+ cells or saline (controls) intravenously into nude rats 48 hours after LAD ligation. At 48 hours, there was histological evidence for angiogenesis in the peri-infarct zone of the animals injected with progenitor cells [1]. Two weeks later, by co-staining with the myocyte-specific marker desmin and terminal deoxynucleotidyl transferase dUTP nick end labeling (TUNEL) , they found a sixfold lower number of apoptotic myocytes in the infarct zone of the LV of rats injected with the CD34+ cells compared with saline controls. The biochemistry of the apoptotic cell death involves activation of the caspase cascade (effector caspases 3 and 7) [54]. Caspase 3 activation both triggers DNA fragmentation (identified by TUNEL staining) and induces cell membrane alterations in cells undergoing apoptosis for phagocytic engulfment. This last pathway has been used to target apoptosis for radioimaging. Phosphatidylserine (PS) is normally restricted to the inner layer of the phospholipid bilayer cell membrane. During apoptosis, PS is flipped to the outer bilayer. The naturally occurring protein annexin V5 avidly binds PS and was labeled first with a fluorescent probe and subsequently with a radiotracer for *in vitro* and *in vivo* imaging of apoptosis [55]. Following coronary artery occlusion with sudden interruption of blood flow, both necrotic and apoptotic death of myocytes occur in the infarct and border zones. Investigators from Maastricht have imaged ischemia-induced myocyte apoptosis occurring in infarct zone in experimental animals and in patients [56,57]. In addition, the number of myocytes undergoing apoptotic cell death in some patients with dilated ischemic cardiomyopathy is sufficient for detection on SPECT imaging following injection of $^{99m}$Tc annexin V5 [57].

## 7.6 Summary

Direct labeling of stem cells with single-photon or PET radiolabels is limited by half-lives of the radioprobes, loss of tracer from cells after injection, and exposure dose to cells but is useful for evaluating the early biodistribution of the cells following injection. By comparison, reporter gene technology offers greater promise for tracking cell engraftment and early survival. In this approach, stem cells are transduced with a reporter gene encoding for intracellular enzymes or transmembrane proteins. The radionuclide probe is designed as either substrate for the enzyme (PET-labeled nucleoside substrate probe for thymidine kinase) or transmembrane protein (free $^{99}$mTc for sodium iodide receptor). The advantage of this approach is that cells can potentially be monitored for longer time periods and with reduced exposure dose to the grafted cells. The duration of gene expression for these methods is limited and not yet fully defined. Radionuclide probes that target beneficial effects of stem cells on the myocardium, such as increases in angiogenesis and reduction in apoptosis, have potential uses for monitoring cell-based therapy.

## References

1. Kocher, A.A., Shuster M., Szaboics MJ. et al., Neovascularization of ischemic myocardium by human bone-marrow-derived angioblasts prevents cardiomyocyte apoptosis, reduces remodeling and improves cardiac function. *Nat Med*, 2001. **7**(4): p. 430–6.
2. Orlic, D., Kajstura J., Chimenti S., et al., Transplanted adult bone marrow cells repair myocardial infarcts in mice. *Ann N Y Acad Sci*, 2001. **938**: p. 221–9; discussion 229–30.
3. Schuster, M.D., Kocher AA., Seki T., et al., Myocardial neovascularization by bone marrow angioblasts results in cardiomyocyte regeneration. *Am J Physiol Heart Circ Physiol*, 2004. **287**(2): p. H525–32.
4. Assmus, B., Schächinger V, Teupe C. et al., Transplantation of Progenitor Cells and Regeneration Enhancement in Acute Myocardial Infarction (TOPCARE-AMI). *Circulation*, 2002. **106**(24): p. 3009–17.
5. Assmus, B., Honold J., Schächinger V. et al., Transcoronary transplantation of progenitor cells after myocardial infarction. *N Engl J Med*, 2006. **355**(12): p. 1222–32.
6. Engelmann, M.G., Theiss HD., Hennig-Theiss C., et al., Autologous bone marrow stem cell mobilization induced by granulocyte colony-stimulating factor after subacute ST-segment elevation myocardial infarction undergoing late revascularization: final results from the G-CSF-STEMI (Granulocyte Colony-Stimulating Factor ST-Segment Elevation Myocardial Infarction) trial. *J Am Coll Cardiol*, 2006. **48**(8): p. 1712–21.
7. Bartunek, J., Vanderheyden M., Vandekerckhove B., et al., Intracoronary injection of CD133-positive enriched bone marrow progenitor cells promotes cardiac recovery after recent myocardial infarction: feasibility and safety. *Circulation*, 2005. **112**(9 Suppl): p. I178–83.
8. Caspi, O., Huber I, Kehat I., et al., Transplantation of human embryonic stem cell-derived cardiomyocytes improves myocardial performance in infarcted rat hearts. *J Am Coll Cardiol*, 2007. **50**(19): p. 1884–93.
9. Cao, F., van der Bogt KE., Sadrzadeh A., et al., Spatial and temporal kinetics of teratoma formation from murine embryonic stem cell transplantation. *Stem Cells Dev*, 2007. **16**(6): p. 883–91.

10. Brenner, W., Aicher A., Eckey T., et al., 111In-Labeled CD34+ Hematopoietic Progenitor Cells in a Rat Myocardial Infarction Model. 2004. **45**(3): p. 512–518.
11. Aicher, A., Brenner W., Zuhayra M., et al., Assessment of the tissue distribution of transplanted human endothelial progenitor cells by radioactive labeling. *Circulation*, 2003. **107**(16): p. 2134–9.
12. Zhou, R., Thomas DH., Qiao H., et al., In vivo detection of stem cells grafted in infarcted rat myocardium. *J Nucl Med*, 2005. **46**(5): p. 816–22.
13. Barbash, I.M., Chouraqui P., Baron J., et al., Systemic delivery of bone marrow-derived mesenchymal stem cells to the infarcted myocardium: feasibility, cell migration, and body distribution. *Circulation*, 2003. **108**(7): p. 863–8.
14. Doyle, B., Kemp BJ., Chareonthaitawee P., et al., Dynamic tracking during intracoronary injection of 18F-FDG-labeled progenitor cell therapy for acute myocardial infarction. *J Nucl Med*, 2007. **48**(10): p. 1708–14.
15. Kang, W.J., Kang HJ., Kim HS., et al., Tissue distribution of 18F-FDG-labeled peripheral hematopoietic stem cells after intracoronary administration in patients with myocardial infarction. *J Nucl Med*, 2006. **47**(8): p. 1295–301.
16. Gambhir, S.S., Bauer E, Black ME., et al., A mutant herpes simplex virus type 1 thymidine kinase reporter gene shows improved sensitivity for imaging reporter gene expression with positron emission tomography. *Proc Natl Acad Sci USA*, 2000. **97**(6): p. 2785–90.
17. Wu, J.C., Inubushi M, Sundaresan G., et al., Positron emission tomography imaging of cardiac reporter gene expression in living rats. *Circulation*, 2002. **106**(2): p. 180–3.
18. Miyagawa, M., Anton M, Wagner B., et al., Non-invasive imaging of cardiac transgene expression with PET: comparison of the human sodium/iodide symporter gene and HSV1-tk as the reporter gene. *Eur J Nucl Med Mol Imaging*, 2005. **32**(9): p. 1108–14.
19. Chang, G.Y., Cao R, Krishnan M., et al., Positron emission tomography imaging of conditional gene activation in the heart. *J Mol Cell Cardiol*, 2007. **43**(1): p. 18–26.
20. Kim, Y.H., Lee DS., Kang JH., et al., Reversing the silencing of reporter sodium/iodide symporter transgene for stem cell tracking. *J Nucl Med*, 2005. **46**(2): p. 305–11.
21. Terrovitis, J., Kwok KF., Lautamaki R., et al., Ectopic expression of the sodium-iodide symporter enables imaging of transplanted cardiac stem cells in vivo by single-photon emission computed tomography or positron emission tomography. *J Am Coll Cardiol*, 2008. **52**(20): p. 1652–60.
22. Wu, J.C., Molecular Imaging: Antidote to Cardiac Stem Cell Controversy. *Journal of the American College of Cardiology*, 2008. **52**(20): p. 1661–1664.
23. Van Tendeloo, V.F., Ponsaerts P., Lardon F., et al., Highly efficient gene delivery by mRNA electroporation in human hematopoietic cells: superiority to lipofection and passive pulsing of mRNA and to electroporation of plasmid cDNA for tumor antigen loading of dendritic cells. *Blood*, 2001. **98**(1): p. 49–56.
24. Wiehe, J.M., Ponsaerts P., Rojewski MT., et al., mRNA-mediated gene delivery into human progenitor cells promotes highly efficient protein expression. *J Cell Mol Med*, 2007. **11**(3): p. 521–30.
25. Ryser, M.F., Ugate F., Lehmann R., et al., mRNA Transfection of CXCR4-GFP Fusion-Simply Generated by PCR-Results in Efficient Migration of Primary Human Mesenchymal Stem Cells. *Tissue Eng Part C Methods*, 2008. **14**(3): p. 179–84.

26. Noiseux, N., Gnecchi M., Lopez-Ilasaca M., et al., Mesenchymal stem cells over-expressing Akt dramatically repair infarcted myocardium and improve cardiac function despite infrequent cellular fusion or differentiation. *Mol Ther*, 2006. **14**(6): p. 840–50.

27. Templin, C., Kotlarz D., Faulhaber J., et al., Ex vivo expanded hematopoietic progenitor cells improve cardiac function after myocardial infarction: role of beta-catenin transduction and cell dose. *J Mol Cell Cardiol*, 2008. **45**(3): p. 394–403.

28. Mangi, A.A., Noiseux N., Kong D., et al., Mesenchymal stem cells modified with Akt prevent remodeling and restore performance of infarcted hearts. *Nat Med*, 2003. **9**(9): p. 1195–201.

29. Kajstura, J., Cheng W., Reiss K., et al., Apoptotic and necrotic myocyte cell deaths are independent contributing variables of infarct size in rats. *Lab Invest*, 1996. **74**(1): p. 86–107.

30. Koyanagi, M., Egashira K, Kitamoto S., et al., Role of monocyte chemoattractant protein-1 in cardiovascular remodeling induced by chronic blockade of nitric oxide synthesis. *Circulation*, 2000. **102**(18): p. 2243–8.

31. Martire, A., Fernandez, B., Strohm C., et al., Cardiac overexpression of monocyte chemoattractant protein-1 in transgenic mice mimics ischemic preconditioning through SAPK/JNK1/2 activation. *Cardiovasc Res*, 2003. **57**(2): p. 523–34.

32. Sun, M., Dawood F, Wen WH., et al., Excessive tumor necrosis factor activation after infarction contributes to susceptibility of myocardial rupture and left ventricular dysfunction. *Circulation*, 2004. **110**(20): p. 3221–8.

33. James, T.N., The variable morphological coexistence of apoptosis and necrosis in human myocardial infarction: significance for understanding its pathogenesis, clinical course, diagnosis and prognosis. *Coron Artery Dis*, 1998. **9**(5): p. 291–307.

34. Gwechenberger, M., Mendoza LH., Youker KA., et al., Cardiac myocytes produce interleukin-6 in culture and in viable border zone of reperfused infarctions. *Circulation*, 1999. **99**(4): p. 546–51.

35. van Amerongen, M.J., Harmsen MC., van Rooijen N., et al., Macrophage depletion impairs wound healing and increases left ventricular remodeling after myocardial injury in mice. *Am J Pathol*, 2007. **170**(3): p. 818–29.

36. Burchfield, J.S. and S. Dimmeler, Role of paracrine factors in stem and progenitor cell mediated cardiac repair and tissue fibrosis. *Fibrogenesis Tissue Repair*, 2008. **1**(1): p. 4.

37. Vandervelde, S., van Luyn MJ., Tio RA, et al., Signaling factors in stem cell-mediated repair of infarcted myocardium. *J Mol Cell Cardiol*, 2005. **39**(2): p. 363–76.

38. Payne, T.R., Oshima H., Okada M., et al., A relationship between vascular endothelial growth factor, angiogenesis, and cardiac repair after muscle stem cell transplantation into ischemic hearts. *J Am Coll Cardiol*, 2007. **50**(17): p. 1677–84.

39. Zhang, M., Mal N., Kiedrowski M., et al., SDF-1 expression by mesenchymal stem cells results in trophic support of cardiac myocytes after myocardial infarction. *Faseb J*, 2007.

39. a　See, F, Seki T, Psaltis PJ., et al., Therapeutic effects of human STRO-3-selected mesenchymal precursor cells and their soluble factors in experimental myocardial ischemia. *J Cell Mol Med* 2010; doi: 10.1111/j.1582-4934.2010.01214x

40. Niagara, M.I., Haider HK., Jiang S., et al., Pharmacologically preconditioned skeletal myoblasts are resistant to oxidative stress and promote angiomyogenesis via release of paracrine factors in the infarcted heart. *Circ Res*, 2007. **100**(4): p. 545–55.
41. Hu, X., Yu SP., Frazer JL., et al., Transplantation of hypoxia-preconditioned mesenchymal stem cells improves infarcted heart function via enhanced survival of implanted cells and angiogenesis. *J Thorac Cardiovasc Surg*, 2008. **135**(4): p. 799–808.
42. Giancotti, F.G. and E. Ruoslahti, Integrin signaling. *Science*, 1999. **285**(5430): p. 1028–32.
43. Senger, D.R., Claffey KP., Benes JE., et al., Angiogenesis promoted by vascular endothelial growth factor: regulation through alpha1beta1 and alpha2beta1 integrins. *Proc Natl Acad Sci USA*, 1997. **94**(25): p. 13612–7.
44. Tsou, R. and F.F. Isik, Integrin activation is required for VEGF and FGF receptor protein presence on human microvascular endothelial cells. *Mol Cell Biochem*, 2001. **224**(1–2): p. 81–9.
45. Bayless, K.J., R. Salazar, and G.E. Davis, RGD-dependent vacuolation and lumen formation observed during endothelial cell morphogenesis in three-dimensional fibrin matrices involves the alpha(v)beta(3) and alpha(5)beta(1) integrins. *Am J Pathol*, 2000. **156**(5): p. 1673–83.
46. Haubner, R., Wester HJ., Burkhart F., et al., Glycosylated RGD-containing peptides: tracer for tumor targeting and angiogenesis imaging with improved biokinetics. *J Nucl Med*, 2001. **42**(2): p. 326–36.
47. Haubner, R., Alphavbeta3-integrin imaging: a new approach to characterise angiogenesis? *Eur J Nucl Med Mol Imaging*, 2006. **33 Suppl 1**: p. 54–63.
48. Giordano, F.J., Ping P., McKirnan MD., et al., Intracoronary gene transfer of fibroblast growth factor-5 increases blood flow and contractile function in an ischemic region of the heart. *Nat Med*, 1996. **2**(5): p. 534–9.
49. Meoli, D.F., Sadeghi M., Krassilnikova S., et al., Noninvasive imaging of myocardial angiogenesis following experimental myocardial infarction. *J Clin Invest*, 2004. **113**(12): p. 1684–91.
50. Dobrucki L.W., Tsutsumi Y, Kalinowski L., et al. Analysis of angiogenesis induced by local IGF-1 expression after myocardial infarction using microSPECT-CT imaging. *J Mol Cell Cardiol*, 2010. **48**: p. 1071–79.
51. van den Borne S.W.M., Isobe S., Zandbergen HR., et al. Molecular imaging of interstitial alterations in remodeling myocardium after myocardial infarction. *J Am Coll Cardiol*, 2008. **52**(24): p. 2017–28.
52. Verjans J., Wolders S, Laufer W., et al. Early molecular imaging of interstitial changes in patients after myocardial infarction: Comparison with delayed contrast-enhanced magnetic resonance imaging. *J Nucl Cardiol* 2010. On-line first July 2010.
53. Johnson, L.L., Schofield L, Donahay T., et al., Radiolabeled arginine-glycine-aspartic acid peptides to image angiogenesis in swine model of hibernating myocardium. *JACC Cardiovasc Imaging*, 2008. **1**(4): p. 500–10.
54. Wolters, S.L., Corsten MR., Reutelingsperger CP., et al., Cardiovascular molecular imaging of apoptosis. *Eur J Nucl Med Mol Imaging*, 2007. **34 Suppl 1**: p. S86–98.
55. Dumont, E.A., Hofstra L, Van Heerde WL., et al., Cardiomyocyte death induced by myocardial ischemia and reperfusion: measurement with recombinant human annexin-V in a mouse model. *Circulation*, 2000. **102**(13): p. 1564–8.

56. Hofstra, L., Liem IH, Dumont EA., et al., Visualisation of cell death in vivo in patients with acute myocardial infarction. *Lancet*, 2000. **356**(9225): p. 209–12.
57. Kietselaer, B.L., Reutelingsperger CP., Boersma HH., et al., Noninvasive detection of programmed cell loss with 99mTc-labeled annexin A5 in heart failure. *J Nucl Med*, 2007. **48**(4): p. 562–7.

# 8

## Fluorescence Imaging of Stem Cells In Vivo: Evolving Technologies and Applications

David E. Sosnovik

## CONTENTS

## 8.1 Introduction

Fluorescence and bioluminescence imaging play a central role in the characterization of stem cells both *in vitro* and *in vivo*. The physical principles and application of bioluminescence imaging are reviewed in Chapter 13. This chapter thus focuses largely on the principles and applications of fluorescence imaging. Particular attention is placed on the recent development of noninvasive fluorescence techniques, such as fluorescence tomography (FMT), and their application in the cardiovascular system. The principles and challenges discussed, however, apply equally to the imaging of other organ systems and regions of the body.

Fluorescence techniques are routinely used to characterize cells *in vitro* (Giepmans et al. 2006). The properties of fluorescence that underlie the power and value of fluorescence techniques *in vitro* also form the basis of fluorescence imaging *in vivo*. The sensitivity of fluorescence imaging is picomolar or better, and the multispectral (multiple wavelengths) nature of the technique allows multiple targets to be imaged simultaneously. The stability of the fluorescence signal over time and the absence of ionizing radiation also contribute significantly to the appeal of fluorescence imaging.

Fluorescence imaging can be divided into those techniques with microscopic, mesoscopic, or macroscopic resolution (Weissleder and Ntziachristos 2003). Microscopic techniques are able to resolve surface and subsurface fluorescence (<1 mm) *in vivo* and generally require a catheter-based, invasive, or semi-invasive approach (Weissleder and Ntziachristos 2003). Macroscopic and mesoscopic approaches can be divided into those that are planar, such as fluorescence reflectance imaging, and those that are tomographic (Ntziachristos et al. 2005; Weissleder and Ntziachristos 2003). The ability to perform FMT *in vivo* has been facilitated by progress in the mathematical modeling of photon propagation in tissue, the expanding availability of biologically compatible near-infrared (NIR) fluorochromes, and the development of highly sensitive photon detection technologies. FMT is quantitative, completely noninvasive, and capable of imaging fluorochromes in deep structures, such as the myocardium, in small animals (Ntziachristos et al. 2005; Weissleder and Ntziachristos 2003).

## 8.2 Physical Basis of Fluorescence Tomography

Fluorescence imaging is based on the absorption of energy from an external light source, which is then almost immediately reemitted at a longer wavelength of lower energy. Many of the fluorochromes used for *in vitro* imaging (i.e., green fluorescent protein [GFP]) absorb light at wavelengths that fall into the visible spectrum. The use of these fluorochromes for *in vivo* imaging, however, is limited by the high absorption coefficient of light within the visible spectrum by body tissues (Frangioni 2003; Ntziachristos et al. 2005; Weissleder and Ntziachristos 2003). In addition, tissue autofluorescence tends to be highest at wavelengths within the visible spectrum, further complicating the use of fluorochromes, such as GFP, for *in vivo* imaging (Frangioni 2003; Ntziachristos et al. 2005; Weissleder and Ntziachristos 2003).

The use of cyanine dyes, with absorption and emission spectra in the NIR range (650–850 nm), has thus been a key advance for fluorescence imaging of deep structures (Figure 8.1) (Frangioni 2003; Mahmood et al. 1999). Hemoglobin (the principal absorber of visible light), water, and lipids (the principal absorbers of infrared light) all have their lowest absorption coefficients in the NIR region of around 650–900 nm. Imaging in the NIR region also has the advantage of minimizing tissue autofluorescence, which is lowest in this range (Figure 8.1). The use of NIR fluorochromes thus allows deeper structures within the body to be imaged *in vivo* and provides high target/background ratios. In fact, penetration depths of 7–14 cm are theoretically possible, depending on tissue type (Ntziachristos et al. 2002a).

The imaging principles involved in FMT are similar to x-ray computed tomography (CT) in that multiple source-detector configurations are used

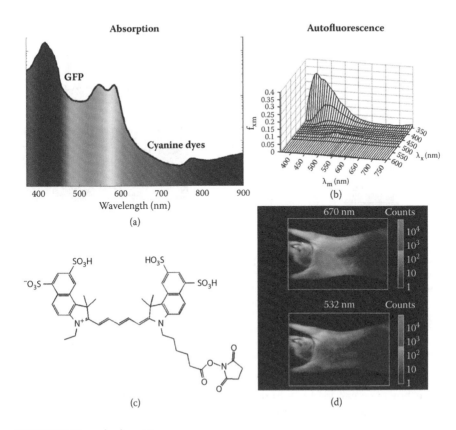

**FIGURE 8.1 (See color insert.)**
(a) The absorption of light by tissue is strongest in the visible spectrum and lowest in the near-infrared portion of the spectrum. (b) Tissue autofluorescence is likewise significantly stronger in the visible portion of the spectrum than in the near infrared. (c) Schematic of Cy5.5, a prototypical near-infrared cyanine fluorochrome. (d) Illumination of a mouse with light in the visible spectrum and the near infrared (670 nm). Transillumination is significantly higher in the near infrared. (Panels a, b, d adapted from Weissleder, R. and V. Ntziachristos. 2003. *Nat Med* 9, no. 1: 123–28. With permission.)

to create a tomographic dataset (Figure 8.2) (Ntziachristos et al. 2002b). FMT, however, requires the use of a theoretical approach that accounts for the diffuse nature of photons in tissues, where both absorption and scattering modulate the signal (Ntziachristos and Weissleder 2001). The propagation of light from a point source through a diffusive medium is described by a Green's function unique to that point (Figure 8.2). The impact of tissue absorption and scattering on the emitted fluorescence signal can thus be formulated as a linear algebraic problem; the detected fluorescence is the product of the fluorochrome distribution and a sensitivity matrix of Green's functions (Ntziachristos and Weissleder 2001). FMT reconstruction involves the solution of the inverse problem (inversion of the sensitivity matrix), which resolves the biodistribution of the fluorescent sources from the detected

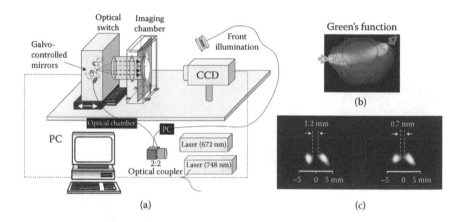

(a)

(b)

(c)

**FIGURE 8.2 (See color insert.)**
(a) Schematic of a second-generation FMT system. The mouse in this system is placed in an imaging chamber containing an optical matching medium. The illumination sources are arranged in a planar slab geometry in front of the mouse. (b) Green's function of light propagation in tissue. The radial width of the function is caused by tissue scattering of light. (c) Two fluorochrome-filled tubes imaged with FMT at variable separation distances. The width of the Green's functions leads to an ill-posed inverse problem and limits the spatial resolution of FMT to the macroscopic range. (Panels A, C adapted with permission from Graves, E.E., J. Ripoll, R. Weissleder, and V. Ntziachristos. 2003. *Med Phys* 30, no. 5: 901–11.)

fluorescence (Ntziachristos and Weissleder 2001). The scattering of light by tissue, however, can lead to a highly ill-posed and computationally challenging inverse problem, limiting the spatial resolution of FMT (Figure 8.2).

At a given time in an FMT experiment, a single-point excitation source illuminates the tissue from a spatially unique position, and the photon field distributes in three dimensions (3D) along isocontour lines within the tissue. In each illumination position, the fluorochromes act as secondary sources, emitting energy at a higher wavelength and with an intensity that depends on the position of the light source. The excitation and fluorescence wavelengths are collected with a charge-coupled device (CCD) camera at multiple spatial points using appropriate filters. Normalization of the emitted fluorescence by the signal at the excitation wavelength produces a measure of the fluorescence signal known as the *normalized Born ratio* (Ntziachristos and Weissleder 2001). This parameter is insensitive to tissue heterogeneity and variation in the optical properties of tissue along the propagation path and is used in the inverse problem to yield a 3D dataset of fluorochrome distribution (Ntziachristos and Weissleder 2001).

FMT of the myocardium (in a mouse model *in vivo*) was initially performed with a second-generation FMT system (Sosnovik et al. 2007) consisting of 46 illumination sources arranged in a slab geometry (Graves et al. 2003). Each source was illuminated over a 5-second time gate, yielding a total acquisition time per mouse of approximately 7 minutes (Sosnovik et al. 2007). Submillimeter resolution could be achieved with this system (Graves et al. 2003); however, the mice needed to be partially immersed in an optical matching medium consisting of 1% intralipid and 150 ppm of India ink. Third-generation FMT systems have been developed and have significantly improved the robustness of FMT, as described next.

## 8.3 Recent Advances in Fluorescence Tomography

Several recent technical advances in FMT have substantially increased the sophistication and scope of the technique. Noncontact systems consisting of 360° source-detector configurations and reconstruction schemes based on early photon propagation have been developed (Deliolanis et al. 2007; Niedre et al. 2008). The high density of third-generation source-detector configurations has also allowed FMT of proteins in the visible spectrum, such as GFP, to be performed (Garofalakis et al. 2007). FMT images are also increasingly fused with anatomical CT and magnetic resonance imaging (MRI) datasets, and hybrid systems are under development.

X-ray CT systems rotate a source-detector configuration around the patient. Noncontact 360° FMT systems, however, employ a rotating stage that rotates the animal in set increments through 360° rather than the detector system (Deliolanis et al. 2007). In addition, at each rotational position the laser beam is scanned over multiple points. In an early implementation of this system, rotation was performed in thirty-six 10° increments, with 36 images (in a 12 × 3 pattern) acquired at each position (Deliolanis et al. 2007). A total of 36 × 36 projections through the animal were thus acquired over 30 minutes. The use of 360° projection systems improves the performance of FMT over prior slab geometry configurations and allows imaging to be performed without the need to immerse the mouse in an optical matching medium. Surface reconstruction of the mouse is performed as it is rotated through 360°, allowing photon propagation in air and tissue to be separated and the effects of refraction at the air-tissue interface to be taken into account (Figure 8.3) (Deliolanis et al. 2007). These systems thus do not require the optical detection device (CCD camera) to be in contact with the tissue and do not require a simple cylindrical or rectangular geometry to be created by immersing the animal in optical matching fluids.

FMT can be performed using continuous-wave, time-resolved, or frequency domain measurements (Eppstein et al. 2002; Ntziachristos et al. 2005). Early photon FMT systems employ a pulsed femtosecond laser in a time-resolved implementation of FMT (Niedre et al. 2008). Fluorescence is detected during a 100-picosecond gate, starting when the incident laser pulse reaches the imaging chamber (Niedre et al. 2008). Only early photons, which undergo minimal radial scattering, are detected during this gate, and approximately 99% of photons are rejected. The exclusive detection of early photons significantly

**Surface Extraction**

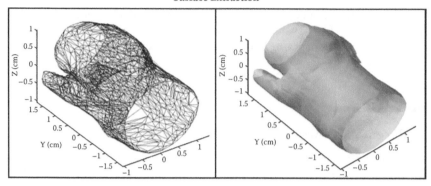

**360° Non-contact FMT Reconstruction**

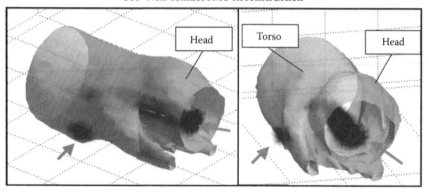

**FIGURE 8.3**

Third-generation 360° noncontact FMT systems. The mouse is rotated in the system and illuminated from 360°. Accurate extraction of surface features allows the effects of refraction at the tissue interface to be calculated. The mouse therefore does not need to be placed in an optical matching medium. Throughput and simplicity are thus significantly improved. Two fluorescent tubes (arrows) implanted in a mouse are accurately resolved with this system. (Adapted with permission from Deliolanis, N., T. Lasser, D. Hyde, A. Soubret, J. Ripoll, and V. Ntziachristos. 2007. *Opt Lett* 32, no. 4: 382–84.)

reduced the width of the Green's functions and caused the inverse problem to become significantly less ill posed. Although the majority of photons were rejected, the accuracy of regular and early-photon FMT systems appears to be similar (Niedre et al. 2008). Early-photon FMT, however, has the potential to improve the spatial resolution of FMT significantly (Niedre et al. 2008). FMT at the mesoscopic scale of small insects and excised tissues has also been described and could become a useful tool to evaluate the 3D structure of engineered tissues (Vinegoni et al. 2008)

The experience with FMT to date has largely involved the imaging of exogenous NIR fluorochromes (Ntziachristos et al. 2002b; Sosnovik et al. 2007). The ability to image fluorescent proteins, such as GFP, with FMT would be a significant advance since GFP can be used as a reporter of stem cell survival and differentiation. The challenges of imaging deep structures in the visible spectrum have been discussed in this chapter. Nevertheless, recent data suggest that FMT of fluorescent proteins may be possible under certain circumstances (Garofalakis et al. 2007). Transgenic mice with GFP-expressing T cells were imaged with a modified FMT system capable of being configured to produce a very high source-detector density over the target of interest (Garofalakis et al. 2007). Measurements were performed in reflection mode, unlike conventional FMT, which uses transillumination to acquire the raw data. A modified solution of the diffusion equation, accounting for the high absorption of light in the visible spectrum, was used to derive the Green's functions for the tomographic reconstruction. With these modifications, GFP-expressing T cells could be successfully imaged in the thymus and spleen (Garofalakis et al. 2007). It should be noted, however, that these organs are fairly superficial and contain large numbers ($>10^7$) of T cells. The ability of this system to detect realistic numbers of GFP-expressing stem cells in deep structures, such as the myocardium, remains unlikely. However, significant effort is being devoted to the generation of red-shifted proteins, which may be more feasible to image in deep tissues (Deliolanis et al. 2008).

## 8.4 Imaging Agents

Fluorescent imaging agents can be broadly divided into small organic fluorochromes, fluorescent proteins, and quantum dots (QDs) (Frangioni 2003; Giepmans et al. 2006). Some imaging agents, such as indocyanine green, can provide a useful readout due to their pharmacokinetic distribution (Rangaraj et al. 2008; Sevick-Muraca et al. 2008). Other agents produce readouts by binding to a specific target (Dumont et al. 2001) or by activation by a specific enzyme of interest (Mahmood et al. 1999). Targeted fluorescence agents are frequently substantially larger than PET-detectable agents but are usually substantially smaller than many MRI-detectable agents, such as magnetic nanoparticles. The differences in the size and physical properties of these constructs can have important implications for probe delivery and pharmacokinetics.

Activatable NIR fluorescent probes have become central to *in vivo* fluorescence imaging and FMT. The NIR fluorochromes on these probes are held in close physical approximation to each other by peptide linkers, such as polylysine chains, quenching the fluorescence (Mahmood et al. 1999). Cleavage of the linker at the recognition site releases the fluorochromes and produces fluorescence. Activatable agents produce no background signal until contact with the enzyme of interest and thus constitute an extremely robust amplification strategy. Increases in signal intensity of up to 100-fold above background have been reported with these agents. Two types of enzyme-activatable NIR agents have been imaged in the myocardium. A cathepsin-activatable fluorochrome has been used to image protease secretion by macrophages infiltrating healing myocardial infarcts (Nahrendorf et al. 2007). Peak cathepsin activity was noted 3–5 days after infarction. Likewise, a matrix metalloproteinase- (MMP-) activatable construct has been used to image MMP activity in healing myocardial infarcts (Chen et al. 2005).

Organic fluorochromes can have several limitations (Frangioni 2003). They are easily bleached *in vitro* (rarely a problem *in vivo*), have narrow excitation spectra, and can have fairly broad emission spectra. A new class of fluorescent agents, QDs, has been developed and does not suffer from these limitations (Dubertret et al. 2002; Frangioni 2003; Giepmans et al. 2006). QDs, which consist of a semiconductor core surrounded by an outer shell, do not bleach, have broad excitation and narrow emission spectra, and are highly tunable (Dubertret et al. 2002; Frangioni 2003; Giepmans et al. 2006). Subtle changes in the size of a QD can significantly change both the absorption and emission wavelengths of the particle, thereby tuning the construct. QDs are also generally more stable than conventional fluorochromes and emit significantly more energy. However, most of the current QD constructs contain highly toxic core substances, such as cadmium, selenide, or arsenic. The development of QDs for deep-tissue imaging will also require these agents both to emit and to be excited in the NIR and thus be subjected to low levels of attenuation at both their excitation and emission wavelengths. The experience with noninvasive QD imaging in the myocardium is thus still fairly limited. Invasive techniques, however, have been used successfully to image QD-labeled mesenchymal stem cells (MSCs) in the myocardium (Rosen et al. 2007). In addition, the highly multispectral nature of QDs makes them an ideal tool for subsurface imaging of multiple stem cell types in experimental settings (Lin et al. 2007). More details of QD labeling can be found in Chapter 14.

## 8.5  Fluorescence Imaging of Stem Cells

FMT could play an important role in several aspects of stem cell imaging in the myocardium. These include fluorescent tracking of labeled stem cells, simultaneous imaging of multiple stem cell types, imaging of the myocardial microenvironment, and reporter imaging of fluorescent proteins.

Tracking of stem cells injected in the myocardium is frequently performed with MRI of magnetically labeled cells (Hill et al. 2003; Kraitchman et al. 2003). The sensitivity of MRI for the detection of these cells in the myocardium *in vivo*, however, is approximately $10^5$ cells. This is substantially worse than the sensitivity achievable with other techniques. When dual-labeled cells (iron oxide and [111]indium oxine) were injected intravenously, the cells could be detected in the myocardium with single-photon emission tomography (SPECT) but not by MRI (Kraitchman et al. 2005). Tracking injected stem cells with FMT, however, could provide substantially higher sensitivity than MRI. NIR fluorochromes are frequently conjugated to magnetic nanoparticles to yield magnetofluorescent nanoparticles (MFNPs) such as cross-linked iron oxide (CLIO)-Tat (Lewin et al. 2000), CLIO-Cy5.5 (Sosnovik et al. 2007), and CLIO-protamine-Cy5.5 (Reynolds et al. 2005). Macrophages labeled endogenously, after intravenous injection of CLIO-Cy5.5, can be robustly imaged in healing myocardial infarctions in mice with both planar techniques (Figure 8.4) and FMT (Figure 8.5) (Sosnovik et al. 2007). It is thus highly likely that stem cells labeled with either magnetofluorescent or fluorescent agents will also

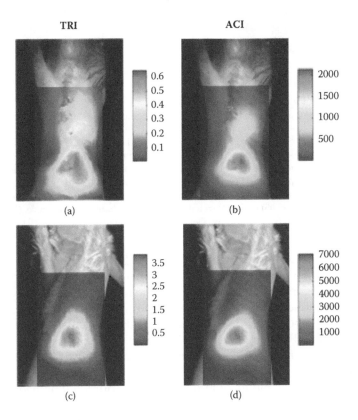

**FIGURE 8.4 (See color insert.)**
Planar transillumination fluorescence images in an infarcted mouse (a, b) and a sham-operated mouse (c, d). The mice have been injected with CLIO-Cy5.5, which is taken up by macrophages infiltrating the healing infarct. The fluorescence images have been superimposed on white light images of the mice. Two postprocessing schemes have been used: TRI (transillumination ratio image) and ACI (attenuation corrected image). Substantial hepatic uptake of CLIO-Cy5.5 is seen in all mice. Thoracic uptake of the agent is seen only in the infarcted mice. (Reproduced from Sosnovik, D.E., M. Nahrendorf, N. Deliolanis, M. Novikov, E. Aikawa, L. Josephson, A. Rosenzweig, R. Weissleder, and V. Ntziachristos. 2007. *Circulation* 115, no. 11: 1384–91.)

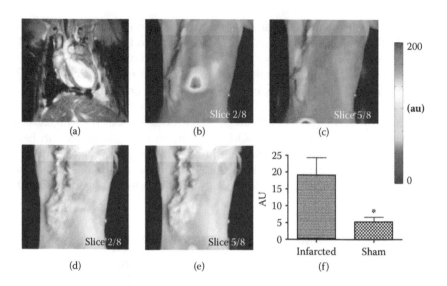

**FIGURE 8.5 (See color insert.)**
FMT of myocardial macrophage infiltration *in vivo*. Reconstructed coronal slices from the 3D FMT dataset have been superimposed on white light images of the mice. Slices 2–4 in the FMT dataset intersected the heart, while slices 5–8 passed posterior to it. (a) Long-axis MRI in an infarcted mouse corresponding to (b) slice 2 from the fluorescence dataset of that mouse, which passed through the heart. (c) Slice 5 from the fluorescence dataset of the infarcted mouse, which passed posterior to the heart. The corresponding slices (d = slice 2, e = slice 5) of a sham-operated mouse are shown. (f) Depth-resolved fluorescence intensity in the heart was significantly greater (*$p < .05$) in the infarcted mice than the sham-operated mice. (Reproduced from Sosnovik, D.E., M. Nahrendorf, N. Deliolanis, M. Novikov, E. Aikawa, L. Josephson, A. Rosenzweig, R. Weissleder, and V. Ntziachristos. 2007. *Circulation* 115, no. 11: 1384–91.)

be highly detectable in the myocardium with FMT. In preliminary studies, cardiac progenitor cells labeled with the magnetofluorescent construct CLIO-protamine-Cy5.5 have shown normal viability, proliferation, differentiation, and resting contraction. It has also been shown that NIR fluorochromes alone can be used to label cytotoxic T cells without adversely affecting their viability and function (Foster et al. 2008; Swirski et al. 2007).

The small size and flexibility of fluorescence probes also allows them to be directed against important targets near injected stem cells. This was demonstrated in a mouse model of radial fracture nonunion. Engineered MSCs injected into the fracture site led to the production of new bone, which could be detected *in vivo* through FMT of the NIR fluorochrome OsteoSense (Visen Medical, Woburn, MA), which binds to hydroxyapatite (Zilberman et al. 2008). FMT of targeted and enzyme-activatable agents can thus provide valuable information on the local microenvironment with which the stem cell interacts. In addition, multispectral NIR FMT of deep tissues is feasible and can allow several processes in the microenvironment to be imaged simultaneously (Figure 8.6) (Nahrendorf et al. 2007). Multispectral FMT could also be used to track different cell populations simultaneously.

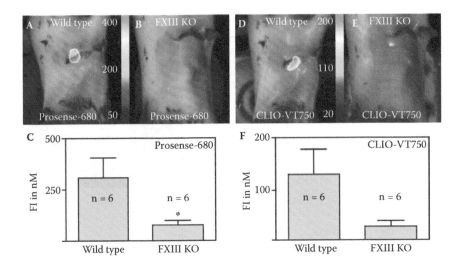

**FIGURE 8.6 (See color insert.)**

Multispectral FMT of the inflammatory response in healing infarcts. Prosense-680 is a protease-activatable Cy5.5-based fluorochrome with an emission wavelength of 680 nm. CLIO-VT750 is a magnetofluorescent nanoparticle with an emission wavelength of 750 nm. Nanoparticle uptake and protease secretion by infiltrating macrophages can be imaged *in vivo* with FMT and spectrally resolved. In the example shown, factor XIII knockout mice with healing infarcts had a significantly lower level of inflammatory activity than wild-type mice. (Reproduced from Nahrendorf, M., D.E. Sosnovik, P. Waterman, F.K. Swirski, A.N. Pande, E. Aikawa, J.L. Figueiredo, M.J. Pittet, and R. Weissleder. 2007. *Circ Res* 100, no. 8: 1218–25.)

QDs in particular are highly suited to simultaneous multiwavelength imaging of several cell populations. This concept was demonstrated in a study involving groups of embryonic stem cells (ESCs) labeled with six spectrally distinct QDs (Lin et al. 2007). At 24 hours, 72% of cells were labeled with the QD, but only 4% remained labeled at 4 days, reflecting the effects of cell division. No differences in ESC proliferation, viability, and differentiation were found between labeled and unlabeled cells. Six groups of ESCs, each containing $10^6$ cells and labeled with a spectrally distinct QD, were implanted subcutaneously in athymic nude mice. Signal from all six groups groups of cells could be detected for 48 hours (Figure 8.7) (Lin et al. 2007). Thereafter, only signal in the infrared (800-nm) range could be detected and remained detectable for 14 days. While several questions, including the potential for chronic toxicity, will require further study, the use of QDs to image stem cells will likely become a valuable adjunct in the imaging armamentarium.

Cell-labeling techniques (whether employing magnetic nanoparticles, organic fluorochromes, QDs, or radioisotopes) report cell position but not cell survival. Reporter techniques, such as the imaging of luciferase and thymidine kinase, are thus needed to image stem cell survival *in vivo* (Cao et al. 2006). MRI-detectable reporter probes have been developed but are still largely experimental (Genove et al. 2005; Gilad et al. 2007). Likewise,

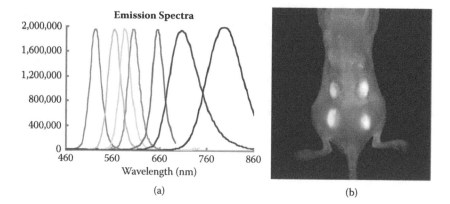

(a)                                                    (b)

**FIGURE 8.7 (See color insert.)**
(a) Emission spectra of an array of quantum dots. The emission spectra of these quantum dots are narrow, supporting highly multispectral imaging approaches. (b) Fluorescence signals from six individual stem cell populations, each labeled with a distinct quantum dot and implanted superficially under the skin of a mouse, can be independently resolved. (Reproduced with permission from Lin, S., X. Xie, M.R. Patel, Y.H. Yang, Z. Li, F. Cao, O. Gheysens, Y. Zhang, S.S. Gambhir, J.H. Rao, and J.C. Wu. 2007. *BMC Biotechnol* 7, no. 1: 67. With permission.)

imaging of fluorescent reporter proteins could conceptually be used to assay cell survival *in vivo*. FMT-based detection of fluorescent reporter proteins such as GFP, however, is likely to be challenging in deep tissues such as the myocardium (Garofalakis et al. 2007). Significant interest thus exists in the development of red-shifted fluorescent reporter proteins, which may be detectable in deep tissues with FMT (Deliolanis et al. 2008).

## 8.6 Translation and Future Challenges

FMT of the myocardium and many other deep tissues is currently limited to mice. Thus, fluorescence imaging of these organs in humans will require invasive or catheter-based approaches. An intraoperative system for NIR fluorescence Near Infrared Fluorescence (NIRF) of exposed tissues has been developed (Ohnishi et al. 2006) and could conceptually be used to guide the surgical implantation of fluorescently labeled stem cells. Moreover, catheter- and endoscopy-based NIRF detectors have been developed and could be used to image stem cells in the cardiovascular, respiratory, urogenital, and gastrointestinal systems. Noninvasive NIR fluorescence imaging of the breast in humans, however, is feasible and is being increasingly studied (Sevick-Muraca et al. 2008).

The translation of QDs will be fairly challenging due to the toxic moieties in these agents. Selected NIR fluorochromes, such as indocyanine green, however, are already approved for human use. In addition, clinical studies with analogous NIR cyanine fluorochromes, such as Cy5.5, are beginning. In one such study, laparoscopic detection of a protease-activatable NIR fluorochrome will be studied in patients with ovarian cancer. In the preclinical setting, the role of *in vivo* fluorescence imaging to study stem cell migration, survival, and differentiation will increase steadily. FMT of small animals will likely prove particularly useful when multiple types of stem cells are injected simultaneously. The impact of fluorescence imaging on stem cell therapy in the clinical area will likely be determined largely by the development of suitable hardware and imaging platforms. A robust and collaborative effort will be required among stem cell scientists, diagnostic pharmaceutical (imaging agent) companies, and the manufacturers of clinical imaging platforms to bring fluorescence imaging of stem cells into the clinical mainstream.

# References

Cao, F., S. Lin, X. Xie, P. Ray, M. Patel, X. Zhang, M. Drukker, S.J. Dylla, A.J. Connolly, X. Chen, I.L. Weissman, S.S. Gambhir, and J.C. Wu. 2006. In vivo visualization of embryonic stem cell survival, proliferation, and migration after cardiac delivery. *Circulation* 113, no. 7: 1005–14.

Chen, J., C.H. Tung, J.R. Allport, S. Chen, R. Weissleder, and P.L. Huang. 2005. Near-infrared fluorescent imaging of matrix metalloproteinase activity after myocardial infarction. *Circulation* 111, no. 14: 1800–05.

Deliolanis, N., T. Lasser, D. Hyde, A. Soubret, J. Ripoll, and V. Ntziachristos. 2007. Free-space fluorescence molecular tomography utilizing 360 degrees geometry projections. *Opt Lett* 32, no. 4: 382–84.

Deliolanis, N.C., R. Kasmieh, T. Wurdinger, B.A. Tannous, K. Shah, and V. Ntziachristos. 2008. Performance of the red-shifted fluorescent proteins in deep-tissue molecular imaging applications. *J Biomed Opt* 13, no. 4: 044008.

Dubertret, B., P. Skourides, D.J. Norris, V. Noireaux, A.H. Brivanlou, and A. Libchaber. 2002. In vivo imaging of quantum dots encapsulated in phospholipid micelles. *Science* 298, no. 5599: 1759–62.

Dumont, E.A., C.P. Reutelingsperger, J.F. Smits, M.J. Daemen, P.A. Doevendans, H.J. Wellens, and L. Hofstra. 2001. Real-time imaging of apoptotic cell-membrane changes at the single-cell level in the beating murine heart. *Nat Med* 7, no. 12: 1352–55.

Eppstein, M.J., D.J. Hawrysz, A. Godavarty, and E.M. Sevick-Muraca. 2002. Three-dimensional, Bayesian image reconstruction from sparse and noisy data sets: near-infrared fluorescence tomography. *Proc Natl Acad Sci USA* 99, no. 15: 9619–24.

Foster, A.E., S. Kwon, S. Ke, A. Lu, K. Eldin, E. Sevick-Muraca, and C.M. Rooney. 2008. In vivo fluorescent optical imaging of cytotoxic T lymphocyte migration using IRDye800cw near-infrared dye. *Appl Opt* 47, no. 31: 5944–52.

Frangioni, J.V. 2003. In vivo near-infrared fluorescence imaging. *Curr Opin Chem Biol* 7, no. 5: 626–34.

Garofalakis, A., G. Zacharakis, H. Meyer, E.N. Economou, C. Mamalaki, J. Papamatheakis, D. Kioussis, V. Ntziachristos, and J. Ripoll. 2007. Three-dimensional in vivo imaging of green fluorescent protein-expressing T cells in mice with noncontact fluorescence molecular tomography. *Mol Imaging* 6, no. 2: 96–107.

Genove, G., U. Demarco, H. Xu, W.F. Goins, and E.T. Ahrens. 2005. A new transgene reporter for in vivo magnetic resonance imaging. *Nat Med* 11, no. 4: 450–54.

Giepmans, B.N., S.R. Adams, M.H. Ellisman, and R.Y. Tsien. 2006. The fluorescent tool-box for assessing protein location and function. *Science* 312, no. 5771: 217–24.

Gilad, A.A., M.T. Mcmahon, P. Walczak, P.T. Winnard, Jr., V. Raman, H.W. Van Laarhoven, C.M. Skoglund, J.W. Bulte, and P.C. Van Zijl. 2007. Artificial reporter gene providing MRI contrast based on proton exchange. *Nat Biotechnol* 25, no. 2: 217–19.

Graves, E.E., J. Ripoll, R. Weissleder, and V. Ntziachristos. 2003. A submillimeter reso-lution fluorescence molecular imaging system for small animal imaging. *Med Phys* 30, no. 5: 901–11.

Hill, J.M., A.J. Dick, V.K. Raman, R.B. Thompson, Z.X. Yu, K.A. Hinds, B.S. Pessanha, M.A. Guttman, T.R. Varney, B.J. Martin, C.E. Dunbar, E.R. Mcveigh, and R.J. Lederman. 2003. Serial cardiac magnetic resonance imaging of injected mesen-chymal stem cells. *Circulation* 108, no. 8: 1009–14.

Kraitchman, D.L., A.W. Heldman, E. Atalar, L.C. Amado, B.J. Martin, M.F. Pittenger, J.M. Hare, and J.W. Bulte. 2003. In vivo magnetic resonance imag-ing of mesenchymal stem cells in myocardial infarction. *Circulation* 107, no. 18: 2290–93.

Kraitchman, D.L., M. Tatsumi, W.D. Gilson, T. Ishimori, D. Kedziorek, P. Walczak, W.P. Segars, H.H. Chen, D. Fritzges, I. Izbudak, R.G. Young, M. Marcelino, M.F. Pittenger, M. Solaiyappan, R.C. Boston, B.M. Tsui, R.L. Wahl, and J.W. Bulte. 2005. Dynamic imaging of allogeneic mesenchymal stem cells trafficking to myocardial infarction. *Circulation* 112, no. 10: 1451–61.

Lewin, M., N. Carlesso, C.H. Tung, X.W. Tang, D. Cory, D.T. Scadden, and R. Weissleder. 2000. Tat peptide-derivatized magnetic nanoparticles allow in vivo tracking and recovery of progenitor cells. *Nat Biotechnol* 18, no. 4: 410–14.

Lin, S., X. Xie, M.R. Patel, Y.H. Yang, Z. Li, F. Cao, O. Gheysens, Y. Zhang, S.S. Gambhir, J.H. Rao, and J.C. Wu. 2007. Quantum dot imaging for embryonic stem cells. *BMC Biotechnol* 7, no. 1: 67.

Mahmood, U., C.H. Tung, A. Bogdanov, Jr., and R. Weissleder. 1999. Near-infrared optical imaging of protease activity for tumor detection. *Radiology* 213, no. 3: 866–70.

Nahrendorf, M., D.E. Sosnovik, P. Waterman, F.K. Swirski, A.N. Pande, E. Aikawa, J.L. Figueiredo, M.J. Pittet, and R. Weissleder. 2007. Dual channel optical tomo-graphic imaging of leukocyte recruitment and protease activity in the healing myocardial infarct. *Circ Res* 100, no. 8: 1218–25.

Niedre, M.J., R.H. De Kleine, E. Aikawa, D.G. Kirsch, R. Weissleder, and V. Ntziachristos. 2008. Early photon tomography allows fluorescence detection of lung carcinomas and disease progression in mice in vivo. *Proc Natl Acad Sci USA* 105, no. 49: 19126–31.

Ntziachristos, V., J. Ripoll, L.V. Wang, and R. Weissleder. 2005. Looking and listening to light: the evolution of whole-body photonic imaging. *Nat Biotechnol* 23, no. 3: 313–20.

Ntziachristos, V., J. Ripoll, and R. Weissleder. 2002a. Would near-infrared fluorescence signals propagate through large human organs for clinical studies? *Opt Lett* 27, no. 5: 333–35.

Ntziachristos, V., C.H. Tung, C. Bremer, and R. Weissleder. 2002b. Fluorescence molecular tomography resolves protease activity in vivo. *Nat Med* 8, no. 7: 757–60.

Ntziachristos, V. and R. Weissleder. 2001. Experimental three-dimensional fluorescence reconstruction of diffuse media by use of a normalized born approximation. *Opt Lett* 26, no. 12: 893–95.

Ohnishi, S., J.L. Vanderheyden, E. Tanaka, B. Patel, A.M. De Grand, R.G. Laurence, K. Yamashita, and J.V. Frangioni. 2006. Intraoperative detection of cell injury and cell death with an 800 nm near-infrared fluorescent annexin V derivative. *Am J Transplant* 6, no. 10: 2321–31.

Rangaraj, A.T., R.K. Ghanta, R. Umakanthan, E.G. Soltesz, R.G. Laurence, J. Fox, L.H. Cohn, R.M. Bolman, 3rd, J.V. Frangioni and F.Y. Chen. 2008. Real-time visualization and quantification of retrograde cardioplegia delivery using near infrared fluorescent imaging. *J Card Surg* 23, no. 6: 701–08.

Reynolds, F., R. Weissleder, and L. Josephson. 2005. Protamine as an efficient membrane-translocating peptide. *Bioconjug Chem* 16, no. 5: 1240–45.

Rosen, A.B., D.J. Kelly, A.J. Schuldt, J. Lu, I.A. Potapova, S.V. Doronin, K.J. Robichaud, R.B. Robinson, M.R. Rosen, P.R. Brink, G.R. Gaudette, and I.S. Cohen. 2007. Finding fluorescent needles in the cardiac haystack: tracking human mesenchymal stem cells labeled with quantum dots for quantitative in vivo three-dimensional fluorescence analysis. *Stem Cells* 25, no. 8: 2128–38.

Sevick-Muraca, E.M., R. Sharma, J.C. Rasmussen, M.V. Marshall, J.A. Wendt, H.Q. Pham, E. Bonefas, J.P. Houston, L. Sampath, K.E. Adams, D.K. Blanchard, R.E. Fisher, S.B. Chiang, R. Elledge, and M.E. Mawad. 2008. Imaging of lymph flow in breast cancer patients after microdose administration of a near-infrared fluorophore: feasibility study. *Radiology* 246, no. 3: 734–41.

Sosnovik, D.E., M. Nahrendorf, N. Deliolanis, M. Novikov, E. Aikawa, L. Josephson, A. Rosenzweig, R. Weissleder, and V. Ntziachristos. 2007. Fluorescence tomography and magnetic resonance imaging of myocardial macrophage infiltration in infarcted myocardium in vivo. *Circulation* 115, no. 11: 1384–91.

Swirski, F.K., C.R. Berger, J.L. Figueiredo, T.R. Mempel, U.H. Von Andrian, M.J. Pittet, and R. Weissleder. 2007. A near-infrared cell tracker reagent for multiscopic in vivo imaging and quantification of leukocyte immune responses. *PLoS ONE* 2, no. 10: e1075.

Vinegoni, C., C. Pitsouli, D. Razansky, N. Perrimon, and V. Ntziachristos. 2008. In vivo imaging of *Drosophila melanogaster* pupae with mesoscopic fluorescence tomography. *Nat Methods* 5, no. 1: 45–47.

Weissleder, R. and V. Ntziachristos. 2003. Shedding light onto live molecular targets. *Nat Med* 9, no. 1: 123–28.

Zilberman, Y., I. Kallai, Y. Gafni, G. Pelled, S. Kossodo, W. Yared, and D. Gazit. 2008. Fluorescence molecular tomography enables in vivo visualization and quantification of nonunion fracture repair induced by genetically engineered mesenchymal stem cells. *J Orthop Res* 26, no. 4: 522–30.

# 9

# Ultrasound Imaging Basics

Howard Leong-Poi

## CONTENTS

## 9.1 Introduction

As the field of stem cells and its applications for tissue regeneration in a wide variety of diseases continues to grow rapidly, there is increasing interest in noninvasive imaging techniques both to evaluate the therapeutic response to cell-based therapies and to track cells after their delivery [1,2]. Ultrasound imaging is one technique that has become a widely utilized modality within medical imaging and clinical cardiology, with important applications in experimental research, due in large part to its many advantages: (a) noninvasive imaging, (b) lack of ionizing radiation, (c) real-time imaging of anatomical structures, (d) visualization and quantification of flow through vascular structures via Doppler techniques, (e) excellent temporal resolution (30–120 Hz with clinical imaging systems and up to 1,000 Hz with small-animal imaging systems), (f) very good spatial resolution (<1 mm with clinical imaging systems and all the way down to 30 μm with ultrahigh-frequency probes using small-animal imaging systems), and (g) portability. Ultrasound, however, is hampered by limited access to certain anatomic structures (e.g., the lungs, bones, and intracerebral structures); susceptibility to imaging artifacts [3]; a limited depth of penetration (especially with higher-frequency probes); and lack of sufficient contrast resolution in grayscale imaging.

Ultrasonic imaging works by the pulse-echo principle: intermittent pulses at varying frequencies generated by ultrasound transducers are propagated through the target tissue, where they are partially absorbed, scattered, and reflected. Based on propagation time and the amplitude of the back-scattered ultrasound pulse wave, the received signal is processed, and the borders between tissues with different echogenic properties are delineated and displayed in real time [4]. Ultrasound contrast agents have now been developed for clinical imaging that result in improved delineation between tissues and the vascular space or blood pool in patients with suboptimal images [5,6]. Contrast-enhanced ultrasound (CEU) imaging relies on the ultrasonic detection of these contrast agents and provides improved endocardial border detection in echocardiography and differentiation between normal and pathological tissue types in medical imaging, such as hepatic or renal ultrasound. More recently, ultrasound contrast agents have been developed that allow the targeted imaging of specific disease-related molecular events that occur at the cellular level [7–9]. It is this approach using CEU molecular imaging and targeted ultrasound contrast agents that yields the highest potential for the noninvasive imaging of stem cells using ultrasonic imaging.

## 9.2  Microbubble Ultrasound Contrast Agents

The contrast agent used for ultrasound imaging consists of gas-filled microbubbles with a shell of biocompatible material, such as lipids, proteins, or biopolymers [10]. With mean diameters less than 6 μm, these bubbles have a similar rheology to red blood cells within the circulation [11,12] and can traverse unimpeded through the pulmonary microcirculation after intravenous administration (Figure 9.1). Unlike most other radiologic contrast agents, microbubble ultrasound contrast agents remain purely intravascular during their transit. While earlier-generation microbubble contrast agents were air or nitrogen filled, later-generation agents used high molecular weight inert gases, such as perfluorocarbons and sulfur hexafluoride, which offer an excellent safety profile, increased bubble stability, and a longer persistence in the systemic circulation. The physical properties of microbubble contrast agents and their responses to insonation are determined by their gas content, peripheral shell composition, the frequency of the ultrasound beam, pulse repetition frequency, and acoustic power employed [13,14]. These microbubbles oscillate when exposed to ultrasound, undergoing compression and expansion in response to the peaks and nadirs of the acoustic wave. The reflected wave is detected, processed, and displayed, forming the basis for CEU imaging.

Commercially available microbubble contrast agents are currently approved to opacify the blood pool, improving endocardial border delineation during echocardiography, and delineating hepatic and renal pathologies [15,16].

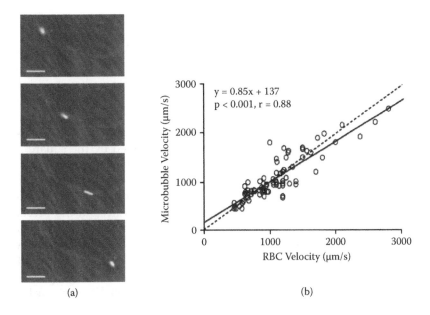

(a)                                         (b)

**FIGURE 9.1**

(a) Examples of sequential intravital microscopy frames of a circulating fluorescent microbubble obtained 30 ms apart (scale bars 20 µm). (b) Graph showing a linear relationship between red blood cell (RBC) velocity and microbubble velocity in capillaries. (Reproduced with permission from Lindner JR, Song J, Jayaweera AR, Sklenar J, Kaul S. *J Am Soc Echocardiogr.* 2002; 15:396–403.)

Imaging algorithms and techniques now allow the ultrasonic assessment and quantification of myocardial and tissue perfusion [17–19], an important endpoint for studies of stem cell delivery for therapeutic angiogenesis. More recently, molecular imaging using CEU has become possible with the development of novel "site-targeted" microbubbles [7,8,20]. Unlike contrast ultrasound methods to opacify the blood pool and evaluate tissue perfusion, which utilize free-flowing microbubbles, this method uses custom-designed microbubbles that are retained in regions of disease by virtue of their shell properties or by the conjugation of specific targeting ligands to their outer shell. Since microbubbles remain entirely within the intravascular space throughout their circulation, the disease states that can be targeted are characterized by events that occur predominantly within the luminal aspect of the vasculature, such as thrombosis [21], endothelial dysfunction [22], angiogenesis [23–25], ischemia [26], and inflammation [27–30]. Therefore, CEU molecular imaging may be ideally suited for the site-targeted imaging of intravascular events, such as progenitor cell incorporation or engraftment in the formation of new vessels. One important limitation of this technique for imaging stem cells may be the inability to target and detect cells that are not accessible via the vascular space, such as cells grafted within the interstitial space or within tissues.

## 9.3 Imaging of Microbubble Contrast Agents

When insonated, microbubbles oscillate, exhibiting strong linear and nonlinear responses that allow their detection during ultrasonic imaging (Figure 9.2). The primary determinant of the microbubble response to insonation is the acoustic power or energy of the emitted ultrasound pulse. The acoustic power is usually expressed by the mechanical index (MI), defined as the peak negative acoustic pressure divided by the square root of the ultrasound frequency. MI is a term generated by regulatory agencies to be used as a safety index to measure the potential for nonthermal bioeffects during insonation. In the absence of attenuation, the MI is highest at the focus of the beam and shifts closer to the transducer in the presence of attenuation. MI ranges between 0.1 and 2.0 on most clinical ultrasound imaging systems and can be adjusted manually on most systems. However, it should be noted that the MI has poor reproducibility on different ultrasound systems, in that various systems may yield distinct acoustic pressures and bioeffects at the same MI. Thus, the MIs displayed on different machines are not precisely comparable.

Several imaging techniques have been developed specifically for contrast microbubble detection (Table 9.1). Imaging strategies can be divided into "nondestructive or low-MI" and "destructive or high-MI" techniques. At very low acoustic powers, the microbubble response is linear, with returning frequencies similar to that of the transmitted signal (Figure 9.2). At a low-to-intermediate MI, microbubbles generate strong harmonic signals (at frequencies that are multiples of the fundamental frequency) due to their nonlinear physical behavior in an ultrasound field. When insonated at a high MI, microbubbles produce a wideband harmonic signal due to microbubble oscillation and destruction (Figure 9.2). Contrast-specific ultrasound imaging techniques use various processing algorithms to register selectively the nonlinear harmonic signals produced by microbubbles and to suppress the signal produced by surrounding tissues, yielding an increased contrast-to-tissue or signal-to-noise ratio. The different contrast-specific techniques can be distinguished by their basic underlying principles (Table 9.1). The main methods to improve signal-to-noise ratio include (a) techniques to selectively detect the nonlinear signals from microbubbles (harmonic, ultraharmonic, subharmonic) with or without Doppler techniques to suppress stationary tissue signal (grayscale harmonic, ultraharmonic, subharmonic, and harmonic power Doppler imaging; all high-MI destructive imaging techniques) [31–33]; (b) techniques that modulate the phase of the emitted ultrasound pulses to eliminate linear tissue signal at lower powers and amplify nonlinear bubble responses (pulse inversion, pulse inversion Doppler, coherent contrast imaging) [34–36]; (c) techniques that modulate the amplitude of the emitted ultrasound pulses to eliminate linear tissue signal at lower powers and amplify nonlinear bubble responses (power modulation angio imaging); and

**A – Very Low Power Imaging**

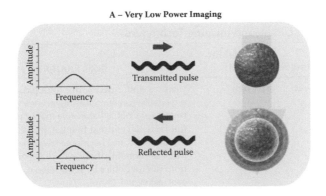

**B – Low Power Imaging**

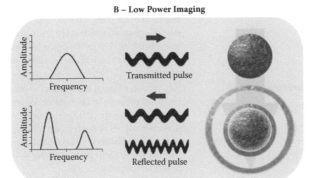

**C – High Power Imaging**

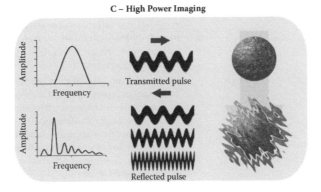

**FIGURE 9.2**

Microbubble responses to ultrasound at different acoustic powers or mechanical index (MI). (A) At a very low MI, the bubbles oscillate, producing a linear acoustic response with the returning spectrum similar to the transmitted pulse, centered on the fundamental transmit frequency. (B) At a low-to-intermediate MI, the bubbles oscillate more vigorously, producing a stronger nonlinear acoustic response, with the returning spectrum consisting of fundamental and harmonic components (multiples of the fundamental transmit frequency). (C) At a high MI, microbubbles are destroyed, producing a very strong nonlinear signal consisting of multiple frequencies, including fundamental, subharmonic, and varying harmonic frequencies.

**TABLE 9.1**

Contrast-Specific Ultrasound Imaging Techniques

| Ultrasound Technique | High/Low MI | Basic Principles |
|---|---|---|
| Grayscale harmonic imaging | High | System is configured with a high-pass filter to selectively receive returning echoes with a frequency double (*harmonic*) that of the fundamental emitted frequency. Overlap between fundamental and harmonic frequency bandwidths leads to contamination of the harmonic response from microbubbles by tissue signal, reducing the contrast signal-to-noise ratio. |
| Harmonic power-Doppler imaging | High | Uses Doppler techniques to separate echoes from microbubbles within the blood and signal from tissue. A high-pass filter cancels signal produced by tissue, selectively detecting the signal from microbubbles. Use of Doppler techniques leads to motion artifact signals, especially from moving tissue. |
| Pulse inversion imaging | High or low | Uses two sequential ultrasound waves that are inverted replicas of each other. When returning echoes are summed, linear echoes from tissue are inverted copies of each other and cancel each other out, while nonlinear microbubble signals are incompletely cancelled, yielding selective signal from microbubbles. |
| Ultraharmonic imaging | High | Uses ultraharmonic signals that occur between higher harmonics (e.g., between second- and third-harmonic frequencies). Tissue produces relatively few ultraharmonic signals, whereas microbubbles generate strong signals in the ultraharmonic range, resulting in an improved signal-to-noise ratio. |
| Subharmonic imaging | High | Utilizes returning signal from echoes with a frequency component at half the insonation fundamental and harmonic frequencies. Subharmonic microbubble signal is greater than second-harmonic signal, and nonlinear responses from tissue do not generate a subharmonic component, yielding a very good microbubble signal-to-tissue noise ratio. |
| Pulse inversion Doppler imaging | Low | Uses a series of pulses (three or more) along each line of ultrasound, with each transmitted pulse an inverted copy of the prior pulse. System processing forms an average of the first and third returning pulses (noninverted) and adds the returning signal from the second pulse (inverted). This leads to cancellation of linear tissue signal, but not of nonlinear microbubble signal. Doppler processing helps to further separate linear tissue signal from nonlinear bubble signal. |

**TABLE 9.1 (*Continued*)**

Contrast-Specific Ultrasound Imaging Techniques

| Ultrasound Technique | High/Low MI | Basic Principles |
|---|---|---|
| Power modulation angio imaging | Low | Utilizes a train of three low acoustic power pulses per ultrasound line. The first and third pulses have half the amplitude of the second pulse. The low acoustic power generates mainly linear tissue responses. The received echoes from the first and third half-amplitude pulses are scaled and subtracted from the second full-amplitude pulse, resulting in effective elimination of linear tissue signal, while the microbubble response continues to be nonlinear to all three pulses, yielding a positive signal even after subtraction. |
| Cadence contrast imaging | Low | Uses one pulse per scan line but inverts the transmitted pulse on alternate scan lines; thus, adjacent lines receive a pulse of alternating phases. Adjacent returning pulses are summed and displayed as a line midway between the returning pulses. This results in cancellation of returning linear signal from tissue, while nonlinear bubble signal remains. The use of single pulses per line yields higher frame rates compared to multipulse techniques. |
| Cadence contrast pulse sequencing | Low | Uses multiple (three) pulses with varying amplitudes and phases, with the first and third being half-amplitude noninverted pulses and the second pulse having a full-amplitude inverted pulse. The system processes returning signal, amplifying the returning signal from the half-amplitude pulses, and then summing the weighted echo sequences. This process eliminates linear fundamental signal from tissue and produces high sensitivity for nonlinear harmonic signal generated by microbubbles. |

(d) techniques that modulate *both* the phase and amplitude of the emitted ultrasound pulses (cadence contrast pulse sequencing) [37].

While contrast-specific ultrasound imaging techniques were developed to image free-flowing microbubbles within the circulation, these techniques have also been adapted to image microbubbles retained within tissue for use in contrast ultrasound molecular imaging. Most imaging techniques used to detect site-targeted microbubbles have employed a high acoustic power or MI, which allows a stronger acoustic response from microbubbles retained within target tissues, compared to low-power imaging. This is of particular importance for contrast ultrasound molecular imaging due to the following factors: (a) given the delay between injection of targeted microbubbles and binding to their targets and subsequent ultrasound imaging (average 5- to

10-minute delay depending on the protocol), adhered bubbles are prone to a reduction in volume/radius over time, which reduces their ultrasound scattering properties; (b) destruction of targeted microbubbles at higher acoustic powers allows their removal from the target ligands and allows repeated studies over short periods of time [38]; (c) the number of microbubbles retained within tissue after their intravenous administration is significantly lower than administered and lower than that observed during microbubble imaging for blood pool opacification or tissue perfusion, thus high-MI imaging maximizes signal from the lower number of retained bubbles; and (d) microbubble oscillation and acoustic responses are dampened to varying degrees by their attachment to target cells, particularly in the event of cellular phagocytosis of microbubbles, in which bubbles are retained wholly within cells [39]. *In vitro* studies have demonstrated that there are reduced oscillations and acoustic signal generation where microbubbles are phagocytosed by leukocytes; however, microbubbles that are simply attached to the surface of cells do not seem to be significantly damped [40]. Thus, while most experimental studies have been performed using high-MI imaging modalities, it is possible to perform CEU molecular imaging using a low-MI nondestructive imaging technique, such as cadence contrast pulse sequencing [41].

## 9.4  Contrast Ultrasound Imaging Algorithms and Protocols

### 9.4.1  Perfusion

Ultrasound imaging techniques have been developed for the assessment and quantification of tissue perfusion, the principles of which are summarized in Figure 9.3. These techniques use free-flowing contrast microbubble agents, which have a similar rheology to red cells within the circulation and hence make ideal flow tracers. Ultrasound probes have a specific beam width or elevation, providing a 2D (two-dimensional) representation of a specified thickness of tissue being insonified (Figure 9.3). For perfusion assessment, microbubbles are administered as a continuous infusion, and after approximately 2–3 minutes, a steady state is achieved when their concentration within the total blood pool is constant, and within any specific tissue, the local concentration is proportional to the blood volume fraction of that local blood pool [42]. When the concentration of microbubbles is within the linear range, the signal from bubbles is proportional to their concentration within the blood pool. Thus, the acoustic signal derived from microbubbles within tissue at steady state is proportional to the blood volume within the tissue being insonified. If all the microbubbles within the tissue are then destroyed with high-energy ultrasound pulses, then subsequent imaging can be performed to measure the rate of microbubble reappearance in tissue within

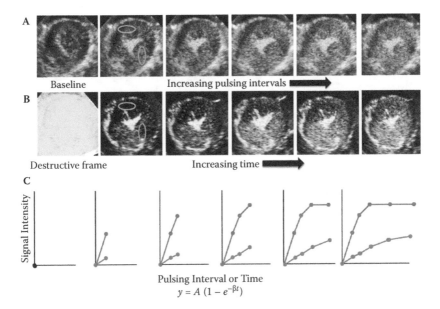

**FIGURE 9.3 (See color insert.)**
Examples of myocardial contrast echocardiographic perfusion images in a canine model of coronary artery stenosis, induced by an occluder on the left anterior descending (LAD) coronary artery. (A) High-power contrast second-harmonic imaging at increasing pulsing intervals, showing a perfusion defect (dark area) in the anterior myocardium. (B) Low-power pulse inversion Doppler contrast imaging at increasing time points after high-power destructive frames (leftmost panel), showing a perfusion defect (dark area) in the anterior myocardium. (C) Progressive plots of background-subtracted signal intensity versus pulsing intervals or time in the ischemic anterior myocardium (LAD territory, red region of interest) and normal posterolateral myocardium (left circumflex territory, blue region of interest), showing the reduced rate of replenishment (reflected by β) and the lower plateau signal intensity (A) in the ischemic LAD bed (red) as compared to the normal left circumflex bed (blue).

the ultrasound beam elevation, which reflects red cell or blood flow velocity (Figure 9.3). The more rapidly microbubbles reappear within tissue, the greater the red cell blood flow velocity is. Areas of reduced blood flow will show late and incomplete replenishment, resulting in relative perfusion defects. Because blood flow constitutes a volume of blood moving at a certain mean velocity, the product of tissue blood volume fraction and blood flow velocity reflects microvascular flow (Figure 9.3) [19].

Perfusion can be quantified using both high-MI "destructive" imaging and low-MI "nondestructive" imaging. Prior to infusion of microbubbles, after ultrasound parameters are optimized, images can be acquired for background subtraction. Newer imaging modalities (see Table 9.1) are able to reduce the background signal from tissue, which is why baseline unenhanced images are relatively dark. For high-MI destructive imaging, during a continuous infusion of microbubbles, imaging is performed at long pulsing intervals (up to 20 seconds depending on the tissue examined) to assess steady-state

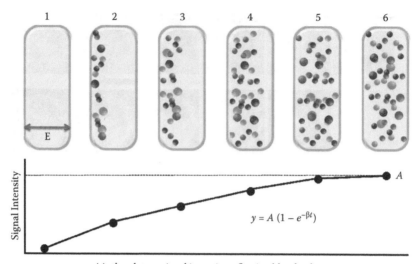

- $A$ is the plateau signal intensity reflecting blood volume
- $\beta$ is the rate constant reflecting mean microbubble velocity
- $A \times \beta$ represents blood flow

**FIGURE 9.4**

Schematic diagram of method to quantify blood flow using contrast-enhanced ultrasound and perfusion microbubbles. During a continuous infusion of microbubbles, after bubble destruction by high-power ultrasound there is progressive filling of the vasculature within the ultrasound-beam thickness (elevation, E) by microbubbles, at increasing time intervals (panels 1 to 6). When the signal intensity from microbubbles is plotted against time, the points can be fitted to the 1-exponential equation $y = A(1 - e^{-\beta t})$, where $A$ is the plateau signal intensity reflecting blood volume, $\beta$ is the rate constant reflecting microbubble or blood flow velocity, and the product of the two $A \times \beta$ represents blood flow.

concentration/signal from microbubbles within the microcirculation. For cardiac imaging, ultrasound imaging is triggered to the electrocardiogram (ECG) to obtain images at similar times within the cardiac cycle (usually end systole) [43] to help facilitate background subtraction. Once steady state has been achieved and microbubble concentration is optimal, the pulsing interval is reduced, and images are then acquired at increasing pulsing intervals until plateau signal intensity is achieved. For quantification, off-line analysis can then be performed where regions of interest are placed, and signal intensity within each region can be determined at each pulsing interval. When signal intensity is plotted against the pulsing interval, the points are fitted to a 1-exponential curve, where measurements of blood volume $A$ and blood flow velocity $\beta$ can be derived (Figure 9.4). The product of $A$ and $\beta$ represents blood flow.

For low-MI nondestructive imaging, the principles remain the same. During a continuous infusion of microbubbles, imaging is performed to assess steady-state concentration/signal from microbubbles within the microcirculation. As there is minimal bubble destruction, triggering is not absolutely required, and imaging can be performed in "real time." Once steady-state microbubble concentration is achieved, several high-power destructive frames of ultrasound are delivered to destroy bubbles within the tissue being insonified (Figure 9.4). After bubble destruction, contrast imaging is performed until plateau signal intensity is again achieved, during which time microbubble replenishment within the tissue can be observed. As before, for quantification, off-line analysis can be performed where the signal intensity within regions of interest can be measured and plotted against time after the destructive pulses. Points are once again fitted to a 1-exponential curve to derive blood volume $A$, blood flow velocity $\beta$, and blood flow $A \times \beta$ (Figure 9.4).

These methods have been used to assess perfusion in multiple organs accessible to ultrasound, including the heart [19,43,44], kidney [17], skeletal muscle [45], brain [46], and liver. Both high-MI destructive imaging and low-MI nondestructive imaging methods for quantitative perfusion have been validated experimentally against radiolabeled microspheres [19,44]. In addition, this technique has been validated clinically in patients, against coronary flow wire measurements [18] and positron emission tomographic (PET) imaging [47]. Moreover, CEU perfusion imaging has been used experimentally to quantify neovascularization responses to growth factor therapy, gene therapy, and cell-based therapies for angiogenesis [48]. Three-dimensional (3D) echocardiographic assessment of perfusion has been studied and holds promise as a method to assess myocardial perfusion [49].

### 9.4.2 Molecular Imaging

Unlike free-flowing microbubbles developed to opacify the blood pool and assess tissue perfusion, targeted microbubbles for molecular imaging are retained within the vasculature at sites of disease, where they can be detected by contrast ultrasound imaging. Targeting can be *nonspecific*: shell properties or composition cause microbubbles to be retained within tissue, such as the binding of albumin or lipid microbubbles to sites of inflammation due to attachment and phagocytosis to activated neutrophils [28,50,51] or retention of anionic microbubbles due to charge coupling [52]. Or, it can be *specific*: monoclonal antibodies or peptides can be attached to the surface of microbubbles, and these microbubbles are retained within tissue, where their specific targeted molecule/ligand is expressed on the luminal surface of the vasculature, such as microbubbles targeted to P-selection by antibody or peptide strategies, binding to sites of inflammation or ischemia where P-selectin is locally upregulated [26,29] (Figure 9.5).

Imaging algorithms have been developed for CEU molecular imaging. After an intravenous injection, site-targeted microbubbles circulate within the vasculature, where they can bind to their specific targets. The kinetics of targeted microbubbles after a bolus injection is illustrated in Figure 9.6. Immediately after the bolus, circulating microbubble concentration is high. As targeted microbubbles bind to their targets, the concentration of bubbles

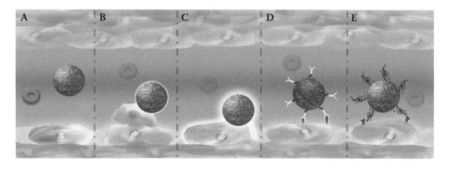

**FIGURE 9.5 (See color insert.)**
Methods of targeting microbubbles. (A) Nontargeted, freely circulating microbubble. (B) Microbubble targeted by shell composition, in this case a leukocyte-avid bubble for the imaging of inflammation. (C) Microbubble targeted by shell surface properties, in this case charge-related binding. (D) Microbubble targeted by attachment of a monoclonal antibody against a specific target. (E) Microbubble targeted by attachment of a targeting peptide or small molecule against a specific intravascular target.

retained at targeted sites gradually increases. In the meantime, free-flowing microbubble concentration gradually falls as they are cleared from the blood pool. To detect retained microbubbles without competing signal from freely circulating microbubbles, the ideal time to image is at the point when retained microbubble concentration is high and levels of circulating bubbles are much lower (Figure 9.6). While imaging following bolus injection

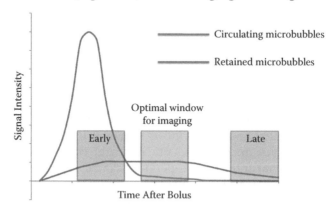

**FIGURE 9.6**
Representative graph of the acoustic signal from targeted and freely circulating microbubbles after an intravenous bolus. While the number and signal from circulating bubble rises rapidly, the number and signal from retained targeted microbubbles rise slower as they accumulate within target tissues and plateaus longer. If ultrasound imaging is performed too early, the signal from circulating bubbles is very high and does not allow easy detection of adhered targeted microbubbles. If ultrasound imaging is performed too late, the number of retained microbubbles is very low. The ideal time for imaging is during the plateau phase of targeted bubble retention, at a time when circulating bubbles have markedly decreased.

provides more time and opportunity for bubbles to bind to their targets and allows free-flowing bubble concentration to fall, this has to be balanced against the progressive degradation and loss of retained microbubbles. The time to image after bolus injection depends on the tissue examined, the regional perfusion and local hydrodynamics within the target tissue, the circulation volume, the targeted ligand affinity to its receptor (on-off rates), and the half-life of circulating and retained microbubbles. Most experimental protocols have waited for at least 5 to 10 minutes after injection to allow maximal binding to targets and clearance of freely circulating bubbles. At the start of imaging, the initial ultrasound pulse detects mostly retained microbubbles within the tissue insonified, along with the few remaining circulating microbubbles. As imaging of targeted microbubbles is usually performed using high-power imaging techniques, the ultrasound pulses result in the destruction of microbubbles within the tissue. Subsequent imaging at long pulsing intervals allows the tissue to replenish with microbubbles, providing the signal from the few remaining freely circulating microbubbles. By using these late images at long pulsing intervals as background frames, one can subtract the signal from circulating microbubbles from the initial frame, yielding an image representing retained targeted microbubbles alone (Figure 9.7). In this fashion, CEU imaging algorithms allow subtraction of tissue signal and signal from any remaining circulating unbound microbubbles, yielding the signal from retained microbubbles alone. The ability to destroy targeted microbubbles or to "null" the signal after image acquisition also allows the repeated administration of targeted microbubbles for serial molecular imaging over short periods of time or the sequential administration of different targeted agents for multiple-target molecular imaging [53].

Figure 9.8 illustrates the potential of CEU molecular imaging of the heart. In this study, CEU molecular imaging of inflammation using targeted microbubbles bearing phosphatidyl serine in the bubble shell was used to serially image the inflammatory response to coronary artery ligation in a canine model of ischemia/infarction-reperfusion. High-power second-harmonic imaging was used in this study to detect retained microbubbles. The regional distribution of the inflammation after ischemia-reperfusion by CEU was comparable to that of a radionuclide tracer and was localized to the infarct zones by postmortem staining. Finally, the first reported study of CEU molecular imaging of progenitor cells was recently published [53]. In this study, site-specific microbubbles were developed to target a genetically engineered cell surface marker, H-2Kk, on endothelial progenitor cells (EPCs). Binding efficiency was first demonstrated using an *in vitro* flow chamber system. *In vivo* studies were then performed in a Matrigel plug model, demonstrating that EPC-targeted microbubbles bind to engrafted EPCs within supplemented plugs and could be detected by their enhancement on CEU imaging using high-MI pulse inversion imaging (Figure 9.9).

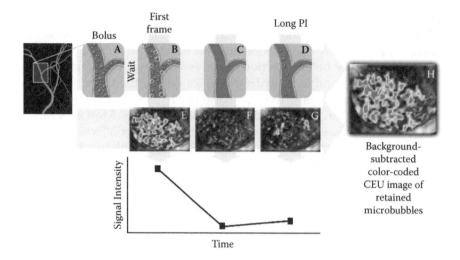

**FIGURE 9.7**

Acquisition protocol for targeted imaging using contrast-enhanced ultrasound molecular imaging in a rodent hind limb. After an intravenous bolus of site-targeted microbubbles, their concentration in the circulation is very high (A). Ultrasound imaging is suspended to allow the microbubbles to continually circulate and bind to their target receptors. After waiting a period of time (5–10 minutes), ultrasound imaging of the hind limb muscle is commenced, where the first frame of ultrasound detects mostly retained microbubbles with a small amount of signal related to circulating microbubbles (B and E). Further high-power ultrasound imaging results in destruction of all bubbles within the muscle (C and F), and the signal within tissue drops. Subsequent imaging at long pulsing intervals detects any remaining freely circulating microbubbles (D and G). When the signal from freely circulating microbubbles is subtracted from the initial frame (panel E minus panel G), this yields the signal from retained targeted microbubbles alone (H). The graph demonstrates the signal intensity measurements at various time points after intravenous site-targeted microbubble bolus.

## 9.5 Conclusions

In summary, CEU imaging allows the quantification of tissue perfusion, an important and relevant endpoint for stem cell therapies in cardiovascular medicine. The recent development of site-specific microbubbles and imaging algorithms to detect retained microbubbles has now allowed molecular imaging using CEU techniques, with many potential opportunities for the detection and quantification of important pathophysiologic events that occur within the vascular compartment. CEU molecular imaging is already being used in the research setting to noninvasively evaluate *in vivo* phenotypes, with clinical applications likely to follow. Finally, CEU molecular imaging techniques for the imaging and tracking of stem cells have begun to be tested, opening up a potential new avenue of research applications for contrast ultrasound imaging.

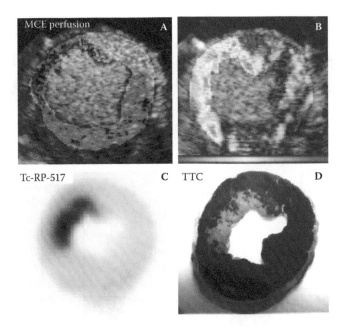

**FIGURE 9.8 (See color insert.)**
Myocardial contrast echocardiographic (MCE) molecular imaging of reperfusion injury/ inflammation using leukocyte-targeted microbubbles in a canine model of ischemia/infarc- tion-reperfusion. (A) Background-subtracted color-coded MCE short-axis image of myocardial perfusion showing a perfusion defect (dark region) in the anterior myocardium, where infarc- tion has occurred. (B) Background-subtracted color-coded short-axis image of signal from retained leukocyte-targeted microbubbles demonstrating the area of inflammation within the myocardium, indicating area of reperfusion injury. This area of reperfusion injury corresponds to that seen in (C), which shows $^{99m}$Tc-RP517-labeled leukocyte accumulation by gamma cam- era imaging, and (D) infarct zone by postmortem 2,3,5-triphenyltetrazolium chloride (TTC) staining. (Reproduced with permission from Christiansen JP, Leong-Poi H, Klibanov AL, Kaul S, Lindner JR. *Circulation* 2002; 105:1764–67.)

## Acknowledgment

I would like to thank Michael A. Kuliszewski for his invaluable help with the illustrations.

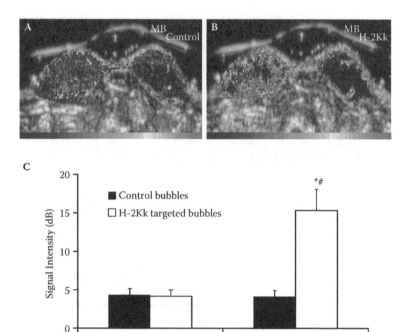

**FIGURE 9.9**

Molecular imaging of endothelial progenitor cell (EPC) engraftment. Background-subtracted color-coded contrast ultrasound images of EPC engraftment in supplemented Matrigel plugs (left plug, EPCs transfected to express the cell surface marker H-2Kk; right plug, control EPC supplemented), after intravenous bolus injection of (A) control microbubbles ($MB_{Control}$) and (B) H-2Kk-targeted bubbles ($MB_{H-2Kk}$). While there was minimal signal after control MB, there was a strong contrast ultrasound signal from H-2Kk-targeted bubbles in the left plug (H-2Kk-EPC supplemented), with only minimal signal in the control EPC-supplemented right plug. (C) EPC-targeted contrast ultrasound imaging data. While there was minimal signal for control bubbles in both control EPC- and H-2Kk-EPC-supplemented plugs (black bars), the signal intensity was greater for H-2Kk-targeted bubbles (white bars) in the H-2Kk-EPC-supplemented, as compared to the control EPC-supplemented right plug. *$p < .001$ versus $MB_{Control}$, #$p < .005$ versus control mock-transfected EPC plug. (Reproduced from Kuliszewski MA, Fujii H, Liao C, et al. *Cardiovasc Res* 2009; 83:653–62. With permission.)

## References

1. Ly HQ, Frangioni JV, Hajjar RJ. Imaging in cardiac cell-based therapy: *In vivo* tracking of the biological fate of therapeutic cells. *Nat Clin Pract Cardiovasc Med* 2008; 5 Suppl 2:S96–S102.

2. Frangioni JV, Hajjar RJ. In vivo tracking of stem cells for clinical trials in cardiovascular disease. *Circulation* 2004; 110:3378–3383.

3. Kremkau FW, Taylor KJ. Artifacts in ultrasound imaging. *J Ultrasound Med* 1986; 5:227–237.

4. Goldstein A. Overview of the physics of US. *Radiographics* 1993; 13:701–704.

5. Kaul S. Myocardial contrast echocardiography: 15 years of research and development. *Circulation* 1997; 96:3745–3760.

6. Cosgrove D. Ultrasound contrast agents: An overview. *Eur J Radiol* 2006; 60:324–330.

7. Lindner JR. Molecular imaging with contrast ultrasound and targeted microbubbles. *J Nucl Cardiol* 2004; 11:215–221.

8. Patel RM. Microbubble: A potential ultrasound tool in molecular imaging. *Curr Pharm Biotechnol* 2008; 9:406–410.

9. Kiessling F, Huppert J, Palmowski M. Functional and molecular ultrasound imaging: concepts and contrast agents. *Curr Med Chem* 2009; 16:627–642.

10. Correas JM, Bridal L, Lesavre A, Mejean A, Claudon M, Helenon O. Ultrasound contrast agents: Properties, principles of action, tolerance, and artifacts. *Eur Radiol* 2001; 11:1316–1328.

11. Keller MW, Segal SS, Kaul S, Duling B. The behavior of sonicated albumin microbubbles within the microcirculation: A basis for their use during myocardial contrast echocardiography. *Circ Res* 1989; 65:458–467.

12. Lindner JR, Song J, Jayaweera AR, Sklenar J, Kaul S. Microvascular rheology of Definity microbubbles after intra-arterial and intravenous administration. *J Am Soc Echocardiogr* 2002; 15:396–403.

13. Stride E. Physical principles of microbubbles for ultrasound imaging and therapy. *Cerebrovasc Dis* 2009; 27 Suppl 2:1–13.

14. Postema M, Schmitz G. Bubble dynamics involved in ultrasonic imaging. *Expert Rev Mol Diagn* 2006; 6:493–502.

15. Honos G, Amyot R, Choy J, Leong-Poi H, Schnell G, Yu E. Contrast echocardiography in Canada: Canadian Cardiovascular Society/Canadian Society of Echocardiography position paper. *Can J Cardiol* 2007; 23:351–356.

16. Mulvagh SL, Rakowski H, Vannan MA, et al. American Society of Echocardiography consensus statement on the clinical applications of ultrasonic contrast agents in echocardiography. *J Am Soc Echocardiogr* 2008; 21:1179–1201; quiz 1281.

17. Wei K, Le E, Bin JP, Coggins M, Thorpe J, Kaul S. Quantification of renal blood flow with contrast-enhanced ultrasound. *J Am Coll Cardiol* 2001; 37:1135–1140.

18. Wei K, Ragosta M, Thorpe J, Coggins M, Moos S, Kaul S. Noninvasive quantification of coronary blood flow reserve in humans using myocardial contrast echocardiography. *Circulation* 2001; 103:2560–2565.

19. Wei K, Jayaweera AR, Firoozan S, Linka A, Skyba DM, Kaul S. Quantification of myocardial blood flow with ultrasound-induced destruction of microbubbles administered as a constant venous infusion. *Circulation* 1998; 97:473–483.

20. Kaufmann BA, Lindner JR. Molecular imaging with targeted contrast ultrasound. *Curr Opin Biotechnol* 2007; 18:11–16.

21. Unger EC, McCreery TP, Sweitzer RH, Shen D, Wu G. In vitro studies of a new thrombus-specific ultrasound contrast agent. *Am J Cardiol* 1998; 81:58G–61G.

22. Villanueva FS, Jankowski RJ, Klibanov S, et al. Microbubbles targeted to intercellular adhesion molecule-1 bind to activated coronary artery endothelial cells. *Circulation* 1998; 98:1–5.

23. Leong-Poi H, Christiansen J, Heppner P, et al. Assessment of endogenous and therapeutic arteriogenesis by contrast ultrasound molecular imaging of integrin expression. *Circulation* 2005; 111:3248–3254.

24. Leong-Poi H, Christiansen J, Klibanov AL, Kaul S, Lindner JR. Noninvasive assessment of angiogenesis by ultrasound and microbubbles targeted to alpha(v)-integrins. *Circulation* 2003; 107:455–460.

25. Weller GE, Wong MK, Modzelewski RA, et al. Ultrasonic imaging of tumor angiogenesis using contrast microbubbles targeted via the tumor-binding peptide arginine-arginine-leucine. *Cancer Res* 2005; 65:533–539.

26. Kaufmann BA, Lewis C, Xie A, Mirza-Mohd A, Lindner JR. Detection of recent myocardial ischaemia by molecular imaging of P-selectin with targeted contrast echocardiography. *Eur Heart J* 2007; 28:2011–2017.

27. Kaufmann BA, Sanders JM, Davis C, et al. Molecular imaging of inflammation in atherosclerosis with targeted ultrasound detection of vascular cell adhesion molecule-1. *Circulation* 2007; 116:276–284.

28. Lindner JR, Dayton PA, Coggins MP, et al. Noninvasive imaging of inflammation by ultrasound detection of phagocytosed microbubbles. *Circulation* 2000; 102:531–538.

29. Lindner JR, Song J, Christiansen J, Klibanov AL, Xu F, Ley K. Ultrasound assessment of inflammation and renal tissue injury with microbubbles targeted to P-selectin. *Circulation* 2001; 104:2107–2112.

30. Christiansen JP, Leong-Poi H, Klibanov AL, Kaul S, Lindner JR. Noninvasive imaging of myocardial reperfusion injury using leukocyte-targeted contrast echocardiography. *Circulation* 2002; 105:1764–1767.

31. Lencioni R, Cioni D, Bartolozzi C. Tissue harmonic and contrast-specific imaging: back to gray scale in ultrasound. *Eur Radiol* 2002; 12:151–165.

32. Becher H, Tiemann K, Schlief R, Luderitz B, Nanda NC. Harmonic power Doppler contrast echocardiography: Preliminary clinical results. *Echocardiography* 1997; 14:637.

33. Shi WT, Forsberg F, Hall AL, et al. Subharmonic imaging with microbubble contrast agents: initial results. *Ultrason Imaging* 1999; 21:79–94.

34. Harvey CJ, Blomley MJ, Eckersley RJ, Heckemann RA, Butler-Barnes J, Cosgrove DO. Pulse-inversion mode imaging of liver specific microbubbles: improved detection of subcentimetre metastases. *Lancet* 2000; 355:807–808.

35. Tiemann K, Lohmeier S, Kuntz S, et al. Real-time contrast echo assessment of myocardial perfusion at low emission power: First experimental and clinical results using power pulse inversion imaging. *Echocardiography* 1999; 16:799–809.

36. Simpson DH, Chin CT, Burns PN. Pulse inversion Doppler: A new method for detecting nonlinear echoes from microbubble contrast agents. *IEEE Trans Ultrason Ferroelectr Freq Control* 1999; 46:372–382.

37. Eckersley RJ, Chin CT, Burns PN. Optimising phase and amplitude modulation schemes for imaging microbubble contrast agents at low acoustic power. *Ultrasound Med Biol* 2005; 31:213–219.

38. Behm CZ, Kaufmann BA, Carr C, et al. Molecular imaging of endothelial vascular cell adhesion molecule-1 expression and inflammatory cell recruitment during vasculogenesis and ischemia-mediated arteriogenesis. *Circulation* 2008; 117:2902–2911.

39. Dayton PA, Chomas JE, Lum AF, et al. Optical and acoustical dynamics of microbubble contrast agents inside neutrophils. *Biophys J* 2001; 80:1547–1556.

40. Lankford M, Behm CZ, Yeh J, Klibanov AL, Robinson P, Lindner JR. Effect of microbubble ligation to cells on ultrasound signal enhancement: Implications for targeted imaging. *Invest Radiol* 2006; 41:721–728.
41. Stieger SM, Dayton PA, Borden MA, et al. Imaging of angiogenesis using Cadence contrast pulse sequencing and targeted contrast agents. *Contrast Media Mol Imaging* 2008; 3:9–18.
42. Kaul S. Myocardial contrast echocardiography: basic principles. *Prog Cardiovasc Dis* 2001; 44:1–11.
43. Leong-Poi H, Le E, Rim SJ, Sakuma T, Kaul S, Wei K. Quantification of myocardial perfusion and determination of coronary stenosis severity during hyperemia using real-time myocardial contrast echocardiography. *J Am Soc Echocardiogr* 2001; 14:1173–1182.
44. Masugata H, Cotter B, Peters B, Ohmori K, Mizushige K, DeMaria AN. Assessment of coronary stenosis severity and transmural perfusion gradient by myocardial contrast echocardiography: Comparison of gray-scale B-mode with power Doppler imaging. *Circulation* 2000; 102:1427–1433.
45. Lindner JR, Womack L, Barrett EJ, et al. Limb stress-rest perfusion imaging with contrast ultrasound for the assessment of peripheral arterial disease severity. *JACC Cardiovasc Imaging* 2008; 1:343–350.
46. Rim SJ, Leong-Poi H, Lindner JR, et al. Quantification of cerebral perfusion with "real-time" contrast-enhanced ultrasound. *Circulation* 2001; 104:2582–2587.
47. Vogel R, Indermuhle A, Reinhardt J, et al. The quantification of absolute myocardial perfusion in humans by contrast echocardiography: Algorithm and validation. *J Am Coll Cardiol* 2005; 45:754–762.
48. Fujii H, Tomita S, Nakatani T, et al. A novel application of myocardial contrast echocardiography to evaluate angiogenesis by autologous bone marrow cell transplantation in chronic ischemic pig model. *J Am Coll Cardiol* 2004; 43:1299–1305.
49. Toledo E, Lang RM, Collins KA, et al. Imaging and quantification of myocardial perfusion using real-time three-dimensional echocardiography. *J Am Coll Cardiol* 2006; 47:146–154.
50. Lindner JR, Song J, Xu F, et al. Noninvasive ultrasound imaging of inflammation using microbubbles targeted to activated leukocytes. *Circulation* 2000; 102:2745–2750.
51. Lindner JR, Coggins MP, Kaul S, Klibanov AL, Brandenburger GH, Ley K. Microbubble persistence in the microcirculation during ischemia/reperfusion and inflammation is caused by integrin- and complement-mediated adherence to activated leukocytes. *Circulation* 2000; 101:668–675.
52. Fisher NG, Christiansen JP, Klibanov A, Taylor RP, Kaul S, Lindner JR. Influence of microbubble surface charge on capillary transit and myocardial contrast enhancement. *J Am Coll Cardiol* 2002; 40:811–819.
53. Kuliszewski MA, Fujii H, Liao C, et al. Molecular imaging of endothelial progenitor cell engraftment using contrast-enhanced ultrasound and targeted microbubbles. *Cardiovasc Res* 2009; 83:653–662.

# 10

## MRI Basics and Principles for Cellular Imaging

Lisa M. Gazdzinski, Paula J. Foster, Eddy S. M. Lee, and Brian K. Rutt

**CONTENTS**

## 10.1 Introduction

Magnetic resonance imaging (MRI) is a powerful medical imaging technique capable of providing exquisite soft tissue contrast without the use of ionizing radiation. Specialized techniques allow MRI to be used to study small-animal models of human diseases, allowing for truly longitudinal studies, thereby reducing the number of animals required. Furthermore, techniques have been developed that allow MRI to visualize processes at the cellular level in live animals. This chapter acts as an introduction to basic MRI principles that will help in understanding MR methods and applications described in other chapters.

## 10.2 Nuclear Magnetic Resonance

All nuclei with an odd number of protons or neutrons can exhibit nuclear magnetic resonance (NMR), as they possess a nonzero angular momentum, or nuclear spin **I**, due to the intrinsic spin of the constituent nucleons. This nuclear spin gives rise to a magnetic moment, $\mu = \gamma \mathbf{I}$, where $\gamma$ is the gyromagnetic ratio (Figure 10.1) and is unique to each nuclear species (Reiser 2008).

Nuclear spin is quantized. The magnitude of the angular momentum vector is limited to discrete values, $|I| = \hbar\sqrt{I(I+1)}$, where $\hbar$ is Planck's constant ($\hbar = 1.05 \times 10^{-34}$ Js), and $I$ is the spin quantum number, which is either integer or half-integer (Figure 10.2). Furthermore, in the presence of a magnetic field, only certain orientations are permitted, defined by $I_z = m\hbar$, where $m$ is the magnetic quantum number ($m = -I, -I + 1, \dots , I - 1, I$) (Reiser 2008).

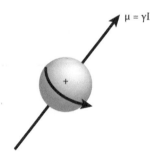

**FIGURE 10.1**
Depiction of the magnetic moment $\mu$ of a nucleus with angular momentum **I**. (From Bernas, L., Preclinical cellular MRI of glioma using iron oxide contrast agents, PhD thesis, Department of Medical Biophysics, University of Western Ontario, 2009. University of Western Ontario.)

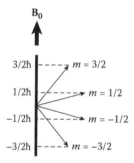

**FIGURE 10.2**
Depiction of the quantization of nuclear spin showing the allowed orientations for a nucleus with $I = 3/2$ when placed in a magnetic field **B₀**. The quantized magnitude of the angular momentum vector is also shown. (From Bernas, L., Preclinical cellular MRI of glioma using iron oxide contrast agents, PhD thesis, Department of Medical Biophysics, University of Western Ontario, 2009. University of Western Ontario.)

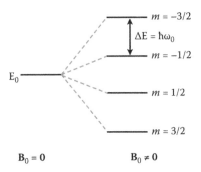

**FIGURE 10.3**
Depiction of Zeeman splitting of spin states for a nucleus with $I = 3/2$ placed in a magnetic field $\mathbf{B}_0$. The energy difference between neighboring states is . (From Bernas, L., Preclinical cellular MRI of glioma using iron oxide contrast agents, PhD thesis, Department of Medical Biophysics, University of Western Ontario, 2009. University of Western Ontario.)

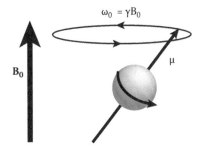

**FIGURE 10.4**
Depiction of a precessing spin in a magnetic field $\mathbf{B}_0$. (From Bernas, L., Preclinical cellular MRI of glioma using iron oxide contrast agents, PhD thesis, Department of Medical Biophysics, University of Western Ontario, 2009. University of Western Ontario.)

In the absence of a magnetic field, the magnetic moments, or spins, are randomly oriented. However, when placed in a magnetic field $\mathbf{B}_0$, Zeeman splitting occurs, resulting in $2I + 1$ spin states with energies $E_m = -\gamma\hbar B_0 m$ (Figure 10.3).

The combined effect of the magnetic field and the nuclear angular momentum is to cause the spins to precess about $\mathbf{B}_0$ at a characteristic frequency called the Larmor frequency $\omega_0$, which is related to the magnitude of $\mathbf{B}_0$ according to the Larmor equation, $\omega_0 = \gamma B_0$ (Figure 10.4).

At thermal equilibrium, the distribution of spins among the allowed states is dictated by the Boltzmann distribution, with lower-energy states having slightly higher occupancy than higher-energy states. In the high-temperature approximation, the magnitude of the equilibrium magnetization $M_0$ (Figure 10.5) resulting from the vector sum of nuclear magnetic moments per unit volume is given by

$$M_0 \cong \frac{\rho I(I+1)\gamma^2 \hbar B_0}{3kT}$$

where $\rho$ is the spin density, $T$ is the absolute temperature of the sample, and $k$ is the Boltzmann constant ($k = 1.38 \times 10^{-23}$ J/K) (Nishimura 1996).

Transitions between spin states can be induced by irradiation with a radio-frequency (RF) field oriented perpendicular to the $B_0$ field and having photon energy equal to the energy difference between neighboring states, $\Delta E = \gamma \hbar B_0 = \hbar \omega_0$. The angular frequency of this field is therefore the Larmor frequency, which reveals the resonance condition and defines the rate at which the disturbed magnetization precesses about the $B_0$ field (Reiser 2008).

While any $I \neq 0$ nucleus can exhibit magnetic resonance, the hydrogen nucleus, which possesses only one proton and has $I = 1/2$, is by far the most often used for MRI and MR spectroscopy. The reasons for this are that it has a high natural abundance (>99%), is extremely prevalent in biological systems, and possesses the highest gyromagnetic ratio of all stable nuclei (42.577 MHz/T), resulting in a relatively high equilibrium magnetization. In the field of cellular and molecular imaging, other nonproton nuclei such as carbon-13, fluorine-19, sodium-23, and phosphorus-31 have also attracted interest. All of these have lower gyromagnetic ratios than the hydrogen nucleus, and therefore lower sensitivity, but they exhibit some advantages in terms of probing specific cellular metabolism or function.

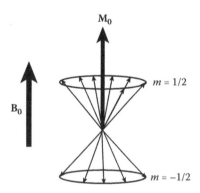

**FIGURE 10.5**
Depiction of the distribution of spins between the two allowed spin states for a nucleus with $I = 1/2$, giving rise to the net magnetization $M_0$ at thermal equilibrium. The excess of spins in the lower-energy ($m = 1/2$) state is extremely small—only about 3 in $10^6$ for hydrogen nuclei at body temperature. (From Bernas, L., Preclinical cellular MRI of glioma using iron oxide contrast agents, PhD thesis, Department of Medical Biophysics, University of Western Ontario, 2009. University of Western Ontario.)

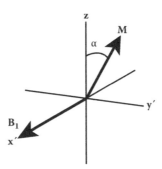

**FIGURE 10.6**
Depiction of rotation of magnetization **M** by application of magnetic field **B**$_1$, oriented along the $x'$-axis. The strength and duration of **B**$_1$ determines the angle $\alpha$ by which the magnetization is rotated. (From Bernas, L., Preclinical cellular MRI of glioma using iron oxide contrast agents, PhD thesis, Department of Medical Biophysics, University of Western Ontario, 2009. University of Western Ontario.)

## 10.3 The Nuclear Magentic Resonance Signal

As mentioned, the effect on individual spins of applying an RF field **B**$_1$ perpendicular to the main magnetic field and rotating at the Larmor frequency is to induce transitions from a lower energy state to a higher energy state, thereby disrupting the equilibrium magnetization. From a classical point of view, the application of **B**$_1$ induces a torque on the net magnetization vector **M**, causing it to rotate away from its equilibrium direction along **B**$_0$. As **M** is made to point away from **B**$_0$, it begins to precess about **B**$_0$, also at the Larmor frequency (Nishimura 1996). Since the magnetization vector and the applied **B**$_1$ field rotate about **B**$_0$ at $\omega_0$, it is useful to define a rotating frame $(x', y', z)$, such that the applied **B**$_1$ field vector remains stationary (Haacke et al. 1999).

In the rotating frame, an RF field applied along the $x'$ axis causes a nutation of **M** about this axis, tipping it down toward the $x'y'$ plane (Figure 10.6). The duration $t$ of the RF pulse dictates the flip angle $\alpha$ by which **M** will be rotated:

$$\omega_1 = \omega_0 = \gamma B_1$$

$$\alpha = \gamma B_1 t$$

After cessation of the RF excitation pulse, the magnetization returns to its thermal equilibrium, **M**$_0$ along the $z$ axis, through a process called *relaxation*. Relaxation is governed by two separate, but simultaneous, processes: T1 relaxation, which describes the return of the longitudinal magnetization to

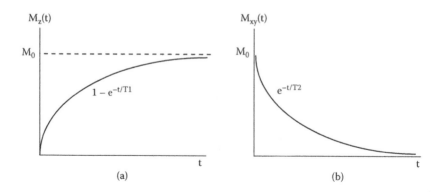

**FIGURE 10.7**
Graphs depicting the exponential return of the excited magnetization to thermal equilibrium. (a) Exponential growth of the longitudinal magnetization is characterized by T1. (b) Exponential decay of the transverse magnetization is characterized by T2. (From Bernas, L., Preclinical cellular MRI of glioma using iron oxide contrast agents, PhD thesis, Department of Medical Biophysics, University of Western Ontario, 2009. University of Western Ontario.)

$M_0$, and T2 relaxation, which describes the decay of the transverse magnetization (Figure 10.7) (Nishimura 1996).

At the atomic level, random fluctuations in the local magnetic field experienced by excited spins are caused by motion of nearby magnetic moments and exchange processes. The component of these fluctuations at the Larmor frequency and perpendicular to $\mathbf{B}_0$ induces transitions between spin states, analogous to the effect of the $\mathbf{B}_1$ field, thereby affecting the occupancy of the spin states and thus the magnitude of the macroscopic longitudinal magnetization. The extra energy carried by the excited spins is dissipated into the surrounding "lattice" until thermal equilibrium is reached; thus, the regrowth of the longitudinal magnetization (T1 relaxation) is also called spin-lattice relaxation. Although the primary effect of this relaxation mechanism is on the longitudinal component of the magnetization, it also disrupts the phase coherence of the excited spins and contributes to transverse relaxation (Reiser 2008).

The component of the magnetic field fluctuations parallel to $\mathbf{B}_0$ affects the local precession frequency of the spins as it affects the local polarizing magnetic field. As these fluctuations are random, the effect is to cause dephasing of the spins, and this lack of coherence causes a decay of the macroscopic transverse magnetization (Figure 10.8). As this mechanism is due to interaction between spins, T2 relaxation is also called spin-spin relaxation (Reiser 2008).

Static inhomogeneities in the magnetic field, caused by tissue boundaries or technical imperfections, for example, also contribute to decay of the transverse magnetization at a rate characterized by T2'. Thus, the true decay rate of the transverse magnetization is faster than T2 and is called T2-star (T2*).

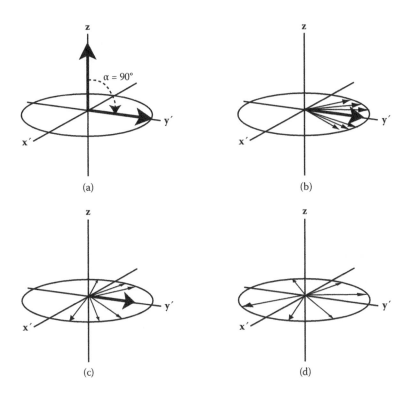

**FIGURE 10.8**
Depiction of spin dephasing in transverse plane. (a) Immediately following a 90° RF pulse along the $x'$ axis, the magnetization vector lies along the $y'$ axis. (b)–(d) Over time, spins contributing to the net magnetization lose phase coherence due to small random differences in the magnetic field experienced by different spins. Spins experiencing stronger magnetic fields precess more quickly, whereas those experiencing weaker magnetic fields precess more slowly. In the rotating frame, this results in some spins rotating clockwise and others rotating counterclockwise. Eventually, total phase coherence is lost, and the transverse magnetization decays to zero. (From Bernas, L., Preclinical cellular MRI of glioma using iron oxide contrast agents, PhD thesis, Department of Medical Biophysics, University of Western Ontario, 2009. University of Western Ontario.)

$$\frac{1}{T2^*} = \frac{1}{T2} + \frac{1}{T2'}$$

As these are static inhomogeneities, they can be compensated for by using the spin echo (SE) pulse sequence, allowing pure T2 relaxation to be observed.

The evolution of the magnetization **M** following an RF excitation pulse can be described mathematically by the phenomenological Bloch equation:

$$\frac{d\mathbf{M}}{dt} = \mathbf{M} \times \gamma \mathbf{B}_0 - \frac{M_x\mathbf{i} + M_y\mathbf{j}}{T2} - \frac{(M_z - M_0)\mathbf{k}}{T1}$$

where $M_x$, $M_y$, and $M_z$ are the components of the magnetization, and $\mathbf{i}$, $\mathbf{j}$, and $\mathbf{k}$, are unit vectors in the $x$, $y$, and $z$ directions, respectively (Nishimura 1996). The cross-product term describes the precession of the magnetization about $\mathbf{B_0}$, whereas other terms describe the relaxation behavior.

It is this evolving magnetization that gives rise to the NMR signal that is used in MRI. Specifically, according to Faraday's law of induction, the precessing magnetization in the transverse plane, a time-varying magnetic field, induces an electric current in a coil of wire oriented perpendicular to it, which may be the same coil as used to apply the RF pulse or a second one. This current, which reflects the evolution of the transverse magnetization, constitutes the NMR signal.

---

## 10.4 Signal-to-Noise Ratio

In MRI, one of the fundamental measures of image quality is the signal-to-noise ratio (SNR). SNR is defined mathematically as

$$SNR = \frac{S}{\sigma_{noise}}$$

where $S$ is the signal amplitude, and $\sigma_{noise}$ is the standard deviation of the noise. The SNR is dependent on a number of factors, including sample characteristics (T1, T2, $\rho$), imaging sequence parameters (resolution, total readout time), and instrumental parameters ($B_0$, receiver coil characteristics).

Sample properties such as relaxation times and proton spin density affect the amount of transverse magnetization and hence the amount of signal. Image resolution, defined by voxel volume ($\Delta x \Delta y \Delta z$) and total image readout time $t_{ro}$, influence SNR according to

$$SNR \propto \Delta x \Delta y \Delta z \sqrt{t_{ro}}$$

The readout time is the product of the number of averages, phase encodes, and the readout time per individual acquisition. The equation predicts that to maintain SNR, image acquisition time must increase quadratically with an increase in resolution (decreasing voxel volume). For example, an increase in resolution by decreasing voxel volume by a factor of two must be accompanied by a quadrupling in acquisition time.

The reduction in SNR at higher resolution and the inefficiencies of increasing acquisition time to compensate for SNR loss is a significant challenge for high-resolution MRI. One common way to improve SNR for this application is by increasing the main magnetic field strength $B_0$. The SNR scales according to

$$SNR \propto B_o \quad \text{or} \quad SNR \propto B_o^{7/4}$$

depending on whether noise is body (specimen) dominated or coil dominated, respectively. Typical MR field strengths for high-resolution MR scanners are in the range of 2–11.7 T.

## 10.5 Contrast Agents

An MRI contrast agent is most simply a substance that will affect the magnetic properties of water protons to an extent that is observable in an image. There are two main classes of contrast agents: paramagnetic and superparamagnetic, both of which shorten the relaxation times of bulk water protons. Paramagnetic contrast agents are usually considered to be T1 shortening because although they shorten both T1 and T2, the T1 effect is dominant. Superparamagnetic agents also shorten both T1 and T2 (as well as T2*), but depending on the size of the nanoparticulate agent, the T2 (T2*) effect is usually much larger than the T1 effect.

### 10.5.1 Paramagnetic Agents

Paramagnetism is a form of magnetism that occurs only in the presence of an externally applied magnetic field. In the absence of an externally applied magnetic field, the net magnetization is zero; thermal motion causes the spins to be randomly oriented. In the presence of an external field, a small magnetization is induced because a small fraction of the spins is oriented by the field. This fraction is proportional to the field strength (Haacke et al. 1999).

Paramagnetic materials that tumble at a rotational rate close to the Larmor frequency transfer energy between the paramagnetic ion and the surrounding water molecules or "lattice," thereby enhancing the longitudinal relaxation rate ($R1 = 1/T1$) or shortening the T1 relaxation time of the water molecules. This transfer of energy occurs through dipole-dipole and scalar interactions, both requiring that the paramagnetic ion be in close contact with the water molecule. Dipole-dipole interactions will also increase the transverse relaxation rate ($R2 = 1/T2$), but the relative, or fractional, change in R1 will dominate over that in R2 in almost all tissues because the initial (precontrast) R1 is less than R2. Local magnetic field inhomogeneities will occur, with areas of higher contrast agent concentration experiencing a greater magnetic field than areas of lower concentration. These inhomogeneities in local magnetic field will cause a loss of phase coherence in magnetic moments, causing a shortening of T2 (Smith and Lange 1998).

## 10.5.2 Superparamagnetic Agents

Superparamagnetism occurs when atoms bearing unpaired electrons interact within a crystal structure such that they form single magnetic domains within the crystal, as is the case for iron oxide contrast agents. Each magnetic domain has a magnetic dipole that is much greater than the sum of the contributing unpaired electrons (Wang et al. 2001). In the absence of an externally applied magnetic field, the magnetic domains of a superparamagnetic substance are randomly oriented, and the substance has no net magnetization. However, when placed in a magnetic field, the magnetic domains orient themselves with the applied field, resulting in strong magnetization (Bean and Livingston 1959; Wood and Hardy 1993).

All iron oxide nanoparticulate contrast agents exhibit superparamagnetic properties at body temperature. T1 shortening will be caused by dipole-dipole interactions and scalar interactions, as with paramagnetic contrast agents. T2 (and T2*) shortening will be caused by local magnetic field inhomogeneities, which are much greater with superparamagnetic agents than paramagnetic agents because coupling forces between magnetic moments cause superparamagnetics to be much stronger than paramagnetics. Superparamagnetic agents will shorten T2 and T2* to a greater extent than T1 and are therefore most often classified as T2 or T2* shortening agents (Vejpravova et al. 2005).

## 10.6  Contrast Agent Concentration and Relaxivity

Longitudinal and transverse relaxation changes will modify MR image intensities, with the image signal modulation dependent on contrast agent concentration: T1 effects dominate at low doses, and T2 effects dominate at higher doses. At higher concentrations, or for stronger magnetic moments, local field inhomogeneities become more significant and cause a greater loss in phase coherence, increasing T2 effects and overpowering T1 effects. T1 shortening will cause an increased signal in T1-weighted MR images, while T2 shortening will cause a decrease in signal intensity on T2-weighted images. Relaxivity $r_i$ is the change in relaxation rate after the introduction of the contrast agent ($\Delta R_i$, i = 1, 2) normalized by the concentration of contrast agent or metal ion [M].

$$r_i = \Delta R_i / [M]; \qquad i = 1, 2$$

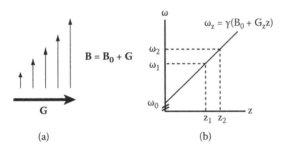

(a)             (b)

**FIGURE 10.9**
Gradient field and slice selection. (a) For imaging, the main magnetic field $\mathbf{B}_0$ is superimposed by field gradients $G_z$, $G_y$, and $G_x$, which impose a linear dependence of the Larmor frequency on location along the gradient directions. (b) The application of an RF field with frequency range $\omega_{z1} < \omega_z < \omega_{z2}$ excites only spins located in a slice $z_1 < z < z_2$. (From Bernas, L., Preclinical cellular MRI of glioma using iron oxide contrast agents, PhD thesis, Department of Medical Biophysics, University of Western Ontario, 2009. University of Western Ontario.)

## 10.7 Magnetic Resonance Imaging

In basic terms, an MRI system consists of three magnets: the main polarizing magnet, which generates the $B_0$ field; the RF coils, which excite the magnetization and detect the NMR signal; and the gradient coils, which instill spatial variation in the magnetic field across the sample, allowing the detected NMR signal to be spatially localized.

The application of a magnetic field gradient across the sample modifies the magnetic field experienced by spins at different locations and thus instills a variation of the precession frequency in space. The simultaneous application of a linear gradient pulse (e.g., along the z direction) and an RF pulse centered at the Larmor frequency allows the selective excitation of spins within a thin slice of the sample (Nishimura 1996). As illustrated in Figure 10.9, the thickness of the excited slice is dictated by the range of frequencies contained in the RF pulse, that is, the excitation bandwidth, and the strength of the gradient ($G_z$).

Following the slice-selective excitation, the application of a gradient along the y direction for a given duration again causes a variation in precession frequency across the sample, which results in a variation in phase across the sample in the y direction when this gradient is removed. The position of the spins along the y axis is therefore encoded in the phase of their precession (Figure 10.10b).

Finally, the application of a third gradient pulse, along the x direction, during the acquisition of the NMR signal, instills a frequency variation in the signal that is directly related to the location of the spins along the x axis (Figure 10.10c). The acquired NMR signal is therefore composed of a range of frequencies and phases defining the location of the contributing spins.

Mathematically, the NMR signal $S(t)$ can be expressed completely by the following signal equation, which combines the aforementioned Bloch equation with the effects of the applied gradient fields as well as characteristics of the receiver coil (Haacke et al. 1999):

$$S(t) = A\omega_0 e^{-\omega_0 t} \int M_0(\vec{r}) D[T1(\vec{r})] e^{\frac{-t}{T2^*(\vec{r})}} B_1(\vec{r}) e^{i2\pi \left[\frac{\gamma}{2\pi}\int_0^t \vec{G}(\vec{r},t')dt'\right]\cdot\vec{r}} e^{i2\pi\left[\frac{\gamma}{2\pi}\int_0^t \Delta B(\vec{r},t')dt'\right]} dV$$

The acquired signal is a summed effect of all spins within the imaging volume $V$. Thus, it depends on the net magnetization $M_0(\vec{r})$; T2* decay, expressed as

$$e^{\frac{-t}{T2^*(\vec{r})}}$$

the T1, if full recovery does not occur between excitation pulses (the $D[T1(\vec{r})]$ term); the receiver coil sensitivity $B_1(\vec{r})$; and the phase acquired due to the applied gradient fields

$$e^{i2\pi\left[\frac{\gamma}{2\pi}\int_0^t \vec{G}(\vec{r},t')dt'\right]\cdot\vec{r}}$$

as well as that due to unintentional field inhomogeneities

$$e^{i2\pi\left[\frac{\gamma}{2\pi}\int_0^t \Delta B(\vec{r},t')dt'\right]}$$

The signal is proportional to the Larmor frequency $\omega_0$ because the electromotive force (emf) induced in the receiver coil is proportional to the rate of change of the magnetic flux through the coil and oscillates at this frequency, as indicated by the $e^{-i\omega_0 t}$ term. The proportionality constant $A$ relates the emf in the receiver coil to the red signal.

A simple pulse sequence diagram, which illustrates the timing of the different gradient pulses mentioned, is shown in Figure 10.11 (Reiser 2008).

MRI data are acquired in spatial frequency space, or *k-space*, and are then subjected to a Fourier transformation to reveal the reconstructed image (Figure 10.12). The effect of the gradient pulses, as they were described, is to traverse k-space line by line, sampling the NMR signal along one line for every phase encode step. The sampling rates along the $k_x$ and $k_y$ directions dictate the field of view (FOV) of the image in the $x$ and $y$ directions, respectively, and the extent of k-space sampled dictates the resulting image

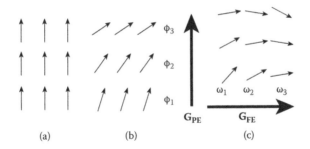

**FIGURE 10.10**
Depiction of phase encoding and frequency encoding. (a) Direction of magnetization within excited slice before phase encoding. (b) Direction of magnetization along phase-encoding direction after application of $G_{PE}$. (c) The effect of the frequency encoding gradient is to impose a linear dependence of the Larmor frequency along the direction of $G_{FE}$. Each position is thereby defined by a unique combination of frequency and phase in the acquired signal. (From Bernas, L. 2009. Preclinical cellular MRI of glioma using iron oxide contrast agents, PhD thesis, Department of Medical Biophysics, University of Western Ontario. University of Western Ontario.)

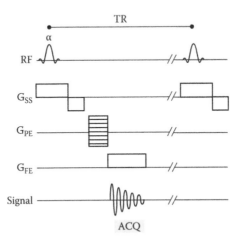

**FIGURE 10.11**
Simple pulse sequence timing diagram. The application of an RF pulse simultaneously with a slice select gradient ($G_{SS}$) excites spins located in a defined slice. Phase encoding ($G_{PE}$) and frequency encoding ($G_{FE}$) gradients are used to assign unique Larmor frequencies and phases to spins at different locations within the excited slice. The entire sequence is repeated after a time TR to acquire all of the data necessary to reconstruct the image. (From Bernas, L., Preclinical cellular MRI of glioma using iron oxide contrast agents, PhD thesis, Department of Medical Biophysics, University of Western Ontario, 2009. University of Western Ontario.)

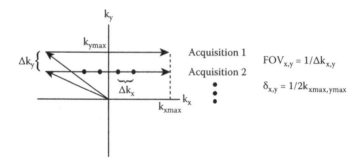

**FIGURE 10.12**
Depiction of a common k-space trajectory. $G_y$ defines the position of the line of k-space to be acquired along $k_y$, and $G_x$ defines the position along $k_x$. The field of view of the final image is related to the sampling rate, and the image resolution is related to the extent of k-space sampled. (From Bernas, L., Preclinical cellular MRI of glioma using iron oxide contrast agents, PhD thesis, Department of Medical Biophysics, University of Western Ontario, 2009. University of Western Ontario.)

resolution (Nishimura 1996). A two-dimensional (2D) Fourier transform of the k-space data yields the image of the selected slice. The combination of a series of 2D slices is one way of obtaining images of a full 3D volume.

Alternatively, true three-dimensional (3D) MRI can be performed by exciting a thick slab, rather than a thin slice, and performing a second phase-encoding step to spatially encode spins along the third dimension. A 3D Fourier transform is then applied to the data to reconstruct the final 3D image (Reiser 2008). This approach has some advantages over the 2D stack approach in that it allows for very thin, contiguous slices and produces datasets that can be reformatted to any orientation. The trade-off is longer scan time and higher sensitivity to motion due to the additional phase-encoding step.

Contrast in MR images is determined by the intrinsic MR characteristics of the sample (proton density [PD], T1, and T2), as well as by the way the magnetization is manipulated by the particular pulse sequence and specific parameters used.

## 10.8 Pulse Sequences

### 10.8.1 Spin Echo

The spin echo (SE) sequence is one of the most commonly used pulse sequences, as it is insensitive to inhomogeneities in the $B_0$ field and can generate images with T1, T2, or PD weighting, depending on the selection of the imaging parameters, TE (echo time), and TR (repetition time). The basic structure of this sequence is shown in Figure 10.13. It consists of a slice-selective

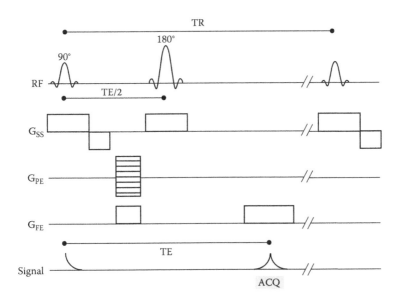

**FIGURE 10.13**

Timing diagram for spin echo pulse sequence. (From Bernas, L., Preclinical cellular MRI of glioma using iron oxide contrast agents, PhD thesis, Department of Medical Biophysics, University of Western Ontario, 2009. University of Western Ontario.)

**TABLE 10.1**

Effect of Selection of TE and TR on Spin Echo Image Contrast

| Contrast | TR | TE |
|---|---|---|
| PD-weighted | >> T1 | << T2 |
| T1-weighted | ≈ T1 | << T2 |
| T2-weighted | >> T1 | ≈ T2 |

90° RF pulse followed after a time TE/2 by a 180° pulse, which reverses the dephasing of the transverse magnetization due to static field inhomogeneities. This leads to the formation of a "spin echo" at a time TE, at which time the signal is acquired. The entire sequence is repeated after some delay (TR) until k-space is fully sampled, allowing the image to be reconstructed using a 2D Fourier transform (Reiser 2008).

The SE signal depends on the PD, T1, and T2 of the sample. The dominant image contrast is determined by the selection of TE and TR, as summarized in Table 10.1 (Reiser 2008).

## 10.8.2 Gradient Echo

A second type of pulse sequence that is commonly used is the gradient echo (GRE) sequence. In this technique, as the name suggests, gradient pulses are

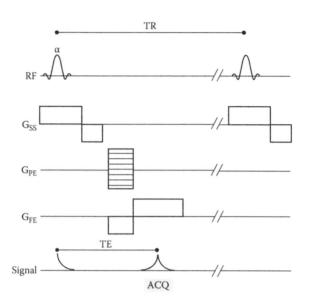

**FIGURE 10.14**
Timing diagram for gradient echo pulse sequence. (From Bernas, L., Preclinical cellular MRI of glioma using iron oxide contrast agents, PhD thesis, Department of Medical Biophysics, University of Western Ontario, 2009. University of Western Ontario.)

used rather than 180° RF pulses to refocus the magnetization and generate echoes. A simple pulse sequence diagram of the GRE sequence is shown in Figure 10.14.

The signal acquired depends on PD, T1, and T2* (the last because the static inhomogeneity effects are not eliminated as they are by SE techniques). In GRE imaging, flip angles smaller than 90° are used, which allows for much shorter TRs, and therefore shorter scan times than are possible using the SE sequence, as full T1 recovery is not necessary before the next RF excitation. After a few excitation pulses, a steady state is reached in which the decrease in the longitudinal magnetization is exactly balanced by T1 recovery (Reiser 2008). The remaining transverse magnetization may either be destroyed through the use of a dephasing gradient, as is the case for the spoiled gradient echo (SPGR) sequence, or be driven to steady state, as in steady-state free precession (SSFP) sequences.

### 10.8.3 Balanced Steady-State Free Precession

Cellular MRI is most commonly performed using T2*-weighted GRE sequences as they are sensitive to the presence of magnetic field-perturbing contrast agents such as superparamagnetic iron oxide (SPIO) nanoparticles. These sequences, however, are limited by relatively poor tissue contrast and a potentially overwhelming "blooming artifact" caused by the SPIO, which may prevent adequate visualization of the surrounding anatomy. T2-weighted

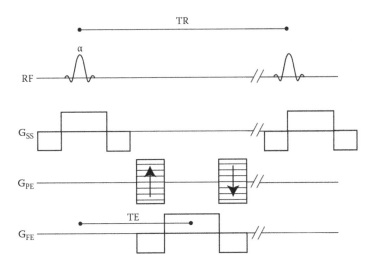

**FIGURE 10.15**
Timing diagram for bSSFP pulse sequence. (From Bernas, L., Preclinical cellular MRI of glioma using iron oxide contrast agents, PhD thesis, Department of Medical Biophysics, University of Western Ontario, 2009. University of Western Ontario.)

techniques are also used as they have better tissue contrast and do not suffer as extensively from the blooming artifact, but these are much less sensitive to SPIO. The balanced steady-state free precession (bSSFP) pulse sequence has also been shown to be extremely sensitive to the SPIO, while reducing the blooming artifact by refocusing much of the background inhomogeneity, similarly to an SE formation (Lebel et al. 2006). This sequence is highly SNR efficient and is therefore attractive for small-animal studies.

The bSSFP pulse sequence is a special type of SSFP sequence in which the net gradient-induced dephasing between repetitions is zero. That is, the gradient pulses are fully balanced by pulses of opposite polarity, as shown in Figure 10.15 (Scheffler and Lehnhardt 2003).

As for other SSFP pulse sequences, the magnetization eventually reaches a steady state, in which the perturbation of the magnetization by the RF excitation is exactly balanced by T1 and T2 recovery by the next excitation. To ensure a smooth approach to the steady state, it is common for the RF pulse train to begin with an $\alpha/2$ pulse followed after a time TR/2 by an $\alpha$ pulse and then by $\alpha$ pulses separated by TR (Scheffler and Lehnhardt 2003). The steady-state signal intensity $M_{SS}$ is given by

$$M_{SS} = M_0 \frac{\sqrt{E_2}(1-E_1)\sin\alpha}{1-(E_1-E_2)\cos\alpha - E_1 E_2}$$

where $E_{1,2} = e^{-TR/T1,2}$, $M_0$ is the thermal equilibrium magnetization, and $\alpha$ is the flip angle.

Where TR << T1 and T2, which is the case for most tissues, this equation (Scheffler and Lehnhardt 2003) simplifies to

$$M_{SS} = M_0 \frac{\sin \alpha}{1 + \cos \alpha + (1 - \cos \alpha) \frac{T1}{T2}}$$

The image contrast is therefore not purely T1 or T2 weighted, but rather depends on differences in the T2/T1 ratio between tissues. Tissues with T2 ≈ T1 appear bright, whereas those with T2 << T1 have low signal. The flip angle that results in the greatest signal is given by

$$\cos \alpha = \frac{\frac{T1}{T2} - 1}{\frac{T1}{T2} + 1}$$

The maximum achievable signal (Haacke et al. 1999) for a given tissue is therefore given by

$$M_{SS} = \frac{1}{2} M_0 \sqrt{\frac{T2}{T1}}$$

For tissues with T1 ≈ T2, such as cerebral spinal fluid and fat, the maximum signal therefore approaches 50% of the thermal equilibrium magnetization. This makes the bSSFP sequence the most SNR efficient of all known sequences, meaning it can achieve very high SNR in short scan times (Haacke et al. 1999; Scheffler and Lehnhardt 2003).

The bSSFP sequence is, however, highly sensitive to local field inhomogeneities, which cause some dephasing of the magnetization between excitation pulses and therefore interfere with the steady state. An example of the steady-state signal as a function of dephasing θ during one TR is shown in Figure 10.16.

The variation in signal due to off-resonance effects gives rise to the characteristic banding artifact that can plague bSSFP images. The artifact can often be minimized by using short TRs, thereby minimizing the dephasing that can occur between excitation pulses. However, this approach is not adequate for some combinations of T1, T2, and α and is not suitable for all applications. Techniques that combine multiple images acquired using RF pulses with linearly varying phase have been developed that aid in removing the banding artifact (Bangerter et al. 2004). The result of varying the phase of the RF pulses, or phase cycling, is to move the banding artifact across the image, resulting in a more uniform signal when multiple images are combined.

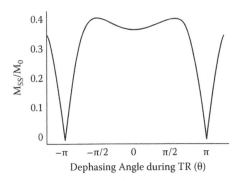

Dephasing Angle during TR (θ)

**FIGURE 10.16**

Example of bSSFP steady-state signal as a function of dephasing due to off-resonance effects. The signal is stable across some range of off-resonance frequencies but displays sharp drops in intensity when the dephasing approaches multiples of $2\pi$. This variation in signal gives rise to the banding artifact often seen in bSSFP images. (From Bernas, L., Preclinical cellular MRI of glioma using iron oxide contrast agents, PhD thesis, Department of Medical Biophysics, University of Western Ontario, 2009. University of Western Ontario.)

Although the sensitivity of bSSFP to off-resonance effects can be challenging, it also explains why the sequence is highly sensitive to the presence of iron oxide contrast agents. The combination of the high SNR efficiency of the bSSFP sequence and its sensitivity to SPIO has enabled the detection of single cells *in vivo* at 1.5 and 3 T (Bernas et al. 2010; Heyn et al. 2006b).

## 10.9 Cellular MRI

Cellular imaging with MRI enables the assessment of dynamic cellular processes that previously could only be studied indirectly using *ex vivo* histology techniques. The ability to study such processes *in vivo* over time has broad application to areas including the study of cancer metastasis, stem cell and other therapeutic cell tracking, and inflammation research (Anderson et al. 2004; Arai et al. 2006; Arbab et al. 2004, 2008; Bernas et al. 2007; Dunn et al. 2005; Oweida et al. 2007). The detection of cells using MRI is best accomplished by high-resolution (on the order of a hundred microns), high-quality imaging, as well as the incorporation of appropriate contrast agents.

The resolution and SNR required for cellular imaging techniques are most often achieved using systems with high field strength, dedicated to small-animal imaging (Arbab et al. 2007; Shapiro et al. 2005; Valable et al. 2007). The higher SNR achieved with the systems with high field strength allows for high-resolution images that are essential for imaging the anatomy of small animals, to be acquired in shorter scan times. There are, however, some

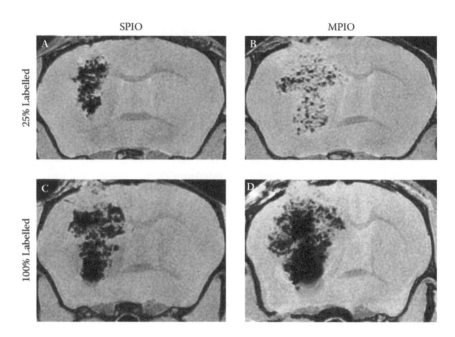

**FIGURE 10.17**

(A)–(D) Examples of bSSFP images of mouse brain showing strong iron oxide-based contrast resulting from glioma cells loaded with SPIOs or microsize paramagnetic iron oxide (MPIO). SPIO-labeled tumors are shown in (A) and (C), and MPIO-labeled tumors are shown in (B) and (D). Cells were labeled with 5 pg iron per cell in all cases; the injected population consisted of either 25% labeled cells (A) and (B) or 100% labeled cells (C) and (D). Even at this low iron/cell level, 100% labeled cells is excessive for both SPIO and MPIO as it leads to large areas of signal loss that prevent structural details from being visualized. At this iron/cell level, 25% labeled cells in the injected population results in a manageable amount of signal loss remaining after 18 days. (From Bernas, L., P. Foster, and B. Rutt. 2010. *Magn Reson Med.*)

challenges associated with the use of high field strengths, including poorer field homogeneity, longer T1 relaxation times, and shorter T2/T2* relaxation times. For this reason, many excellent cellular imaging results have been obtained using clinical field strength scanners (1.5 and 3 T), sometimes with the addition of custom-built, high-performance, insertable gradient coils that allow for high-resolution, fast imaging and effectively converting a clinical system into a small-animal imaging system without changing the main magnetic field strength. High SNR can still be obtained with these clinical systems with lower field strength through the use of highly optimized RF coils, specifically designed for each type of sample or animal, and by utilizing highly SNR-efficient pulse sequences such as the bSSFP pulse sequence (Figure 10.17) (Bernas et al. 2010; Foster-Gareau et al. 2003; Heyn et al. 2006a).

Contrast agents that have been used for cellular MRI include paramagnetic agents based on gadolinium or manganese (Aoki et al. 2006; Brekke et al. 2007; Giesel et al. 2006; Modo et al. 2002; Strijkers et al. 2007), as well as agents based on SPIO (Arbab et al. 2003; Bulte and Kraitchman 2004; Corot et al. 2006; Foster-Gareau et al. 2003; Frank et al. 2004). Paramagnetic contrast agents are attractive because they can be used to generate positive contrast on T1-weighted images. However, the sensitivity of these agents is several orders of magnitude lower than that achievable using iron oxide agents, and studies have been limited to visualizing tens of thousands of cells (Brekke et al. 2007; Giesel et al. 2006; Modo et al. 2002). Iron oxide agents are available in a wide range of sizes and surface coatings, which dictate their biological characteristics (Corot et al. 2006; Thorek et al. 2006). Standard dextran-coated SPIO agents (80–200 nm in diameter), such as Feridex I.V.® (AMAG Pharmaceuticals, Inc., Cambridge, MA), are biodegradable and used clinically for liver and spleen imaging as they are quickly taken up by the cells of the reticuloendothelial system following intravenous injection, highlighting cancerous lesions (Modo et al. 2005; Yoshikawa et al. 2006). SPIO agents have also been used extensively in the research setting for labeling macrophages via intravenous injection to track their involvement in the inflammatory response in diseases, including multiple sclerosis (Dousset et al. 1999; Oweida et al. 2007) and spinal cord injury (Dunn et al. 2005). Other cell types, including stem cells and cancer cells, can be labeled *in vitro* with SPIO.

The use of transfection agents or other techniques is sometimes necessary for efficient labeling of nonphagocytic cells (Arbab et al. 2003; Montet-Abou et al. 2005; Walczak et al. 2005) and is covered in more detail in Chapter 15. Ultrasmall agents (ultrasmall superparamagnetic iron oxide [USPIO], 10–40 nm in diameter) have longer circulation times than SPIO and are therefore used as blood pool agents for MR angiography as well as for lymph node imaging (Corot et al. 2006). Larger, micron-size particles (micron-sized iron oxide [MPIO]), such as Bangs beads (Bangs Laboratories, Inc., Fishers, IN) and SiMag beads (Chemicell, Berlin, Germany), are coated with biologically inert substances, such as polystyrene or silica, and are used exclusively in the research setting (Hinds et al. 2003). MPIO particles are efficiently taken up by both phagocytic and nonphagocytic cells without the need for transfection agents (Bernas et al. 2007; Shapiro et al. 2005; Williams et al. 2007); and since they are larger and contain more iron per particle, they have the potential to load more iron into cells for higher detection sensitivity.

The strong magnetization of SPIO particles within a cell causes a strong perturbation in the magnetic field surrounding the iron-loaded cell. The T2 relaxation is greatly enhanced in this region as protons diffusing through this inhomogeneous magnetic field become quickly dephased. This leads to areas of strong signal loss in T2*-weighted images. Since the effect on the magnetic field extends beyond the cell borders, the areas of signal loss are much larger than the cells themselves (the blooming effect), allowing for sensitive cell detection. The use of iron oxide contrast agents has enabled the detection of single iron-labeled cells both *in vitro* and *in vivo* by several groups. *In vitro* experiments were carried out using SPIO-labeled

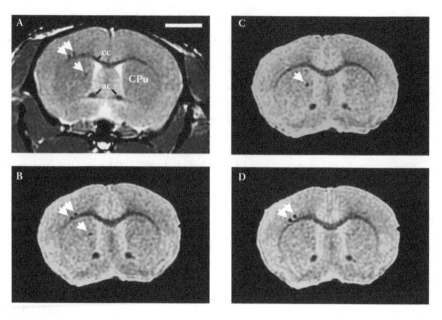

**FIGURE 10.18**

*In vivo* and *ex vivo* MRI of a mouse brain from an animal injected with SPIO-labeled macrophages. (A) *In vivo* MRI showing three signal voids (100 × 100 µm² in plane, 200-µm slice thickness). (B) *Ex vivo* MRI at the same resolution and slice location as (A) showing the same three signal voids. (C), (D) High-resolution *ex vivo* MRI (100 × 100 µm² in plane, 100-µm slice thickness). The three signal voids in (A) and (B) now appear in two different MR slices. cc, corpus callosum; ac, anterior commissure; CPu, caudate putamen. (From Heyn, C., J.A. Ronald, L.T. Mackenzie, I.C. Macdonald, A.F. Chambers, B.K. Rutt, and P.J. Foster. 2006. *Magn Reson Med* 55, no. 1: 23–29.)

macrophages embedded in gelatin, allowing for optical validation of single-cell detection (Dodd et al. 1999; Foster-Gareau et al. 2003). In a common hepatocyte transplantation model, single cells labeled with 1.63-µm MPIO particles were detected at 7 T in the livers of mice following injection into the spleens. Careful validation using confocal microscopy confirmed that single SPIO-labeled macrophages (~50 pg Fe/cell) were detected at 1.5 T in the brains of mice following intracardiac injection (Figure 10.18) (Heyn et al. 2006a).

## 10.10  Limitations of Iron Oxide-Based Cellular MRI

There are certain limitations to the use of iron oxide-based reagents that must be taken into account when analyzing MRI image data. Contrast agents are diluted or lost when cells divide or die; as such, the time frame for imaging cells labeled using these agents must be within days to several weeks, depending on the rates of cell proliferation and death (Frank et al. 2004).

Another limitation is that signal loss in MR images sensitive to iron can be caused by sources other than iron-labeled cells. This includes anatomical features (such as air in sinuses, bone, blood vessels, and hemorrhage), which lead to artifacts caused by local magnetic field inhomogeneities. This biology has the potential to lead to the incorrect identification of regions of signal loss as iron-labeled cells (van den Bos et al. 2006). Similarly, the nonspecific uptake of iron particles by macrophages that engulf dead labeled cells can be problematic (Amsalem et al. 2007; Pawelczyk et al. 2008, 2009; Terrovitis et al. 2008).

## 10.11 Positive Contrast Cellular MRI

To overcome these limitations, other contrast agents are being developed. There are a few reports of cells that have been successfully labeled with gadolinium-based contrast agents for detection by MRI (Crich et al. 2004; Klasson et al. 2008; Shen et al. 2009). $^{19}$F MRI has been used to image perfluoropolyether (PFPE)-labeled dendritic cells *in vivo* with minimal adverse effects on cell function (Ahrens et al. 2005). Due to the negligible endogenous $^{19}$F background signal, this approach allows for more specific imaging of the labeled cells (Helfer et al. 2010). The image obtained with the $^{19}$F signal can be overlaid on the anatomical information derived from a conventional $^{1}$H image to obtain a more detailed image. The main limitation of gadolinium or PFPE labels is the much lower sensitivity compared with iron nanoparticles.

Another active and promising area of research has been the development of a number of methods that create positive contrast from cells labeled with iron oxide. The first general class of methods is based on the design of pulse sequences that are sensitized to the susceptibility-induced, off-resonance spins that surround the labeled cells. Selective imaging of the off-resonance signals around cells has been proposed (Cunningham et al. 2005). As an extension of this method, an inversion recovery on-resonance water suppression (IRON) sequence places a saturation pulse to suppress on-resonance water signal leaving only the signal from off-resonant protons presumably due to iron oxide-labeled cells (Stuber et al. 2007). Positive contrast can also be generated using a diffusion-mediated off-resonance saturation (ORS) sequence (Zurkiya and Hu 2006), in which bulk water protons are imaged with and without the ORS pulse, and the diffusion of water molecules around the labeled cells acts to generate the relevant contrast. The "white marker" or gradient echo acquisition for superparamagnetic particle (GRASP) technique (Mani et al. 2006; Seppenwoolde et al. 2003) uses a dephasing gradient on the slice-select axis to spoil (suppress) the signal across the sample at all locations except where iron-loaded cells act to generate local gradients to cancel the externally imposed gradient.

The second general class of methods takes a different approach by deriving positive contrast through postprocessing of images (Posse 1992). The local magnetic field gradients induced by magnetic susceptibilities lead to echo shifts in k-space with GRE imaging. The effect is exploited by applying a shifted reconstruction window in k-space (Bakker et al. 2006). A susceptibility gradient mapping (SGM) technique has been proposed that generates susceptibility vectors from a regular complex GRE dataset to develop positive contrast in 3D (Dahnke et al. 2008). A third technique is to exploit the advantage of ultrashort echo time (UTE) imaging (Gold et al. 1995) in which Short-T2 species, such as iron-labeled cells, will appear bright. A second Long-TE image is acquired and subtracted from the short-TE image to yield difference images in which the background is suppressed, including fat and blood, while iron-labeled cells appear bright.

Some of these positive contrast methods have been compared by imaging labeled glioma cells transplanted to the flanks of nude rats (Liu et al. 2008). Positive contrast was generated from the presence of labeled cells, but it remains to be determined whether positive contrast yielded better cellular detection. One disadvantage of positive contrast images is that they lack anatomical detail and therefore are best supplemented with traditional negative contrast images, which may in fact demonstrate higher sensitivity to cellular detection.

---

# References

Ahrens, E.T., R. Flores, H. Xu, and P.A. Morel. 2005. In vivo imaging platform for tracking immunotherapeutic cells. *Nat Biotechnol* 23, no. 8: 983–87.

Amsalem, Y., Y. Mardor, M.S. Feinberg, N. Landa, L. Miller, D. Daniels, A. Ocherashvilli, R. Holbova, O. Yosef, I.M. Barbash, and J. Leor. 2007. Iron-oxide labeling and outcome of transplanted mesenchymal stem cells in the infarcted myocardium. *Circulation* 116, no. 11 Suppl: I38–I45.

Anderson, S.A., J. Shukaliak-Quandt, E.K. Jordan, A.S. Arbab, R. Martin, H. Mcfarland, and J.A. Frank. 2004. Magnetic resonance imaging of labeled T-cells in a mouse model of multiple sclerosis. *Ann Neurol* 55, no. 5: 654–59.

Aoki, I., Y. Takahashi, K.H. Chuang, A.C. Silva, T. Igarashi, C. Tanaka, R.W. Childs, and A.P. Koretsky. 2006. Cell labeling for magnetic resonance imaging with the T1 agent manganese chloride. *NMR Biomed* 19, no. 1: 50–59.

Arai, T., T. Kofidis, J.W. Bulte, J. De Bruin, R.D. Venook, G.J. Berry, M.V. Mcconnell, T. Quertermous, R.C. Robbins, and P.C. Yang. 2006. Dual in vivo magnetic resonance evaluation of magnetically labeled mouse embryonic stem cells and cardiac function at 1.5 T. *Magn Reson Med* 55, no. 1: 203–9.

Arbab, A.S., L.A. Bashaw, B.R. Miller, E.K. Jordan, J.W. Bulte, and J.A. Frank. 2003. Intracytoplasmic tagging of cells with ferumoxides and transfection agent for cellular magnetic resonance imaging after cell transplantation: Methods and techniques. *Transplantation* 76, no. 7: 1123–30.

Arbab, A.S., B. Janic, R.A. Knight, S.A. Anderson, E. Pawelczyk, A.M. Rad, E.J. Read, S.D. Pandit, and J.A. Frank. 2008. Detection of migration of locally implanted AC133+ stem cells by cellular magnetic resonance imaging with histological findings. *FASEB J* 22, no. 9: 3234–46.

Arbab, A.S., E.K. Jordan, L.B. Wilson, G.T. Yocum, B.K. Lewis, and J.A. Frank. 2004. In vivo trafficking and targeted delivery of magnetically labeled stem cells. *Hum Gene Ther* 15, no. 4: 351–60.

Arbab, A.S., A.M. Rad, A.S. Iskander, K. Jafari-Khouzani, S.L. Brown, J.L. Churchman, G. Ding, Q. Jiang, J.A. Frank, H. Soltanian-Zadeh, and D.J. Peck. 2007. Magnetically-labeled sensitized splenocytes to identify glioma by MRI: a preliminary study. *Magn Reson Med* 58, no. 3: 519–26.

Bakker, C.J., J.H. Seppenwoolde, and K.L. Vincken. 2006. Dephased MRI. *Magn Reson Med* 55, no. 1: 92–97.

Bangerter, N.K., B.A. Hargreaves, S.S. Vasanawala, J.M. Pauly, G.E. Gold, and D.G. Nishimura. 2004. Analysis of multiple-acquisition SSFP. *Magn Reson Med* 51, no. 5: 1038–47.

Bean, C.P. and J.D. Livingston. 1959. Superparamagnetism. *J Appl Phys* 30, no. 4: 120S–29S.

Bernas, L., P. Foster and B. Rutt. 2007. Magnetic resonance imaging of in vitro glioma cell invasion. *J Neurosurg* 106, no. 2: 306–13.

Bernas, L., P. Foster, and B. Rutt. 2010. Imaging iron-loaded mouse glioma tumors with bSSFP at 3 T. *Magn Reson Med* 64(1): 23–31.

Brekke, C., S.C. Williams, J. Price, F. Thorsen, and M. Modo. 2007. Cellular multiparametric MRI of neural stem cell therapy in a rat glioma model. *Neuroimage* 37, no. 3: 769–82.

Bulte, J.W. and D.L. Kraitchman. 2004. Iron oxide MR contrast agents for molecular and cellular imaging. *NMR Biomed* 17, no. 7: 484–99.

Corot, C., P. Robert, J.M. Idee, and M. Port. 2006. Recent advances in iron oxide nanocrystal technology for medical imaging. *Adv Drug Deliv Rev* 58, no. 14: 1471–1504.

Crich, S.G., L. Biancone, V. Cantaluppi, D. Duo, G. Esposito, S. Russo, G. Camussi, and S. Aime. 2004. Improved route for the visualization of stem cells labeled with a Gd-/Eu-chelate as dual (MRI and fluorescence) agent. *Magn Reson Med* 51, no. 5: 938–44.

Cunningham, C.H., T. Arai, P.C. Yang, M.V. Mcconnell, J.M. Pauly, and S.M. Conolly. 2005. Positive contrast magnetic resonance imaging of cells labeled with magnetic nanoparticles. *Magn Reson Med* 53, no. 5: 999–1005.

Dahnke, H., W. Liu, D. Herzka, J.A. Frank, and T. Schaeffter. 2008. Susceptibility gradient mapping (SGM): A new postprocessing method for positive contrast generation applied to superparamagnetic iron oxide particle (SPIO)-labeled cells. *Magn Reson Med* 60, no. 3: 595–603.

Dodd, S.J., M. Williams, J.P. Suhan, D.S. Williams, A.P. Koretsky, and C. Ho. 1999. Detection of single mammalian cells by high-resolution magnetic resonance imaging. *Biophys J* 76, no. 1 Pt 1: 103–9.

Dousset, V., C. Delalande, L. Ballarino, B. Quesson, D. Seilhan, M. Coussemacq, E. Thiaudiere, B. Brochet, P. Canioni, and J.M. Caille. 1999. In vivo macrophage activity imaging in the central nervous system detected by magnetic resonance. *Magn Reson Med* 41, no. 2: 329–33.

Dunn, E.A., L.C. Weaver, G.A. Dekaban, and P.J. Foster. 2005. Cellular imaging of inflammation after experimental spinal cord injury. *Mol Imaging* 4, no. 1: 53–62.

Foster-Gareau, P., C. Heyn, A. Alejski, and B.K. Rutt. 2003. Imaging single mammalian cells with a 1.5 T clinical MRI scanner. *Magn Reson Med* 49, no. 5: 968–71.

Frank, J.A., S.A. Anderson, H. Kalsih, E.K. Jordan, B.K. Lewis, G.T. Yocum, and A.S. Arbab. 2004. Methods for magnetically labeling stem and other cells for detection by in vivo magnetic resonance imaging. *Cytotherapy* 6, no. 6: 621–25.

Giesel, F.L., M. Stroick, M. Griebe, H. Troster, C.W. Von Der Lieth, M. Requardt, M. Rius, M. Essig, H.U. Kauczor, M.G. Hennerici, and M. Fatar. 2006. Gadofluorine m uptake in stem cells as a new magnetic resonance imaging tracking method: an in vitro and in vivo study. *Invest Radiol* 41, no. 12: 868–73.

Gold, G.E., J.M. Pauly, A. Macovski, and R.J. Herfkens. 1995. MR spectroscopic imaging of collagen: Tendons and knee menisci. *Magn Reson Med* 34, no. 5: 647–54.

Haacke, E.M., R.W. Brown, M.R. Thompson, and R. Venkatesan. 1999. *Magnetic Resonance Imaging Physical Principles and Sequence Design*. New York: Wiley.

Helfer, B.M., A. Balducci, A.D. Nelson, J.M. Janjic, R.R. Gil, P. Kalinski, I.J. De Vries, E.T. Ahrens, and R.B. Mailliard. 2010. Functional assessment of human dendritic cells labeled for in vivo (19)F magnetic resonance imaging cell tracking. *Cytotherapy* 12, no. 2: 238–50.

Heyn, C., J.A. Ronald, L.T. Mackenzie, I.C. Macdonald, A.F. Chambers, B.K. Rutt, and P.J. Foster. 2006a. In vivo magnetic resonance imaging of single cells in mouse brain with optical validation. *Magn Reson Med* 55, no. 1: 23–29.

Heyn, C., J.A. Ronald, S.S. Ramadan, J.A. Snir, A.M. Barry, L.T. Mackenzie, D.J. Mikulis, D. Palmieri, J.L. Bronder, P.S. Steeg, T. Yoneda, I.C. Macdonald, A.F. Chambers, B.K. Rutt, and P.J. Foster. 2006b. In vivo MRI of cancer cell fate at the single-cell level in a mouse model of breast cancer metastasis to the brain. *Magn Reson Med* 56, no. 5: 1001–10.

Hinds, K.A., J.M. Hill, E.M. Shapiro, M.O. Laukkanen, A.C. Silva, C.A. Combs, T.R. Varney, R.S. Balaban, A.P. Koretsky, and C.E. Dunbar. 2003. Highly efficient endosomal labeling of progenitor and stem cells with large magnetic particles allows magnetic resonance imaging of single cells. *Blood* 102, no. 3: 867–72.

Klasson, A., M. Ahren, E. Hellqvist, F. Soderlind, A. Rosen, P.O. Kall, K. Uvdal, and M. Engstrom. 2008. Positive MRI contrast enhancement in THP-1 cells with $Gd_2O_3$ nanoparticles. *Contrast Media Mol Imaging* 3, no. 3: 106–11.

Lebel, R.M., R.S. Menon, and C.V. Bowen. 2006. Relaxometry model of strong dipolar perturbers for balanced-SSFP: application to quantification of SPIO loaded cells. *Magn Reson Med* 55, no. 3: 583–91.

Liu, W., H. Dahnke, E.K. Jordan, T. Schaeffter, and J.A. Frank. 2008. In vivo MRI using positive-contrast techniques in detection of cells labeled with superparamagnetic iron oxide nanoparticles. *NMR Biomed* 21, no. 3: 242–50.

Mani, V., K.C. Briley-Saebo, V.V. Itskovich, D.D. Samber, and Z.A. Fayad. 2006. Gradient echo acquisition for superparamagnetic particles with positive contrast (GRASP): Sequence characterization in membrane and glass superparamagnetic iron oxide phantoms at 1.5 T and 3 T. *Magn Reson Med* 55, no. 1: 126–35.

Modo, M., D. Cash, K. Mellodew, S.C. Williams, S.E. Fraser, T.J. Meade, J. Price, and H. Hodges. 2002. Tracking transplanted stem cell migration using bifunctional, contrast agent-enhanced, magnetic resonance imaging. *Neuroimage* 17, no. 2: 803–11.

Modo, M., M. Hoehn, and J.W. Bulte. 2005. Cellular MR imaging. *Mol Imaging* 4, no. 3: 143–64.

Montet-Abou, K., X. Montet, R. Weissleder, and L. Josephson. 2005. Transfection agent induced nanoparticle cell loading. *Mol Imaging* 4, no. 3: 165–71.

Nishimura, D.G. 1996. *Principles of Magnetic Resonance Imaging*: Stanford, CA: Stanford University Press.

Oweida, A.J., E.A. Dunn, S.J. Karlik, G.A. Dekaban, and P.J. Foster. 2007. Iron-oxide labeling of hematogenous macrophages in a model of experimental autoimmune encephalomyelitis and the contribution to signal loss in fast imaging employing steady state acquisition (FIESTA) images. *J Magn Reson Imaging* 26, no. 1: 144–51.

Pawelczyk, E., A.S. Arbab, A. Chaudhry, A. Balakumaran, P.G. Robey, and J.A. Frank. 2008. In vitro model of bromodeoxyuridine or iron oxide nanoparticle uptake by activated macrophages from labeled stem cells: implications for cellular therapy. *Stem Cells* 26, no. 5: 1366–75.

Pawelczyk, E., E.K. Jordan, A. Balakumaran, A. Chaudhry, N. Gormley, M. Smith, B.K. Lewis, R. Childs, P.G. Robey, and J.A. Frank. 2009. In vivo transfer of intracellular labels from locally implanted bone marrow stromal cells to resident tissue macrophages. *PLoS One* 4, no. 8: e6712.

Posse, S. 1992. Direct imaging of magnetic field gradients by group spin-echo selection. *Magn Reson Med* 25, no. 1: 12–29.

Reiser, M.F. 2008. *Magnetic Resonance Tomography*. New York: Springer.

Scheffler, K. and S. Lehnhardt. 2003. Principles and applications of balanced SSFP techniques. *Eur Radiol* 13, no. 11: 2409–18.

Seppenwoolde, J.H., M.A. Viergever, and C.J. Bakker. 2003. Passive tracking exploiting local signal conservation: The white marker phenomenon. *Magn Reson Med* 50, no. 4: 784–90.

Shapiro, E.M., S. Skrtic, and A.P. Koretsky. 2005. Sizing it up: cellular MRI using micron-sized iron oxide particles. *Magn Reson Med* 53, no. 2: 329–38.

Shen, J., X.M. Zhong, X.H. Duan, L.N. Cheng, G.B. Hong, X.B. Bi, and Y. Liu. 2009. Magnetic resonance imaging of mesenchymal stem cells labeled with dual (MR and fluorescence) agents in rat spinal cord injury. *Acad Radiol* 16, no. 9: 1142–54.

Smith, R.C. and Lange, R.C. 1998. Relaxation times and mechanisms. In *Understanding Magnetic Resonance Imaging*, 45–57. Boca Raton, FL: CRC Press.

Strijkers, G.J., W.J. Mulder, G.A. Van Tilborg, and K. Nicolay. 2007. MRI contrast agents: Current status and future perspectives. *Anticancer Agents Med Chem* 7, no. 3: 291–305.

Stuber, M., W.D. Gilson, M. Schar, D.A. Kedziorek, L.V. Hofmann, S. Shah, E.J. Vonken, J.W. Bulte, and D.L. Kraitchman. 2007. Positive contrast visualization of iron oxide-labeled stem cells using inversion-recovery with ON-resonant water suppression (IRON). *Magn Reson Med* 58, no. 5: 1072–77.

Terrovitis, J., M. Stuber, A. Youssef, S. Preece, M. Leppo, E. Kizana, M. Schar, G. Gerstenblith, R.G. Weiss, E. Marban, and M.R. Abraham. 2008. Magnetic resonance imaging overestimates ferumoxide-labeled stem cell survival after transplantation in the heart. *Circulation* 117, no. 12: 1555–62.

Thorek, D.L., A.K. Chen, J. Czupryna, and A. Tsourkas. 2006. Superparamagnetic iron oxide nanoparticle probes for molecular imaging. *Ann Biomed Eng* 34, no. 1: 23–38.

Valable, S., E.L. Barbier, M. Bernaudin, S. Roussel, C. Segebarth, E. Petit, and C. Remy. 2007. In vivo MRI tracking of exogenous monocytes/macrophages targeting brain tumors in a rat model of glioma. *Neuroimage* 37 Suppl 1: S47–S58.

Van Den Bos, E.J., T. Baks, A.D. Moelker, W. Kerver, R.J. Van Geuns, W.J. Van Der Giessen, D.J. Duncker, and P.A. Wielopolski. 2006. Magnetic resonance imaging of haemorrhage within reperfused myocardial infarcts: possible interference with iron oxide-labelled cell tracking? *Eur Heart J* 27, no. 13: 1620–26.

Vejpravova, J.P., Sechovsky, V., Niznansky, D., Plocek, J., Hutlova, A., and Rehspringer, J.-L. 2005. Superparamagnetism of Co-ferrite nanoparticles. *WDS'05*, Prague, Mattyzpress. June, 519–23.

Walczak, P., D.A. Kedziorek, A.A. Gilad, S. Lin, and J.W. Bulte. 2005. Instant MR labeling of stem cells using magnetoelectroporation. *Magn Reson Med* 54, no. 4: 769–74.

Wang, Y.X., S.M. Hussain, and G.P. Krestin. 2001. Superparamagnetic iron oxide contrast agents: Physicochemical characteristics and applications in MR imaging. *Eur Radiol* 11, no. 11: 2319–31.

Williams, J.B., Q. Ye, T.K. Hitchens, C.L. Kaufman, and C. Ho. 2007. MRI detection of macrophages labeled using micrometer-sized iron oxide particles. *J Magn Reson Imaging* 25, no. 6: 1210–18.

Wood, M.L. and P.A. Hardy. 1993. Proton relaxation enhancement. *J Magn Reson Imaging* 3, no. 1: 149–56.

Yoshikawa, T., D.G. Mitchell, S. Hirota, Y. Ohno, K. Oda, T. Maeda, M. Fujii, and K. Sugimura. 2006. Gradient- and spin-echo T2-weighted imaging for SPIO-enhanced detection and characterization of focal liver lesions. *J Magn Reson Imaging* 23, no. 5: 712–19.

Zurkiya, O. and X. Hu. 2006. Off-resonance saturation as a means of generating contrast with superparamagnetic nanoparticles. *Magn Reson Med* 56, no. 4: 726–32.

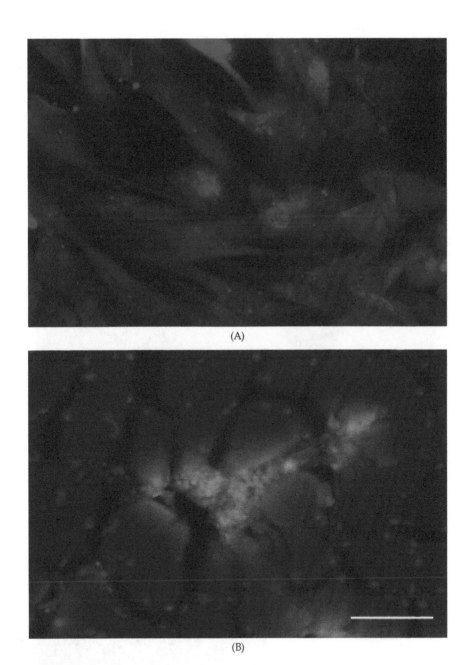

**FIGURE I.1**
See details on p. xviii.

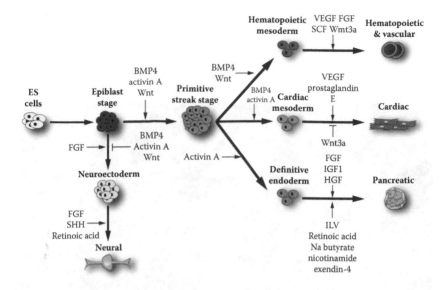

**FIGURE 1.3**
See details on p. 6.

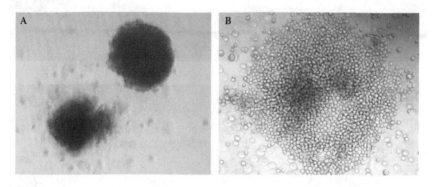

**FIGURE 1.5**
See details on p. 10.

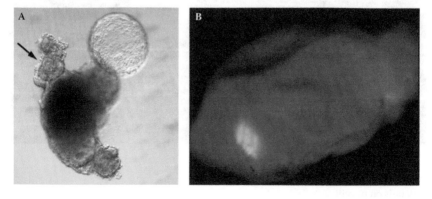

**FIGURE 1.6**
See details on p. 12.

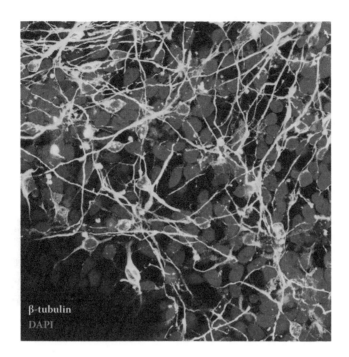

**FIGURE 1.7**
See details on p. 16.

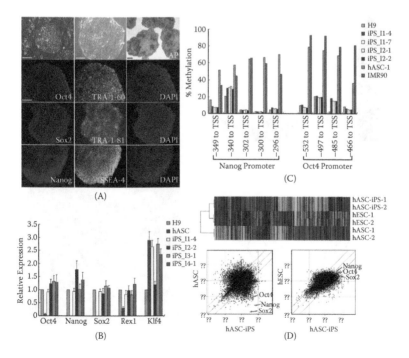

**FIGURE 4.1**
See details on p. 82.

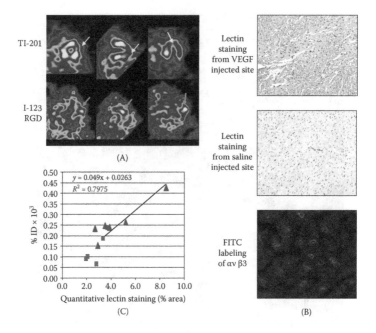

TI-201

I-123 RGD

(A)

$y = 0.049x + 0.0263$
$R^2 = 0.7975$

% ID × 10³

Quantitative lectin staining (% area)

(C)

Lectin staining from VEGF injected site

Lectin staining from saline injected site

FITC labeling of αv β3

(B)

**FIGURE 7.5**
See details on p. 132.

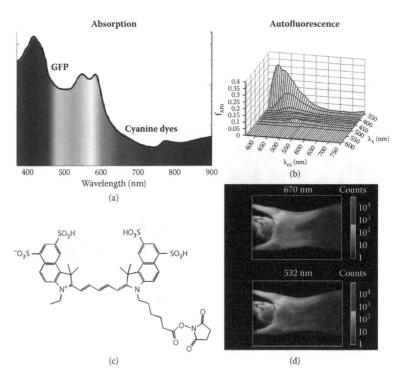

Absorption

GFP

Cyanine dyes

Wavelength (nm)

(a)

Autofluorescence

$f_{xm}$

$\lambda_m$ (nm)

$\lambda_x$ (nm)

(b)

(c)

670 nm    Counts

532 nm    Counts

(d)

**FIGURE 8.1**
See details on p. 141.

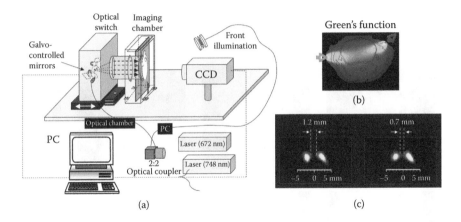

**FIGURE 8.2**
See details on p. 142.

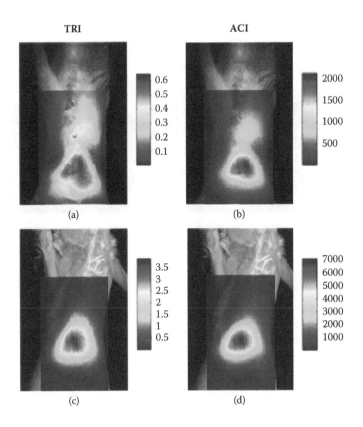

**FIGURE 8.4**
See details on p. 147.

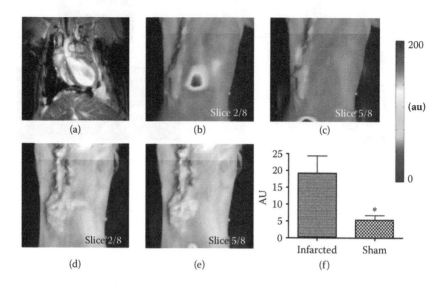

**FIGURE 8.5**
See details on p. 148.

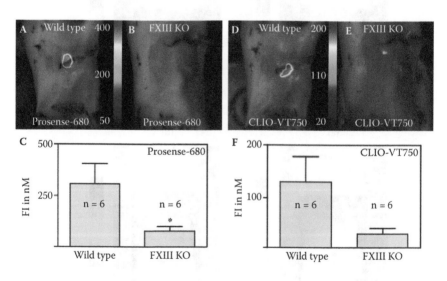

**FIGURE 8.6**
See details on p. 149.

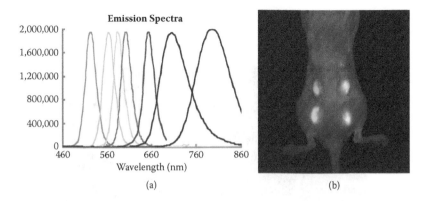

(a)  (b)

**FIGURE 8.7**
See details on p. 150.

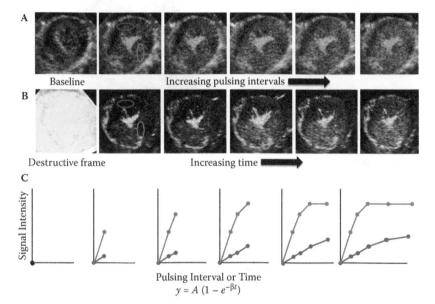

$$y = A (1 - e^{-\beta t})$$

**FIGURE 9.3**
See details on p. 163.

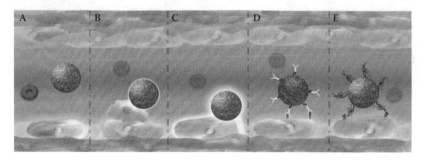

**FIGURE 9.5**
See details on p. 166.

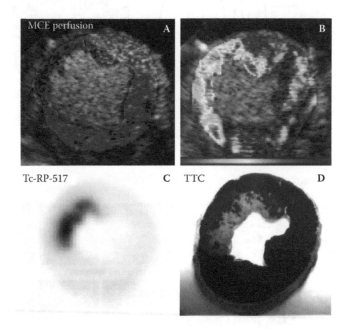

**FIGURE 9.8**
See details on p. 169.

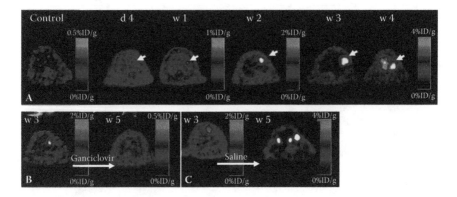

**FIGURE 12.3**
See details on p. 223.

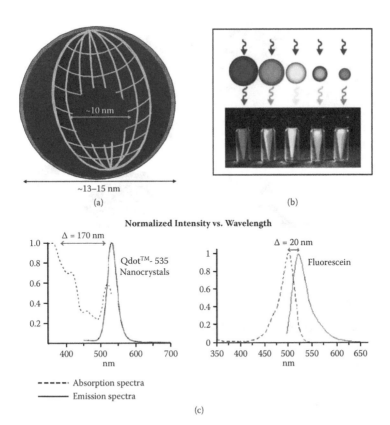

**FIGURE 14.1**
See details on p. 257.

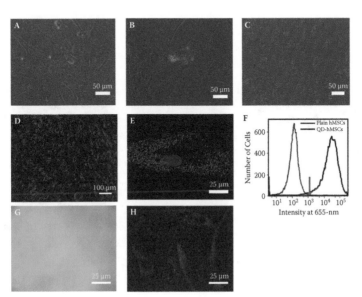

**FIGURE 14.3**
See details on p. 260.

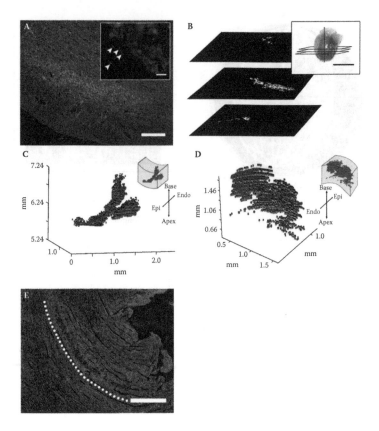

**FIGURE 14.4**
See details on p. 261.

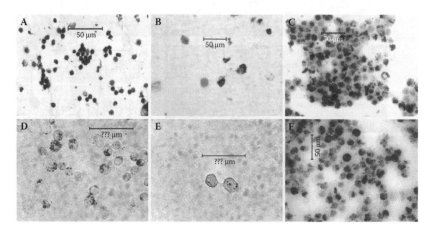

**FIGURE 15.4**
See details on p. 279.

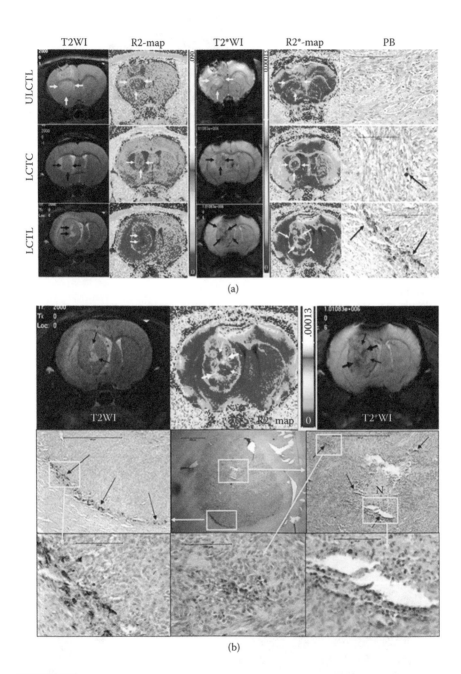

**FIGURE 15.5**
See details on pp. 282 and 283.

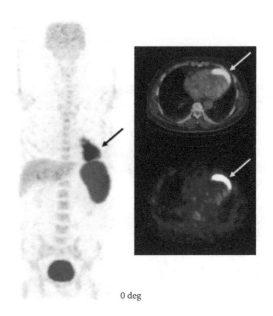

0 deg

**FIGURE 18.2**
See details on p. 322.

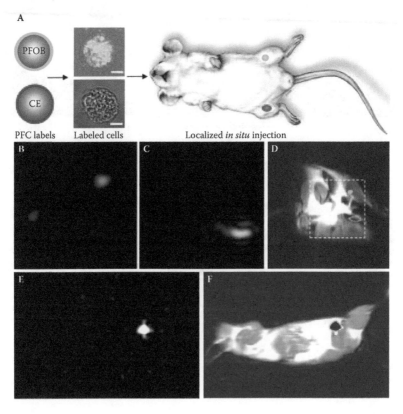

**A**

PFOB

CE

PFC labels    Labeled cells    Localized *in situ* injection

B

C

D

E

F

**FIGURE 18.5**
See details on p. 326.

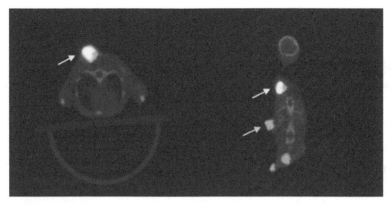

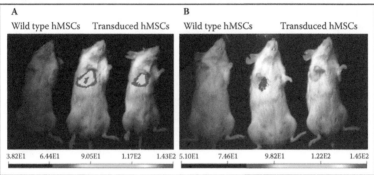

| A | | | B | | |
| Wild type hMSCs | Transduced hMSCs | | Wild type hMSCs | Transduced hMSCs | |

| 3.82E1 | 6.44E1 | 9.05E1 | 1.17E2 | 1.43E2 | 5.10E1 | 7.46E1 | 9.82E1 | 1.22E2 | 1.45E2 |

**FIGURE 18.6**
See details on p. 327.

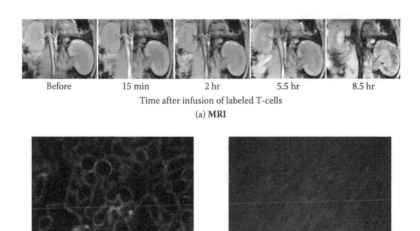

| Before | 15 min | 2 hr | 5.5 hr | 8.5 hr |

Time after infusion of labeled T-cells

(a) **MRI**

| Transplanted kidney | Native kidney |

100 µm

(b) **Fluorescence**

**FIGURE 19.2**
See details on p. 341.

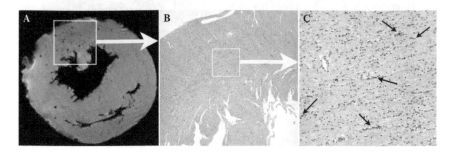

**FIGURE 19.9**
See details on p. 347.

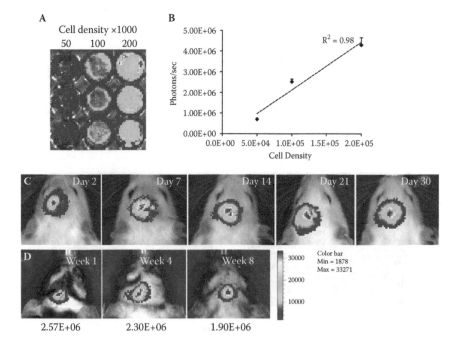

**FIGURE 20.1**
See details on p. 355.

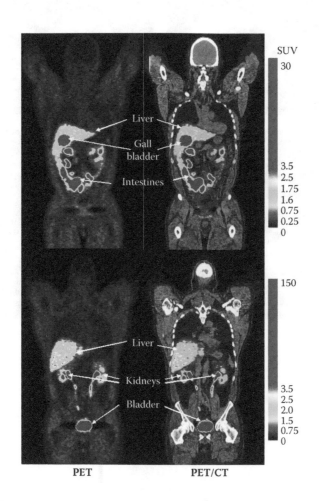

**FIGURE 21.2**
See details on p. 375.

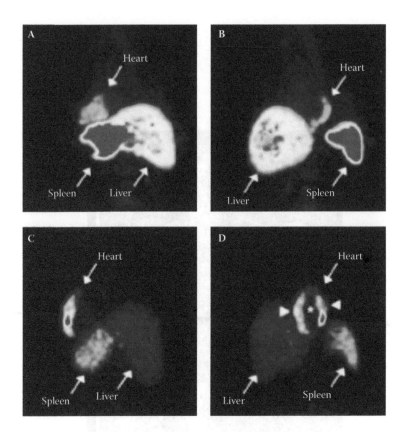

**FIGURE 22.1**
See details on p. 396.

(A)

**FIGURE 22.2**
See details on p. 397.

# 11

## X-Ray Imaging Basics

Jochen Cammin and Katsuyuki Taguchi

**CONTENTS**

## 11.1 Introduction

X-ray images are acquired by transmitting X-rays through an object and recording the transmitted rays on film or with electronic detectors. X-rays are attenuated as they travel through an object, similar to the attenuation of visible light as it travels through a semiopaque material. The magnitude of attenuation depends on both the material and the distance the X-rays have to traverse in the material. The recorded intensity of the acquired image, therefore, provides information about the internal structure of the radiographed object. In medical imaging, X-rays have a wide range of applications and provide complementary information to other imaging modalities like magnetic resonance imaging (MRI), positron emission tomography (PET), single-photon emission computed tomography (SPECT), and ultrasound.

X-ray imaging is used during all stages of medical procedures: (a) it provides preoperative images useful to plan a procedure; (b) it can guide in performing the procedure itself, for example, positioning a catheter at the target blood vessel segment; (c) it can help monitor progress during a procedure; and (d) it can be used to assess the results and success of a procedure.

The most distinguished features of X-ray imaging are the short imaging time (ranging from less than 10 ms for a single-frame image to 30 s for a full-body scan), the wide view angle, and the quantitative nature of the images that helps to identify specific materials. Disadvantages of X-ray imaging include the radiation dose (especially for computed tomographic [CT] scans) and the toxicity of some of the contrast agents.

In the following sections, we first review the basics of X-ray physics: what X-rays are; how they can be produced; and how they interact with matter. Then, we describe planar (projectional) imaging with a focus on C-arm systems. Finally, we discuss CT, which provides cross-sectional and three-dimensional (3D) images. References to further reading are given at the end of the chapter.

## 11.2  What Are X-Rays?

X-rays are electromagnetic waves with a wavelength in vacuum in the range of 0.01 to 10 nm, corresponding to frequencies between $3 \times 10^{16}$ and $3 \times 10^{18}$ Hz. By comparison, the wavelength of visible light ranges from 380 to 750 nm. Typically, X-rays are characterized by the energy of their field quanta, the *photons*, in units of electron volts (eV). Their energy is linked to their wavelength by the following relation:

$$\text{Energy in eV} = \frac{1239.84}{\text{wavelength in nm}}$$

Because X-rays are about 50 to 50,000 times more energetic than visible light, they can penetrate the body of an animal or a human. X-rays are a form of ionizing radiation, meaning that they have enough energy to strip electrons from atoms and leave them ionized.

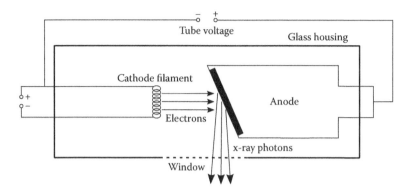

**FIGURE 11.1**
Schematic view of an X-ray tube.

## 11.3 Production of X-Rays and X-Ray Spectra

X-rays for diagnostic purpose are produced by shooting high-speed electrons onto a target material where the kinetic energy of the electrons is converted into energetic X-ray photons. The main components of an X-ray tube are shown in Figure 11.1. Electrons are produced by heating a filament through a low-voltage electric current. A large electric potential difference in the range of 60–140 kV between the cathode filament and the target anode produces a strong electric field that accelerates the electrons toward the anode. The inside of the tube is evacuated so that the electrons do not collide with gas molecules. The electrons hit the anode at high speed and release most of their energy as heat. About 1% of the electrons produce X-ray photons through one of two processes known as *bremsstrahlung* and *ionization*. Bremsstrahlung ("breaking radiation") occurs when an electron is deflected by the electric field of the atomic nucleus in the anode material, causing it to emit a photon (Figure 11.2a). The resulting photon energy spectrum is continuous with an upper boundary given by the cathode-anode potential difference (*tube voltage*). Ionization occurs when the incident electron knocks out an inner-shell electron from an anode atom (Figure 11.2b). The vacant space is filled by an electron from an outer shell. Since electrons in higher shells have more energy than electrons in lower shells, the energy difference is emitted as a photon. The energy spectrum in ionization is discrete because electrons in atomic orbits have quantized energies that are specific to the anode material ("characteristic spectrum"). The choice of the anode material, typically tungsten, molybdenum, or copper, is driven by the following considerations: (a) the probability for bremsstrahlung production grows quadratically with the atomic number Z of the anode material, making heavy elements desirable for effective X-ray production; (b) the differences in

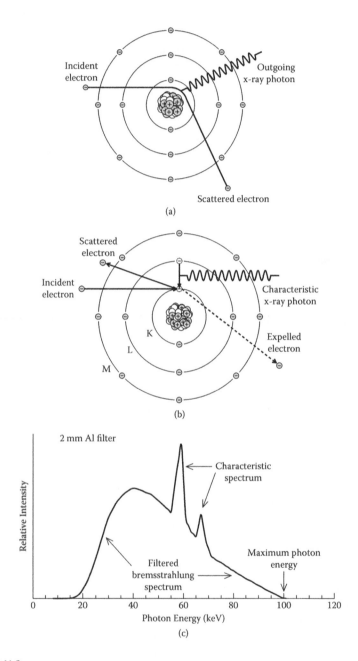

**FIGURE 11.2**
Production of X-ray photons through interaction of electrons with the atoms in the target material: (a) bremsstrahlung, (b) ionization by removal of an inner electron, (c) resulting X-ray spectrum as a function of the photon energy. The tube voltage was set to 100 keV. Low-energy photons (below 20 keV) are missing because of attenuation within the anode material and a 2-mm aluminum filter.

electron energy levels have to lie in the energy range of X-ray photons; and (c) the material must exhibit good heat conductivity. In addition, a filter is placed between the X-ray tube and the patient to absorb low-energy photons that would not reach the detectors. A typical X-ray spectrum emitted to the patient is shown in Figure 11.2c.

## 11.4 Interaction of X-Rays with Matter

When X-rays pass through matter, they can change direction and deposit part or all of their energy in the material. X-rays with energies appropriate for medical imaging interact with matter mainly through three processes: Rayleigh scattering, Compton scattering, and photoelectric absorption (Figure 11.3).

### 11.4.1 Rayleigh Scattering

The incident photon is absorbed by the electron shell, causing all electrons to oscillate in phase. This results in the emission of a photon of the same energy as the incident photon, but at a different angle (Figure 11.3a). Rayleigh scattering is more likely to occur for photon energies between 10 and 50 keV, and it is rare compared to the other two processes.

### 11.4.2 Compton Scattering

In Compton scattering, the incident photon strikes an outer-shell electron, ejecting it from the atom. The photon is emitted at an angle with respect to the incident direction and with lower energy since a fraction of the original energy is carried by the ejected electron (Figure 11.3b). The higher their original energy, the more likely the photons will be scattered in the forward direction. Compton scattering dominates most of the energy range relevant for medical imaging.

### 11.4.3 Photoelectric Absorption

When photoelectric absorption occurs, the incident photon is completely absorbed by the atom, and its energy is used to free an electron, typically from an inner shell (Figure 11.3c). For this to happen, the photon energy has to be larger than the electron-binding energy. The vacancy in the electron shell is filled by an electron from an outer shell, and the energy difference between the shells is emitted as a photon. The newly created vacancy in the outer shell is filled by an electron from another shell and so on. This cascade effect results in the emission of *characteristic* photons, which usually

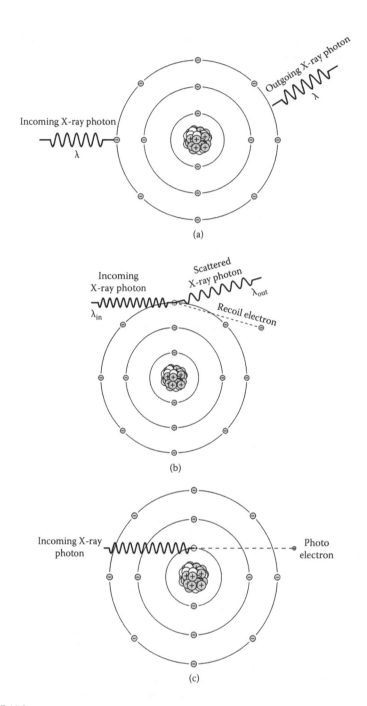

**FIGURE 11.3**
Interaction of X-rays with matter: (a) Rayleigh scattering; (b) Compton scattering; (c) photoelectric absorption.

are absorbed by the surrounding material. Photoelectric absorption is the dominant process in tissue for photons with energy below approximately 30 keV.

### 11.4.4 Implications for Medical Imaging

Compton scattering and photoelectric absorption dominate the interaction of diagnostic X-rays with organic matter such as soft tissue. The probabilities for these effects to occur depend on the incident X-ray energy and the atomic number Z of the penetrated material. To be able to distinguish different materials, the attenuation of the incident X-ray beam needs to be significantly different for different materials. If the X-ray energy is too high, most of the Compton-scattered photons traverse the body almost unchanged, leading to images with low contrast. Lower-energetic X-rays may not penetrate dense material, like bones, or thick materials, but they are suitable for imaging thin, soft tissues. Here, one can take advantage of the fact that absorption due to the photoelectric effect shows a sharp increase at energies around the element-specific energy level of the innermost electron orbit (i.e., K-edge imaging). Furthermore, the absorption probability for the photoelectric effect increases with the cube of the atomic number Z, so that small differences in tissue composition are amplified in the attenuation. The low-energy part of the X-ray spectrum is not useful for medical imaging since it is completely absorbed by the patient's body. To avoid this unnecessary radiation dose, a filter is placed between the X-ray tube and the patient. For planar or C-arm systems, the filter consists of a thin layer of copper or aluminum. For diagnostic CT systems (discussed in the section 11.6 on X-ray CT), a bow-tie-shaped filter is used.

### 11.4.5 Attenuation Coefficients

The attenuation of the intensity that an X-ray beam encounters when traveling through a uniform material depends exponentially on the thickness of the material:

$$I(x) = I_0 \, e^{-\alpha_{E_\gamma} x}$$

where $I_0$ is the incident intensity, $x$ is the path length of the X-rays through the object, and $\alpha_{E_\gamma}$ is the *linear attenuation coefficient*. The attenuation coefficient is specific for a material and depends on the photon energy as indicated by the subscript $E_\gamma$. Consequently, the transmitted X-ray spectrum contains information about both the material composition and the thickness of the structures inside the object. The linear attenuation coefficient of a material is determined by its specific interaction with X-rays as described. Therefore, it can be written as follows:

$$\alpha = \alpha_{\text{Rayleigh}} + \alpha_{\text{Compton}} + \alpha_{\text{Photoelectric}}$$

## 11.5  Planar X-Ray Imaging and C-arm Systems

The original and still widely used method to acquire an X-ray image consists of irradiating the object with X-rays from one side and capturing a single transmission image with a suitable recording device on the other side of the object. This technique produces a two-dimensional (2D) projection of the 3D object without depth information along the direction of projection. Traditional detectors consist of an image-intensifying phosphorescent layer and a photographic or charge-coupled device (CCD) camera. These bulky detectors have been replaced by much lighter and more compact *flat-panel detectors*. Planar (projection) X-ray imaging is frequently used to display single images of anatomical structures (bones, lungs), as well as for real-time monitoring of interventional or surgical procedures (*fluoroscopy*). A common application for fluoroscopy is the visualization of blood vessels and the perfusion of tissues through digital subtraction angiography (DSA). In this procedure, first a catheter is inserted through a blood vessel (artery) with the help of a guidewire. Projection X-ray images of 3–60 frames per second (fps) at a low dose allow the doctor to monitor the correct placement of the guidewire in real time. Next, a reference image with a higher dose is recorded, immediately followed by the injection of a radiopaque contrast agent through the catheter and the acquisition of a series of image frames. The vessels filled with the contrast agent become clearly visible in the X-ray images. To enhance the quality of the image, the reference image is automatically subtracted for each frame. This masks obstructing anatomical structures like bones while leaving volumes filled with the contrast agent visible (Figure 11.4). A radiologist can then identify regions of reduced blood flow and their cause (e.g., stenoses).

C-arm systems (Figure 11.5) acquire a series of planar images and are often used for interventional procedures when real-time imaging is required. Examples include angiography, needle biopsy, and transarterial chemoembolization. Due to their high spatial resolution and real-time image acquisition, C-arm systems are also well suited for monitoring the delivery and tracking of stem cells. C-arm systems enable X-ray imaging with large flexibility in spatial orientation. The X-ray source and detector are mounted on opposite ends of a C-shaped support frame, which can rotate around both its isocenter and its diameter. The C-arm can also be moved along and rotated around the patient table. This allows for almost any positioning of the imaging system around the patient and an optimal choice of the projection angle. The open design of C-arm system offers easy

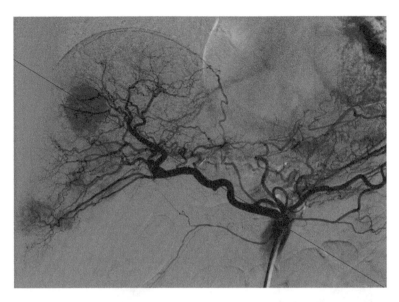

**FIGURE 11.4**
Digital subtraction angiography (DSA) for imaging neuroendocrine tumors in the liver. (Image courtesy of MingDe Lin, PhD, Philips Research, Briarcliff Manor, NY, and the Johns Hopkins Hospital, Baltimore, MD.)

access to the patient during the imaging process, which is important for interventional procedures. Modern C-arm systems are equipped with flat-panel detectors with up to 1,920 by 2,480 pixels on an area of 30 by 38 cm. Flat-panel detectors have a higher *detective quantum efficiency* (DQE) compared to image-intensifying detectors. This reduces the required patient dose. At full resolution, they can be read out at about 6 fps. In 2 × 2 (4 × 4) binning mode, the number of effective pixels is reduced by a factor of four (16), but allows a readout at 30 (60) fps. A combination of two C-arm systems can be used to produce simultaneous images from two different projection angles and provide better 3D orientation (biplane imaging). This setup is often used for interventional cardiovascular procedures.

An increasing use for C-arm systems is CT (C-arm CT or cone-beam CT), which introduces 3D imaging to interventional and surgical procedures. Similar to diagnostic CT (discussed in the next section), the 3D imaging is achieved by recording numerous projections while rotating the C-arm around. Due to the design, the maximum rotation angle is about 220° with a scan time of 5–8 s (Video 11.1 on web site editors). The size of the reconstructed volume is governed by the size of the flat-panel detector since the system is not translated along the table axis during a scan. The advantages of C-arm CT over diagnostic CT include higher spatial resolution, open access to the patient, and the ability to scan the patient in the procedure room without relocating the patient. However, the image quality of C-arm CT has not

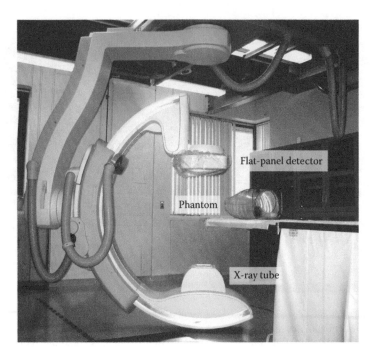

**FIGURE 11.5**
A modern X-ray C-arm system. The flat-panel detector is protected by a plastic cover. (Image taken by the authors at the Johns Hopkins Hospital, Baltimore, MD, with assistance from MingDe Lin, PhD, Philips Research, Briarcliff Manor, NY.)

yet reached that of diagnostic CT in terms of temporal resolution, contrast resolution, noise from scattered X-rays, and truncation artifacts due to the limited detector size.

## 11.6 X-Ray Computed Tomography

The lack of volume information in planar, 2D projection images can be overcome by measuring projections at many positions around an object (Figure 11.6a). From the projection data, cross-sectional images of the object can be reconstructed, commonly via *filtered backprojection*. This process is known as *X-ray computed tomography* (CT). Repeating this step for several cross-sectional slices yields a 3D image. In contemporary CT scanners, the detector elements are arranged in a single arc opposite the X-ray source, and the X-rays are emitted from the source in a fan-like (truncated *cone-beam*) geometry. Both the X-ray tube and the detector rotate around the object. The cross section of the human body roughly resembles an ellipse; therefore, X-rays in the boundary of the fan-beam traverse less material than those in the center. This results in a nonuniform intensity distribution (and hence nonuniform noise) along the detector. To reduce this nonuniformity and the necessary dose on the edges of the patient's body, a filter made of bow-tie-shaped aluminum is placed between the X-ray source and the patient.

In *helical CT*, the projections are acquired continuously while the table with the patient moves through the scanner, resulting in a helical path of the scanner-detector system around the patient (Figure 11.6b). Several slices can be projected simultaneously if the detector system consists of several detector arrays. Modern helical CT scanners with at least 64 detector arrays can achieve three revolutions per second with a table speed of 15 cm/s. This makes it possible to perform a complete heart scan in less than 1 s.

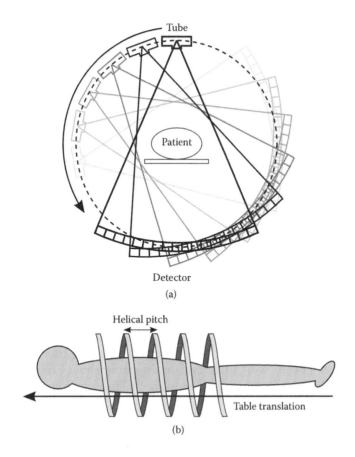

**FIGURE 11.6**
Principles of tomographic imaging. (a) Rotation of the X-ray tube and detector around the patient. (b) Combined with the table translation, the X-ray tube describes a helical motion around the patient.

### 11.6.1 Representation of CT Images

X-ray CT images typically are reconstructed with 512 by 512 pixels by 100–1,000 slices. As opposed to other imaging modalities, the pixel value of a CT image itself represents an absolute, physical quantity: the linear attenuation coefficient (i.e., the degree of opacity against the X-ray). By convention, the pixel values $\mu(x, y)$ are normalized to the attenuation coefficient of water $\mu_{water}$ and multiplied by a factor of a thousand:

$$I_{CT}(x, y) = 1,000 \times \frac{\mu(x, y) - \mu_{water}}{\mu_{water}}$$

Using this definition, water has a *CT number* or *Hounsfield unit (HU)* $I_{CT}$ of zero. Air has a value of about –1,000, soft tissues around 0–100, and bone on the order of 1,000. The display for structures of interest can be enhanced by restricting the display to a subrange of CT numbers and assigning the full range of available grayscales to this subrange. The range is defined by the central value ($L$ = level) and width ($W$ = window). Values of $L$ = 0 HU, $W$ = 500 HU are typically suitable for cardiac imaging; $L$ = 1,600 HU, $W$ = 400 HU for bone; and $L$ = –700 HU, $W$ = 2,000 HU for lung, although there are individual variations.

Usually, a CT scan is performed for a particular diagnostic purpose and focuses on a limited set of materials or organs (e.g., bone or brain). The visibility of the material of interest can be enhanced during the image reconstruction process by applying specific *kernels* (or reconstruction filters). Bone kernels increase the sharpness of the image by accentuating high (spatial) frequencies at the cost of additional noise. This is tolerable because of the high contrast present in bone images. Soft tissue kernels aim at reducing the noise by suppressing high frequencies.

### 11.6.2 Scan Parameters, Image Quality, and Artifacts

The operational parameters for a CT scan (and for planar imaging) are laid out in a *protocol*. The protocol defines the setting of the tube voltage, tube current, table speed, scan time, and other parameters. The tube voltage determines the X-ray energy and is expressed as the maximum (*peak*) energy that occurs in the X-ray spectrum (*peak kilovoltage*, kVp). The tube current sets the beam intensity and hence the radiation dose per time. It is defined in milliamperes (mA); however, the time-integrated intensity, expressed as milliampere seconds (mAs), is more common and defines the total dose. Many protocols exist, each optimized for a particular imaging goal. Typical values for tube voltage and current for a CT scan are 80 to 140 kVp and 50 to 800 mA, respectively.

Several factors have an impact on the quality of the reconstructed image. The spatial resolution is driven by the size of the detector elements, the detector collimation height, the number of projections per revolution, the helical pitch, the reconstruction kernel, and the image slice thickness. The contrast is determined by the X-ray energy and the patient size. In addition, several artifacts may degrade the accuracy of the reconstructed image. The most important ones are patient motion (e.g., from breathing) while taking projections, leading to inconsistent data for reconstruction and beam hardening due to the energy dependence of the attenuation coefficients. In the latter case, the incident X-ray beam contains photons from a large energy range. Since the attenuation decreases with the photon energy, the transmitted energy spectrum is shifted toward higher, "harder," energies. The reconstructed attenuation coefficient then depends on the length of the object and effectively appears to be smaller the longer the path is through the object. This leads to artifacts in the reconstructed image. Sophisticated mechanisms are implemented in commercial CT scanners to suppress these and other artifacts as they may lead to misinterpretation of physiological structures.

### 11.6.3 Additional X-Ray CT Applications

#### 11.6.3.1 Dual Energy

Scanners have been developed that contain two X-ray sources and two detectors that are displaced by 90°. In dual-energy mode, the two X-ray tubes operate at different peak kilovoltages. Since the linear attenuation coefficient is energy and material dependent, this mode allows determination of the absolute content of two *basis materials* in the object, such as bone and soft tissue. Handling a third basis material with high atomic number such as iodine or barium (thus with a K-edge in the energy absorption curve) requires additional assumptions and is the subject of ongoing research.

#### 11.6.3.2 CT Fluoroscopy

Multislice CT scans with up to 12 fps make it possible to perform thin-slab 3D imaging of interventional procedures in real time. These scans require special setups, speed-optimized reconstruction algorithms, and low-dose protocols. A common application for CT fluoroscopy is needle aspiration biopsy.

#### 11.6.3.3 Perfusion CT

Studies of regional blood flow are performed after intravenous injection of a contrast agent and sequential imaging using a standard CT scanner. Postprocessing software applications calculate blood flow and volume, for example, to help determine ischemic regions. The most common applications of this technique are cerebral and liver perfusion imaging.

## 11.7 Conclusions

X-rays provide an important and ubiquitous imaging modality that can be applied in planning, guiding, monitoring, and assessing medical procedures. Both planar and CT imaging are frequently used for fast, noninvasive diagnostics, interventional radiology, therapy, and many other medical applications. The introduction of radiopaque materials for stem cell labeling also makes it possible to use the most advanced X-ray imaging devices, such as the emerging C-arm CT technique to track stem cells.

> Video 11.1: Video of C-arm CT using a flat-panel X-ray angiographic unit (Axiom Artis dFA, Siemens AG, Forchheim, Germany) taken using standard manufacturer presets to obtain a dynamic subtraction angiogram. An initial 8-s, 220° mask run is obtained, followed by a second rotational angiogram after an injection of iodinated contrast using a power injector. (Courtesy of Dara L. Kraitchman, VMD, PhD.)

## Suggested Reading

J. T. Bushberg et al. 2002. *The Essential Physics of Medical Imaging* (2nd edition). Lippincott Williams & Wilkins, Philadelphia.

J. Hsieh. 2009. *Computed Tomography: Principles, Design, Artifacts, and Recent Advances.* (2nd edition). SPIE Press monograph Vol. PM 188. John Wiley & Sons, Inc., New York.

K. Iniewski. 2009. *Medical Imaging. Principles, Detectors, and Electronics.* Wiley-Interscience, Hoboken, NJ.

M. Mahesh. 2009. *MDCT Physics: The Basics—Technology, Image Quality and Radiation Dose.* Lippincott Williams & Wilkins, Philadelphia.

G. V. Pavlinsky. 2008. *Fundamentals of X-Ray Physics.* Cambridge International Science, Cambridge, UK.

# 12

## Radionuclide Cell-Labeling Methods

**Rong Zhou and Hui Qiao**

### CONTENTS

## 12.1 Introduction

Positron emission tomography (PET) and single-photon emission tomography (SPECT) detect radionuclides and their interactions with biochemical processes in living subjects. These radionuclide imaging modalities are routinely used in the clinic for a variety of diagnostic purposes. Due to their exquisite subnanomolar ($10^{-11}$ to $10^{-10}$ M) sensitivity, PET and SPECT are able to measure biological processes at very low concentrations. The mass of radionuclides can be injected in tracer quantities and generally does not affect the biological system of the host.

Technological developments of both PET and SPECT have led to the implementation of specialized systems for small-animal imaging, with much higher spatial resolution (<2 mm) [1–11]. Such systems, often referred to as μPET and μSPECT, have dramatically advanced the *in vivo* imaging in small-animal models and facilitated clinical translation of diagnostic or therapeutic strategies developed in these models.

The high sensitivity of radionuclide imaging makes it suitable for noninvasive tracking of stem cells. In this chapter, methods available to detect stem cells by radionuclide imaging are discussed. In general, they can be divided into *two* categories: (a) the direct labeling method and (b) the reporter gene approach [12]. The former involves loading radionuclides into stem cells, which are imaged for short-term (hours or days) tracking of cell distribution. The latter involves transfecting stem cells with a reporter gene whose expression is imaged and associated with cell survival or proliferation over a relatively long period of time (weeks or months). The two approaches are compared for their sensitivity, specificity, and feasibility for clinical implementations.

## 12.2 Direct Labeling of Stem Cells with Radionuclides

As a natural extension of the clinical practice in which autologous white blood cells are labeled with [$^{111}$In]oxyquinoline (oxine) [13,14] or [T-99m]hexamethylprophylene amine oxime (HMPAO) and are reinfused in the patient for localization of inflammations, a variety of radionuclides has been used for labeling stem cells as summarized in Table 12.1. [$^{18}$F]FDG (fluorodeoxyglucose) ($t_{1/2}$ = 110 minutes) has been applied in a number of labeling studies likely because this radionuclide is approved by the Food and Drug Administration (FDA) for diagnostic imaging of cancers, and as such, the regulation hurdle of using it for cell tracking might be more manageable. In general, radioisotopes with a relatively long decay half-life, such as, $^{111}$In ($t_{1/2}$ = 2.8 days) and $^{64}$Cu ($t_{1/2}$ = 12.7 hours) are preferred for tracking cells over a period of days.

The direct labeling method does not involve extensive manipulation of the stem cells beyond a short incubation period; thus, it is preferred for clinical implementation. The radionuclide is carried into the cells via a lipophilic chelator (in the case of [$^{111}$In]oxine) or by a transporter (in the case of $^{18}$F fluorodexoyglucose [FDG]), which governs the initial extraction of the tracer into the cells. Once inside the cells, a trapping mechanism reduces the lipophilicity of the molecule, and the isotope is retained. After incubation, the cells are washed to remove any unbound activity and are injected into the

**TABLE 12.1**

Radionuclides Used for Direct Labeling of Stem Cells

| | Half-Life | Detection Modality | References |
|---|---|---|---|
| [$^{111}$In]oxine | 2.8 days | SPECT | 15–17, 22, 24, 26, 27, 29, 95–97 |
| [$^{111}$In]tropolone | 2.8 days | SPECT | 20 |
| [$^{99m}$Tc]exametazime | 6.0 hours | SPECT | 98 |
| [$^{18}$F]FDG | 109.8 minutes | PET | 18, 21, 23, 25, 28, 99 |
| [$^{18}$F]SFB | 109.8 minutes | PET | 100 |
| [$^{18}$F]HFB | 109.8 minutes | PET | 101 |
| [$^{64}$Cu]PTSM | 12.7 hours | PET | 21 |

*Note:* FDG, 2-deoxy-2-fluoro-D-glucose; SFB, *N*-succinimidyl-4-fluorobenzoate; HFB, hexadecyl-4-fluorobenzoate; PTSM, pyruvaldehyde-bis(N4-methylthio semicarbazone).

host. Labeling efficiencies are closely dependent on the type of stem cell, the radionuclide and its concentration, and incubation time.

To localize the stem cells in target tissues, a dual-isotope study is usually performed to overlay the signals from labeled stem cells to tissue anatomies delineated by the second isotope [15–17]. As shown in Figure 12.1A, [111]In signals from cells were overlaid on myocardial wall anatomy delineated by [99mTc] sestamibi [18]. The autoradiographic confirmation of [111]In signal with the histological staining of stem cells is presented in Figure 12.1B–D. To facilitate the localization of stem cells in tissue anatomies, the SPECT or PET signals from labeled cells can be fused with high-resolution anatomical images from computed tomography (CT) [19] when a SPECT/CT or PET/CT hybrid scanner is available or coregistered with magnetic resonance (MR) images acquired sequentially from a stand-alone MR scanner as shown in Figure 12.1A.

A high intracellular concentration of the radionuclide can be obtained by adjusting the incubation time and concentration of the radionuclide in the labeling media. As such, the direct labeling method can achieve high detection sensitivity. A phantom study suggested that 10,000, 5,000, and

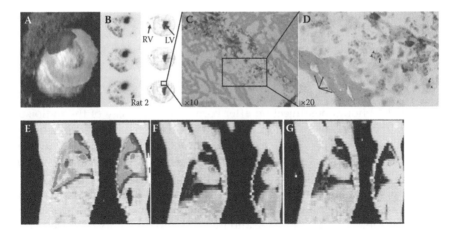

**FIGURE 12.1**
(A) Dual-isotope SPECT images and coregistration with MR images: [111]In-labeled stem cells were implanted in the infarcted rat heart; a dual-energy window detects simultaneously [99mTc] sestamibi (pseudocolored in yellow) and [111]In (blue) signals; the SPECT images were coregistered on MRI image (gray). (B) Autoradiographs of heart sections obtained after transplantation of [111]In and Feridex double-labeled stem cells in the infarcted myocardium of a rat. (C), (D) Prussian blue staining of iron for localization of the stem cells. LV, left ventricle; RV, right ventricle. (Reprinted from Zhou R, Acton PD, Ferrari VA. *Journal of American College of Cardiology* 2006;48:2094–2106 with permission.) (E)–(G) Homing of [111]In]-labeled BMSCs to the infarcted heart (in dogs) after intravenous injection at day 1 (E), day 2 (F), and day 7 (G) by SPECT imaging. For each panel, sagittal (left) and coronal (right) view of fused SPECT (color) and CT (gray) images are shown. Initial retention of cells in the lung is indicated by strong [111]In signal in the lung. (Reprinted from (Kraitchman DL, Tatsumi M, Gilson WD, Ishimori T, Kedziorek D, Walczak P, Segars WP, Chen H, et. al. Dynamic imaging of allogeneic mesenchymal stem cells trafficking to myocardial infarction, *Circulation*, 2005 112: 1454–1464.)

2,900 [$^{111}$In]-labeled cells could be detected, respectively, in a 16-, 32-, and 64-minute scan with appropriate background correction [20]. For cells labeled with [$^{64}$Cu]PTSM (PTSM) [pyruvaldehyde-bis(N4-methylthiosemi-carbazone], μPET detection of a few hundred cells has been reported [21].

The high sensitivity mediated by the direct labeling method allows one to probe the homing [17,22,23] and biodistribution [16–18,24,25] of stem cells. It has been reported that after intravenous, intra-left ventricular [26], intramyo-cardial, intracoronary, or interstitial retrograde coronary venous delivery [27], the majority of labeled stem cells were not retained in the heart. However, myocardial infarction increases the homing of endothelial progenitor cells to the heart significantly compared to nonischemic heart [26]. An elegant hom-ing study that demonstrated a higher sensitivity of radionuclide imaging (compared to Magnetic Resonance Imaging [MRI]) is shown in Figure 12.1E–G: bone marrow-derived mesenchymal stem cells were doubly labeled with [$^{111}$In]oxine and Feridex® and then injected intravenously into a dog model of myocardial infarction; the homing of these cells to the heart was detected by a clinical SPECT scanner. MRI, however, was not able to detect the small number of homed-in cells.

The direct labeling method has two inherent limitations: (a) dilution of labels on cell division may make some labeled cells invisible; and (b) efflux of labels from labeled cells may lead to invisibility of these cells. Consequently, this method is unable to distinguish whether reduction of signal is due to label loss from viable cells (efflux or on cell proliferation) or is due to removal of dead cells. Using [$^{111}$In]oxine, one-third of the original [$^{111}$In] signal was retained at 72 hours postinjection after correction for decay of the radioiso-tope [15]. In comparison, significant label loss occurs rapidly within 2 hours for [$^{18}$F]FDG labeling [21,28]. Therefore, the direct labeling method is suitable for short-term tracking to answer questions such as whether the delivery is successful and where the cells go.

Potential radiation damage to the cell posts an additional concern of direct radionuclide labeling method. In a study in which hematopoietic progenitor cells were labeled with [$^{111}$In]oxine and injected into the cavity of the left ven-tricle heart in a rat model of myocardial infarct [29], while gamma imaging revealed homing of the progenitor cells to infarcted myocardium, signifi-cant impairment of proliferation and function of labeled cells was observed. However, such damage can be minimized by adjusting the radioactivity during incubation, as demonstrated in a study by Jin et al., which suggested that 100% cell viability and 80% labeling efficiency can be achieved when the incubation activity of [$^{111}$In] does not exceed 0.9 MBq, corresponding to 0.14 Bq per cell [20].

In conclusion, the direct labeling approach is feasible for implementation in the clinic and is suitable for short-term tracking of the location and dis-tribution of injected cells with high sensitivity. However, dilution and efflux of the label would reduce the specificity of this method and prevent it from tracking the number of surviving cells over time. The reporter gene approach discussed next would allow monitoring of cell survival over a long period of time.

## 12.3 Tracking Stem Cells by Reporter Genes Detected by Radionuclide Imaging

Classic reporter genes include LacZ (encoding the enzyme β-galactosidase) and green fluorescence protein (GFP), both of which are detected on histological sections; *in vivo* detection of GFP is limited in small animals due to limited tissue penetration of the fluorescent light. While many investigators were involved in the development of reporter genes detectable by radionuclide imaging, the pioneering work and extensive development that established the utility of herpes simplex type 1 virus thymidine kinase (HSV1-tk) reporter gene by radionuclide imaging was performed by the Gambhir laboratory [30–43] and the Tjuvajev and Blasberg group [44–50].

The utility of radionuclide reporter gene for labeling stem cells was first demonstrated by Wu and colleagues for tracking stem cells injected in the heart [51,52], and this approach has proved successful in other laboratories as well [53,54]. While radionuclide imaging modalities are highly sensitive, for small animals reporter genes such as firefly luciferase have a higher sensitivity, and the luminescence imaging modality is more cost efficient compared to PET/SPECT. Therefore, optical reporter genes were applied for tracking stem cell survival in many studies by the group that pioneered the radionuclide reporter gene approach [55–62]. It should be noted, however, that due to limited tissue penetration, optical imaging is *not* suitable for humans or large animals except via endoscopic or intraoperative access.

Products of reporter gene expression generally can be divided into three categories: enzymes, receptors, and transporters. Diagrams in Figure 12.2 illustrate how these three reporters work for radionuclide imaging modalities using a representative in each category.

*HSV1-tk* expresses a viral thymidine kinase (*tk* is used to represent the gene, while *TK* represents the protein) (Figure 12.2A). The viral TK enzyme phosphorylates a range of substrates, including thymidine (the natural substrate), analogs of pyrimidine and acycloguanosines. The monophosphates that result from TK reaction are converted by cellular enzymes to di- and triphosphates, which are trapped inside the cells; in contrast, cells not expressing TK will not metabolize or retain the tracer to an appreciable degree. The metabolites of viral TK, if present at a sufficiently high concentration, result in cell death; therefore, HSV1-tk has been used as a suicide gene in gene therapy protocols [63–65] or as a negative selection marker to eliminate TK-expressing cells *in vitro* [66] when a pharmacological concentration of the substrate was reached. For *imaging* HSV1-tk expression, the substrate

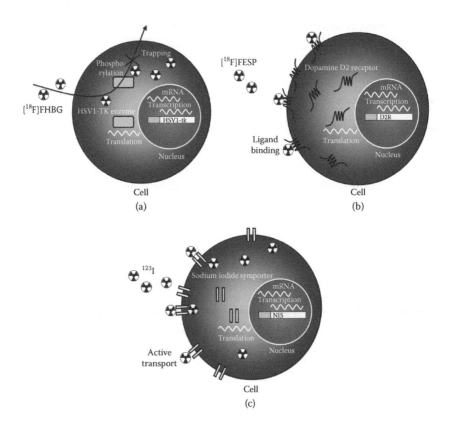

**FIGURE 12.2**

Schematic diagrams of three different types of reporter genes. (a) HSV1-tk expression results in the production of the enzyme HSV1-TK. This enzyme phosphorylates the radiotracer [18F] FHBG, causing it to become trapped inside the cells. (b) The D2R reporter gene produces the dopamine D2 receptor, which migrates to the cell surface. This seven-transmembrane receptor binds radioactive ligands, such as 18F-FESP. (c) The sodium-iodide symporter actively transports radioactive iodine, and other analogous tracers, into the cell. However, without subsequent organification, the trapping is incomplete, and some iodine may leak out. (Reprinted from Zhou R, Acton PD, Ferrari VA. *Journal of American College of Cardiology* 2006;48:2094–2106.)

is administered at tracer quantities to avoid toxicity to TK-expressing cells. To improve the detection sensitivity and specificity, a variety of pyrimidine analogs and acycloguanosine derivatives (which are substrates of HSV1-TK) have been designed that result in better tracer retention (see review article, Reference 67), and a mutant form of the wild-type HSV1-tk gene, HSV1-sr39tk, has also been obtained [68] that utilizes acycloguanosine substrates such as 18F FHBG more efficiently [38,69].

The reporter and suicidal properties of HSV1-tk have been utilized to visualize the embryonic stem cells (ESCs) grafted in the heart and to eliminate them when they were forming teratoma, respectively [52,70]. As shown in Figure 12.3, the PET signal was observed at week 1 in animals receiving 10

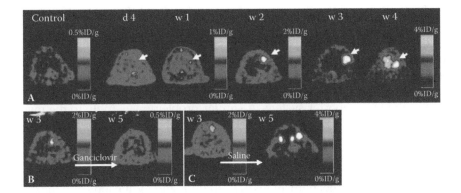

**FIGURE 12.3 (See color insert.)**
(A) *In vivo* tracking of stem cell survival and proliferation using the reporter gene method. Ten million ($10^7$) murine embryonic stem cells transfected with a truncated version of HSV1-tk were injected into the myocardium of a noninfarcted nude rat; PET was performed at day 4 and weeks 1, 2, 3, and 4. Approximately 1 mCi [$^{18}$F]9-[3-fluoro-1-hydroxy-2-(propoxymethyl)]guanine was injected intravenously for visualization of HSV1-tk-expressing cells. Positive signal was observed at 1 week after injection of cells, but control animals showed background activities only. Quantification of imaging signals showed a drastic increase of thymidine kinase activity from week 2 to week 4, corresponding to proliferation of embryonic stem cells into teratoma. (B), (C) After administration of a pharmacological dose of ganciclovir (50 mg/kg twice daily for 2 weeks), [$^{18}$F] PET signal disappeared, suggesting elimination of sr39tk-expressing cells, whereas [$^{18}$F] signal became stronger in saline-treated rats. (Reprinted from Cao F, Lin S, Xie X, Ray P, Patel M, Zhang X, Drukker M, Dylla SJ, Connolly AJ, Chen X, Weissman IL, Gambhir SS, Wu JC. *Circulation* 2006;113(7):1005–1014. With permission.)

million HSV1-sr39tk-expressing ESCs. Quantification of imaging signals showed a dramatic increase from week 2 to week 4, corresponding to proliferation of ESCs into cardiac teratoma. Treating the rat with a pharmacological dose of ganciclovir for 2 weeks apparently eliminated the teratoma because no [$^{18}$F] PET signal was detected thereafter.

To validate that the number of transplanted cells surviving in the heart is correlated with the [$^{18}$F] PET signal, Qiao et al. [71] utilized a sex-mismatched transplantation scheme in which male ESCs were injected into female recipient hearts; in this scheme, the number of transplanted cells can be quantified by estimation of the male Sry gene using the real-time polymerase chain reaction (RT-PCR), an accurate but terminal method. The results showed a strong correlation between the *in vivo* [$^{18}$F] PET signal versus the number of surviving ESCs in the heart estimated by RT-PCR.

*Dopamine type 2 receptor (D2R)* (Figure 12.2B) is one of the few reporter genes that could be used for cell tracking in the brain because the reporter probe, for example, [$^{18}$F]fluoroethylspiperone (FESP), readily crosses the intact blood–brain barrier [72]. FESP binds to the D2R with high affinity and has been shown to provide a quantitative measure of D2R expression in living animals [73]. To overcome the potential problems created by the occupancy of the ectopic D2R by the endogenous agonist, which may increase levels of cellular cyclic adenosine monophosphate (cAMP) and lead to physiologic consequences on the target tissue/cells [74], a mutant strain of the D2R, D2R80A, has been developed [75]. D2R80A uncouples FESP binding from intracellular signaling, eliminating physiological responses to FESP. In the brain, the presence of endogenous D2R, particularly in the striatum, would introduce a high background signal, making it difficult to resolve the presence of the D2R reporter system. A similar problem may apply to the myocardium, in which expression of a detectable level of D2R is reported [76].

The *sodium-iodide symporter (NIS)* is a membrane glycoprotein and expresses in high concentrations in the thyroid and lower concentrations in salivary glands, stomach, thymus, breast, and other tissues. The symporter provides an active transport mechanism to carry sodium and iodine ions into the cell. As shown in Figure 12.2C, the NIS reporter gene is detected by radionuclide imaging using [$^{123}$I]iodine or [$^{99m}$Tc]pertechnetate for SPECT and [$^{124}$I]iodine for PET [77–82]. PET imaging of stem cells transfected by NIS reporter has been reported [53]. The endogenous expression of NIS is limited to a few tissues, allowing it to be ectopically expressed in a variety of other tissues and organs, such as the heart. However, the use of NIS as a reporter system could be hampered by the efflux of the radiotracers from nonthyroid tissues. In the thyroid, organification of the iodine occurs, catalyzed by the enzyme thyroperoxidase, which effectively traps the iodine after it is transported into the tissue. Coexpression of thyroperoxidase with the NIS reporter gene may improve the retention of radioactive iodine in the transfected cells [83].

*Other reporters detected by radionuclide imaging* include the following: (a) human mitochondrial thymidine kinase 2 (hTK2), which is detected using the same radionuclides as for HSV1-tk; (b) human somatostatin receptors (hSSTRs), which are detected by radiolabeled derivatives of somatostatin receptor agonist and antagonists, such as [$^{64}$Cu]TETA-octreotide [84]; and (c) human norepinephrine transporter (hNET), which is detected by [$^{124}$I] metaiodobenzylguanine (MIBG) [85] or [$^{11}$C]m-hydroxyephedrine [86]. These reporters can potentially be used for labeling cells; for instance, T-lymphocytes expressing hNET have been detected by *in vivo* PET imaging [87].

*Detection sensitivity of reporter gene labeled cells* is determined by factors that affect the concentration of the radiotracer retained by the cell as the result of reporter gene expression. These factors include (a) the type of reporter gene, (b) the level of reporter gene expression, and (c) availability of radiotracer to the stem cells after the tracer is injected systemically. Tracer availability will not be limited if the reporter gene-expressing cells are preincubated with the tracer *in vitro* before cell injection [88].

Whether the reporter gene is expressed as an enzyme, receptor, or transporter (Figure 12.2) could affect its sensitivity. In general, enzyme-based reporter genes have the advantage of signal amplification since the enzyme can continuously phosphorylate (in case of TK) the tracer, leading to accumulation of radioactivity inside the cell. This advantage may not be applicable to a transporter-based reporter gene if the cell lacks mechanisms to retain the tracer, as in the case of NIS. Receptor-based reporter genes could also have a low sensitivity if the receptor is capable of binding only one reporter probe. However, other receptors, such as low-density lipoprotein receptor, on binding to their ligands, undergo endocytosis and recycle to cell membrane in several minutes to internalize more ligands.

The level of reporter gene expression ultimately determines its detection sensitivity. Since gene expression is closely regulated by the strength of the promoter, to enhance the sensitivity of reporter gene detection, a strong, constitutively active promoter is often used. A reporter gene driven by such a promoter is minimally regulated by physiological processes in the cell and is always expressed as long as the cell is alive. In comparison, a tissue-specific promoter is usually weaker. A tissue-specific reporter gene can therefore monitor stem cell differentiation *in vitro* [89], but its relatively low sensitivity may not permit monitoring of differentiation *in vivo*. A much higher level of the reporter gene expression is obtained in transiently transfected stem cells than stably transfected cells. For the former, multiple copies of the transgene are present in the cytosol, whereas for the latter, usually a single copy of the reporter gene construct integrates into the cell genome. However, stable transfection is necessary to ensure that the reporter gene will not be lost over time or on cell division [90].

The availability of the radiotracer to the cell is also important. For example, transporting of radiolabeled substrate into HSV1-tk-expressing cells is mediated by nucleoside transporters, and this is the rate-limiting step for intracellular accumulation of the tracer. For many applications, stem cells are directly injected into the target tissue or organ, and as such, how well the injection site is perfused (blood flow and capillary density) will decide how many tracer molecules will be available to injected cells. For example, stem cells injected into poorly perfused (ischemic) myocardium could compete with each other for the limited amount of tracer that diffuses from nearby capillaries. The demand for tracer by a large number of stem cells may exceed what is available at the injection site.

To estimate the sensitivity of PET detection of reporter gene-expressing stem cells in the heart, 5 million ESCs stably transfected with HSV1-sr39tk were injected into the myocardial wall of rats, and the survival kinetics of these cells was studied by serial PET imaging and, after each imaging time point, by RT-PCR to estimate the number of surviving cells [71]. Figure 12.4A shows the time course of the image contrast, which was well below 1 on days 0 (3–5 hours after cell injection), 1, 2, and 3 but sharply increased to 1.3 on day 4, when a hot spot in the heart was first visualized on [$^{18}$F] PET images. The hot spot was indeed localized in the myocardium, as revealed by overlaying the [$^{18}$F]FHBG and [$^{13}$N]ammonia images (Figure 12.4B and 12.4C). RT-PCR estimated the number of surviving cells to be 3.4 million at this time. Phantom studies further suggested that the activity ratio between the hot spot and surrounding myocardium could reach or exceed 8:1 (Figure 12.4F). While such a high activity ratio would ensure robust detection, a threshold

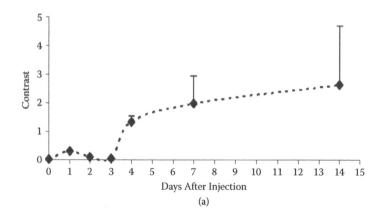

(a)

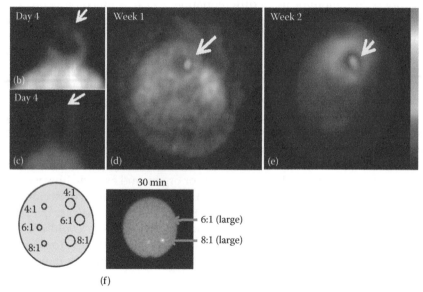

**FIGURE 12.4**

(a) Contrast of [$^{18}$F] images acquired at days 0–14 was estimated by , where *VOIh* is the volume of interest encompassing the hot spot, while *VOIref* was placed on reference (septum) region of the myocardial wall. Images were shown for day 4 (b) and (c), week 1 (c), and week 2 (e) after injection of 5 million embryonic stem cells stably expressing sr39tk. [$^{13}$N]ammonia image (shown in grayscale) provides the anatomical reference of the heart (myocardial wall in particular) to facilitate the localization of the $^{18}$F signals (color scale). (f) A phantom study was performed to aid the estimation of the *in vivo* uptake ratio. Three small capillaries (ID = 0.75 mm) and three large capillaries (ID = 1.75 mm) containing the activity concentration of 4:1, 6:1, and 8:1, respectively, were scanned for 30 min. The contrast value of 1.3 was obtained from the large capillary with 8:1 concentration activity; the contrast matched that on day 4 [$^{18}$F] PET images acquired with the same scanning time, suggesting that activity ratio between the hot spot and surrounding myocardium could be 8:1 or higher. (Reprinted from Qiao H, Surti S, Choi SR, Raju K, Zhang H, Ponde DE, Kung HF, Karp JS, Zhou R. *Molecular Imaging and Biology* 2009;11(6):408–14. With permission.)

of 3.4 million suggested relatively low detection sensitivity, which could be attributed to stable transfection of the reporter gene and the limited availability of tracer at the injection site containing a large number of cells. The observation that [$^{18}$F] signals were detected only 4 days after injection was consistent with previous studies performed under similar conditions [52,54].

## 12.4 Limitations of Reporter Gene Approach

First, the inevitable molecular manipulation of the stem cells is a major concern for the reporter gene approach. Since the reporter gene is usually randomly inserted into the cell genome, the consequence of its expression is unpredictable. A proteomic study reported that stable expression of a trifusion protein in ESCs had no significant changes in protein quantity compared to the untransfected counterparts and had no effect on cell viability, proliferation, and capability of cardiac differentiation [91]. However, downregulation of genes involved in cell cycling, cell death, and nucleic acid metabolism and upregulation of genes for homeostasis and antiapoptosis were also observed [92]. Second, the immunogenicity of cells expressing a xenogeneic reporter gene such as HSV1-tk limits their survival and clinical utility. Human reporter genes such as hTK2 and hNIS are expected to play a leading role in clinical applications of the reporter gene method in gene and cell therapy [93]. Third, epigenetic modulations of reporter gene may attenuate its expression over time [94].

## 12.5 Conclusion

As a complement to the direct labeling method, the reporter gene approach is capable of tracking cell survival and may provide a useful tool (in the case of HSV1-tk) for eliminating the cell when necessary. While apparently safe in many animal studies, the long-term safety in humans is still a major concern that is expected to limit the clinical implementation of the reporter gene approach in the near future.

## References

1. Cherry SR, Shao Y, Silverman RW, Meadors K, Siegel S, Chatziioannou A, Young JW, Jones WF, Moyers JC, Newport D, Boutefnouchet A, Farquhar TH, Andreaco M, Paulus MJ, Binkley DM, Nutt R, Phelps ME. MicroPET: a high resolution PET scanner for imaging small animals. *IEEE Trans Nucl Sci* 1997; 44(3):1161–1166.
2. Weber DA, Ivanovic M. Pinhole SPECT—ultra-high-resolution imaging for small animal studies. *J Nucl Med* 1995; 36(12):2287–2289.
3. Ishizu K, Mukai T, Yonekura Y, Pagani M, Fujita T, Magata Y, Nishizawa S, Tamaki N, Shibasaki H, Konishi J. Ultra-high resolution SPECT system using four pinhole collimators for small animal studies. *J Nucl Med* 1995; 36(12):2282–2287.
4. Chatziioannou AF, Cherry SR, Shao Y, Silverman RW, Meadors K, Farquhar TH, Pedarsani M, Phelps ME. Performance evaluation of microPET: a high-resolution lutetium oxyorthosilicate PET scanner for animal imaging. *J Nucl Med* 1999; 40(7):1164–1175.
5. MacDonald LR, Patt BE, Iwanczyk JS, Tsui BMW, Wang YC, Frey EC, Wessell DE, Acton PD, Kung HF. Pinhole SPECT of mice using the LumaGEM gamma camera. *IEEE Trans Nucl Sci* 2001; 48(3):830–836.
6. Green MV, Seidel J, Vaquero JJ, Jagoda E, Lee I, Eckelman WC. High resolution PET, SPECT and projection imaging in small animals. *Comput Med Imaging Graph* 2001; 25(2):79–86.
7. McElroy DP, MacDonald LR, Beekman FJ, Wang YC, Patt BE, Iwanczyk JS, Tsui BMW, Hoffman EJ. Performance evaluation of A-SPECT: a high resolution desktop pinhole SPECT system for imaging small animals. *IEEE Trans Nucl Sci* 2002; 49(5):2139–2147.
8. Surti S, Karp JS, Perkins AE, Freifelder R, Muehllehner G. Design evaluation of A-PET: a high sensitivity animal PET camera. *IEEE Trans Nucl Sci* 2003; 50(5, pt.2):1357–1363.
9. Acton PD, Kung HF. Small animal imaging with high resolution single photon emission tomography. *Nucl Med Biol* 2003; 30(8):889–895.
10. Yang YF, Tai YC, Siegel S, Newport DF, Bai B, Li QZ, Leahy RM, Cherry SR. Optimization and performance evaluation of the microPET II scanner for in vivo small-animal imaging. *Phys Med Biol* 2004; 49(12):2527–2545.
11. Beekman FJ, van der Have F, Vastenhouw B, van der Linden AJA, van Rijk PP, Burbach JPH, Smidt MP. U-SPECT-I: a novel system for submillimeter-resolution tomography with radiolabeled molecules in mice. *J Nucl Med* 2005; 46(7):1194–1200.
12. Zhou R, Acton PD, Ferrari VA. Imaging stem cells implanted in infarcted myocardium. *J Am Coll Cardiol* 2006; 48:2094–2106.
13. Peters AM, Saverymuttu SH. The value of indium-labelled leucocytes in clinical practice. *Blood Rev* 1987; 1(1):65–76.
14. Dutton JA, Bird NJ, Skehan SJ, Peters AM. Evaluation of a 3-hour indium-111 leukocyte image as a surrogate for a technetium-99m nanocolloid marrow scan in the diagnosis of orthopedic infection. *Clin Nucl Med* 2004; 29(8):469–474.
15. Zhou R, Thomas DH, Qiao H, Bal HS, Choi S-R, Alavi A, Ferrari VA, Kung HF, Acton PD. In vivo detection of stem cells grafted in infarcted rat myocardium. *J Nucl Med* 2005; 46(5):816–822.

16. Tran N, Franken PR, Maskali F, Nloga J, Maureira P, Poussier S, Groubatch F, Vanhove C, Villemot J-P, Marie P-Y. Intramyocardial implantation of bone marrow-derived stem cells enhances perfusion in chronic myocardial infarction: Dependency on initial perfusion depth and follow-up assessed by gated pinhole SPECT. *J Nucl Med* 2007; 48(3):405–412.

17. Caveliers V, De Keulenaer G, Everaert H, Van Riet I, Van Camp G, Verheye S, Roland J, Schoors D, Franken PR, Schots R. In vivo visualization of [111]In labeled CD133[+] peripheral blood stem cells after intracoronary administration in patients with chronic ischemic heart disease. *Q J Nucl Med Mol Imaging* 2007; 51(1):61–66.

18. Shen D, Liu D, Cao Z, Acton PD, Zhou R. Co-registration of MR and SPECT images for non-invasive localization of stem cells grafted in the infarcted rat myocardium. *Mol Imaging Biol* 2007; 9(1):24–31.

19. Kang WJ, Kang HJ, Kim HS, Chung JK, Lee MC, Lee DS. Tissue distribution of 18F-FDG-labeled peripheral hematopoietic stem cells after intracoronary administration in patients with myocardial infarction. *J Nucl Med* 2006; 47(8):1295–1301.

20. Jin Y, Kong H, Stodilka RZ, Wells RG, Zabel P, Merrifield PA, Sykes J, Prato FS. Determining the minimum number of detectable cardiac-transplanted [111]In-tropolone-labelled bone-marrow-derived mesenchymal stem cells by SPECT. *Phys Med Biol* 2005; 50(19):4445–4455.

21. Adonai N, Nguyen KN, Walsh J, Iyer M, Toyokuni T, Phelps ME, McCarthy T, McCarthy DW, Gambhir SS. Ex vivo cell labeling with [64]Cu-pyruvaldehyde-bis(N4-methylthiosemicarbazone) for imaging cell trafficking in mice with positron-emission tomography. *Proc Natl Acad Sci USA* 2002; 99(5):3030–3035.

22. Kraitchman DL, Tatsumi M, Gilson WD, Ishimori T, Kedziorek D, Walczak P, Segars WP, Chen HH, Fritzges D, Izbudak I, Young RG, Marcelino M, Pittenger MF, Solaiyappan M, Boston RC, Tsui BM, Wahl RL, Bulte JW. Dynamic imaging of allogeneic mesenchymal stem cells trafficking to myocardial infarction. *Circulation* 2005; 112(10):1451–1461.

23. Blocklet D, Toungouz M, Berkenboom G, Lambermont M, Unger P, Preumont N, Stoupel E, Egrise D, Degaute JP, Goldman M, Goldman S. Myocardial homing of nonmobilized peripheral-blood CD34[+] cells after intracoronary injection. *Stem Cells* 2006; 24(2):333–336.

24. Gao J, Dennis JE, Muzic RF, Lundberg M, Caplan AI. The dynamic in vivo distribution of bone marrow-derived mesenchymal stem cells after infusion. *Cells Tissues Organs* 2001; 169(1):12–20.

25. Qian H, Yang Y, Huang J, Gao R, Dou K, Yang G, Li J, Shen R, He Z, Lu M, Zhao S. Intracoronary delivery of autologous bone marrow mononuclear cells radiolabeled by 18F-fluoro-deoxy-glucose: tissue distribution and impact on postinfarct swine hearts. *J Cell Biochem* 2007; 102(1):64–74.

26. Aicher A, Brenner W, Zuhayra M, Badorff C, Massoudi S, Assmus B, Eckey T, Henze E, Zeiher AM, Dimmeler S. Assessment of the tissue distribution of transplanted human endothelial progenitor cells by radioactive labeling. *Circulation* 2003; 107(16):2134–2139.

27. Hou D, Youssef EA-S, Brinton TJ, Zhang P, Rogers P, Price ET, Yeung AC, Johnstone BH, Yock PG, March KL. Radiolabeled cell distribution after intramyocardial, intracoronary, and interstitial retrograde coronary venous delivery: Implications for current clinical trials. *Circulation* 2005; 112(9suppl):I-150–I-156.

28. Doyle B, Kemp BJ, Chareonthaitawee P, Reed C, Schmeckpeper J, Sorajja P, Russell S, Araoz P, Riederer SJ, Caplice NM. Dynamic tracking during intracoronary injection of [18]F-FDG-labeled progenitor cell therapy for acute myocardial infarction. *J Nucl Med* 2007; 48(10):1708–1714.

29. Brenner W, Aicher A, Eckey T, Massoudi S, Zuhayra M, Koehl U, Heeschen C, Kampen WU, Zeiher AM, Dimmeler S, Henze E. [111]In-labeled CD34[+] hematopoietic progenitor cells in a rat myocardial infarction model. *J Nucl Med* 2004; 45(3):512–518.

30. Gambhir SS, Barrio JR, Wu L, Iyer M, Namavari M, Satyamurthy N, Bauer E, Parrish C, MacLaren DC, Borghei AR, Green LA, Sharfstein S, Berk AJ, Cherry SR, Phelps ME, Herschman HR. Imaging of adenoviral-directed herpes simplex virus type 1 thymidine kinase reporter gene expression in mice with radiolabeled ganciclovir. *J Nucl Med* 1998; 39(11):2003–2011.

31. Gambhir SS, Barrio JR, Herschman HR, Phelps ME. Assays for noninvasive imaging of reporter gene expression. *Nucl MediBiol* 1999; 26(5):481–490.

32. Gambhir SS, Herschman HR, Cherry SR, Barrio JR, Satyamurthy N, Toyokuni T, Phelps ME, Larson SM, Balatoni J, Finn R, Sadelain M, Tjuvalev J, Blasberg R. Imaging transgene expression with radionuclide imaging technologies. *Neoplasia* 2000; 2(1–2):118–138.

33. Yaghoubi SS, Wu L, Liang Q, Toyokuni T, Barrio JR, Namavari M, Satyamurthy N, Phelps ME, Herschman HR, Gambhir SS. Direct correlation between positron emission tomographic images of two reporter genes delivered by two distinct adenoviral vectors. *Gene Ther* 2001; 8(14):1072–1080.

34. Sun X, Annala AJ, Yaghoubi SS, Barrio JR, Nguyen KN, Toyokuni T, Satyamurthy N, Namavari M, Phelps ME, Herschman HR, Gambhir SS. Quantitative imaging of gene induction in living animals. *Gene Ther* 2001; 8(20):1572–1579.

35. Wu JC, Inubushi M, Sundaresan G, Schelbert HR, Gambhir SS. Positron emission tomography imaging of cardiac reporter gene expression in living rats. *Circulation* 2002; 106(2):180–183.

36. Inubushi M, Wu JC, Gambhir SS, Sundaresan G, Satyamurthy N, Namavari M, Yee S, Barrio JR, Stout D, Chatziioannou AF, Wu L, Schelbert HR. Positron-emission tomography reporter gene expression imaging in rat myocardium. *Circulation* 2003; 107(2):326–332.

37. De A, Lewis XZ, Gambhir SS. Noninvasive imaging of lentiviral-mediated reporter gene expression in living mice. *Mol Ther* 2003; 7(5 Pt 1):681–691.

38. Min JJ, Iyer M, Gambhir SS. Comparison of [(18)F]FHBG and [(14)C]FIAU for imaging of HSV1-tk reporter gene expression: adenoviral infection vs stable transfection. *Eur J Nucl Med Mol Imaging* 2003; 10:10.

39. Chen IY, Wu JC, Min JJ, Sundaresan G, Lewis X, Liang Q, Herschman HR, Gambhir SS. Micro-positron emission tomography imaging of cardiac gene expression in rats using bicistronic adenoviral vector-mediated gene delivery. *Circulation* 2004; 109(11):1415–1420.

40. Green LA, Nguyen K, Berenji B, Iyer M, Bauer E, Barrio JR, Namavari M, Satyamurthy N, Gambhir SS. A tracer kinetic model for [18]F-FHBG for quantitating herpes simplex virus type 1 thymidine kinase reporter gene expression in living animals using PET. *J Nucl Med* 2004; 45(9):1560–1570.

41. Wu JC, Chen IY, Wang Y, Tseng JR, Chhabra A, Salek M, Min J-J, Fishbein MC, Crystal R, Gambhir SS. Molecular imaging of the kinetics of vascular endothelial growth factor gene expression in ischemic myocardium. *Circulation* 2004; 110(6):685–691.

42. Rodriguez-Porcel M, Brinton TJ, Chen IY, Gheysens O, Lyons J, Ikeno F, Willmann JK, Wu L, Wu JC, Yeung AC, Yock P, Gambhir SS. Reporter gene imaging following percutaneous delivery in swine moving toward clinical applications. *J Am Coll Cardiol* 2008; 51(5):595–597.

43. Willmann JK, Paulmurugan R, Rodriguez-Porcel M, Stein W, Brinton TJ, Connolly AJ, Nielsen CH, Lutz AM, Lyons J, Ikeno F, Suzuki Y, Rosenberg J, Chen IY, Wu JC, Yeung AC, Yock P, Robbins RC, Gambhir SS. Imaging gene expression in human mesenchymal stem cells: From small to large animals. *Radiology* 2009:2513081616.

44. Tjuvajev JG, Stockhammer G, Desai R, Uehara H, Watanabe K, Gansbacher B, Blasberg RG. Imaging the expression of transfected genes in vivo. *Cancer Res* 1995; 55(24):6126–6132.

45. Tjuvajev JG, Finn R, Watanabe K, Joshi R, Oku T, Kennedy J, Beattie B, Koutcher J, Larson S, Blasberg RG. Noninvasive imaging of herpes virus thymidine kinase gene transfer and expression: A potential method for monitoring clinical gene therapy. *Cancer Res* 1996; 56(18):4087–4095.

46. Tjuvajev JG, Avril N, Oku T, Sasajima T, Miyagawa T, Joshi R, Safer M, Beattie B, DiResta G, Daghighian F, Augensen F, Koutcher J, Zweit J, Humm J, Larson SM, Finn R, Blasberg R. Imaging herpes virus thymidine kinase gene transfer and expression by positron emission tomography. *Cancer Res* 1998; 58(19):4333–4341.

47. Tjuvajev JG, Chen SH, Joshi A, Joshi R, Guo ZS, Balatoni J, Ballon D, Koutcher J, Finn R, Woo SL, Blasberg RG. Imaging adenoviral-mediated herpes virus thymidine kinase gene transfer and expression in vivo. *Cancer Res* 1999; 59(20):5186–5193.

48. Tjuvajev JG, Joshi A, Callegari J, Lindsley L, Joshi R, Balatoni J, Finn R, Larson SM, Sadelain M, Blasberg RG. A general approach to the non-invasive imaging of transgenes using cis-linked herpes simplex virus thymidine kinase. *Neoplasia* 1999; 1(4):315–320.

49. Blasberg RG, Tjuvajev JG. Herpes simplex virus thymidine kinase as a marker/reporter gene for PET imaging of gene therapy. *Q J Nucl Med* 1999; 43(2):163–169.

50. Koehne G, Doubrovin M, Doubrovina E, Zanzonico P, Gallardo HF, Ivanova A, Balatoni J, Teruya-Feldstein J, Heller G, May C, Ponomarev V, Ruan S, Finn R, Blasberg RG, Bornmann W, Riviere I, Sadelain M, O'Reilly RJ, Larson SM, Tjuvajev JG. Serial in vivo imaging of the targeted migration of human HSV-TK-transduced antigen-specific lymphocytes. *Nat Biotechnol* 2003; 21(4):405–413.

51. Wu JC, Chen IY, Sundaresan G, Min JJ, De A, Qiao JH, Fishbein MC, Gambhir SS. Molecular imaging of cardiac cell transplantation in living animals using optical bioluminescence and positron emission tomography. *Circulation* 2003; 108(11):1302–1305.

52. Cao F, Lin S, Xie X, Ray P, Patel M, Zhang X, Drukker M, Dylla SJ, Connolly AJ, Chen X, Weissman IL, Gambhir SS, Wu JC. In vivo visualization of embryonic stem cell survival, proliferation, and migration after cardiac delivery. *Circulation* 2006; 113(7):1005–1014.

53. Terrovitis J, Kwok KF, Lautamäki R, Engles JM, Barth AS, Kizana E, Miake J, Leppo MK, Fox J, Seidel J, Pomper M, Wahl RL, Tsui B, Bengel F, Marbán E, Abraham MR. Ectopic expression of the sodium-iodide symporter enables imaging of transplanted cardiac stem cells in vivo by single-photon emission computed tomography or positron emission tomography. *J Am Coll Cardiol* 2008; 52(20):1652–1660.

54. Qiao H, Zhang H, Zheng Y, Ponde DE, Shen D, Gao F, Bakken AB, Schmitz A, Kung HF, Ferrari VA, Zhou R. Embryonic stem cell grafting in normal and infarcted myocardium: serial assessment with MR imaging and PET dual detection. *Radiology* 2009; 250(3):821–829.

55. Sheikh AY, Lin SA, Cao F, Cao Y, van der Bogt KE, Chu P, Chang CP, Contag CH, Robbins RC, Wu JC. Molecular imaging of bone marrow mononuclear cell homing and engraftment in ischemic myocardium. *Stem Cells* 2007; 25(10):2677–2684.

56. van der Bogt KEA, Sheikh AY, Schrepfer S, Hoyt G, Cao F, Ransohoff KJ, Swijnenburg R-J, Pearl J, Lee A, Fischbein M, Contag CH, Robbins RC, Wu JC. Comparison of different adult stem cell types for treatment of myocardial ischemia. *Circulation* 2008; 118(14 suppl 1):S121–S129.

57. Swijnenburg RJ, Schrepfer S, Cao F, Pearl JI, Xie X, Connolly AJ, Robbins RC, Wu JC. In vivo imaging of embryonic stem cells reveals patterns of survival and immune rejection following transplantation. *Stem Cells Dev* 2008; 17(6):1023–1029.

58. Swijnenburg RJ, Schrepfer S, Govaert JA, Cao F, Ransohoff K, Sheikh AY, Haddad M, Connolly AJ, Davis MM, Robbins RC, Wu JC. Immunosuppressive therapy mitigates immunological rejection of human embryonic stem cell xenografts. *Proc Natl Acad Sci USA* 2008; 105(35):12991–12996.

59. Li Z, Wu JC, Sheikh AY, Kraft D, Cao F, Xie X, Patel M, Gambhir SS, Robbins RC, Cooke JP, Wu JC. Differentiation, survival, and function of embryonic stem cell derived endothelial cells for ischemic heart disease. *Circulation* 2007; 116(11 suppl):I-46–I-54.

60. Li Z, Suzuki Y, Huang M, Cao F, Xie X, Connolly AJ, Yang PC, Wu JC. Comparison of reporter gene and iron particle labeling for tracking fate of human embryonic stem cells and differentiated endothelial cells in living subjects. *Stem Cells* 2008; 26(4):864–873.

61. Cao F, Sadrzadeh Rafie AH, Abilez OJ, Wang H, Blundo JT, Pruitt B, Zarins C, Wu JC. In vivo imaging and evaluation of different biomatrices for improvement of stem cell survival. *J Tissue Eng Regen Med* 2007; 1(6):465–468.

62. Cao F, van der Bogt KE, Sadrzadeh A, Xie X, Sheikh AY, Wang H, Connolly AJ, Robbins RC, Wu JC. Spatial and temporal kinetics of teratoma formation from murine embryonic stem cell transplantation. *Stem Cells Dev* 2007; 16(6):883–891.

63. Eck SL, Alavi JB, Alavi A, Davis A, Hackney D, Judy K, Mollman J, Phillips PC, Wheeldon EB, Wilson JM. Treatment of advanced CNS malignancies with the recombinant adenovirus H5.010RSVTK: A phase I trial. *Hum Gene Ther* 1996; 7(12):1465–1482.

64. Shand N, Weber F, Mariani L, Bernstein M, Gianella-Borradori A, Long Z, Sorensen AG, Barbier N. A phase 1–2 clinical trial of gene therapy for recurrent glioblastoma multiforme by tumor transduction with the herpes simplex thymidine kinase gene followed by ganciclovir. GLI328 European-Canadian Study Group. *Hum Gene Ther* 1999; 10(14):2325–2335.

65. Alauddin MM, Shahinian A, Gordon EM, Bading JR, Conti PS. Preclinical evaluation of the penciclovir analog 9-(4-[(18)F]fluoro-3-hydroxymethylbutyl)guanine for in vivo measurement of suicide gene expression with PET. *J Nucl Med* 2001; 42(11):1682–1690.

66. Anderson D, Self T, Mellor IR, Goh G, Hill SJ, Denning C. Transgenic enrichment of cardiomyocytes from human embryonic stem cells. *Mol Ther* 2007; 15(11):2027–2036.

67. Tjuvajev JG, Doubrovin M, Akhurst T, Cai S, Balatoni J, Alauddin MM, Finn R, Bornmann W, Thaler H, Conti PS, Blasberg RG. Comparison of radiolabeled nucleoside probes (FIAU, FHBG, and FHPG) for PET imaging of HSV1-tk gene expression. *J Nucl Med* 2002; 43(8):1072–1083.

68. Gambhir SS, Bauer E, Black ME, Liang Q, Kokoris MS, Barrio JR, Iyer M, Namavari M, Phelps ME, Herschman HR. A mutant herpes simplex virus type 1 thymidine kinase reporter gene shows improved sensitivity for imaging reporter gene expression with positron emission tomography. *Proc Natl Acad Sci USA* 2000; 97(6):2785–2790.

69. Miyagawa M, Anton M, Haubner R, Simoes MV, Stadele C, Erhardt W, Reder S, Lehner T, Wagner B, Noll S, Noll B, Grote M, Gambhir SS, Gansbacher B, Schwaiger M, Bengel FM. PET of cardiac transgene expression: comparison of 2 approaches based on herpes viral thymidine kinase reporter Gene. *J Nucl Med* 2004; 45(11):1917–1923.

70. Cao F, Drukker M, Lin S, Sheikh AY, Xie X, Li Z, Connolly AJ, Weissman IL, Wu JC. Molecular imaging of embryonic stem cell misbehavior and suicide gene ablation. *Cloning Stem Cells* 2007; 9(1):107–117.

71. Qiao H, Surti S, Choi SR, Raju K, Zhang H, Ponde DE, Kung HF, Karp JS, Zhou R. Death and proliferation time course of stem cells transplanted in the myocardium. *Mol Imaging Biol* 2009; 11(6):408–414.

72. Barrio JR, Satyamurthy N, Huang SC, Keen RE, Nissenson CH, Hoffman JM, Ackermann RF, Bahn MM, Mazziotta JC, Phelps ME. 3-(2'-[18F]fluoroethyl)spiperone: In vivo biochemical and kinetic characterization in rodents, nonhuman primates, and humans. *J Cereb Blood Flow Metab* 1989; 9(6):830–839.

73. MacLaren DC, Gambhir SS, Satyamurthy N, Barrio JR, Sharfstein S, Toyokuni T, Wu L, Berk AJ, Cherry SR, Phelps ME, Herschman HR. Repetitive, non-invasive imaging of the dopamine D2 receptor as a reporter gene in living animals. *Gene Ther* 1999; 6(5):785–791.

74. MacLaren DC, Toyokuni T, Cherry SR, Barrio JR, Phelps ME, Herschman HR, Gambhir SS. PET imaging of transgene expression. *Biol Psychiatry* 2000; 48(5):337–348.

75. Liang Q, Satyamurthy N, Barrio JR, Toyokuni T, Phelps MP, Gambhir SS, Herschman HR. Noninvasive, quantitative imaging in living animals of a mutant dopamine D2 receptor reporter gene in which ligand binding is uncoupled from signal transduction. *Gene Ther* 2001; 8(19):1490–1498.

76. Amenta F, Gallo P, Rossodivita A, Ricci A. Radioligand binding and auto-radiographic analysis of dopamine receptors in the human heart. *Naunyn Schmiedebergs Arch Pharmacol* 1993; 347(2):147–154.

77. Chung JK. Sodium iodide symporter: Its role in nuclear medicine. *J Nucl Med* 2002; 43(9):1188–1200.

78. Shin JH, Chung JK, Kang JH, Lee YJ, Kim KI, Kim CW, Jeong JM, Lee DS, Lee MC. Feasibility of sodium/iodide symporter gene as a new imaging reporter gene: comparison with HSV1-tk. *Eur J Nucl Med Mol Imaging* 2004; 31(3):425–432.

79. Niu G, Gaut AW, Ponto LL, Hichwa RD, Madsen MT, Graham MM, Domann FE. Multimodality noninvasive imaging of gene transfer using the human sodium iodide symporter. *J Nucl Med* 2004; 45(3):445–449.

80. Miyagawa M, Beyer M, Wagner B, Anton M, Spitzweg C, Gansbacher B, Schwaiger M, Bengel FM. Cardiac reporter gene imaging using the human sodium/iodide symporter gene. *Cardiovasc Res* 2005; 65(1):195–202.

81. Miyagawa M, Anton M, Wagner B, Haubner R, Souvatzoglou M, Gansbacher B, Schwaiger M, Bengel FM. Non-invasive imaging of cardiac transgene expression with PET: Comparison of the human sodium/iodide symporter gene and HSV1-tk as the reporter gene. *Eur J Nucl Med Mol Imaging* 2005; 32(9):1108–1114.

82. Kang JH, Lee DS, Paeng JC, Lee JS, Kim YH, Lee YJ, Hwang do W, Jeong JM, Lim SM, Chung JK, Lee MC. Development of a sodium/iodide symporter (NIS)-transgenic mouse for imaging of cardiomyocyte-specific reporter gene expression. *J Nucl Med* 2005; 46(3):479–483.

83. Huang M, Batra RK, Kogai T, Lin YQ, Hershman JM, Lichtenstein A, Sharma S, Zhu LX, Brent GA, Dubinett SM. Ectopic expression of the thyroperoxidase gene augments radioiodide uptake and retention mediated by the sodium iodide symporter in non-small cell lung cancer. *Cancer Gene Ther* 2001; 8(8):612–618.

84. Wang M, Caruano AL, Lewis MR, Meyer LA, Vander Waal RP, Anderson CJ. Subcellular localization of radiolabeled somatostatin analogues: Implications for targeted radiotherapy of cancer. *Cancer Res* 2003; 63(20):6864–6869.

85. Moroz MA, Serganova I, Zanzonico P, Ageyeva L, Beresten T, Dyomina E, Burnazi E, Finn RD, Doubrovin M, Blasberg RG. Imaging hNET reporter gene expression with 124I-MIBG. *J Nucl Med* 2007; 48(5):827–836.

86. Buursma AR, Beerens AMJ, de Vries EFJ, van Waarde A, Rots MG, Hospers GAP, Vaalburg W, Haisma HJ. The human norepinephrine transporter in combination with 11C-m-hydroxyephedrine as a reporter gene/reporter probe for PET of gene therapy. *J Nucl Med* 2005; 46(12):2068–2075.

87. Doubrovin MM, Doubrovina ES, Zanzonico P, Sadelain M, Larson SM, O'Reilly RJ. In vivo imaging and quantitation of adoptively transferred human antigen-specific T cells transduced to express a human norepinephrine transporter gene. *Cancer Res* 2007; 67(24):11959–11969.

88. Zanzonico P, Koehne G, Gallardo HF, Doubrovin M, Doubrovina E, Finn R, Blasberg RG, Riviere I, O'Reilly RJ, Sadelain M, Larson SM. [131I]FIAU labeling of genetically transduced, tumor-reactive lymphocytes: Cell-level dosimetry and dose-dependent toxicity. *Eur J Nucl Med Mol Imaging* 2006; 33(9):988–997.

89. Xie X, Chan KS, Cao F, Huang M, Li Z, Lee A, Weissman IL, Wu JC. Imaging of STAT3 signaling pathway during mouse embryonic stem cell differentiation. *Stem Cells Dev* 2009; 18(2):205–214.

90. Acton PD, Zhou R. Imaging reporter genes for cell tracking with PET and SPECT. *Q J Nucl Med Mol Imaging* 2005; 49(4):349–360.

91. Wu JC, Cao F, Dutta S, Xie X, Kim E, Chungfat N, Gambhir S, Mathewson S, Connolly AJ, Brown M, Wang EW. Proteomic analysis of reporter genes for molecular imaging of transplanted embryonic stem cells. *Proteomics* 2006; 6(23):6234–6249.

92. Wu JC, Spin JM, Cao F, Lin S, Xie X, Gheysens O, Chen IY, Sheikh AY, Robbins RC, Tsalenko A, Gambhir SS, Quertermous T. Transcriptional profiling of reporter genes used for molecular imaging of embryonic stem cell transplantation. *Physiol Genomics* 2006; 25(1):29–38.

93. Serganova I, Ponomarev V, Blasberg R. Human reporter genes: potential use in clinical studies. *Nucl Med Biol* 2007; 34(7):791–807.

94. Krishnan M, Park JM, Cao F, Wang D, Paulmurugan R, Tseng JR, Gonzalgo ML, Gambhir SS, Wu JC. Effects of epigenetic modulation on reporter gene expression: Implications for stem cell imaging. *FASEB J* 2006; 20(1):106–108.

95. Chin BB, Nakamoto Y, Bulte JW, Pittenger MF, Wahl R, Kraitchman DL. [111]In oxine labeled mesenchymal stem cell SPECT after intravenous administration in myocardial infarction. *Nucl Med Commun* 2003; 24(11):1149–1154.

96. Tran N, Poussier S, Franken PR, Maskali F, Groubatch F, Vanhove C, Antunes L, Karcher G, Villemot JP, Marie PY. Feasibility of in vivo dual-energy myocardial SPECT for monitoring the distribution of transplanted cells in relation to the infarction site. *Eur J Nucl Med Mol Imaging* 2006; 33(6):709–715.

97. Stodilka RZ, Blackwood KJ, Kong H, Prato FS. A method for quantitative cell tracking using SPECT for the evaluation of myocardial stem cell therapy. *Nucl Med Commun* 2006; 27(10):807–813.

98. Barbash IM, Chouraqui P, Baron J, Feinberg MS, Etzion S, Tessone A, Miller L, Guetta E, Zipori D, Kedes LH, Kloner RA, Leor J. Systemic delivery of bone marrow-derived mesenchymal stem cells to the infarcted myocardium: Feasibility, cell migration, and body distribution. *Circulation* 2003; 108(7):863–868.

99. Hofmann M, Wollert KC, Meyer GP, Menke A, Arseniev L, Hertenstein B, Ganser A, Knapp WH, Drexler H. Monitoring of bone marrow cell homing into the infarcted human myocardium. *Circulation* 2005; 111(17):2198–2202.

100. Olasz EB, Lang L, Seidel J, Green MV, Eckelman WC, Katz SI. Fluorine-18 labeled mouse bone marrow-derived dendritic cells can be detected in vivo by high resolution projection imaging. *J Immunol Methods* 2002; 260(1–2):137–148.

102. Ma B, Hankenson KD, Dennis JE, Caplan AI, Goldstein SA, Kilbourn MR. A simple method for stem cell labeling with fluorine 18. *Nucl Med Biol* 2005; 32(7):701–705.

# 13

## Principles of Bioluminescence Imaging

Maarten A. Lijkwan, Ernst Jan Bos, and Joseph C. Wu

## CONTENTS

## 13.1 Introduction

For centuries, sailors and those at sea have reported seeing seas aglow with a dim light or light seen on disturbing the seawater by oars. These "legends" of the sea are now thought to be caused by bioluminescent bacteria or dino-flagellates [1]. Christopher Columbus reported lights in the sea near the coast of San Salvador, and Sir Frances Drake described tropical fireflies in the East Indies. In addition, Shakespeare's Hamlet addressed the glowworm's luminescence capabilities [2]. Partly due to the mysterious nature of these phenomena, bioluminescence has attracted the interest of researchers as long ago as Aristotle's time; he was the first to record self-luminosity of bioluminescent organisms, and in 1555, Conrad Gesner published the first book on bioluminescence and chemiluminescence [2,3]. About 40 years ago, characterization of the various luciferases began [4,5], which has led to the broad applications of *in vivo* bioluminescence imaging (BLI) modalities available to date. Many research groups worldwide contributed to the development of BLI applications following the initial characterization, when hurdles such as the proper instrumentation for detecting weak optical signals [6], the understanding of the optical properties of mammalian tissue [7], and the design of specific reporter genes with the ability to produce detectable optical signals in living animals [8,9] needed to be overcome.

The traditional application of the luciferase-luciferin bioluminescence reaction was as an ATP (adenosine triphosphate) indicator [10]. Intracellular/cytoplasmic ATP concentration changes could be measured by microinjection of recombinant luciferase protein into single cells followed by incubation with luciferin [11,12]. Other initial applications of BLI included the detection of microbial contamination for environmental health assays and luciferase imaging in the food contamination field using a bioluminescent reporter strain of *Escherichia coli*.

Beyond the initial usage of the luciferase-luciferin bioluminescence reaction, these days BLI has a wide applicability in many fields, such as monitoring protein-protein interactions via bioluminescence resonance energy transfer (BRET) analogous to fluorescence resonance energy transfer (FRET), monitoring transcriptional and posttranscriptional regulation of functionally important genes, analyzing cell injury-induced expression and regulation, monitoring free $Ca^{2+}$ concentrations, and visualizing immune responses [10,13]. In stem cell research and oncology research, BLI techniques have been particularly useful for the real-time noninvasive tracking of cell fate in animal models.

## 13.2 Principles of Bioluminescence

Although every bioluminescent and chemiluminescent reaction is called a "luciferase-luciferin" reaction, the molecular structures and the specific chemistry involved in this reaction differ among organisms. These differences, caused by independent evolution of the different species, have resulted in several forms of luciferases and substrates with different colors of emission. The basic chemiluminescent reaction requires energy, oxygen, and luciferase as a substrate. As mentioned, the specific luciferase substrate differs depending on the species, with some reactions requiring additional cofactors for light production [14]. Five basic luciferase-luciferin systems can be distinguished in different organisms as listed in Table 13.1. Of these

**TABLE 13.1**

Overview of the Luciferase-Luciferin Systems

| Luciferase-Luciferin System | Substrate (Luciferin) | Cofactors | Luciferases (kDa) | Maximum Emission (nm) |
|---|---|---|---|---|
| Bacteria | $FMNH_2$ | Aldehydes | 80 | 490–500 |
| Dinoflagellates | Tetrapyrrole | $H^+$ | 60–130 | 475 |
| Coelenterazine (Renilla) | Coelenterazine | $Ca^{2+}$ | 35 | 460–490 |
| Vargulae | Imidazopyrazine (Vargulin) | — | 60 | 465 |
| Firefly | Benzothiazole | $ATP; Mg^{2+}; O_2$ | 62 | 560 |

systems, three luciferases have been characterized and cloned for routine laboratory use:

1. The bacterial chemiluminescent reaction, found mostly in luminescent bacteria and certain fishes, involves the oxidation of a reduced riboflavin phosphate (Reduced Flavin mononucleotide [$FMNH_2$]) by a two-subunit luciferase in association with a long-chain aldehyde. This reaction produces a blue-green light with a peak emission at 490 nm [15]. Since the genetic properties and the luminescence conditions are well understood, this system is used in a wide variety of biotechnological applications, such as infection dynamics and pathogen distribution [16,17]. The main advantage of the bacterial chemiluminescent reaction is that this system does not require additional exogenous substrate to generate light emission because of the ability to express the enzymes for substrate synthesis. The main disadvantage of the bacterial luciferase-luciferin system is the bacterial enzyme, which is limited to use in prokaryotic systems [15].

2. Coelenterazine is the most widely known luciferase next to firefly luciferase and can be found in marine organisms like jellyfish (*Aequorea*) and sea pansies (*Renilla*). In contrast to both bacterial luciferase and firefly luciferase, this system does not require a cellular energy source since the substrate provides the essential energy necessary for the reaction [15]. This reaction produces a blue light with a peak emission at 475 nm, and $Ca^{2+}$ as a cofactor is required. In the presence of the accessory green fluorescence protein (GFP), the *Renilla* chemiluminescent reaction emits a green light with an emission peak at approximately 509 nm; in the absence of GFP, it emits a blue light with an emission peak at approximately 475 nm [18]. Similar effects on light emission are seen in the *Aequorea* chemiluminescent reaction in the presence or absence of *Aequorea* GFP.

Since the initial report of GFP as a marker for gene expression and protein localization [19], GFP has proven to be an extremely valuable tool in biomedical research, with numerous applications available to date [20]. One of the applications in combination with the chemiluminescence reaction is BRET, a technique now used for the detection of protein-protein interactions both *in vitro* and in living cells [10]. Resonance energy transfer occurs during protein interaction, which can be measured as differences in light emission [21]. Initially, this technique utilizes the coelenterazine luminescence system in combination with GFP, but these days the firefly luciferase luminescence system is also used. One of the main advantages of the coelenterazine luminescence system is that the enzyme does not require cofactors provided by the host cell, ameliorating potential problems of substrate transportation into the cytoplasm of the cells [15]. Two of the main limiting factors for the use of coelenterazine for *in vivo* BLI are the limited light emission spectrum (below 590 nm), which prevents deep tissue transmission, and the relatively unstable substrate, which tends to bind to serum proteins [15].

3. The firefly luciferase is the most widely used chemiluminescent reaction for *in vivo* BLI. This system is mostly found in fireflies, click beetles, and railroad worms. The chemiluminescent reaction among these species is similar, as all emit light with the luciferin substrate [18]. Two superfamilies of bioluminescent beetles are found, with fireflies belonging to the Lampyridae family of the Cantharoidea superfamily and click beetles being members of the Elateroidea superfamily [22]. The chemiluminescent reaction here uses benzothiazole as a substrate and requires ATP (in combination with $Mg^{2+}$ and $O_2$) as a cofactor, resulting in a yellow-green light with an emission peak at 560 nm [14]. Because of this unique property, as mentioned, this chemiluminescent reaction was used as a standard ATP assay. Currently, the firefly luciferase chemiluminescent reaction is widely used for *in vivo* bioluminescent imaging. Contag et al. were the first to show the detection of light emission in a mouse model using *Salmonella* strains expressing bacterial luciferase [23]. Firefly luciferase seems to be ideal for *in vivo* use because of the cellular and biodistribution characteristics of its luciferin substrate [15]. For instance, luciferin passes freely though cell membranes and is rapidly distributed [24]. Luciferin can also pass through both the blood-brain barrier [25] as well as the placental barrier [26], and produces no adverse reactions from the host animal [15]. Because of these advantages, most *in vivo* bioluminescent imaging models available for the real-time noninvasive tracking of cell fate use the firefly luciferase-luciferin system. Table 13.2 shows the three most widely used luciferase-luciferin

**TABLE 13.2**

Molecular Structure of the Most Widely Used Luciferin Substrates

| Organism | Substrate | Luciferin (Substrate) Structure |
|---|---|---|
| Bacteria, certain fish | FMNH$_2$ | |
| Jellyfish (*Aequorea*), sea pansies (*Renilla*) | Coelenterazine | |
| Fireflies, click beetles, railroad worms | Benzothiazole | |

systems with the organism from which each system originated and the molecular structure of the substrate.

## 13.3 Tracking Cell Survival Using Bioluminescence Imaging

Bioluminescence imaging has been used extensively in the tracking of cells, giving real-time information on survival, growth, and proliferation. The concept of reporter gene imaging is shown in Figure 13.1. This multistep process requires the creation of a reporter gene construct that contains a promoter and the reporter (luciferase) gene. A viral or nonviral vector delivers the construct into the cell nucleus, to be transcribed to mRNA (messenger RNA). Subsequent translation of the mRNA results in luciferase proteins. The luciferase proteins generate a signal once they interact with the reporter probe and the necessary cofactors. This signal is detectable by a charge-coupled device (CCD) camera. Since the initial reports of the reporter gene imaging concept, multiple groups have used this concept to track stem cell fate. Li et al. [27] tracked the fate of embryonic stem cell-derived endothelial cells (ES-ESCs) in a murine heart model to investigate induction of angiogenesis. Van der Bogt et al. [28] used BLI to compare the therapeutic efficacy of four different cell types (bone marrow mononuclear cells, mesenchymal stem cells, skeletal myoblasts, and fibroblasts) in a murine myocardial infarction model monitoring cell fate for

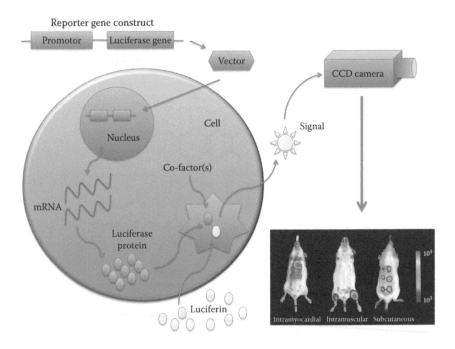

**FIGURE 13.1**

The concept of reporter gene imaging. A viral or nonviral vector delivers the reporter gene construct into the cell nucleus, to be transcribed to mRNA. Subsequent translation of the mRNA results in luciferase proteins. The luciferase proteins generate a signal once they interact with the reporter probe and the necessary cofactors. This signal is detectable by a charge-coupled device (CCD) camera. Stem cells containing the reporter gene can be tracked *in vivo* using a variety of cell transplantation models: intramyocardial, intramuscular, or subcutaneous.

up to 6 weeks. Using a similar concept, Li et al. [29] evaluated the survival and function of cardiac stem cells as shown in Figure 13.2.

## 13.4 Monitoring Stem Cell Behavior

Tracking cell behavior has also proven invaluable in cancer research by providing repetitive and long-term information on tumor formation and metastasis, with the first report of a high-throughput model dating to 1999 [30]. Various models of increasing complexity and ingenuity have been developed since that time [31]. Rauch et al. [32] developed a mouse model of spontaneous lymphoma in which they coupled expression of the oncogen Tax with luciferin activation, thus enabling the visualization of malignant transformation.

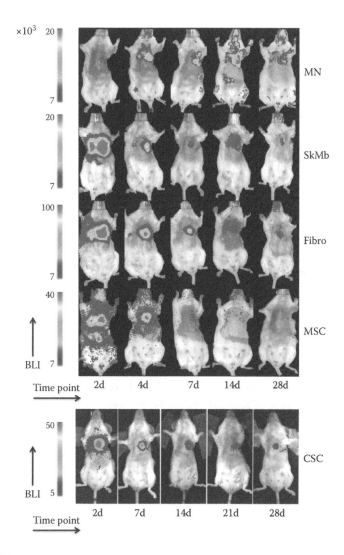

**FIGURE 13.2**
Tracking stem cell proliferation and survival *in vivo* using bioluminescence imaging. *In vivo* longitudinal survival BLI of mononuclear stem cells (MN), skeletal myoblasts (SkMb), fibroblasts (Fibro), mesenchymal stem cells (MSC), and cardiac stem cells (CSC) transplanted from L2G transgenic mice to FVB mice. Scale bars represent signal intensity in photons/second/square centimeter/sr. (Images adapted from van der Bogt, K.E., et al., *Circulation* 2008, 118(14 Suppl): S121–29 with permission from Wolters Kluwer; and Li, Z., et al., *J Am Coll Cardiol* 2009, 53(14): 1229–40 with permission from Elsevier.)

## 13.5 Monitoring Drug/Stem Cell Distribution

BLI is also widely applied in drug research, for which the same principles for cell tracking apply [33,34]. BLI has proven to be a versatile and high-throughput technique to monitor drug distribution and interaction *in vivo* [34]. Applications range from antiviral therapy [35] to anticancer agents [36]. McMillin et al. [33] have demonstrated the value of BLI for the preclinical evaluation of anticancer agents developed for an *in vitro* model that mimics the microenvironment of a tumor to assess stroma-mediated drug resistance. By using the transcriptional promoters from genes to drive luciferase expression, researchers can study gene expression *in vivo* [24,37]. Chen et al. [38], for example, monitored the induction of nonislet β-cells into insulin-producing cells in a murine model using rat insulin promoter to drive luciferase expression.

## 13.6 Monitoring Cell Biology and Gene Expression

As mentioned, another aspect of cell biology to which BLI is applied is the detection of protein-protein interactions. Several techniques have been described [10]. Major approaches include (a) luciferase complementation imaging, a technique in which luciferase is fragmented into two inactive sites and fused to the proteins of interest, and on the subsequent interaction of which the inactive fragments of luciferase are brought together to reconstitute bioluminescence activity [39,40]; and (b) BRET, a technique based on the transfer of energy between a luminescent energy donor and a fluorescent energy acceptor [41].

## 13.7 Advantages and Limits of Bioluminescence Imaging

Although there are several other noninvasive imaging modalities, such as magnetic resonance imaging (MRI) [42] and positron emission tomography (PET) [43], BLI carries several advantages. First, the method is cheap and facile compared to most systems [15,44], making BLI suitable as a high-throughput imaging modality [45]. Furthermore, BLI is noninvasive and can be conducted repeatedly in an experimental animal, allowing longitudinal observation [37]. BLI has proven useful in multiple constructs in a variety of important fields of investigation as discussed in this chapter. Unfortunately, BLI also has a number of shortcomings, discussed next.

### 13.7.1 Image Limitations

Since BLI is based on photographic principles using a highly sensitive CCD camera to capture the emitted light, it creates only planar images, lacking information on signal depth [46]. Tomographic imaging reconstructing the data to a three-dimensional image provides a solution but has so far proven difficult to implement reliably [47].

### 13.7.2 Penetration Depth

The most notable drawback of BLI is the lack of tissue penetration by the bioluminescent signal, limiting its application to small-animal models. The diffusion of light through tissue is impaired by a number of factors, such as density, composition, and thickness. Essential to penetration is the wavelength of emitted light. Blue-to-green light has a wavelength between 400 and 600 nm and is strongly absorbed by hemoglobin and skin pigment, significantly impairing its signal and limiting its use to a depth of only a few millimeters [48,49]. Orange-to-red light with a wavelength greater than 600 nm is the candidate of choice for imaging of deeper tissue (>1 cm). As such, the emission spectra of most applied luciferases contain wavelengths of 600 nm and above, with firefly and click beetle luciferase containing the highest proportion of red light [49]. Several factors have been investigated to optimize photon emission. Zhao et al. [49] demonstrated that at body temperature, the brightness of several luciferases increases, and firefly luciferase expresses higher emission wavelengths, ranging from 578 nm at 25°C to 612 nm at 37°C.

### 13.7.3 Light Scattering

Besides limiting penetration, tissue also hampers light transmission by scattering the emitted photons. The cause of this phenomenon lies in variations in the refractive index of cell membranes and intracellular structures. The consequence of this scattering is a diffuse light emission that results in lower spatial resolution of the acquired data [48]. Despite these drawbacks, techniques such as optical clearing and others may eventually improve BLI imaging. Attempts have been made to reduce photon scattering through optical tissue clearing by injecting glycerol into mouse skin [50]. Although partially effective, this technique is not yet practicable.

### 13.7.4 Autoluminescence

Another factor to take into account is autoluminescence of experimental animals themselves. This background signal is present in mice and probably all living creatures, and it intervenes with the measurement of low-intensity signals. This phenomenon is not well understood; although rather minimal, it definitely exceeded the lower detection limit of 100 photons/second/centimeter$^2$/steradian photons/second/centimeter2/sr (photons/s/cm$^2$/sr) from the popular IVIS® imaging system (Xenogen, Alameda, CA) in an experiment by Troy et al. [51]. Taking sample images of white furred mice and nude mice, they detected a background signal of 1,600 photons/s/cm$^2$/sr and 1,000 photons/s/cm$^2$/sr, respectively. Shaving white mice showed no difference, but the background signal for black mice increased due to the higher absorptive properties of black fur. They further noted that after injection, luciferase substrate also contributed to an increased background signal, further reducing the signal-to-noise ratio. This effect is especially strong for coelenterazine [52].

### 13.7.5 Luciferase Substrate

The luciferase substrate has also been the subject of examination. As mentioned, light with higher wavelengths offers better tissue penetration, and successful attempts have been made to improve luciferases by increasing the emission of red light [53,54]. Because active metabolism is required for signal emission, BLI gives highly accurate data on cell viability. As such, it is also dependent on cofactors. Coelenterazine substrate requires only the presence of $O_2$, while the firefly luciferase substrate also requires ATP and $Mg^{2+}$ [52], as discussed in this chapter. Factors influencing the availability of these cofactors, such as hypoxia and necrosis, can thus affect BLI accuracy [30,36]. Distribution of the substrate also plays a role in BLI sensitivity. Berger et al. [55] investigated tissue distribution in mice using radiolabeled D-luciferin, showing heterogeneous tissue distribution following the natural elimination routes by the organ system. A comparison between intraperitoneal and intravenous injection showed lower blood levels and organ activity but longer duration of activity after intraperitoneal injection. Uptake in brain and muscle was generally lower. In conclusion, they suggested that depending on the site of investigation, intravenous injection is preferred for early and high-uptake values and intraperitoneal injection for prolonged organ uptake. Naturally, sufficient substrate and timing are also crucial for success. In an effort to overcome the limitations of tissue distribution and clearing, especially for imaging of short time processes, Gross et al. [56] implanted micro-osmotic pumps that continuously provided D-luciferin in a nude mouse model.

### 13.7.6 Sensitivity

Troy et al. [51] also assessed BLI sensitivity of cell detection *in vivo* and *in vitro*. BLI proved capable of detecting as little as 100 cells *in vitro* using PC-3M-luc-C6 (human prostate adenocarcinoma) cells. The same type of cells implanted subcutaneously in a nude mouse required 400 cells for sufficient signal, and injecting the cells in the mouse flank required a minimum of 900 cells. Rabinovich et al. created an improved firefly luciferase transfection vector capable of generating more than 100 times more light compared with the normal retroviral vector-based transfection of T cells through a combination of codon optimization, removal of cryptic splice sites, and retroviral modification [57], making detection of fewer than 10 autologous T-cell numbers possible in B6 albino mice. The number of cells necessary for sufficient signal not only depends on the types of vector and promoter for driving the transgene but also on the cell type used. In the same article, human peripheral blood mononuclear cells (PBMCs) transfected with the enhanced vector only showed a tenfold light increase on comparison.

## 13.8 Comparing Bioluminescence Imaging to Other Imaging Modalities

Despite several disadvantages, BLI provides an excellent imaging modality for *in vivo* small-animal research compared with other imaging modalities. The method is reliable and safe, the range of applications is broad, and its viability has been proven in numerous studies [28,45,58,59]. Fluorescent imaging (FLI) using either reporter genes or fluorescent dyes carries a broader range of wavelengths, allowing deeper tissue imaging without requiring the administration of a substrate that is dependent on excitation light to visualize the fluorochrome of interest. Despite several advantages, high levels of autofluorescence limit this method, making FLI generally less sensitive and with a lower spatial resolution than BLI [51]. Other imaging modalities such as MRI, PET, and single-photon emission computed tomography (SPECT) are expensive and often lack the simplicity BLI offers. Table 13.3 gives an overview of the different imaging systems with information on signal penetration depth, spatial resolution, image dimensions, availability of reporter genes, and clinical applicability. Despite its advantages as a noninvasive, convenient, and reliable imaging modality, the translation of BLI to large-animal models and clinical application has been seriously hampered by the limited

**TABLE 13.3**

Overview of Imaging Systems

| Modality | Penetration | Resolution | Sensitivity | 3D | Clinic |
|----------|-------------|------------|-------------|-----|--------|
| PET | Unlimited | 1 mm | $10^{-11}$ to $10^{-12}$ mole/L | + | + |
| SPECT | Unlimited | 1 mm | $10^{-10}$ to $10^{-11}$ mole/L | + | + |
| MRI | Unlimited | 10–100 µm | $10^{-3}$ to $10^{-5}$ mole/L | + | + |
| BLI | <1 cm | >1 mm | Approximately $10^{-15}$ to $10^{-17}$ mole/L[a] | Developing | – |
| FLI | <10 cm | 1 mm | Approximately $10^{-9}$ to $10^{-12}$ mole/L[a] | Developing | Developing |

[a] Not well characterized, for more information please refer to Massoud et al. 2003 [64].

penetration of the bioluminescent signal. In contrast, MRI, PET, and SPECT are not limited by lack of tissue penetration and therefore seem more attractive options for clinical application. Although recent developments have been made [60,61], MRI reporter genes have not yet been as widely applied as PET and SPECT, mostly due to concerns regarding their detection sensitivity threshold (see Table 13.3). This is less the case with PET and SPECT, which have been applied in various studies [43,62]; moreover, reporter gene imaging using PET has been applied in a human case [63]. Further in-depth comparisons of the different (molecular) imaging modalities can be found in a review by Massoud et al. [64]. Despite recent advances, it is unlikely that BLI will overcome its major drawback of limited tissue penetration anytime soon. Nevertheless, it is clear that for a high-throughput imaging modality in cell and small-animal models, BLI is an indispensable tool with numerous applications in the field of modern research.

## 13.9 Conclusion

Since the characterization of multiple luciferase substrates some 40 years ago, the applications of BLI have been widespread, ranging from intracellular/cytoplasmic ATP concentration measurements and the detection of microbial contamination for environmental health assays to monitoring protein-protein interactions via BRET and the real-time noninvasive tracking of (stem) cell fate in animal models. The firefly luciferase-luciferin system is most commonly used for this tracking of cells, giving real-time information on survival, growth, and proliferation. Despite several disadvantages as discussed in this chapter, BLI provides an excellent imaging modality for *in vivo* small-animal research compared with other imaging modalities. This method is reliable and safe, the range of applications is broad, and its viability has been proven in numerous studies. Taken together, BLI is a valuable tool with numerous applications in the field of modern research.

## Acknowledgments

This work was supported in part by Burroughs Wellcome Foundation, NIH R01 EB009689, R01 HL093172, RC1 HL099117, and the Netherlands Organization for Health Research and Development."

## References

1. Miller, S.D., et al., Detection of a bioluminescent milky sea from space. *Proc Natl Acad Sci USA*, 2005. 102(40): 14181–84.
2. Lee, J., Bioluminescence: The first 3000 years. *J Sib Fed U*, 2008. Biology 3(1): 194–205.
3. Lee, J. and Smith C.K. A history of bioluminescence. 2010 http://www.photobiology.info.
4. McElroy, W.D., H.H. Seliger, and E.H. White, Mechanism of bioluminescence, chemiluminescence and enzyme function in the oxidation of firefly luciferin. *Photochem Photobiol*, 1969. 10(3): 153–70.
5. Johnson, F.H. and O. Shimomura, Enzymatic and nonenzymatic bioluminescence. *Photophysiology*, 1972(7): 275–334.
6. Oshiro, M., Cooled CCD versus intensified cameras for low-light video—applications and relative advantages. *Methods Cell Biol*, 1998. 56: 45–62.
7. Cheong, W.F., A review of the optical properties of biological tissues. *IEEE J Quantum Electron*, 1990. 26(12): 2166–85.
8. Contag, P.R., et al., Bioluminescent indicators in living mammals. *Nat Med*, 1998. 4(2): 245–47.
9. Padera, T.P., et al., Conventional and high-speed intravital multiphoton laser scanning microscopy of microvasculature, lymphatics, and leukocyte-endothelial interactions. *Mol Imaging*, 2002. 1(1): 9–15.
10. Welsh, D.K. and S.A. Kay, Bioluminescence imaging in living organisms. *Curr Opin Biotechnol*, 2005. 16(1): 73–8.
11. Dumollard, R., et al., Sperm-triggered [Ca2+] oscillations and $Ca^{2+}$ homeostasis in the mouse egg have an absolute requirement for mitochondrial ATP production. *Development*, 2004. 131(13): 3057–67.
12. Allue, I., et al., Evidence for rapid consumption of millimolar concentrations of cytoplasmic ATP during rigor-contracture of metabolically compromised single cardiomyocytes. *Biochem J*, 1996. 319(Pt 2): 463–69.
13. Greer, L.F., 3rd and A.A. Szalay, Imaging of light emission from the expression of luciferases in living cells and organisms: a review. *Luminescence*, 2002. 17(1): 43–74.
14. Hastings, J.W., Chemistries and colors of bioluminescent reactions: a review. *Gene*, 1996. 173(1 Spec No): 5–11.
15. Doyle, T.C., S.M. Burns, and C.H. Contag, In vivo bioluminescence imaging for integrated studies of infection. *Cell Microbiol*, 2004. 6(4): 303–17.
16. Sharma, P.K., et al., Spatiotemporal progression of localized bacterial peritonitis before and after open abdomen lavage monitored by in vivo bioluminescent imaging. *Surgery*, 2010. 147(1): 89–97.
17. Wiles, S., et al., Bioluminescent monitoring of in vivo colonization and clearance dynamics by light-emitting bacteria. *Methods Mol Biol*, 2009. 574: 137–53.
18. Wilson, T. and J.W. Hastings, Bioluminescence. *Annu Rev Cell Dev Biol*, 1998. 14: 197–230.
19. Chalfie, M., et al., Green fluorescent protein as a marker for gene expression. *Science*, 1994. 263(5148): 802–5.
20. Shaner, N.C., G.H. Patterson, and M.W. Davidson, Advances in fluorescent protein technology. *J Cell Sci*, 2007. 120(Pt 24): 4247–60.

21. Bacart, J., et al., The BRET technology and its application to screening assays. *Biotechnol J*, 2008. 3(3): 311–24.
22. Wood, K.V., Y.A. Lam, and W.D. McElroy, Introduction to beetle luciferases and their applications. *J Biolumin Chemilumin*, 1989. 4(1): 289–301.
23. Contag, C.H., et al., Photonic detection of bacterial pathogens in living hosts. *Mol Microbiol*, 1995. 18(4): 593–603.
24. Contag, C.H., et al., Visualizing gene expression in living mammals using a bioluminescent reporter. *Photochem Photobiol*, 1997. 66(4): 523–31.
25. Rehemtulla, A., et al., Rapid and quantitative assessment of cancer treatment response using in vivo bioluminescence imaging. *Neoplasia*, 2000. 2(6): 491–95.
26. Lipshutz, G.S., et al., In utero delivery of adeno-associated viral vectors: Intraperitoneal gene transfer produces long-term expression. *Mol Ther*, 2001. 3(3): 284–92.
27. Li, Z., et al., Differentiation, survival, and function of embryonic stem cell derived endothelial cells for ischemic heart disease. *Circulation*, 2007. 116(11 Suppl): I46–I54.
28. van der Bogt, K.E., et al., Comparison of different adult stem cell types for treatment of myocardial ischemia. *Circulation*, 2008. 118(14 Suppl): S121–29.
29. Li, Z., et al., Imaging survival and function of transplanted cardiac resident stem cells. *J Am Coll Cardiol*, 2009. 53(14): 1229–40.
30. Edinger, M., et al., Noninvasive assessment of tumor cell proliferation in animal models. *Neoplasia*, 1999. 1(4): 303–10.
31. Lyons, S.K., Advances in imaging mouse tumour models in vivo. *J Pathol*, 2005. 205(2): 194–205.
32. Rauch, D., et al., Imaging spontaneous tumorigenesis: Inflammation precedes development of peripheral NK tumors. *Blood*, 2009. 113(7): 1493–500.
33. McMillin, D.W., et al., Tumor cell-specific bioluminescence platform to identify stroma-induced changes to anticancer drug activity. *Nat Med*, 2010. 16(4): 483–89.
34. Rudin, M. and R. Weissleder, Molecular imaging in drug discovery and development. *Nat Rev Drug Discov*, 2003. 2(2): 123–31.
35. Luker, K.E. and G.D. Luker, Applications of bioluminescence imaging to antiviral research and therapy: Multiple luciferase enzymes and quantitation. *Antiviral Res*, 2008. 78(3): 179–87.
36. Zhang, C., et al., Advancing bioluminescence imaging technology for the evaluation of anticancer agents in the MDA-MB-435-HAL-Luc mammary fat pad and subrenal capsule tumor models. *Clin Cancer Res*, 2009. 15(1): 238–46.
37. Wu, J.C., et al., Noninvasive optical imaging of firefly luciferase reporter gene expression in skeletal muscles of living mice. *Mol Ther*, 2001. 4(4): 297–306.
38. Chen, X., et al., In vivo detection of extrapancreatic insulin gene expression in diabetic mice by bioluminescence imaging. *PLoS One*, 2010. 5(2): e9397.
39. Paulmurugan, R., Y. Umezawa, and S.S. Gambhir, Noninvasive imaging of protein-protein interactions in living subjects by using reporter protein complementation and reconstitution strategies. *Proc Natl Acad Sci USA*, 2002. 99(24): 15608–13.
40. Luker, K.E., et al., Kinetics of regulated protein-protein interactions revealed with firefly luciferase complementation imaging in cells and living animals. *Proc Natl Acad Sci USA*, 2004. 101(33): 12288–93.

41. Perroy, J., et al., Real-time monitoring of ubiquitination in living cells by BRET. *Nat Methods*, 2004. 1(3): 203–8.

42. Kraitchman, D.L. and J.W. Bulte, Imaging of stem cells using MRI. *Basic Res Cardiol*, 2008. 103(2): 105–13.

43. Zhang, Y., et al., Tracking stem cell therapy in the myocardium: Applications of positron emission tomography. *Curr Pharm Des*, 2008. 14(36): 3835–53.

44. Contero, A., et al., High-throughput quantitative bioluminescence imaging for assessing tumor burden. *Methods Mol Biol*, 2009. 574: 37–45.

45. Paroo, Z., et al., Validating bioluminescence imaging as a high-throughput, quantitative modality for assessing tumor burden. *Mol Imaging*, 2004. 3(2): 117–24.

46. Ntziachristos, V., et al., Looking and listening to light: the evolution of whole-body photonic imaging. *Nat Biotechnol*, 2005. 23(3): 313–20.

47. Wang, G., et al., Overview of bioluminescence tomography—a new molecular imaging modality. *Front Biosci*, 2008. 13: 1281–93.

48. Rice, B.W., M.D. Cable, and M.B. Nelson, In vivo imaging of light-emitting probes. *J Biomed Opt*, 2001. 6(4): 432–40.

49. Zhao, H., et al., Emission spectra of bioluminescent reporters and interaction with mammalian tissue determine the sensitivity of detection in vivo. *J Biomed Opt*, 2005. 10(4): 41210.

50. Jansen, E.D., et al., Effect of optical tissue clearing on spatial resolution and sensitivity of bioluminescence imaging. *J Biomed Opt*, 2006. 11(4): 041119.

51. Troy, T., et al., Quantitative comparison of the sensitivity of detection of fluorescent and bioluminescent reporters in animal models. *Mol Imaging*, 2004. 3(1): 9–23.

52. Zhao, H., et al., Characterization of coelenterazine analogs for measurements of *Renilla* luciferase activity in live cells and living animals. *Mol Imaging*, 2004. 3(1): 43–54.

53. Loening, A.M., A.M. Wu, and S.S. Gambhir, Red-shifted *Renilla reniformis* luciferase variants for imaging in living subjects. *Nat Methods*, 2007. 4(8): 641–43.

54. Branchini, B.R., et al., Red-emitting luciferases for bioluminescence reporter and imaging applications. *Anal Biochem*. 396(2): 290–97.

55. Berger, F., et al., Uptake kinetics and biodistribution of 14C-D-luciferin—a radiolabeled substrate for the firefly luciferase catalyzed bioluminescence reaction: Impact on bioluminescence based reporter gene imaging. *Eur J Nucl Med Mol Imaging*, 2008. 35(12): 2275–85.

56. Gross, S., et al., Continuous delivery of D-luciferin by implanted micro-osmotic pumps enables true real-time bioluminescence imaging of luciferase activity in vivo. *Mol Imaging*, 2007. 6(2): 121–30.

57. Rabinovich, B.A., et al., Visualizing fewer than 10 mouse T cells with an enhanced firefly luciferase in immunocompetent mouse models of cancer. *Proc Natl Acad Sci USA*, 2008. 105(38): 14342–46.

58. Santos, E.B., et al., Sensitive in vivo imaging of T cells using a membrane-bound *Gaussia princeps* luciferase. *Nat Med*, 2009. 15(3): 338–44.

59. Ray, P. and S.S. Gambhir, Noninvasive imaging of molecular events with bioluminescent reporter genes in living subjects. *Methods Mol Biol*, 2007. 411: 131–44.

60. Genove, G., et al., A new transgene reporter for in vivo magnetic resonance imaging. *Nat Med*, 2005. 11(4): 450–54.

61. Cohen, B., et al., Ferritin as an endogenous MRI reporter for noninvasive imaging of gene expression in C6 glioma tumors. *Neoplasia*, 2005. 7(2): 109–17.
62. Pomper, M.G., et al., Serial imaging of human embryonic stem-cell engraftment and teratoma formation in live mouse models. *Cell Res*, 2009. 19(3): 370–79.
63. Yaghoubi, S.S., et al., Noninvasive detection of therapeutic cytolytic T cells with [18]F-FHBG PET in a patient with glioma. *Nat Clin Pract Oncol*, 2009. 6(1): 53–58.
64. Massoud, T.F. and S.S. Gambhir, Molecular imaging in living subjects: Seeing fundamental biological processes in a new light. *Genes Dev*, 2003. 17(5): 545–80.

# 14

## Quantum Dot–Labeling Methods

Amy R. Kontorovich, Glenn R. Gaudette,
Michael R. Rosen, Peter R. Brink, and Ira S. Cohen

### CONTENTS

## 14.1 Challenges of Tracking Stem Cells in Mammalian Tissue

The qualities of an ideal agent for tracking stem cells in biological tissue have been outlined by Frangioni and Hajjar (2004) and include the criteria that the agent (a) be biocompatible, safe, and nontoxic; (b) not require any genetic modification or perturbation to the stem cell; (c) permit single-cell detection at any anatomic location; (d) allow quantification of cell number; (e) have minimal or no dilution with cell division; (f) have minimal or no transfer to non-stem cells; (g) permit noninvasive imaging in the living subject over months to years; and (h) require no injectable contrast agent for visualization. This chapter provides a brief overview of traditional and newer cell-labeling agents, highlighting some of their pitfalls and the way in which quantum dot (QD) nanoparticles may represent a unique solution to the challenge of stem cell tracking by satisfying many of the criteria proposed.

Tracking cells can be confounded by high levels of endogenous fluorescence in the host tissue, a property known as *autofluorescence*. There is a multitude of biochemical sources for autofluorescence in mammalian tissue that can be excited by light in the wavelength range of 360–480 nm. These chemicals are emitters of both green and red light and include nicotinamide-adenine dinucleotide phosphate [NAD(P)H, emitting at ~509 nm], lipofuscins (emitting at 540–560 nm and rampant in postmitotic cells such as myocytes), and collagen and elastin (emitting at >515 nm) (Billinton and Knight 2001). Emission spectra of all of these substances are broad and overlap with many of the fluorescent reagents traditionally used to label cells. Furthermore, when studying damaged tissue, as in infarction models, autofluorescence becomes even more significant due to the presence of collagenous scar tissue and inflammatory cells with high levels of oxidative enzymes such as NAD(P)H (Laflamme and Murry 2005).

Techniques for tracking cells delivered to tissue have been discussed in previous chapters, and a brief list includes (a) transfecting cells *in vitro* with either fluorescent (i.e., green fluorescent protein [GFP]) (Potapova et al. 2004) or nonfluorescent (i.e., β-galactosidase) proteins (Jackson et al. 2001); (b) employing fluorescence in situ hybridization (FISH) to identify delivered cells by their unique chromosomal content (i.e., male cells delivered to a female host) (Kajstura et al. 2005); (c) using species-specific surface markers to identify delivered cells in a xenographic host (Potapova et al. 2004); (d) using commercially available fluorescent dyes to label cell cytoplasm (Kraitchman et al. 2003) or nuclei (Thompson et al. 2003); and (e) labeling cells with inorganic particles (i.e., iron oxide particles) (Amado et al. 2005; Dick et al. 2003; Kraitchman et al. 2003) or radiotracers (Barbash et al. 2003; Hofmann et al. 2005; Kraitchman et al. 2005). These can be grouped into two categories: those techniques requiring secondary staining to visualize cells in histological sections (a–c)* and those not requiring secondary staining (d and e). Although each of these approaches has advantages, all have generally focused on identifying the location and fate of a small minority of the delivered cells or produced low-resolution representations of *in vivo* stem cell distribution.

Within the first category of techniques, one of the pitfalls of transfection is that it often does not label 100% of the cells. Loading cells with exogenous DNA also requires use of viruses, lipofectamine vehicle, or electroporation; each of these processes can cause some cell death. Immunostaining—employed when tracking cells with FISH or by surface markers—can potentially generate false positives (Laflamme and Murry 2005), and interpretation of staining outcomes can be difficult in the absence of strict controls. Stem cells are also dynamic and may change their surface marker expression over time if they differentiate *in vivo*. Furthermore, staining results and image features are typically heterogeneous across a given tissue section and among sections, making it important to sample the specimens carefully and making quantitation difficult.

The second category of cell-tracking approaches—those not requiring secondary staining—represents a potentially higher-throughput avenue to identify delivered cells. However, fluorescent dyes often must be taken up by cells via lipid vehicles, which can be cytotoxic. Moreover, they are only retained for short periods of time, are prone to photobleaching when imaged

---

* GFP emission can sometimes be visualized directly under fluorescence microscopy. However, for many tissue types and conditions, autofluorescence precludes identification of GFP-positive cells. In these cases, histological methods must be used to identify the cells using an antibody directed against GFP.

(Parish 1999), and have narrow excitation spectra, and their emission profiles have long tails into the red range (spectral cross talk), making it a challenge to identify multiple probes simultaneously (Bruchez et al. 1998). The emission spectra of fluorescent dyes also overlap with those of host tissue autofluorescence (Billinton and Knight 2001; Laflamme and Murry 2005). Some groups have labeled cells with the nuclear dye DAPI (4′,6-diamidino-2-phenylindole; Kraitchman et al. 2003; Thompson et al. 2003), but this agent intercalates into DNA and can therefore interfere with normal cell function (Parish 1999). Others have loaded stem cells with magnetic nanoparticles (superparamagnetic iron oxide, SPIO) (Kraitchman et al. 2003) or radiotracers (Hofmann et al. 2005) and imaged delivered cells *in vivo* with magnetic resonance imaging (MRI) or positron emission tomographic (PET) scanning, respectively. This offers the advantage of identifying clusters of labeled cells noninvasively but does not afford single-cell resolution. If, however, secondary staining techniques are utilized (i.e., Prussian blue reaction), SPIO-labeled cells can be identified in histological sections at the single-cell level.

In summary, although the methods described are useful for some aspects of stem cell tracking, none offers the ability to unequivocally identify delivered cells *in vivo* with single-cell resolution using relatively high-throughput approaches (i.e., no secondary staining). Recalling the criteria for ideal cell-tracking agents discussed, we also propose that the agent must (i) be detectable above autofluorescence and (j) label nearly 100% of cells. QD nanoparticles are relatively new labeling agents that appear to satisfy most of the criteria outlined and therefore represent an alternative for long-term stem cell tracking *in vivo*.

## 14.2 Special Properties of Quantum Dots

Quantum dots are semiconductor nanoparticles that were discovered in the early 1980s in the laboratories of Louis Brus at Bell Laboratories in New Jersey and of Alexander Efros and A. I. Ekimov of the Yoffe Institute in St. Petersburg, Russia. Semiconductors are composed of elements in the II–VI, III–V, or IV–VI columns of the periodic table. On the macroscopic scale, these materials are important constituents of modern electronic devices because their conductivity can be altered by an external stimulus. As nanoparticles, however, their quantum properties change dramatically such that energy levels no longer are spaced closely and therefore no longer considered continuous. Discrete energy levels establish a scenario called quantum confinement that confers unique absorptive and emissive behavior onto the fluorescent QDs. The semiconductor at the core of a QD is responsible for the impressive optical properties discussed in this section. Several years after Brus' discovery, two of his colleagues at Bell Laboratories, Dr. Moungi

Bawendi and Dr. Paul Alivisatos, found that coating the dots with a passivating inorganic shell further stabilized their fluorescence; they also developed methods to make QDs water soluble (Bruchez et al. 1998). Thus, most of the QDs that are used for biological applications consist of a cadmium selenide or cadmium tellurium semiconductor core, a zinc sulfide inner shell, and an outer polymer coating. The result is a water-soluble particle 13–15 nm in diameter (Figure 14.1a).

Similar to organic fluorophores like fluorescein, QDs absorb photons of light of one wavelength and emit light of a different wavelength. Traditional fluorophores use absorbed energy to transfer electrons to excited states and energy is released (in the form of fluorescent light) when these electrons return to their resting states. When electrons move to different energy levels in QDs, they behave analogously, generating electron holes called excitons. The quantum system of excitons makes QD fluorescence much brighter and more photostable (less prone to photobleaching) than traditional fluorophores that use conjugated double-bond systems of energy transfer. The energy state of an exciton dictates the wavelength of light emitted by a particular QD after excitation.

QDs have a unique property known as tunability; the physical size of the QD determines the wavelength of emitted light. Smaller dots emit blue fluorescent light, and as the core size of the dots increases, emitted light becomes redder. Another important feature that distinguishes QDs from conventional fluorescent dyes is the large distance between the wavelength of excitation and emission light (Figure 14.1b). This energy difference, known as the Stokes shift, means that QDs can be excited by ultraviolet light at a wavelength much lower than the peak emission wavelength. In fact, QDs can be excited by any wavelength lower than their emission wavelength. Thus, both excitation of the particles and collection of emitted light are efficient processes. Furthermore, they have a narrow emitting window resembling an impulse function. These special characteristics of QDs suit them for use in biological imaging, specifically in the highly autofluorescent milieu of the myocardium. Because of the large Stokes shift, there is practically no overlap between absorption and emission spectra, facilitating collection of QD-specific emitted light. Tissue can be excited by light that is less likely to induce autofluorescence. Also, since autofluorescent emission tends to be more significant in the green range, red QDs can be preferentially used. Their extreme brightness allows for very low exposure times when imaging QDs in biological tissues, decreasing the likelihood of simultaneously inducing autofluorescence emission. The narrow emitting window allows for easy identification with spectral analysis, an option available on most confocal microscopes currently available. Finally, as multiplex imaging of several markers is often desirable in biology, the broad excitation spectrum of QDs allows simultaneous detection of different size QDs by the same incident light (Figure 14.1c).

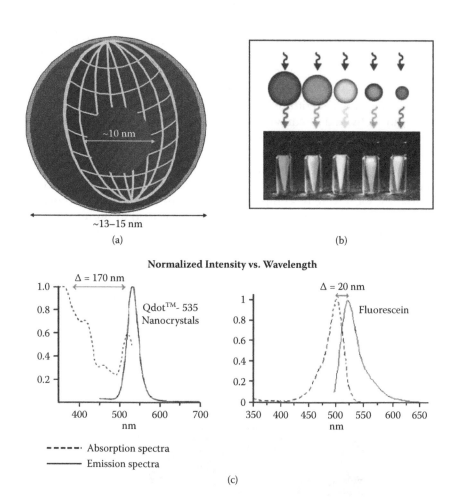

**FIGURE 14.1 (See color insert.)**
(a) Quantum dot (QD) nanoparticles are composed of a core semiconductor material (i.e., CdSe) and coated with a different semiconductor shell (i.e., ZnS). The particle is made biocompatible by virtue of an inert passivating coating. (b) QD particles emit unique wavelengths of fluorescent light on excitation with a common incident light source (in this case, a handheld UV lamp), depending on the size of the QD core, a property known as *tunability*. (c) The Stokes shift, or large difference between the wavelength of excitation and emission light, makes QD detection highly sensitive and superior to conventional fluorescent dyes. (Adapted with permission from Life Technologies Corporation.)

## 14.3 Quantum Dots for Cell Tracking

Quantum dots have been demonstrated as safe and effective labeling agents for tracking a range of cell types, including stem cells. Two landmark studies appeared in *Science* in September 1998, documenting the first biological uses for QDs (Bruchez et al. 1998; Chan and Nie 1998). Since then, several groups have emerged with novel applications. Dubertret et al. (2002) reported the encapsulation of QDs within micelle-forming hydrophilic polymer-grafted lipids and delivered them via microinjection to single cells of *Xenopus* embryos. Major findings from this study were that only the originally injected cells and their progeny retained QDs, labeled cells showed no signs of toxicity, all embryonic cell types were able to arise from labeled cells, the QD fluorescence was detectable above high levels of autofluorescence, and most importantly, QDs were biocompatible (see Figure 14.2). This group was also the first to report the pitfall that intracellular QDs tend to aggregate around the nucleus over time (Dubertret et al. 2002). A similar study published in

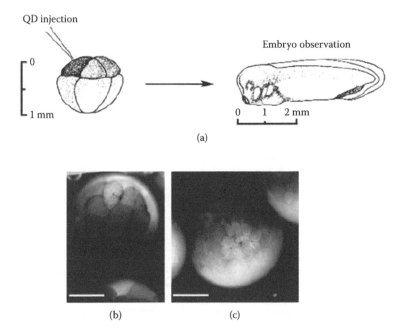

(a)

(b)                    (c)

**FIGURE 14.2**
*Xenopus* lineage tracing (a) QDs microinjected into a single blastomere of the *Xenopus* embryo. The embryos were then cultured until they reached various stages of development, at which point they were imaged. (b) One cell of an eight-cell embryo is labeled with QDs. (c) Same embryo 1 hour after injection, depicting labeled daughter cells. (Reprinted with permission from Dubertret, B., P. Skourides, D.J. Norris, V. Noireaux, A.H. Brivanlou, and A. Libchaber. 2002. *Science* 298, no. 5599: 1759–62.)

2005 demonstrated the use of QDs as lineage tracers by microinjecting them into single cells of zebrafish embryos. Here it was shown that the dots were nontoxic, retained their emission spectra regardless of microenvironment, could be induced to avoid perinuclear aggregation by surface modification with streptavidin, did not pass through gap junctions, and could be imaged in aldehyde-fixed tissue (Rieger et al. 2005). Both of these studies showed that the presence of intracellular QDs did not affect proliferation or differentiation of cells or preclude formation of a fully grown organism; however, these studies only examined labeling by microinjection of single cells.

In 2003, two articles appeared in *Nature Biotechnology* demonstrating that populations of cells could be labeled with QDs. Jaiswal et al. (2003) reported a receptor-mediated and generalizable endocytotic method for introducing QDs into the intracellular space for live cell imaging. Unfortunately, the images in this article again revealed the problem of perinuclear aggregation. Others have compared loading populations of cells via commonly used approaches, including receptor-mediated transfection with a host of proteins (Gao et al. 2004; Hanaki et al. 2003; Silver and Ou 2005; So et al. 2006; Zhang and Huang 2006), lipoprotein-mediated* transfection, and electroporation (Derfus et al. 2004). Some of the problems associated with these early methods of QD loading included inducing cell death (lipid-mediated transfection and electroporation), nonuniformity of labeling, and perinuclear aggregation.

Our group has defined the conditions under which optimal QD uptake occurs (Figure 14.3) (Rosen et al. 2007). Uptake depends on QD surface charge, composition of cell media and the substrate on which the cells are seeded. Using this optimal method of labeling, it is now possible to track QD-labeled stem cells for up to 8 weeks after delivery *in vivo* and to track the location and fate of delivered cells. Furthermore, by directly injecting QD-labeled stem cells into rodent hearts, we created a complete three-dimensional (3D) reconstruction of their distribution in the host tissue (Figure 14.4) (Rosen et al. 2007).

Beyond vital labeling of cells with QDs, one can capitalize on their other unique features. As mentioned, due to the tight band of emission wavelength for a given size particle, it is possible to image multiple QD particles in a single tissue sample. Therefore, it is theoretically possible to deliver two populations of cells to an animal model, each labeled with QDs of different sizes (and therefore emitting at two separate wavelengths) and track their location in time and space. Furthermore, for tracking the fate of delivered cells in histological sections, tissue can be stained with antibodies conjugated to QD particles rather than traditional fluorophores (Rosen et al. 2007). Wu et al. highlighted the ability of QDs for use in multiplex immunostaining of fixed cells (Wu et al. 2003).

---

* Indeed, peptide-conjugated QDs are commercially available and advertised as cell-tracking agents. In our experience, these kits do not permit optimal labeling (Rosen et al. 2007).

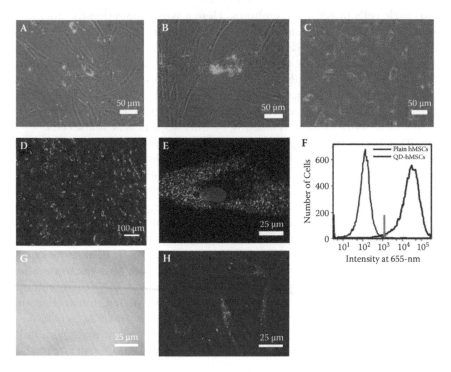

**FIGURE 14.3 (See color insert.)**
Quantum dots (QDs) were loaded into human mesenchymal stem cells (hMSCs) by three approaches. Panels A–C show images of QD fluorescence (655 nm, red) with phase contrast overlays. (A) Electroporation of cells in media containing carboxylated-655-nm QDs resulted in perinuclear aggregation of the dots, nonuniform labeling, and cell death. (B) Use of the receptor-mediated-based Qtracker kit (Quantum Dot, Inc.) again resulted in nonuniform cellular loading with perinuclear aggregation but with better cell survival than with electroporation. (B) In contrast, passively incubating hMSCs in naked QD medium resulted in nearly 100% loading with a pattern that extends to the cell borders. (D) The field in (C) is imaged for QD fluorescence without the phase overlay to demonstrate homogeneity and brightness. The intracellular QD cluster distribution is diffusely cytoplasmic (D) and (E) and largely excludes the nucleus (blue, Hoechst 33342 dye). (F) QD loading efficiency was analyzed using flow cytometry. The threshold for plain hMSCs (gray line) was set such that the intensity range encompassed at least 98% of the control cells (red arrow indicates upper bound of control range). QD-positive status was designated for all cells in the QD-hMSC sample having intensities above the range set for the control group. In four experiments, QD-positive cells (black line) were found in 96% of over 17,000 viable cells. (Panels C–F adapted with permission from Rosen, A.B., D.J. Kelly, A.J. Schuldt, J. Lu, I.A. Potapova, S.V. Doronin, K.J. Robichaud, R.B. Robinson, M.R. Rosen, P.R. Brink, G.R. Gaudette, and I.S. Cohen. 2007. *Stem Cells* 25, no. 8: 2128–38.) (G) GFP-transfected and (H) QD-labeled hMSCs were each seeded onto a 100-μm thick extracellular matrix patch (porcine bladder) and imaged for their respective emissions. Only intracellular QD fluorescence (H) is detectable above autofluorescence, whereas GFP-labeled cells cannot be seen above autofluorescence in (G).

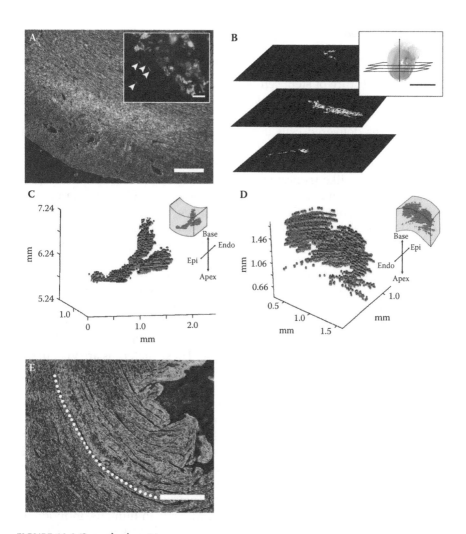

**FIGURE 14.4 (See color insert.)**
QDs can be used to identify single cells (hMSCs) after injection into the rat heart and further used to reconstruct the three-dimensional (3D) distribution of all delivered cells. Rat hearts were injected with QD hMSCs and subsequently excised and processed for tissue analysis. Tissue sections were imaged for QD fluorescence emission (655 nm) with phase overlay. QD hMSCs can be visualized at (A) low power and (inset) high power (Hoechst 33342 dye used to stain nuclei blue). In (A), inset, endogenous nuclei can be seen adjacent to the delivered cells in the midmyocardium (arrows). (B) Binary masks (where white pixels depict QD-positive zones in the image) for all QD-positive sections were used to generate the 3D reconstruction of delivered cells in the tissue. (C) QD hMSC reconstruction in an animal that was terminated 1 hour after injection. (D) Reconstruction from an animal euthanized 1 day after injection with orientation noted in inset. (E) One day after injection into the heart, the pattern of QD hMSCs is well organized and appears to mimic the endogenous myocardial orientation (dotted white line highlights myofibril alignment). (Adapted with permission from Rosen, A.B., D.J. Kelly, A.J. Schuldt, J. Lu, I.A. Potapova, S.V. Doronin, K.J. Robichaud, R.B. Robinson, M.R. Rosen, P.R. Brink, G.R. Gaudette, and I.S. Cohen. 2007. *Stem Cells* 25, no. 8: 2128–38.)

## 14.4  Advances in Noninvasive Imaging Using Quantum Dots

The long-term future of stem cell therapies will depend in part on the ability to track noninvasively and ensure the safety and efficacy of treatments. What might be considered the "holy grail" of stem cell tracking is a label that visualizes delivered cells *in vivo* noninvasively with high resolution. Existing techniques for noninvasive tracking of stem cells have been discussed in previous chapters and in short include loading cells with radioactive substances, like truncated thymidine kinase for PET detection (Cao et al. 2006; MacLaren et al. 2000) or radioisotopes for PET (Hofmann et al. 2005) or single-photon emission computed tomography (SPECT) (Barbash et al. 2003; Kraitchman et al. 2005). Concerns have arisen, however, over the uptake of the label by host tissue, endogenous tissue photon attenuation, and high levels of label needed for detection (Chemaly et al. 2005). Another approach is the transfection of cells with the gene for luciferase to enable visualization using bioluminescence imaging. Major problems with this optical approach include the absorption and scatter of visible light and use of nonhuman genetic material (Chemaly et al. 2005). More commonly used for noninvasive cell tracking are metals like iron (SPIO) (Hill et al. 2003; Kraitchman et al. 2003) and gadolinium [34]. These materials are visualized using MRI. For gadolinium, difficulties arise in loading the cells with sufficient concentrations to permit T1 contrast. In addition, dilution effects will be pronounced as cells divide (Chemaly et al. 2005). SPIOs, like ferumoxides (Feridex, Berlex Laboratories), are most frequently used, but conflicting studies exist on whether these particles interfere with differentiation of stem cells, specifically chondrogenesis of mesenchymal stem cells (Arbab et al. 2004; Bulte et al. 2004; Henning et al. 2009; Kostura et al. 2004). If true, this would suggest they are not a "stealth" particle within the cell and could potentially interfere with other important physiologic functions. Further, should the technique be extended to clinical trials in humans, individuals with electronic pacemakers or implantable defibrillators may be excluded from MRI studies (Martin 2005; Martin et al. 2004). This would isolate a potentially needy patient population, especially in trials using stem cells for cardiac repair.

One promising avenue for noninvasive stem cell tracking with QDs is the use of those nanoparticles emitting near-infrared (NIR) fluorescence (NIR QDs). Kim et al. used NIR QDs as a tracer for sentinel lymph node mapping in rodents and detected the label in cancerous nodes noninvasively after intradermal injection (Kim et al. 2004). While in principle this should suggest that NIR QDs can be used as noninvasive stem cell-tracking agents *in vivo*, the method is currently limited to detection at approximately 1 cm below the skin. For the purposes of tracking stem cells in deeper organs, this is likely insufficient depth penetration. However, newer detection modalities using frequency domain photon migration (Ntziachristos et al. 2003) may permit detection as deep as 10 cm. Therefore, NIR QDs may soon prove useful for noninvasive stem cell tracking.

Another approach toward using QDs for noninvasive stem cell tracking is to capitalize on their inherent features and use existing technologies for detection. Both the core of the QD and the passivating shell contain metal ions (cadmium and zinc, respectively). Metals are radiopaque and can, therefore, be imaged using X-ray technology. Traditional X-rays are too large to interfere with nano- or microscale metals, but it might be possible to use micro-computed tomography (μCT) scanning to image QDs *in vivo*. Based on theoretical data* and preliminary experiments in our laboratory, we have found that μCT can detect QD-labeled stem cells in both explanted tissue samples and in living animals. Still, further experiments will be necessary before this approach can be applied in large animals.

Finally, it may also soon be possible to track QD-labeled stem cells noninvasively using MRI. Several laboratories have attempted to create paramagnetic QDs by coating the particles with designer lipids (Mulder et al. 2006) or doping with metals such as manganese (Yong 2009). QDs may represent a more favorable option for this approach than SPIOs as they will likely load more easily into the cells and avoid the potential problem of interfering with differentiation. While this technology is not yet optimized, paramagnetic QDs should permit noninvasive detection of populations of labeled stem cells in deep tissue and the ability to identify single cells via QD fluorescence in histological sections of explanted tissue samples without secondary staining (Bakalova and Zheler 2007).

## 14.5 Summary and Limitations

QDs appear to satisfy many, but not all, of the criteria of an "ideal" tracking agent as listed in Section 14.1. They can be loaded into cells without genetic or mechanical perturbation, and once inside the cell they are biocompatible, safe, and nontoxic; they do not interfere with cell proliferation or differentiation or transfer to adjacent (unlabeled) cells. In at least the case of labeling human mesenchymal stem cells, nearly 100% of the cells uniformly take up QDs (Rosen et al. 2007). QD-labeled cells can be delivered to animals, subsequently identified in the host tissue with single-cell resolution (in histological sections of explanted tissue), and the number of cells can be quantified. However, there are some important limitations to their usage. First, as with any exogenous label inside a cell, the threshold of detection is based in part on the dilution of that label over multiple cell divisions. In our hands,

---

* Data are based primarily on the difference in density between QD-labeled stem cells and surrounding mammalian tissue.

QD-labeled human mesenchymal stem cells (hMSCs) can be tracked over 8 weeks *in vitro* and after delivery to the canine heart *in vivo*. The QD label is retained by these cells even after differentiation to various phenotypes *in vitro* and *in vivo* (Rosen et al. 2007). However, one concern is the nonuniform dilution of the label that may occur within a population of cells; those dividing more slowly will appear brighter in the tissue, whereas those that proliferate rapidly will more quickly fall below the threshold of detection. Further studies—perhaps using techniques such as flow cytometry to characterize the fluorescence intensity of populations of labeled stem cells over multiple passages—will be necessary to define the time frame over which QD-labeled stem cells can be studied. In addition, with rare exception, the utility of the method is currently limited to studying explanted tissue samples from animals in which QD-labeled cells have been delivered. For the method to be truly noninvasive, future work is needed in developing technologies to detect QDs within living host organs. For this, applications using NIR QDs, µCT, or MRI may represent the future of noninvasive stem cell tracking.

# References

Aime, S., L. Calabi, C. Cavallotti, E. Gianolio, G.B. Giovenzana, P. Losi, A. Maiocchi, G. Palmisano and M. Sisti. 2004. [Gd-AAZTA]-: A new structural entry for an improved generation of MRI contrast agents. *Inorg Chem* 43, no. 24: 7588-90.

Amado, L.C., A.P. Salrais, K.H. Schuleri, M.E. St. John, J. Xie, S.M. Cattaneo, D.J. Durand, T. Fitton, J.Q. Kuang, G.C. Stewart, S. Lehrke, W.A. Baumgartner, B.J. Martin, A.W. Heldman, and J.M. Hare. 2005. Cardiac repair with intramyocardial injection of allogeneic mesenchymal stem cells after myocardial infarction. *Proc Natl Acad Sci USA* 102, no. 32: 11474–79.

Arbab, A.S., G.T. Yocum, H. Kalish, E.K. Jordan, S.A. Anderson, A.Y. Khakoo, E.J. Read, and J.A. Frank. 2004. Efficient magnetic cell labeling with protamine sulfate complexed to ferumoxides for cellular MRI. *Blood* 104, no. 4: 1217–23.

Bakalova, R. and Z. Zhelev. 2007. Designing quantum-dot probes. *Nature Photonics* 1: 487-89.

Barbash, I.M., P. Chouraqui, J. Baron, M.S. Feinberg, S. Etzion, A. Tessone, L. Miller, E. Guetta, D. Zipori, L.H. Kedes, R.A. Kloner, and J. Leor. 2003. Systemic delivery of bone marrow-derived mesenchymal stem cells to the infarcted myocardium: Feasibility, cell migration, and body distribution. *Circulation* 108, no. 7: 863–68.

Billinton, N. and A.W. Knight. 2001. Seeing the wood through the trees: A review of techniques for distinguishing green fluorescent protein from endogenous autofluorescence. *Anal Biochem* 291, no. 2: 175–97.

Bruchez, M., Jr., M. Moronne, P. Gin, S. Weiss, and A.P. Alivisatos. 1998. Semiconductor nanocrystals as fluorescent biological labels. *Science* 281, no. 5385: 2013–16.

Bulte, J.W., D.L. Kraitchman, A.M. Mackay, M.F. Pittenger, A.S. Arbab, G.T. Yocum, H. Kalish, E.K. Jordan, S.A. Anderson, A.Y. Khakoo, E.J. Read, and J.A. Frank. 2004. Chondrogenic differentiation of mesenchymal stem cells is inhibited after magnetic labeling with ferumoxides. *Blood* 104, no. 10: 3410–13.

Cao, F., S. Lin, X. Xie, P. Ray, M. Patel, X. Zhang, M. Drukker, S.J. Dylla, A.J. Connolly, X. Chen, I.L. Weissman, S.S. Gambhir, and J.C. Wu. 2006. In vivo visualization of embryonic stem cell survival, proliferation, and migration after cardiac delivery. *Circulation* 113, no. 7: 1005–14.

Chan, W.C. and S. Nie. 1998. Quantum dot bioconjugates for ultrasensitive nonisotopic detection. *Science* 281, no. 5385: 2016–18.

Chemaly, E.R., R. Yoneyama, J.V. Frangioni, and R.J. Hajjar. 2005. Tracking stem cells in the cardiovascular system. *Trends Cardiovasc Med* 15, no. 8: 297–302.

Derfus, A.M., W.C.W. Chan, and S.R. Bhatia. 2004. Intracellular delivery of quantum dots for live cell labeling and organelle tracking. *Adv Mater* 16, no. 12: 961–66.

Dick, A.J., M.A. Guttman, V.K. Raman, D.C. Peters, B.S. Pessanha, J.M. Hill, S. Smith, G. Scott, E.R. Mcveigh, and R.J. Lederman. 2003. Magnetic resonance fluoroscopy allows targeted delivery of mesenchymal stem cells to infarct borders in swine. *Circulation* 108, no. 23: 2899–2904.

Dubertret, B., P. Skourides, D.J. Norris, V. Noireaux, A.H. Brivanlou, and A. Libchaber. 2002. In vivo imaging of quantum dots encapsulated in phospholipid micelles. *Science* 298, no. 5599: 1759–62.

Frangioni, J.V. and R.J. Hajjar. 2004. In vivo tracking of stem cells for clinical trials in cardiovascular disease. *Circulation* 110, no. 21: 3378–83.

Gao, X., Y. Cui, R.M. Levenson, L.W. Chung, and S. Nie. 2004. In vivo cancer targeting and imaging with semiconductor quantum dots. *Nat Biotechnol* 22, no. 8: 969–76.

Hanaki, K., A. Momo, T. Oku, A. Komoto, S. Maenosono, Y. Yamaguchi, and K. Yamamoto. 2003. Semiconductor quantum dot/albumin complex is a long-life and highly photostable endosome marker. *Biochem Biophys Res Commun* 302, no. 3: 496–501.

Henning, T.D., E.J. Sutton, A. Kim, D. Golovko, A. Horvai, L. Ackerman, B. Sennino, D. Mcdonald, J. Lotz, and H.E. Daldrup-Link. 2009. The influence of ferucarbotran on the chondrogenesis of human mesenchymal stem cells. *Contrast Media Mol Imaging* 4, no. 4: 165–73.

Hill, J.M., A.J. Dick, V.K. Raman, R.B. Thompson, Z.X. Yu, K.A. Hinds, B.S. Pessanha, M.A. Guttman, T.R. Varney, B.J. Martin, C.E. Dunbar, E.R. Mcveigh, and R.J. Lederman. 2003. Serial cardiac magnetic resonance imaging of injected mesenchymal stem cells. *Circulation* 108, no. 8: 1009–14.

Hofmann, M., K.C. Wollert, G.P. Meyer, A. Menke, L. Arseniev, B. Hertenstein, A. Ganser, W.H. Knapp, and H. Drexler. 2005. Monitoring of bone marrow cell homing into the infarcted human myocardium. *Circulation* 111, no. 17: 2198–2202.

Jackson, K.A., S.M. Majka, H. Wang, J. Pocius, C.J. Hartley, M.W. Majesky, M.L. Entman, L.H. Michael, K.K. Hirschi, and M.A. Goodell. 2001. Regeneration of ischemic cardiac muscle and vascular endothelium by adult stem cells. *J Clin Invest* 107, no. 11: 1395–1402.

Jaiswal, J.K., H. Mattoussi, J.M. Mauro, and S.M. Simon. 2003. Long-term multiple color imaging of live cells using quantum dot bioconjugates. *Nat Biotechnol* 21, no. 1: 47–51.

Kajstura, J., M. Rota, B. Whang, and E. Al. 2005. Bone marrow cells differentiate in cardiac cell lineages after infarction independently of cell fusion. *Circ Res* 96, no. 1: 127–37.

Kim, S., Y.T. Lim, E.G. Soltesz, A.M. De Grand, J. Lee, A. Nakayama, J.A. Parker, T. Mihaljevic, R.G. Laurence, D.M. Dor, L.H. Cohn, M.G. Bawendi, and J.V. Frangioni. 2004. Near-infrared fluorescent type II quantum dots for sentinel lymph node mapping. *Nat Biotechnol* 22, no. 1: 93–97.

Kostura, L., D.L. Kraitchman, A.M. Mackay, M.F. Pittenger, and J.W. Bulte. 2004. Feridex labeling of mesenchymal stem cells inhibits chondrogenesis but not adipogenesis or osteogenesis. *NMR Biomed* 17, no. 7: 513–17.

Kraitchman, D.L., A.W. Heldman, E. Atalar, L.C. Amado, B.J. Martin, M.F. Pittenger, J.M. Hare, and J.W. Bulte. 2003. In vivo magnetic resonance imaging of mesenchymal stem cells in myocardial infarction. *Circulation* 107, no. 18: 2290–93.

Kraitchman, D.L., M. Tatsumi, W.D. Gilson, T. Ishimori, D. Kedziorek, P. Walczak, W.P. Segars, H.H. Chen, D. Fritzges, I. Izbudak, R.G. Young, M. Marcelino, M.F. Pittenger, M. Solaiyappan, R.C. Boston, B.M. Tsui, R.L. Wahl, and J.W. Bulte. 2005. Dynamic imaging of allogeneic mesenchymal stem cells trafficking to myocardial infarction. *Circulation* 112, no. 10: 1451–61.

Laflamme, M.A. and C.E. Murry. 2005. Regenerating the heart. *Nat Biotechnol* 23, no. 7: 845–56.

Maclaren, D.C., T. Toyokuni, S.R. Cherry, J.R. Barrio, M.E. Phelps, H.R. Herschman, and S.S. Gambhir. 2000. PET imaging of transgene expression. *Biol Psychiatry* 48, no. 5: 337–48.

Martin, E.T. 2005. Can cardiac pacemakers and magnetic resonance imaging systems co-exist? *Eur Heart J* 26, no. 4: 325–27.

Martin, E.T., J.A. Coman, F.G. Shellock, C.C. Pulling, R. Fair, and K. Jenkins. 2004. Magnetic resonance imaging and cardiac pacemaker safety at 1.5-Tesla. *J Am Coll Cardiol* 43, no. 7: 1315–24.

Mulder, W.J., R. Koole, R.J. Brandwijk, G. Storm, P.T. Chin, G.J. Strijkers, C. De Mello Donega, K. Nicolay, and A.W. Griffioen. 2006. Quantum dots with a paramagnetic coating as a bimodal molecular imaging probe. *Nano Lett* 6, no. 1: 1–6.

Ntziachristos, V., C. Bremer, and R. Weissleder. 2003. Fluorescence imaging with near-infrared light: new technological advances that enable in vivo molecular imaging. *Eur Radiol* 13, no. 1: 195–208.

Parish, C.R. 1999. Fluorescent dyes for lymphocyte migration and proliferation studies. *Immunol Cell Biol* 77, no. 6: 499–508.

Potapova, I., A. Plotnikov, Z. Lu, and et al. 2004. Human mesenchymal stem cells as a gene delivery system to create cardiac pacemakers. *Circ Res* 94, no. 7: 952–59.

Rieger, S., R.P. Kulkarni, D. Darcy, S.E. Fraser, and R.W. Koster. 2005. Quantum dots are powerful multipurpose vital labeling agents in zebrafish embryos. *Dev Dyn* 234, no. 3: 670–81.

Rosen, A.B., D.J. Kelly, A.J. Schuldt, J. Lu, I.A. Potapova, S.V. Doronin, K.J. Robichaud, R.B. Robinson, M.R. Rosen, P.R. Brink, G.R. Gaudette, and I.S. Cohen. 2007. Finding fluorescent needles in the cardiac haystack: Tracking human mesenchymal stem cells labeled with quantum dots for quantitative in vivo three-dimensional fluorescence analysis. *Stem Cells* 25, no. 8: 2128–38.

Silver, J. and W. Ou. 2005. Photoactivation of quantum dot fluorescence following endocytosis. *Nano Lett* 5, no. 7: 1445–49.

So, M.K., C. Xu, A.M. Loening, S.S. Gambhir, and J. Rao. 2006. Self-illuminating quantum dot conjugates for in vivo imaging. *Nat Biotechnol* 24, no. 3: 339–43.

Thompson, R.B., S.M. Emani, B.H. Davis, E.J. Van Den Bos, Y. Morimoto, D. Craig, D. Glower, and D.A. Taylor. 2003. Comparison of intracardiac cell transplantation: autologous skeletal myoblasts versus bone marrow cells. *Circulation* 108 Suppl 1: II-264–71.

Wu, X., H. Liu, J. Liu, K.N. Haley, J.A. Treadway, J.P. Larson, N. Ge, F. Peale, and M.P. Bruchez. 2003. Immunofluorescent labeling of cancer marker Her2 and other cellular targets with semiconductor quantum dots. *Nat Biotechnol* 21, no. 1: 41–46.

Yong, K.T. 2009. Mn-doped near-infrared quantum dots as multimodal targeted probes for pancreatic cancer imaging. *Nanotechnology* 20, no. 1: 15102.

Zhang, Y. and N. Huang. 2006. Intracellular uptake of CdSe-ZnS/polystyrene nanobeads. *Mater Res Part B: Appl Biomater* 76B: 161–68.

# 15

# *Magnetic Resonance Imaging Cell-Labeling Methods*

**Ali S. Arbab and Joseph A. Frank**

**CONTENTS**

## 15.1 Introduction

Personalized treatment using stem cells that are modified or genetically engineered is becoming a reality in the field of medicine. Allogeneic or autologous cells can be used for treatment and possibly for early diagnosis of diseases. Hematopoietic, stromal, and organ-specific stem cells are under evaluation for cell-based therapies for cardiac, neurological, autoimmune, and other disorders (Alessandri et al. 2004; Ben-Hur et al. 2005; Blank et al. 2009; Chachques et al. 2007; Clavel and Verfaillie 2008; Cleland et al. 2008; Heinrich et al. 2009; Manganas and Maletic-Savatic 2005; Walker et al. 2009). However, to advance stem cell therapy from its current state to clinical practice, an investigator has to know the status of the administered cells by an efficient and noninvasive imaging method to monitor the transplanted cells. Currently, different imaging modalities, such as positron emission tomography (PET), single-photon emission computed tomography (SPECT), optical imaging, and magnetic resonance imaging (MRI) are being utilized to track the administered cells (Acton and Zhou 2005; Arbab and Frank 2008; Dingli et al. 2006; Gilad et al. 2008; Hoshino et al. 2007; Politi et al. 2007; Wu et al. 2006). For tracking administered cells by any modality, cells are to be manipulated *ex vivo* with compatible imaging probes.

Because MRI is noninvasive and offers high spatial resolution, it provides an excellent means for *in vivo* cell tracking. To be able to track the administered cells by MRI, cells must be manipulated *ex vivo* by incorporating different MRI contrast agents that do not exhibit toxic effects in labeled cells, enable sufficient threshold detection by MRI, and do not elicit side effects in human recipients. To date, paramagnetic, superparamagnetic, and fluorinated MR contrast agents have been used to label different cell types (Arbab et al. 2006a; Gilad et al. 2008; Partlow et al. 2007; Wilhelm and Gazeau 2008). Superparamagnetic iron oxide (SPIO) nanoparticles are extensively used to label cells for *in vivo* tracking by MRI because they are readily available and biodegradable and have a higher threshold for toxicity (Bulte 2006). In this chapter, different MR contrast agents that are being used to label cells and the different methods that are employed to label cells are discussed. Details of a step-by-step labeling method using ferumoxides and protamine sulfate are also discussed.

## 15.2 Types of Magnetic Nanoparticles Used in Cell Labeling for MRI

MR contrast agents used to label cells can exhibit properties of either paramagnetism (T1 shortening agent) or superparamagnetism (T2 or T2* shortening agent). These agents alter the nuclear magnetic resonance (NMR) relaxation times of the water protons in solution or tissue known as T1, T2, and T2*. Fluorine-based ($^{19}$F) contrast agents are also used to label cells. Instead of targeting water protons, MR coils are tuned to receive signals from $^{19}$F. The advantage of $^{19}$F is that there will be no background noise as there is no natural $^{19}$F available in the biological tissues.

### 15.2.1 Paramagnetic Agents and Methods of Cell Labeling

*Paramagnetism* refers to the ability of a metal such as manganese, gadolinium, or iron to interact with water protons through dipole-dipole interaction with direct inner-sphere effects resulting in a shortening of NMR relaxation times; it is usually associated with enhancement (increase in signal intensity) on T1-weighted images. Gadolinium chelates (i.e., gadolinium diethylenetriaminepentaacetic acid [Gd-DTPA], gadolinium tetraazacyclododecanetetraacetic acid [Gd-DOTA], or [Gd-DO3A]) and manganese chloride are paramagnetic contrast agents used in experimental and clinical studies. These agents tend to shorten T1 relaxation times greater than the T2 and T2* of tissues. The following methods are used to label cells using paramagnetic agents: simple incubation and mechanical methods.

#### 15.2.1.1 Simple Incubation

Gadolinium (Gd) chelate-based contrast agents for cell labeling have been used to label cells *ex vivo* with limited results. Reports indicated modest proton relaxation enhancement or in some cases no T1 enhancement when gadolinium-chelated agents were used for cell labeling (Crich et al. 2004, 2005a; Daldrup-Link et al. 2004; Giesel et al. 2006 ; Himmelreich et al. 2006; Modo et al. 2002; Su et al. 2007). Thus, the need remains to identify an agent that will exert a strong T1 effect, allowing detection of cells in disease models in which the cells need to be conspicuous versus surrounding tissues, labeled cell numbers are low, or the concentration of gadolinium is low.

Manganese (Mn) chloride was the first paramagnetic contrast agent used in MRI, and it has been shown that it can be taken up by cells *in vivo* through calcium channels in the cell membrane (Aoki et al. 2006; Mendonca-Dias et al. 1983; Wolf et al. 1985). Aoki et al. reported that lymphocytes could be labeled following incubation with manganese chloride. MRI of cells in gelatin demonstrated increased signal intensity on T1-weighted images; however, it is not clear whether there would be sufficient contrast enhancement to detect Mn-labeled cells *in vivo* by MRI. Cells have been labeled with paramagnetic manganese oxide (MnO) nanoparticles, and the enhancement could be detected using standard T1-weighted imaging, although further work is needed to fully understand the uptake and safety of this agent in stem cells (Na et al. 2007). The major drawback to the use of Mn as an MR contrast agent is its narrow therapeutic window and potential toxicity.

Giesel et al. (2006) were able to label mesenchymal stem cells (MSCs) using a bifunctional gadofluorine M-Cy3.5 for both MRI and optical imaging. Gadofluorine M-Cy3.5 is designed with a hydrophilic tail that allows the agent to insert in the cell wall and then become internalized into cytosol. Intracerebral implantation of $10^6$ gadofluorine M-Cy3.5-labeled MSC allowed for clear visualization of cells in the rat brain on T1-weighted imaging at clinically relevant 1.5-T field strength that was confirmed by fluorescent microscopy. Anderson et al. (2006) used gadolinium fullerenol, which has higher relativities than conventional gadolinium chelates, to label MSCs. Gadolinium fullerenol-labeled MSCs could be detected on 7-T MRI following direct injection of $10^6$ cells into the rat thigh. Gadolinium fullerenol labeling decreased the stem cell proliferation initially, suggesting that the agent may be altering mitochondrial function. Brekke et al. (2007) used a combination gadolinium chelate with fluorescent tag to label cells and noted a significant decrease in proliferation and an increase in reactive oxygen species with 24 hours of incubation. The transient negative effect of a gadolinium-based agent on cell proliferation used for cellular and molecular imaging will require further evaluation to ensure there is no long-term toxicity or ability of the cells to repair damage.

### 15.2.1.2 Mechanical Methods

Direct injection of high concentrations of gadolinium chelates into *Xenopus laevis* eggs enabled tracking of the labeled cell proliferation and migration during development using MRI and optical imaging (Jacobs and Fraser 1994). However, this approach is not a practical or efficient method for labeling mammalian cells with MR contrast agents. Electroporation has been used to label cells with gadolinium chelates and SPIO nanoparticles (Crich et al. 2005a; Walczak et al. 2005). There is relatively little experience using this approach with MRI contrast agents to label cells, and the long-term effects on cell viability when using this method are unknown. Electroporation is commonly used to introduce DNA into the cell genome, but the associated cell stress due to chemical imbalances and efflux or influx of chemicals from within the cell and surrounding medium often alters cell viability and survival. The type, size, and number of cells; media conditions; the magnitude and duration of the electric pulse; and the handling of cells after electroporation may all be factors that influence cell viability and survival following electroporation with an MR contrast agent. It has been shown that a significant amount of cell lysis and death occurred during electroporation and following labeling with contrast agents (Terreno et al. 2006).

The electroporation method has also been used to label rat glioma cells using MnO nanoparticles for effective labeling and tracking in rat brain after implantation using a 9.4-T animal imaging system (Gilad et al. 2008). However, the MnO-labeled cells were not clearly visualized after 3 days of implantation. Shapiro and Koretsky also used $MnCO_3$ and $MnO_3$ nanoparticles to label cells by simple incubation. However, *in vivo* tracking has not been reported by this group. Similarly, Sotak et al. have reported effective labeling of murine hepatocytes using $Mn^{+3}$-transferrin, but *in vivo* studies have not been performed (Sotak et al. 2008). Investigators are actively working on making engineered nanoparticles containing different metals that can elicit both T1 and T2 effects (Zhang et al. 2008).

## 15.2.2 Superparamagnetic Agents and Methods of Cell Labeling

Superparamagnetic iron oxide nanoparticles are a family of MRI contrast agents presently used to efficiently label cells for cellular imaging. There are various methods used to prepare SPIO nanoparticles, resulting in a wide range of physiochemical differences, including core size (e.g., ultrasmall SPIO [USPIO] vs. SPIO), shape, mono- or oligocrystalline composition, and outer coating that may alter the ability to use these agents to label cells. The coating molecules will contribute to the surface charge or zeta potential of the USPIO/SPIO in water. The zeta potential is the average potential difference in millivolts existing between the surface of the USPIO/SPIO nanoparticles immersed in a conduction liquid (water) and the bulk of the liquid. Dextran-coated ferumoxides, an SPIO, have a zeta potential of −32 mV, while ferumoxtran-10, a USPIO with a shorter chain dextran, has a measured zeta potential of −2.0 to 0 mV (Kalish et al. 2003), with the near-neutral surface potential of ferumoxtran-10 possibly contributing to the long blood half-life compared to the larger SPIO nanoparticles. Zeta potentials have not been reported for other USPIO nanoparticles characterized in the literature as anionic or cationic because of the coating. The SPIO nanoparticles have been characterized as either anionic or cationic (negative or positive zeta potential), which will in turn determine the ability of the contrast agent to interact with cell/plasma membrane. Cationic USPIO nanoparticles have been used to label primary and cloned cell populations in culture through the cell surface membrane, such as HeLa cells for *in vitro* studies. HeLa cells have a negative zeta potential (Walliser and Redmann 1978), and the USPIO agent initially interacts with the plasma membrane through electrostatic interactions followed by endocytosis. The following methods can be used to label cells using SPIOs: simple incubation, mechanical, antigen-antibody mediated, and viral shell mediated.

### 15.2.2.1 Simple Incubation

Both modified and unmodified SPIOs have been used to label cells by the simple incubation method. As mentioned, the surface of most of the commercially available USPIO nanoparticles is negatively charged. Since the electric charge of the cell surface is also negative, the surface charge of the nanoparticles should be modified to enable efficient nanoparticle uptake by the cells. Investigators have modified the surface charge of the nanoparticles by coating them with cationic materials or modified the surface of the coating by attaching membrane-penetrable peptides. Commercially available transfection agents are commonly used to modify the surface charge of the nanoparticles, thereby facilitating the uptake by cells through various endocytosis mechanisms (Arbab et al. 2003a, 2004c, 2006a; Frank et al. 2003). Modified or nonmodified SPIOs are added to the cell cultures at varying concentrations and incubated for a few hours to a few days for adequate cell labeling. However, due to the potential adverse effect of iron oxides and transfection agents, long-term incubation is not usually recommended.

Investigators are continually trying to optimize the labeling of different types of cells based on incubation time and concentration of SPIOs. Our group has developed a technique to create complexes of ferumoxides (an agent approved by the Food and Drug Administration [FDA]) with transfection agents to facilitate cellular uptake by endocytosis (Arbab et al. 2003a, 2003b, 2004a, 2004c; Frank et al. 2003). We introduced the use of protamine sulfate, an FDA-approved agent, instead of commonly used cellular transfection agents such as lipofectamine, to generate ferumoxide-protamine sulfate (FePro) complexes for efficient labeling of different mammalian cells, including stem cells and T lymphocytes (Montet-Abou et al. 2007). These labeled cells have been used in different animal models and tracked by both high-field-strength research and clinical-strength MRI systems (Arbab and Frank 2008; Arbab et al. 2006b, 2007). One of the advantages of labeling cells using FDA-approved agents is the possibility of clinical trial without facing major regulatory hurdles related to contrast and transfection agents. We have reported modification of our method to efficiently label cells using ferumoxides and protamine sulfate within 4 hours (Janic et al. 2009). The following are the step-by-step methods to label suspensions as well as adherent cells using our modified method. For suspension cells (such as hematopoietic stem cells, T cells, etc.), Figure 15.1 shows the schematic representation of the labeling procedures.

Materials needed:

1. RPMI-1640 medium with L-glutamine, MEM (minimum essential medium) nonessential amino acid, and sodium pyruvate
2. Ferumoxide (Feridex, 11.2 mg/Fe per mL, Berlex Laboratory, NJ)
3. Protamine sulfate (American Pharmaceuticals Partners, IL)
4. 15- or 50-mL centrifuge tubes
5. Six-well plates (each well has around 10 cm$^2$ growth surface)

Procedures:

1. Count the cells and put into 15- or 50-mL tubes (based on cell numbers).
2. Centrifuge and decant the supernatant.
3. Add serum-free RPMI-1640 to the cell suspension, mix, and centrifuge it again. (This is important to get rid of all serum.)
4. Decant the supernatant and add serum-free medium again to make $4 \times 10^6$ cells per milliliter of the serum-free medium.
5. Add 100 μg of Feridex (9 μL from the bottle) for every milliliter of cell suspension and mix well by gentle pipetting.

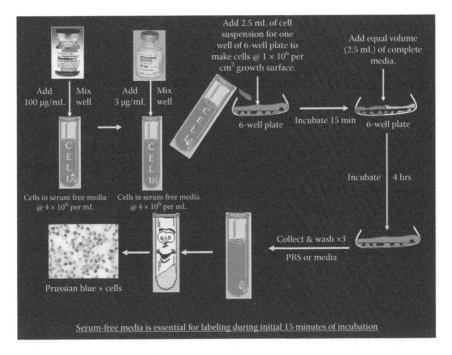

**FIGURE 15.1**
Schematic representation of the superparamagnetic iron oxide labeling procedure for suspension cells.

6. Then, add 3 µg of freshly prepared (in distilled water) protamine sulfate for every milliliter of cell suspension and mix well by gentle pipetting.

7. Transfer the cells to a six-well plate at a concentration of $1 \times 10^6$ per 1-cm$^2$ area. For example, add 2.5 mL cell suspension ($10 \times 10^6$ cells) to a single well in a six-well plate. (The number of cells per square centimeter is important. Please see Figure 15.2 for the importance of cell number per square centimeter of growth surface.)

8. Incubate at 37°C for 15 minutes.

9. Then, add an equal volume of complete medium (containing serum) to the cell suspension. Complete medium should contain serum and other growth factors for respective cells.

10. Incubate further for 4 hours.

11. Collect and wash the cells.

12. Determine the labeling efficiency and functional status of labeled cells.

Note that if more iron per cell is required, the initial concentration of ferumoxides per milliliter of cell suspension may be increased to 200 µg/mL (18

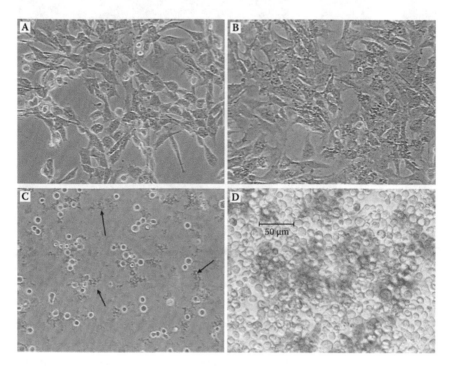

**FIGURE 15.2**
Importance of cell number during incubation. (A), (B) Neural stem cell labeling (adherent cells) using ferumoxide and protamine sulfate. Ferumoxide was added to the cells (after getting rid of old medium and washing with serum-free medium) at 100 μg/mL, mixed well with the serum-free medium, and then freshly prepared (making a stock solution of 1 mg/1 mL of distilled water). Protamine sulfate was added to the medium at 3 μg/mL. The images were obtained with an inverted microscope at ×40 magnification. Note there are no definite complexes seen 5 minutes after addition of ferumoxide and protamine sulfate (A), but small complexes (black dots) are seen attached to the cells at 4 hours (B). However, no large complexes are seen as all of the complexes are distributed equally and become attached to the cells; there are no free complexes to make larger chunks. (C) and (D) Hematopoietic stem cell labeling (suspension cells) using ferumoxide and protamine sulfate. Ferumoxide was added to the cells (after discarding old medium and washing with serum-free medium) at 100 μg/mL, mixed well with the serum-free medium, and then freshly prepared (making a stock solution of 1 mg/1 mL of distilled water) protamine sulfate was added to the medium at 3 μg/mL. Then, the cells were plated in a six-well plate at a cell density of $2 \times 10^5$ and $1 \times 10^6$ cells per square centimeter of growth surface. Note the formation of chunks of complexes (arrows in C) within 5 minutes after incubation. There are no large complexes seen even after 4 hours of incubation (D) when the density of cells is maintained at $1 \times 10^6$ per square centimeter.

µL of Feridex). However, the ratio of protamine sulfate must be kept the same. For 200 µg/mL of ferumoxide, 6 µg of protamine sulfate should be added.

For adherent cells (such as MSCs, neural stem cells, etc.), Figure 15.3 shows a schematic representation of the labeling procedures.

Materials needed:

1. RPMI-1640 medium with L-glutamine, MEM nonessential amino acid, and sodium pyruvate

2. Ferumoxide (Feridex, 11.2 mg/Fe per milliliter, Berlex Laboratory)

3. Protamine sulfate (American Pharmaceuticals Partners)

4. 15- or 50-mL centrifuge tubes

5. Cell culture flasks or plates

Procedures: Adherent cell labeling (MSCs) in T75 culture flask

1. Culture cells to 80–90% confluence.

2. Remove old culture medium completely, add 10 mL serum-free RPMI-1640, and wash the cells.

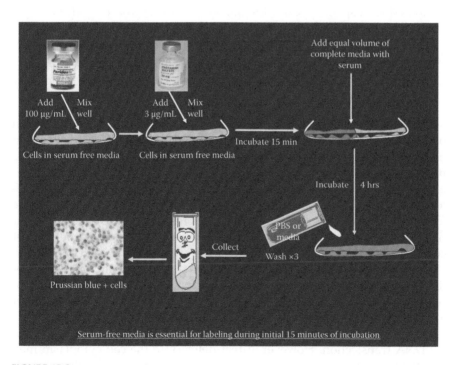

**FIGURE 15.3**
Schematic representation of superparamagnetic iron oxide labeling procedure for adherent cells.

3.  Remove the washing serum-free RPMI 1640 and add 5 mL of serum-free medium.

4.  Add 500 µg (45 µL of Feridex from the bottle) of Feridex (100 µg/mL) to the cells (in the flask containing 5 mL of serum-free medium) and mix thoroughly.

5.  Add 15 µg of freshly prepared protamine sulfate (3 µg/mL) to the cells (in the flask containing 5 mL medium plus Feridex) and mix thoroughly.

6.  Incubate at 37°C for 15 minutes.

7.  Then, add an equal volume (5 mL) of complete medium (containing serum) to the cell. Complete medium should contain serum and other growth factors for respective cells.

8.  Incubate further for 4 hours.

9.  Collect and wash the cells.

10. Determine the labeling efficiency and functional status of labeled cells.

Note that the number of cells in a T75 flask may not be 20 million. Therefore, compared to suspension cells, the added amount of iron per cell is larger, which will in turn facilitate uptake of more iron to the adherent cells.

Figure 15.4 shows different types of labeled cells using ferumoxide-protamine sulfate and the new 4-hour labeling method.

### 15.2.2.2 Mechanical Methods

Mechanical approaches, such as the gene gun or electroporation, have been used effectively to introduce MRI contrast agents into cells. The gene gun fires nanoparticles or magnetic beads directly into cells in culture, driving the particles through the cell membrane or directly into the nucleus. However, it is unknown what the long-term effects are on functional, metabolic, and differential capabilities of the cell (Zhang et al. 2003). Moreover, this technique for labeling cells has its own limitations with respect to the efficiency, potential tissue damages created by the impact of the particles, and small area of coverage (Davidson et al. 2000).

Magnetofection is a technique that utilizes strong magnetic force to introduce SPIO nanoparticles or a desired genome attached with magnetic nanoparticles within the cells (Bhattarai et al. 2008; Gersting et al. 2004; Huth et al. 2004; Mykhaylyk et al. 2007; Plank et al. 2003; Scherer et al. 2002). This technique delivers nanoparticles directly to the cytoplasm, and it is effective for DNA transfection. However, direct delivery to the cell cytoplasm may be a deterrent for magnetic cell labeling because of possible cytotoxicity following the release of iron into the cytoplasm or nucleus. This technique is useful for rapid labeling only in adherent cells. Magnetofection applicability

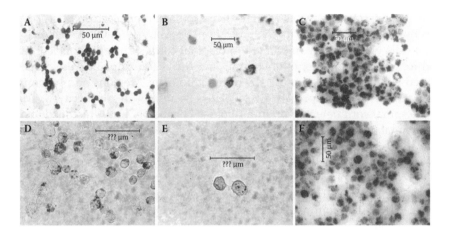

**FIGURE 15.4 (See color insert.)**
Representative images of different types of magnetically labeled cells. (A) Cytotoxic T cells; 3,3'-diaminobenzidine (DAB)-enhanced Prussian blue staining; bar represents 50 μm. (B) Hematopoietic stem cells (CD34+/AC133+); Prussian blue staining; bar represents 20 μm. (C) U251 human glioma cells; Prussian blue staining; bar represents 50 μm. (D) Neural stem cells (C17.2); Prussian blue staining; bar represents 100 μm. (E) Umbilical cord-derived mesenchymal stem cells; Prussian blue staining; bar represents 100 μm. (F) Umbilical cord blood-derived immature dendritic cells; Prussian blue staining; bar represents 50 μm.

in labeling suspension cells or cells with a small cytoplasm-to-nuclear ratio (such as T cells and hematopoietic stem cells) has not been verified. Moreover, details on toxicity and nuclear uptake have not been described.

Electroporation has been used to label neural progenitor cells and MSCs with SPIO nanoparticles without alterations to cell viability, proliferative capacity, or mitochondrial metabolic rate (Walczak et al. 2005). *In vivo* labeling of DNA in muscle using a Gd-based agent as a reporter has also been performed using electroporation (Crich et al. 2005b). It has been reported that the magnetic labeling of embryonic stem cells (ESCs) by electroporation resulted in a significant decrease in the percentage of viable cells compared to labeling cells with transfection agents complexed to ferumoxides (Suzuki et al. 2007). In addition, labeling ESCs by electroporation inhibited differentiation into cardiac progenitor cells. Thus, this method may not be a clinically useful approach for ESCs (Suzuki et al. 2007).

### 15.2.2.3 Antigen-Antibody Mediated

Conjugating antigen-specific internalizing monoclonal antibodies (MoAbs) to the dextran coat of USPIO nanoparticles has facilitated the magnetic labeling of cells by clathrin-mediated endocytosis. (Ahrens et al. 2003; Berry et al. 2004; Bulte et al. 1992, 1999; Moore et al. 1998, 2001). The MoAb (OX-26) to the rat transferrin receptor was covalently attached to USPIO nanoparticles (MION-46L) and used to label rat progenitor oligodendrocytes (CG-4). Iron oxide particles were demonstrated in the CG-4 cells using transmission electron microscopy and by histology using Prussian blue staining. MION-46L-OX-26-labeled rat CG-4 cells were directly implanted into spinal cords of myelin-deficient rats, and ex vivo MR images, obtained on day 10–14 after implantation, demonstrated excellent correlation between the hypointense regions and blooming artifacts caused by the presence of labeled cells and the degree of myelination in the spinal cord detected on immunohistochemistry. These results demonstrated that magnetically labeled cells would not interfere with cell differentiation, migration along an area of pathology, or the formation of myelin wraps around axons. Ahrens et al. (2003) labeled dendritic cells by biotinylating anti-CD-11 MoAb in conjunction with strepavidin attached to dextran-coated iron oxide nanoparticles. The advantage of this two-stage labeling approach is that the CD11 MoAb is commercially available, and the biotin-strepavidin interaction is a commonly used molecular biology procedure resulting in the SPIO being incorporated in endosomes. While using target-specific MoAb attached to SPIO provides efficient cellular labeling, a major drawback of this approach is the need for a specific MoAb for each cell type and species.

### 15.2.2.4 *Viral Shell Mediated*

Viruses and viral shells are being explored as carriers for MRI contrast agents. Using the hemagglutination virus of Japan (HVJ) envelope, encapsulated SPIO nanoparticles were found to label microglial cells in culture (Miyoshi 2005; Song et al. 2006; Toyoda et al. 2004). The HVJ SPIO-labeled microglial cells were injected intracardially, and clusters of cells could be seen within 1 day following transplantation in the brains of mice. The HVJ SPIO particles were reportedly more efficient at labeling cells than combining dextran-coated SPIO with the transfection agent lipofectamine (Toyoda et al. 2004). However, since HVJ envelopes are not commercially available, the use of this agent for labeling cells is limited.

### 15.2.3 Labeling Cells with Perfluorocarbons

Intracellular tagging of cells with $^{19}F$ perfluorocarbons provides researchers with the ability to monitor the migration of cells *in vivo* using $^{19}F$ MRI because of a similar gyromagnetic ratio as proton MRI (fluorine 40.05 MHz/T vs. proton 42 MHz/T) and 100% natural abundance of $^{19}F$. The main advantage of $^{19}F$ MRI is that it allows imaging of the cells without any background, so-called hot-spot imaging, since there are no endogenous fluorine atoms present in the body. Cationic perfluorocarbons have been introduced as contrast agents for direct cell labeling via simple incubation (Ahrens et al. 2005; Partlow et al. 2007; Srinivas et al. 2007). Current reports indicate that viability, proliferation, and cell function are not altered when cells are directly labeled with perfluorocarbons. Ahrens et al. incubated dendritic cells with perfluoropolyether (PEPE) for 18 hours, and following local injection of $4 \times 10^6$ labeled cells in the foot pad of a mouse were able to monitor cell migration to the lymph nodes with *in vivo* [19]F imaging at 11.7 T (Ahrens et al. 2005). Partlow et al. injected mononuclear hematopoietic stem cells labeled for 12 hours with perfluoro-octylbromide (PFOB) or perfluoro-15-crown-5 ether (CE) nanoparticles into the left and right hind limbs of a mouse and were able to image the two populations separately with selected excitation [19]F imaging. These authors found that $4 \times 10^6$ locally injected CE-labeled cells were detected by [19]F MRI at 1.5 T. They also injected $6 \times 10^6$ labeled cells intravenously and were able to localize the cells in the mouse liver with [19]F spectroscopy at 11.7 T. These authors estimated the detection limit in a single voxel was 6,100 cells with [19]F MRI of a cell pellet. In comparison, Anderson et al. (2005) indicated that one could detect as few as 50 SPIO-labeled cells/voxel in experimental systems at 7 T and about 500–2,000 at 3 T (Dahnke and Schaeffter 2005; de Vries et al. 2005).

Ruiz-Cabello et al. (2008) labeled neural stem cells with cationic CE and directly injected cells into the mouse brain; they were able to show that the [19]F MRI signal did not change over a 2-week period. However, these authors could not demonstrate whether the CE label remained in the stem cell or was taken up by microglia in the area of direct implantation. These authors also indicated that they needed to use modified polystyrene culture dishes coated with both carboxylic and amino groups rather than conventional carboxyl-coated dishes to directly label the adherent stem cells. There are several other limitations of using perfluorocarbon nanoparticles for cell labeling, including the need for high concentrations of [19]F to achieve a minimal detection threshold, separate [1]H images for anatomical localization of [19]F-detected cells, and the creation of statistical threshold maps of [19]F MRI to minimize extraneous signal as well as the relatively long scanning times compared to proton MRI. In addition, single-cell imaging of perfluorocarbon-labeled cells has not been demonstrated.

## 15.2.4 Methods to Determine the Optimized Labeling of Stem Cells

### 15.2.4.1 Validity of Labeling Status

Magnetic labeling of cells can be assessed qualitatively using histological Prussian blue stain for iron, the identification of electron dense material in endosomes by electron microscopy, or T2*-weighted MRI for determining the presence of iron in cells (Figure 15.5). Confocal microscopy is another method that can be utilized to determine the localization of magnetic particles colabeled with fluorescent dye or proteins. Multiphoton confocal microscopy can determine the depth as well as location of iron particles in the cytoplasm or nucleus. Quantitative approaches commonly used to determine the amount of iron taken up by the cells include NMR relaxometry, spectrophotometric assays, and inductively coupled plasma mass spectrometry. Determination of intracellular iron is important to validate MRI techniques used in a semi-quantitative manner to determine the number of labeled cells in a region of

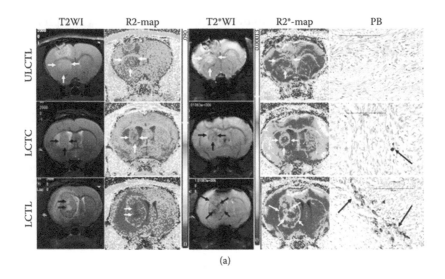

(a)

**FIGURE 15.5 (See color insert.)**
MRI and Prussian blue-positive cells in tumors. (a) T2-weighted image (T2WI) and T2\*-weighted image (T2\*WI) and their corresponding R2 and R2\* maps and DAB enhanced Prussian blue staining from representative animals that received unlabeled cytotoxic T cell CTL (ULCTL, upper row), labeled control T cells (LCTC, middle row), and CTL (LCTL, lower row). Both T2W and T2\*W images show well-established tumors in the brain; however, areas with low signal intensity were only seen in tumors that received LCTC and LCTL. Corresponding R2\* maps show areas with high signal intensity. Animals that received LCTL showed areas with high signal intensity at both the peripheral and central part of the tumors (arrows). Corresponding DAB-enhanced Prussian blue staining shows multiple Prussian blue-positive cells in tumors that received LCTL (arrows). There are a few Prussian blue-positive cells seen in tumor that received LCTC (arrow). No definite Prussian blue-positive cells are seen in tumor that received ULCTL. Areas of necrosis can easily be identified by comparing T2WI and R2 maps (thick arrows) in tumor that received LCTL (lower row). Bars on the images measure 100 μm.

*(Continued)*

interest (Arbab et al. 2003b; Daldrup-Link et al. 2005; Foster-Gareau et al. 2003; Rad et al. 2007). The preservation of cell viability, cell function, and differential capacity of cells is important and needs to be monitored following labeling cells with MRI contrast agents because soluble iron (intracellular) is toxic to the cells and may damage DNA.

### 15.2.4.2 Viability Assay

Iron and the transfection agents alone, which are used to label cells, are potentially toxic to the cells. Ferrous forms of iron in the cytoplasm can stimulate increased production of $H_2O_2$ and hydroxyl free radicals, which can cause denaturation of DNA (Emerit et al. 2001; Gutteridge and Halliwell 1982; Gutteridge and Toeg 1982). Therefore, it is essential to determine the short- and long-term toxicity of labeling cells with iron to translate from

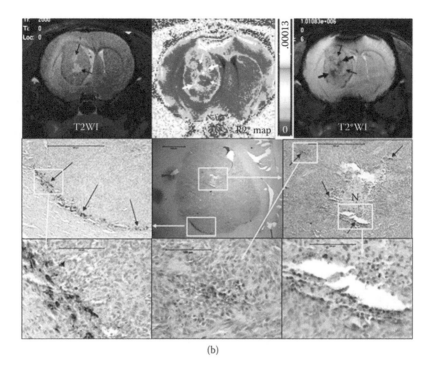

(b)

**FIGURE 15.5** (*Continued*) (See color insert.)
MRI and Prussian blue-positive cells in tumors. (b) Detailed histological analysis of the tumor that received LCTL. Bars on the images measure 100 μm. Upper panel: T2WI and T2*WI show areas of necrosis (high signal intensity on T2WI and T2*WI and low signal intensity on R2* map). Thin arrows show the sites of necrosis, and thick arrows show possible site of accumulated iron-positive cells. Middle panel: Representative histological section with similar tumor orientation (within the constraints of the experimental limitations, i.e., 1-mm thick MRI slices vs. 10-μ thick histological section) show central necrosis (N) in the tumor with areas of iron-positive cells (black arrows) seen in the central and peripheral part of the tumor that received iron-labeled CTLs. Lower panel: Enlarged view of the boxed areas.

bench to bedside. Cell viability is usually determined at time of cell counting using a Trypan blue dye exclusion test. Propidium iodide (PI) can also be used to detect dead cells by fluorescent microscopy or flow cytometry. Changes in the rate of apoptosis, a long-term indication of (U)SPIO labeling of cells can be accomplished using incubating cells with fluorescent-labeled annexin V and PI combined with analysis using flow cytometry. In general, investigators have found that cell viability or rate of apoptosis is not different from unlabeled control cells when cells are labeled using any of the methods described. Short- and long-term proliferation capacity of the labeled cells can be assessed by tritiated-thymidine uptake, Bromodeoxyuridine (BrdU) incorporation, or an MTT (3-[4,5-dimethylthiazol-2-yl]-2,5-diphenyl tetrazolium bromide) assay (Roche Molecular Biochemicals, Indianapolis, IN). Proliferative assays are usually performed as a pulse chase labeling experiment with the (U)SPIO, and determinations are made at different times (Arbab et al. 2003b). The dissolution of (U)SPIO nanoparticles will release ferrous ions in the cells or nucleus, which could initiate a Fenton reaction and through Haber-Weiss chemistry (Emerit et al. 2001) result in the development of free radicals, which could cause damage to DNA. Reactive oxygen species (ROS) can be detected using the fluorescent probe CM-H$_2$DCFDA (Molecular Probes Inc., Eugene, OR). CM-H$_2$DCFDA is a nonfluorescent agent that forms fluorescent esters when reacted with ROS inside cells, which can be analyzed by flow cytometry, fluorescent microscopy, or confocal microscopy.

Pawelczyk et al. (2006) demonstrated following labeling of MSCs and macrophages with ferumoxide-protamine sulfate that a transient decrease in gene and protein expression of the transferrin receptor occurred; gene and protein levels of ferritin either remained stable or increased in response to the iron load. These results indicate that ferumoxide-protamine sulfate labeling of cells elicited appropriate and expected physiological changes of iron metabolism or storage. To date, little or no alterations in cell viability, proliferative capacity, metabolic activity, apoptotic rate, ROS formation, or functional and differential capacity have been reported for cells labeled with clinically approved (U)SPIO MR contrast agents when compared to unlabeled cells. Although previous studies (Kruman et al. 1998; Stauber et al. 1998; Stauber and Pavlakis 1998) indicated toxicity of Trans-activator of Transcription (TaT) protein, a publication indicated no short-term loss of cell viability or accumulation of cross-linked iron oxide (CLIO)-Tat in the nucleus (Kaufman et al. 2003). However, the authors did not show the functional or activation markers of splenocytes after labeling with CLIO-Tat.

### 15.2.4.3 Functional Assay

It is also important that labeled cells should not deviate from their original function or become activated (in the case of T cells). A cellular functional assay should be performed after labeling and compared with that of control cells. Cell activation markers as well as phenotypical expression should be assessed for primary cell populations. Labeling lymphocytes (T cells) with SPIO nanoparticles may change one population of T cells (such as T helper cells) to another (such as cytotoxic T cells). Analysis of different phenotypical cell surface marker expression (i.e., CD markers, CD4, CD8, CD11a, CD19, CD25) of T cells and production of cytokines (interferons, interleukins, tumor necrosis factor [TNF], etc.) should be performed before and after labeling immune cells. For stem cells, primary as well as lineage markers should be assessed before and after labeling with MR contrast agents.

### 15.2.4.4 Determination of Cell Differentiation Capacity

Determination of differential capacity of labeled cells is especially important for nonhematopoietic stem cells, such as MSCs and neural stem cells (NSCs).

MSCs should be differentiated to adipogenic, chondrogenic, and osteogenic lineages. The potential for alterations in chondrogenic differentiation of MSCs *in vitro* have been seen for various transfection agents and may be dose dependent or incubation time dependent (Arbab et al. 2004b; Bulte et al. 2004; Henning et al. 2009). For HSCs, differentiation to colony-forming units should be determined for labeled and unlabeled cells. Endothelial progenitor cells (EPCs) can be differentiated to colony-forming units, endothelial cells, and cord-like structures in a Matrigel system. NSCs should be differentiated to neuron and glial cell lineages. The differential capacity of magnetically labeled cells can also be determined by tracking the *in vivo* migration and targeted incorporation or labeled stem cells in different organs (Watson et al. 2006).

## 15.3 Important Note

The production and marketing of ferumoxides have been discontinued. Although there are alternative SPIO agents that can be used for cell labeling, none of them are approved by the FDA or made in a cGMP facility and available in the United States. Amag Pharmaceuticals (http://www.amag-pharma.com) will be introducing ferumoxytol, a 30-nm carboxyl methyl-dextran-coated SPION as a clinically approved iron supplement in patients with renal failure, and presently ferumoxytol is being evaluated in clinical trials (http://clinicaltrials.gov/ct2/results?term=ferumoxytol) as an MRI contrast agent. It is also not clear whether ferumoxytol can be used to label nonphagocytic adherent cells or cells grown in suspension. According to a review by Neuwelt et al. (2008), who have been running clinical trials with intravenous fermoxytol, combining the agent with protamine *in vivo* does not result in leukocyte labeling as compared to ferumoxides and protamine. Daldrup-Link et al. (2003) did label HSCs with P7228 (similar to ferumoxytol) plus lipofectin, but Prussian blue images were not provided of the labeled cells. Until such a time as a protocol can be developed for labeling stem cells with ferumoxytol, the cell-labeling community will need to use non-USP- (U.S. Pharmaceutical) grade material for cell labeling. For experimental animal studies for SPIO labeling of stem cells or other nonphagocytic cells, the ultimate removal of ferumoxides from the marketplace is probably not an issue since other manufacturers synthesize SPIOs (e.g., Bangs, http://www.bangslabs.com; BioPAL Inc., http://www.biopal.com) that can be used to label nonphagocytic cells magnetically. For planned clinical trials in the United States and possibly in other countries, investigators will need to seek approval or postpone studies until it can be determined whether ferumoxytol can be substituted for ferumoxides for cell labeling.

# References

Acton, P.D. and R. Zhou. 2005. Imaging reporter genes for cell tracking with PET and SPECT. *Q J Nucl Med Mol Imaging* 49, no. 4: 349–60.

Ahrens, E.T., M. Feili-Hariri, H. Xu, G. Genove, and P.A. Morel. 2003. Receptor-mediated endocytosis of iron-oxide particles provides efficient labeling of dendritic cells for in vivo MR imaging. *Magn Reson Med* 49, no. 6: 1006–13.

Ahrens, E.T., R. Flores, H. Xu, and P.A. Morel. 2005. In vivo imaging platform for tracking immunotherapeutic cells. *Nat Biotechnol* 23, no. 8: 983–87.

Alessandri, G., C. Emanueli, and P. Madeddu. 2004. Genetically engineered stem cell therapy for tissue regeneration. *Ann NY Acad Sci* 1015: 271–84.

Anderson, S.A., J. Glod, A.S. Arbab, M. Noel, P. Ashari, H.A. Fine, and J.A. Frank. 2005. Noninvasive MR imaging of magnetically labeled stem cells to directly identify neovasculature in a glioma model. *Blood* 105, no. 1: 420–25.

Anderson, S.A., K.K. Lee, and J.A. Frank. 2006. Gadolinium-fullerenol as a paramagnetic contrast agent for cellular imaging. *Invest Radiol* 41, no. 3: 332–38.

Aoki, I., Y. Takahashi, K.H. Chuang, A.C. Silva, T. Igarashi, C. Tanaka, R.W. Childs, and A.P. Koretsky. 2006. Cell labeling for magnetic resonance imaging with the T1 agent manganese chloride. *NMR Biomed* 19, no. 1: 50–59.

Arbab, A.S., L.A. Bashaw, B.R. Miller, E.K. Jordan, J.W. Bulte, and J.A. Frank. 2003a. Intracytoplasmic tagging of cells with ferumoxides and transfection agent for cellular magnetic resonance imaging after cell transplantation: Methods and techniques. *Transplantation* 76, no. 7: 1123–30.

Arbab, A.S., L.A. Bashaw, B.R. Miller, E.K. Jordan, B.K. Lewis, H. Kalish, and J.A. Frank. 2003b. Characterization of biophysical and metabolic properties of cells labeled with superparamagnetic iron oxide nanoparticles and transfection agent for cellular MR imaging. *Radiology* 229, no. 3: 838–46.

Arbab, A.S. and J.A. Frank. 2008. Cellular MRI and its role in stem cell therapy. *Regen Med* 3, no. 2: 199–215.

Arbab, A.S., E.K. Jordan, L.B. Wilson, G.T. Yocum, B.K. Lewis, and J.A. Frank. 2004a. In vivo trafficking and targeted delivery of magnetically labeled stem cells. *Hum Gene Ther* 15, no. 4: 351–60.

Arbab, A.S., W. Liu, and J.A. Frank. 2006a. Cellular magnetic resonance imaging: Current status and future prospects. *Expert Rev Med Devices* 3, no. 4: 427–39.

Arbab, A.S., S.D. Pandit, S.A. Anderson, G.T. Yocum, M. Bur, V. Frenkel, H.M. Khuu, E.J. Read, and J.A. Frank. 2006b. Magnetic resonance imaging and confocal microscopy studies of magnetically labeled endothelial progenitor cells trafficking to sites of tumor angiogenesis. *Stem Cells* 24, no. 3: 671–78.

Arbab, A.S., A.M. Rad, A.S. Iskander, K. Jafari-Khouzani, S.L. Brown, J.L. Churchman, G. Ding, Q. Jiang, J.A. Frank, H. Soltanian-Zadeh, and D.J. Peck. 2007. Magnetically-labeled sensitized splenocytes to identify glioma by MRI: A preliminary study. *Magn Reson Med* 58, no. 3: 519–26.

Arbab, A.S., G.T. Yocum, H. Kalish, E.K. Jordan, S.A. Anderson, A.Y. Khakoo, E.J. Read, and J.A. Frank. 2004b. Efficient magnetic cell labeling with protamine sulfate complexed to ferumoxides for cellular MRI. *Blood* 104, no. 4: 1217–23.

Arbab, A.S., G.T. Yocum, L.B. Wilson, A. Parwana, E.K. Jordan, H. Kalish, and J.A. Frank. 2004c. Comparison of transfection agents in forming complexes with ferumoxides, cell labeling efficiency, and cellular viability. *Mol Imaging* 3, no. 1: 24–32.

Ben-Hur, T., O. Einstein, and J.W. Bulte. 2005. Stem cell therapy for myelin diseases. *Curr Drug Targets* 6, no. 1: 3–19.

Berry, C.C., S. Charles, S. Wells, M.J. Dalby, and A.S. Curtis. 2004. The influence of transferrin stabilised magnetic nanoparticles on human dermal fibroblasts in culture. *Int J Pharm* 269, no. 1: 211–25.

Bhattarai, S.R., S.Y. Kim, K.Y. Jang, K.C. Lee, H.K. Yi, D.Y. Lee, H.Y. Kim, and P.H. Hwang. 2008. Laboratory formulated magnetic nanoparticles for enhancement of viral gene expression in suspension cell line. *J Virol Methods* 147, no. 2: 213–18.

Blank, A.C., T.A. Van Veen, M.K. Jonsson, J.S. Zelen, J.L. Strengers, T.P. De Boer, and M.A. Van Der Heyden. 2009. Rewiring the heart: Stem cell therapy to restore normal cardiac excitability and conduction. *Curr Stem Cell Res Ther* 4, no. 1: 23–33.

Brekke, C., S.C. Morgan, A.S. Lowe, T.J. Meade, J. Price, S.C. Williams, and M. Modo. 2007. The in vitro effects of a bimodal contrast agent on cellular functions and relaxometry. *NMR Biomed* 20, no. 2: 77–89.

Bulte, J.W. 2006. Intracellular endosomal magnetic labeling of cells. *Methods Mol Med* 124: 419–39.

Bulte, J.W., Y. Hoekstra, R.L. Kamman, R.L. Magin, A.G. Webb, R.W. Briggs, K.G. Go, C.E. Hulstaert, S. Miltenyi, and T.H. The. 1992. Specific MR imaging of human lymphocytes by monoclonal antibody-guided dextran-magnetite particles. *Magn Reson Med* 25, no. 1: 148–57.

Bulte, J.W., D.L. Kraitchman, A.M. Mackay, M.F. Pittenger, A.S. Arbab, G.T. Yocum, H. Kalish, E.K. Jordan, S.A. Anderson, A.Y. Khakoo, E.J. Read, and J.A. Frank. 2004. Chondrogenic differentiation of mesenchymal stem cells is inhibited after magnetic labeling with ferumoxides. *Blood* 104, no. 10: 3410–3.

Bulte, J.W., S. Zhang, P. Van Gelderen, V. Herynek, E.K. Jordan, I.D. Duncan, and J.A. Frank. 1999. Neurotransplantation of magnetically labeled oligodendrocyte progenitors: magnetic resonance tracking of cell migration and myelination. *Proc Natl Acad Sci USA* 96, no. 26: 15256–61.

Chachques, J.C., J.C. Trainini, N. Lago, O.H. Masoli, J.L. Barisani, M. Cortes-Morichetti, O. Schussler, and A. Carpentier. 2007. Myocardial assistance by grafting a new bioartificial upgraded myocardium (MAGNUM clinical trial): One year follow-up. *Cell Transplant* 16, no. 9: 927–34.

Clavel, C. and C.M. Verfaillie. 2008. Bone-marrow-derived cells and heart repair. *Curr Opin Organ Transplant* 13, no. 1: 36–43.

Cleland, J.G., A.P. Coletta, A.T. Abdellah, D. Cullington, A.L. Clark, and A.S. Rigby. 2008. Clinical trials update from the American Heart Association 2007: CORONa, RethinQ, mASCOT, AF-CHF, HART, MASTER, POISE and stem cell therapy. *Eur J Heart Fail* 10, no. 1: 102–8.

Crich, S.G., L. Biancone, V. Cantaluppi, D. Duo, G. Esposito, S. Russo, G. Camussi, and S. Aime. 2004. Improved route for the visualization of stem cells labeled with a Gd-/Eu-chelate as dual (MRI and fluorescence) agent. *Magn Reson Med* 51, no. 5: 938–44.

Crich, S.G., S. Lanzardo, A. Barge, and S. Aime. 2005a. Visualization through magnetic resonance imaging of DNA internalized following "in vivo" electroporation. *Mol Imaging* 4, no. 1: 7–17.

Crich, S.G., S. Lanzardo, A. Barge, G. Esposito, L. Tei, G. Forni, and S. Aime. 2005b. Visualization through magnetic resonance imaging of DNA internalized following "in vivo" electroporation. *Mol Imaging* 4, no. 1: 7–17.

Dahnke, H. and T. Schaeffter. 2005. Limits of detection of SPIO at 3.0 T using T2 relaxometry. *Magn Reson Med* 53, no. 5: 1202–6.

Daldrup-Link, H.E., M. Rudelius, S. Metz, G. Piontek, B. Pichler, M. Settles, U. Heinzmann, J. Schlegel, R.A. Oostendorp, and E.J. Rummeny. 2004. Cell tracking with gadophrin-2: A bifunctional contrast agent for MR imaging, optical imaging, and fluorescence microscopy. *Eur J Nucl Med Mol Imaging* 31, no. 9: 1312–21.

Daldrup-Link, H.E., M. Rudelius, R.A. Oostendorp, M. Settles, G. Piontek, S. Metz, H. Rosenbrock, U. Keller, U. Heinzmann, E.J. Rummeny, J. Schlegel, and T.M. Link. 2003. Targeting of hematopoietic progenitor cells with MR contrast agents. *Radiology* 228, no. 3: 760–67.

Daldrup-Link, H.E., M. Rudelius, G. Piontek, S. Metz, R. Brauer, G. Debus, C. Corot, J. Schlegel, T.M. Link, C. Peschel, E.J. Rummeny, and R.A. Oostendorp. 2005. Migration of iron oxide-labeled human hematopoietic progenitor cells in a mouse model: *In vivo* monitoring with 1.5-T MR imaging equipment. *Radiology* 234, no. 1: 197–205.

Davidson, J.M., T. Krieg, and S.A. Eming. 2000. Particle-mediated gene therapy of wounds. *Wound Repair Regen* 8, no. 6: 452–59.

De Vries, I.J., W.J. Lesterhuis, J.O. Barentsz, P. Verdijk, J.H. Van Krieken, O.C. Boerman, W.J. Oyen, J.J. Bonenkamp, J.B. Boezeman, G.J. Adema, J.W. Bulte, T.W. Scheenen, C.J. Punt, A. Heerschap, and C.G. Figdor. 2005. Magnetic resonance tracking of dendritic cells in melanoma patients for monitoring of cellular therapy. *Nat Biotechnol* 23, no. 11: 1407–13. Epub 2005 Oct 30.

Dingli, D., B.J. Kemp, M.K. O'Connor, J.C. Morris, S.J. Russell, and V.J. Lowe. 2006. Combined I-124 positron emission tomography/computed tomography imaging of NIS gene expression in animal models of stably transfected and intravenously transfected tumor. *Mol Imaging Biol* 8, no. 1: 16–23.

Emerit, J., C. Beaumont, and F. Trivin. 2001. Iron metabolism, free radicals, and oxidative injury. *Biomed Pharmacother* 55, no. 6: 333–39.

Foster-Gareau, P., C. Heyn, A. Alejski, and B.K. Rutt. 2003. Imaging single mammalian cells with a 1.5 T clinical MRI scanner. *Magn Reson Med* 49, no. 5: 968–71.

Frank, J.A., B.R. Miller, A.S. Arbab, H.A. Zywicke, E.K. Jordan, B.K. Lewis, L.H. Bryant, Jr., and J.W. Bulte. 2003. Clinically applicable labeling of mammalian and stem cells by combining superparamagnetic iron oxides and transfection agents. *Radiology* 228, no. 2: 480–87.

Gersting, S.W., U. Schillinger, J. Lausier, P. Nicklaus, C. Rudolph, C. Plank, D. Reinhardt, and J. Rosenecker. 2004. Gene delivery to respiratory epithelial cells by magnetofection. *J Gene Med* 6, no. 8: 913–22.

Giesel, F.L., M. Stroick, M. Griebe, H. Troster, C.W. Von Der Lieth, M. Requardt, M. Rius, M. Essig, H.U. Kauczor, M.G. Hennerici, and M. Fatar. 2006. Gadofluorine M uptake in stem cells as a new magnetic resonance imaging tracking method: An *in vitro* and *in vivo* study. *Invest Radiol* 41, no. 12: 868–73.

Gilad, A.A., P. Walczak, M.T. Mcmahon, H.B. Na, J.H. Lee, K. An, T. Hyeon, P.C. Van Zijl, and J.W. Bulte. 2008. MR tracking of transplanted cells with "positive contrast" using manganese oxide nanoparticles. *Magn Reson Med* 60, no. 1: 1–7.

Gutteridge, J.M. and B. Halliwell. 1982. The role of the superoxide and hydroxyl radicals in the degradation of DNA and deoxyribose induced by a copper-phenanthroline complex. *Biochem Pharmacol* 31, no. 17: 2801–5.

Gutteridge, J.M. and D. Toeg. 1982. Iron-dependent free radical damage to DNA and deoxyribose. Separation of TBA-reactive intermediates. *Int J Biochem* 14, no. 10: 891–93.

Heinrich, A.C., S.A. Patel, B.Y. Reddy, R. Milton, and P. Rameshwar. 2009. Multi- and inter-disciplinary science in personalized delivery of stem cells for tissue repair. *Curr Stem Cell Res Ther* 4, no. 1: 16–22.

Henning, T.D., E.J. Sutton, A. Kim, D. Golovko, A. Horvai, L. Ackerman, B. Sennino, D. McDonald, J. Lotz, and H.E. Daldrup-Link. 2009. The influence of ferucarbotran on the chondrogenesis of human mesenchymal stem cells. *Contrast Media Mol Imaging* 4, no. 4: 165–73.

Himmelreich, U., S. Aime, T. Hieronymus, C. Justicia, F. Uggeri, M. Zenke, and M. Hoehn. 2006. A responsive MRI contrast agent to monitor functional cell status. *Neuroimage* 32, no. 3: 1142–49.

Hoshino, K., H.Q. Ly, J.V. Frangioni, and R.J. Hajjar. 2007. *In vivo* tracking in cardiac stem cell-based therapy. *Prog Cardiovasc Dis* 49, no. 6: 414–20.

Huth, S., J. Lausier, S.W. Gersting, C. Rudolph, C. Plank, U. Welsch, and J. Rosenecker. 2004. Insights into the mechanism of magnetofection using PEI-based magnetofectins for gene transfer. *J Gene Med* 6, no. 8: 923–36.

Jacobs, R.E. and S.E. Fraser. 1994. Magnetic resonance microscopy of embryonic cell lineages and movements. *Science* 263, no. 5147: 681–84.

Janic, B., A.M. Rad, E.K. Jordan, A.S. Iskander, M.M. Ali, N.R. Varma, J.A. Frank, and A.S. Arbab. 2009. Optimization and validation of FePro cell labeling method. *PLoS ONE* 4, no. 6: e5873.

Kalish, H., A.S. Arbab, B.R. Miller, B.K. Lewis, H.A. Zywicke, J.W. Bulte, L.H. Bryant, Jr., and J.A. Frank. 2003. Combination of transfection agents and magnetic resonance contrast agents for cellular imaging: relationship between relaxivities, electrostatic forces, and chemical composition. *Magn Reson Med* 50, no. 2: 275–82.

Kaufman, C.L., M. Williams, L.M. Ryle, T.L. Smith, M. Tanner, and C. Ho. 2003. Superparamagnetic iron oxide particles transactivator protein-fluorescein isothiocyanate particle labeling for in vivo magnetic resonance imaging detection of cell migration: Uptake and durability. *Transplantation* 76, no. 7: 1043–46.

Kruman, I., A. Nath, and M.P. Mattson. 1998. HIV-1 protein TaT induces apoptosis of hippocampal neurons by a mechanism involving caspase activation, calcium overload, and oxidative stress. *Exp Neurol* 154, no. 2: 276–88.

Manganas, L.N. and M. Maletic-Savatic. 2005. Stem cell therapy for central nervous system demyelinating disease. *Curr Neurol Neurosci Rep* 5, no. 3: 225–31.

Mendonca-Dias, M.H., E. Gaggelli, and P.C. Lauterbur. 1983. Paramagnetic contrast agents in nuclear magnetic resonance medical imaging. *Semin Nucl Med* 13, no. 4: 364–76.

Miyoshi, S., J. Flexman, D.J. Cross, K.R. Maravilla, Y. Kim, Y. Anzai, J. Oshima, and S. Minoshima, S. 2005. Transfection of neuroprogenitor cells with iron nanoparticles for magnetic resonance imaging tracking: cell viability, differentiation, and intracellular localization. *Mol Imaging Biol* 4: 1–10.

Modo, M., D. Cash, K. Mellodew, S.C. Williams, S.E. Fraser, T.J. Meade, J. Price, and H. Hodges. 2002. Tracking transplanted stem cell migration using bifunctional, contrast agent-enhanced, magnetic resonance imaging. *Neuroimage* 17, no. 2: 803–11.

Montet-Abou, K., X. Montet, R. Weissleder, and L. Josephson. 2007. Cell internalization of magnetic nanoparticles using transfection agents. *Mol Imaging* 6, no. 1: 1–9.

Moore, A., J.P. Basilion, E.A. Chiocca, and R. Weissleder. 1998. Measuring transferrin receptor gene expression by NMR imaging. *Biochim Biophys Acta* 1402, no. 3: 239–49.

Moore, A., L. Josephson, R.M. Bhorade, J.P. Basilion, and R. Weissleder. 2001. Human transferrin receptor gene as a marker gene for MR imaging. *Radiology* 221: 244–50.

Mykhaylyk, O., Y.S. Antequera, D. Vlaskou, and C. Plank. 2007. Generation of magnetic nonviral gene transfer agents and magnetofection in vitro. *Nat Protocols* 2, no. 10: 2391–2411.

Na, H.B., J.H. Lee, K. An, Y.I. Park, M. Park, I.S. Lee, D.H. Nam, S.T. Kim, S.H. Kim, S.W. Kim, K.H. Lim, K.S. Kim, S.O. Kim, and T. Hyeon. 2007. Development of a T1 contrast agent for magnetic resonance imaging using MnO nanoparticles. *Angew Chem Int Ed Engl* 46, no. 28: 5397–401.

Neuwelt, E.A., B.E. Hamilton, C.G. Varallyay, W.R. Rooney, R.D. Edelman, P.M. Jacobs, and S.G. Watnick. 2009. Ultrasmall superparamagnetic iron oxides (USPIOs): A future alternative magnetic resonance (MR) contrast agent for patients at risk for nephrogenic systemic fibrosis (NSF)? *Kidney Int* 75, 75: 465–74

Partlow, K.C., J. Chen, J.A. Brant, A.M. Neubauer, T.E. Meyerrose, M.H. Creer, J.A. Nolta, S.D. Caruthers, G.M. Lanza, and S.A. Wickline. 2007. 19F magnetic resonance imaging for stem/progenitor cell tracking with multiple unique perfluorocarbon nanobeacons. *FASEB J* 21, no. 8: 1647–54.

Pawelczyk, E., A.S. Arbab, S. Pandit, E. Hu, and J.A. Frank. 2006. Expression of transferrin receptor and ferritin following ferumoxides-protamine sulfate labeling of cells: Implications for cellular magnetic resonance imaging. *NMR Biomed* 19, no. 5: 581–92.

Plank, C., F. Scherer, U. Schillinger, C. Bergemann, and M. Anton. 2003. Magnetofection: Enhancing and targeting gene delivery with superparamagnetic nanoparticles and magnetic fields. *J Liposome Res* 13, no. 1: 29–32.

Politi, L.S., M. Bacigaluppi, E. Brambilla, M. Cadioli, A. Falini, G. Comi, G. Scotti, G. Martino, and S. Pluchino. 2007. Magnetic-resonance-based tracking and quantification of intravenously injected neural stem cell accumulation in the brains of mice with experimental multiple sclerosis. *Stem Cells* 25, no. 10: 2583–92.

Rad, A.M., B. Janic, A. Iskander, H. Soltanian-Zadeh, and A.S. Arbab. 2007. Measurement of quantity of iron in magnetically labeled cells: Comparison among different UV/VIS spectrometric methods. *Biotechniques* 43, no. 5: 627–36.

Ruiz-Cabello, J., P. Walczak, D.A. Kedziorek, V.P. Chacko, A.H. Schmieder, S.A. Wickline, G.M. Lanza and J.W. Bulte. 2008. *In Vivo* "hot spot" MR imaging of neural stem cells using fluorinated nanoparticles. *Magn Reson Med* 60(6): 1506–11.

Scherer, F., M. Anton, U. Schillinger, J. Henke, C. Bergemann, A. Kruger, B. Gansbacher, and C. Plank. 2002. Magnetofection: Enhancing and targeting gene delivery by magnetic force in vitro and in vivo. *Gene Ther* 9, no. 2: 102–9.

Shapiro, E.M. and A.P. Koretsky. 2008. Convertible manganese contrast for molecular and cellular MRI. *Magn Reson Med* 60, no. 2: 265–69.

Song, Y., S. Morikawa, M. Morita, T. Inubushi, T. Takada, R. Torii, and I. Tooyama. 2006. Magnetic resonance imaging using hemagglutinating virus of Japan-envelope vector successfully detects localization of intra-cardially administered microglia in normal mouse brain. *Neurosci Lett* 395, no. 1: 42–45.

Sotak, C.H., K. Sharer, and A.P. Koretsky. 2008. Manganese cell labeling of murine hepatocytes using manganese(III)-transferrin. *Contrast Media Mol Imaging* 3, no. 3: 95–105.

Srinivas, M., P.A. Morel, L.A. Ernst, D.H. Laidlaw, and E.T. Ahrens. 2007. Fluorine-19 MRI for visualization and quantification of cell migration in a diabetes model. *Magn Reson Med* 58, no. 4: 725–34.

Stauber, R.H., E. Afonina, S. Gulnik, J. Erickson, and G.N. Pavlakis. 1998. Analysis of intracellular trafficking and interactions of cytoplasmic HIV-1 Rev mutants in living cells. *Virology* 251, no. 1: 38–48.

Stauber, R.H. and G.N. Pavlakis. 1998. Intracellular trafficking and interactions of the HIV-1 TaT protein. *Virology* 252, no. 1: 126–36.

Su, W., R. Mishra, J. Pfeuffer, K.H. Wiesmuller, K. Ugurbil, and J. Engelmann. 2007. Synthesis and cellular uptake of a MR contrast agent coupled to an antisense peptide nucleic acid—cell-penetrating peptide conjugate. *Contrast Media Mol Imaging* 2, no. 1: 42–49.

Suzuki, Y., S. Zhang, P. Kundu, A.C. Yeung, R.C. Robbins, and P.C. Yang. 2007. In vitro comparison of the biological effects of three transfection methods for magnetically labeling mouse embryonic stem cells with ferumoxides. *Magn Reson Med* 57, no. 6: 1173–79.

Terreno, E., S. Geninatti Crich, S. Belfiore, L. Biancone, C. Cabella, G. Esposito, A.D. Manazza, and S. Aime. 2006. Effect of the intracellular localization of a Gd-based imaging probe on the relaxation enhancement of water protons. *Magn Reson Med* 55, no. 3: 491–97.

Toyoda, K., I. Tooyama, M. Kato, H. Sato, S. Morikawa, Y. Hisa, and T. Inubushi. 2004. Effective magnetic labeling of transplanted cells with HVJ-e for magnetic resonance imaging. *Neuroreport* 15, no. 4: 589–93.

Walczak, P., D.A. Kedziorek, A.A. Gilad, S. Lin, and J.W. Bulte. 2005. Instant MR labeling of stem cells using magnetoelectroporation. *Magn Reson Med* 54, no. 4: 769–74.

Walker, P.A., S.K. Shah, M.T. Harting, and C.S. Cox. 2009. Progenitor cell therapies for traumatic brain injury: Barriers and opportunities in translation. *Dis Model Mech* 2, no. 1–2: 23–38.

Walliser, S. and K. Redmann. 1978. Effect of 5-fluorouracil and thymidine on the transmembrane potential and zeta potential of HeLa cells. *Cancer Res* 38, no. 10: 3555–59.

Watson, D.J., R.M. Walton, S.G. Magnitsky, J.W. Bulte, H. Poptani, and J.H. Wolfe. 2006. Structure-specific patterns of neural stem cell engraftment after transplantation in the adult mouse brain. *Hum Gene Ther* 17, no. 7: 693–704.

Wilhelm, C. and F. Gazeau. 2008. Universal cell labelling with anionic magnetic nanoparticles. *Biomaterials* 29, no. 22: 3161–74.

Wolf, G.L., K.R. Burnett, E.J. Goldstein, and P.M. Joseph. 1985. Contrast agents for magnetic resonance imaging. *Magn Reson Annu* 231–66.

Wu, J.C., J.M. Spin, F. Cao, S. Lin, X. Xie, O. Gheysens, I.Y. Chen, A.Y. Sheikh, R.C. Robbins, A. Tsalenko, S.S. Gambhir, and T. Quertermous. 2006. Transcriptional profiling of reporter genes used for molecular imaging of embryonic stem cell transplantation. *Physiol Genomics* 25, no. 1: 29–38.

Zhang, H.T., J. Ding, G.M. Chow, and Z.L. Dong. 2008. Engineering inorganic hybrid nanoparticles: Tuning combination fashions of gold, platinum, and iron oxide. *Langmuir* 24, no. 22: 3197–202.

Zhang, R.L., L. Zhang, Z.G. Zhang, D. Morris, Q. Jiang, L. Wang, L.J. Zhang, and M. Chopp. 2003. Migration and differentiation of adult rat subventricular zone progenitor cells transplanted into the adult rat striatum. *Neuroscience* 116, no. 2: 373–82.

# 16

## Ultrasound Cell-Labeling Methods

Flordeliza S. Villanueva

### CONTENTS

## 16.1 Introduction

Ultrasound imaging is a noninvasive approach to *in vivo* visualization of exogenously delivered stem cells, which may have unique advantages for clinical application. Therapeutically administered stem cells are acoustically indistinguishable from surrounding tissue and cannot be uniquely identified using standard ultrasound imaging. An ultrasound approach to imaging stem cells is based on the premise that stem cell attachment to, or internalization of, ultrasound contrast agents renders the cells detectable by clinical ultrasound imaging systems. An acoustic approach to imaging stem cells has two fundamental requirements: an ultrasound contrast agent that links to the stem cell and a detection strategy that senses a unique acoustic signature from the contrast agent-stem cell construct, which can thus be anatomically coregistered on a two-dimensional ultrasound image. This chapter outlines the current work in achieving these requirements. A brief review of ultrasound contrast agent technology and detection is provided in the context of defining the ideal stem cell contrast agent. This is followed by a description of labeling strategies that have been thus far developed, their strengths and limitations, and requirements for moving forward with this technology.

## 16.2 Ultrasound Contrast Agents and Detection Methods

This discussion focuses on gas-filled microsphere or "microbubble" contrast agents, measuring 1–4 µm in diameter, which are typically comprised of gases, such as nitrogen or perfluorocarbons, encapsulated by a shell made of a biocompatible material such as phospholipids, albumin, or a polymer [1]. In the presence of ultrasound delivered at their natural resonance frequency, microbubbles expand and contract (oscillate) symmetrically, and at higher acoustic powers, the oscillations become asymmetric (nonlinear behavior) [2,3]. Such oscillations, occurring at frequencies that are multiples of the transmitted frequencies (harmonics), result in ultrasound emissions from the microbubbles themselves, which can be received by ultrasound transducers designed to detect harmonic frequencies or nonlinear microbubble behavior. Because tissue does not oscillate in response to medical ultrasound frequencies, it does not generate a harmonic signal, allowing the selective detection of the microbubble signal (which appears as contrast enhancement) as distinct from tissue (which appears black) on the two-dimensional ultrasound image. This approach is currently used to opacify the blood pool during echocardiographic imaging, for which the contrast agents are used as red blood cell tracers [1]. Two echo contrast agents are approved by the Food and Drug Administration (FDA) in the United States and are comprised of a lipid shell (Definity, Lantheus Medical Imaging) or albumin shell (Optison, General Electric) encapsulating a perfluorocarbon gas [1].

The microbubbles used in clinical practice and research applications undergo nonlinear oscillation at ultrasound frequencies in the range of 1–2 MHz. Numerous microbubble detection strategies have been incorporated into commercially available ultrasound imaging systems, having in common the ability to initiate and detect microbubble oscillations or induce microbubble rupture at high acoustic pressures. These include strategies to detect the second-harmonic signal (second-harmonic imaging) and multipulse cancellation techniques designed to subtract linear fundamental frequency tissue backscatter from nonlinear microbubble signals via amplitude or phase modulation of the acoustic pulse [2,4]. Other approaches that capitalize on the acoustic responsiveness of microbubbles utilize high acoustic pressures to induce microbubble rupture, resulting in broadband frequency ultrasound emissions [2,4].

Ultrasound contrast agents in conjunction with the ultrasound imaging modalities described above have been used as red blood cell tracers to define the blood space during echocardiography. Because of their small size, intravenously injected microbubbles transit the pulmonary circulation, resulting in left ventricular cavity opacification and improved delineation of left ventricular endocardial borders, leading to more accurate assessment of left ventricular function [5]. The red cell tracer attributes of ultrasound contrast agents have also been used in echocardiography to evaluate regional myocardial blood flow, and as such, the contrast agents have been useful in evaluating risk area during coronary occlusion, collateral perfusion, and infarct size and detecting coronary stenosis through assessment of flow reserve [1,6–9].

## 16.3 Ultrasound Contrast Agents for Molecular Imaging

Ultrasound contrast agents can be chemically modified such that their intravascular behavior is uncoupled from red blood cell kinetics, allowing them to function as molecular and cellular imaging probes. The microbubble surface can be modified to bear targeting ligands on the shell, conferring adhesion of the microbubbles to specific molecular targets on the endothelial cell surface. Unlike the transient tissue contrast enhancement resulting from microbubbles used as red cell tracers, persistent contrast enhancement occurs due to microbubble adhesion to function-specific endothelial markers, thus confirming the presence and localization of the function-specific epitope of interest [10,11]. Targeting ligands coupled to the shell of microbubbles have included the naturally occurring ligands (e.g., vascular endothelial growth factor 121 [$VEGF_{121}$], sialyl Lewis$^x$) [12,13], peptide sequences [14,15], and monoclonal antibodies [16–18]. Targeted microbubbles have been used to detect inflammatory overexpression of leukocyte adhesion molecules such as vascular cell adhesion molecule 1 (VCAM1), intercellular adhesion molecule 1 (ICAM1), and P-selectin [13–18]. Microbubble targeting to leukocyte adhesion molecules, in combination with two-dimensional ultrasound, has successfully detected heart transplant rejection (via ICAM1 targeting) [17], myocardial ischemic memory (via P-selectin targeting) [13], and early atherosclerotic plaque (via VCAM targeting) [18] in animal models. Tumor angiogenesis has been ultrasonically imaged using microbubbles targeted to bind to VEGF receptors (via $VEGF_{121}$) [12], $\alpha_v\beta_3$ integrin (via echistatin) [19], as well as microbubbles bearing peptides found by phage display to bind to tumor neovasculature [14].

## 16.4 Ultrasound Contrast Agents for Cellular Imaging

Stem cell tracking with ultrasound is an extension of the molecular imaging concept in that microbubbles are complexed to cells, rendering them acoustically active. The use of ultrasound contrast agents for imaging the fate of exogenously delivered cell therapies can be approached in several ways, having the common general requirement to couple a microbubble to the delivered cell, but differing in the mechanism by which the coupling is achieved (Figure 16.1).

In one approach, the cells can be labeled with microbubbles *in vitro*, then they are administered to the organism, and the cells can be localized using the microbubble-specific ultrasound imaging modalities described. The *ex vivo* coupling of microbubbles to a stem cell can be achieved using molecular targeting as described; the microbubble carries a stem cell-specific ligand on its shell that allows microbubble attachment to the stem cell surface. This would result in a microbubble-stem cell complex in which the microbubbles

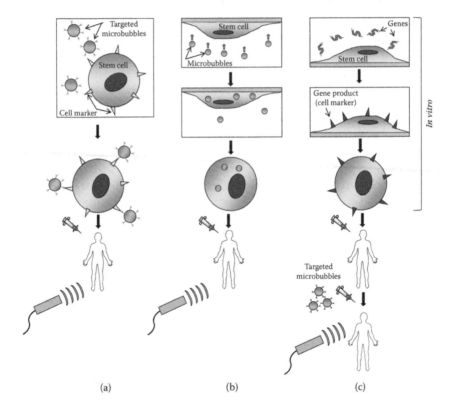

(a)                          (b)                          (c)

**FIGURE 16.1**

Schematic representation of potential approaches to labeling stem cells with microbubbles to enable ultrasound detection. (a) Microbubbles can be attached to the surface of suspended stem cells *in vitro* via a targeting ligand on the microbubble shell that binds to a specific stem cell marker. The labeled cells are then delivered, bearing the microbubbles on the cell surface. (b) Stem cells are labeled *in vitro* by internalization of the microbubbles. In the approach shown, culture plates are inverted, and microbubble interaction with stem cells is achieved by virtue of the buoyant properties of the microbubbles. The labeled stem cells are lifted and delivered, bearing the microbubbles within the cytoplasm. (c) Stem cells are labeled *in vivo*, after delivery, using microbubbles targeted to bind to a cell-specific marker. Prior to delivery, the stem cells are genetically modified *in vitro* to express a unique marker that serves as the microbubble target. Figure not drawn to scale.

are extracellular (Figure 16.1A). The success of this approach is contingent on maintenance of microbubble attachment to the cell during handling and *in vivo* transit of the microbubble-cell complex, which may be technically difficult to achieve. Another disadvantage of this approach is that attachment of multiple microbubbles 1–3 µm in diameter on the surface of a cell would functionally increase cell size, possibly resulting in vascular occlusion if injected intravascularly, or inhibiting motility or engraftment potential.

Another approach to *ex vivo* microbubble labeling of cells would be to promote internalization of the microbubbles by the stem cells, resulting in a microbubble-cell complex in which the microbubbles are *intracellular* (Figure 16.1B). This approach requires that the microbubble remain acoustically active after internalization; that is, the microbubble must retain its gas content. It has been shown that activated leukocytes can internalize microbubbles, and that the leukocytes subsequently become acoustically visible with ultrasound, implying that the microbubbles retain their gas and remain compressible by ultrasound even when located within the cytoplasm of a cell (Figure 16.2) [20]. It has been reported that bone marrow-derived mesenchymal stem cells internalize polymer microbubbles through mechanisms that are incompletely understood [21,22]. It has been shown that following microbubble internalization by mesencyhmal stem cells, the cells can be imaged *in vitro* using nonlinear ultrasound imaging techniques as described, indicating that microbubble oscillation in an ultrasound field is retained even when the microbubbles are within the cytoplasm of the mesenchymal stem cell [21]. Furthermore, spectral analyses of the microbubble-cell

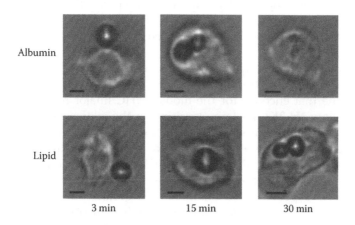

Albumin

Lipid

3 min 15 min 30 min

**FIGURE 16.2**
Evidence that microbubbles can be internalized by cells and remain intact. The cells shown are activated leukocytes incubated with microbubbles composed of an albumin (upper panels) or lipid (lower panels) shell. Microscopic imaging up to 30 minutes after internalization demonstrated persistence of the lipid but not albumin microbubbles. (From Lindner, J.R., Dayton, P.A., Coggins, M.P., et al. 2000. *Circulation* 102:531–38. With permission.)

complexes confirmed that they emit harmonic frequencies, whereas stem cells alone do not exhibit harmonic behavior [22].

Microbubble internalization would be an attractive cell-tracking strategy if several criteria were met. The ideal microbubble should be nontoxic to the cell and "survive" internalization to retain acoustic activity when located within the cell (i.e., retain its gas content). The microbubble should track viability; that is, if the cell dies, the microbubble should also "die," which in acoustic terms means that microbubble shell integrity must be disrupted such that the gas content of the bubble is released. The ideal microbubble would be passed on to daughter cells if cell division occurs, rendering the daughter cells acoustically visible during ultrasound imaging. A polymer microbubble may be best suited to achieve these attributes because of its potential to be more resilient to degradation by lyososomal enzymes, but this will need to be tested.

Rather than labeling the stem cells *ex vivo* prior to injection, another strategy would be to use a molecular imaging approach whereby microbubbles are pretargeted to bind to engrafted stem cells *in vivo* via a ligand on the microbubble shell that binds to a unique marker expressed by the engrafted cells (Figure 16.1C). In this approach, microbubbles targeted to bind to a unique stem cell surface marker are intravenously injected and subsequently "find" the cells *in vivo, after* the cells have already been delivered. This approach is challenged by the fact that, on culture or engraftment, stem cells tend to lose early stem cell markers as they develop a more mature cellular phenotype. For example, bone marrow-derived endothelial progenitor cells (EPCs) evolve an endothelial phenotype in culture, expressing markers such as vascular endothelial group factor receptor-2 (VEGFR2) and CD31 [23], which, if chosen as the microbubble targets, would make it difficult to distinguish between exogenously delivered engrafted cells and endogenous endothelial cells.

Hence, in a study involving exogenous delivery of bone marrow-derived EPCs, prior to administration, the cells were initially transfected with a marker gene that encodes for the mouse MHC (major histocompatibility complex) class 1H-2Kk protein, resulting in the expression of the H-2Kk marker solely on the cell surfaces of the exogenously delivered EPCs [24]. Lipid microbubbles were synthesized bearing anti-H-2Kk antibody on the shell to allow recognition of the EPCs by the microbubble. In parallel plate flow chamber studies using either H-2Kk-targeted or nontargeted microbubbles exposed to EPCs, there was greater adhesion of the targeted microbubbles to H-2Kk-transfected cells compared to control EPCs or mock-transfected EPCs. Subsequent *in vivo* ultrasound imaging studies were performed in rats using subcutaneously implanted Matrigel plugs supplemented with H-2Kk- or mock-transfected EPCs. Signal intensity within the H-2Kk-transfected EPC Matrigel plugs was higher after injection of H-2Kk-targeted microbubbles compared to control microbubbles (Figure 16.3). Furthermore, there was minimal contrast enhancement by

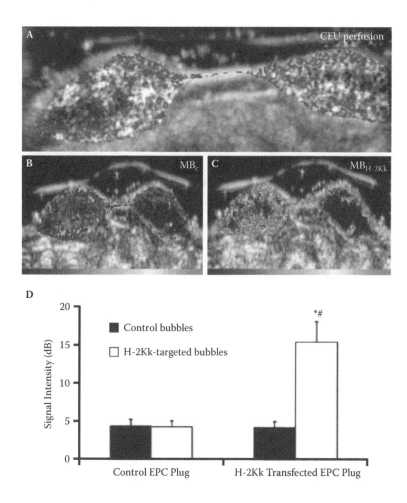

**FIGURE 16.3**
Ultrasound cell tracking using *in vivo* labeling of endothelial progenitor cells (EPCs) that were modified *in vitro*, prior to injection, by transfection with the H-2Kk gene. Matrigel plugs subcutaneously injected at two sites in rats were supplemented with H-2Kk- transfected EPCs (left plug) or control EPCs (right plug). (A) Background-subtracted perfusion imaging of the plugs with nontargeted microbubbles. Delayed imaging of the Matrigel plugs after intravenous injection of (B) control microbubbles or (C) H-2Kk-targeted microbubbles. There was persistent contrast enhancement of the plug supplemented with H-2Kk-transfected EPCs after injection of the H2Kk-targeted microbubbles. (D) Videointensity measurements of the Matrigel plugs under the various experimental conditions. (From Kuliszewski, M.A., Fugii, H., Liao, C., et al. 2009. *Cardiovasc Res* 83:653–62. With permission.)

either microbubble type within plugs supplemented with mock-transfected EPCs. These data suggest the potential to track EPCs using a molecular imaging approach to labeling the cells *in vivo*, subsequent to engraftment. A disadvantage of this technique, however, is the requirement for genetic modification of the EPCs to effect expression of a unique surface marker. Furthermore, because the microbubbles are strictly intravascular probes by virtue of their size, only cell engraftment on the luminal side of the vasculature can be detected.

## 16.5  Advantages of Ultrasound Contrast Agents for Cell Tracking

There are several potential unique advantages of using ultrasound contrast agents for tracking stem cells. Unlike optical methods that require fluorescent labeling, ultrasound imaging of contrast agents can be performed in clinical populations. The spatial resolution of ultrasound exceeds that of nuclear imaging and optical methods [25]. Furthermore, it is a tomographic imaging technique that allows three-dimensional interrogation of target tissue, not possible using optical methods that summate signals emanating along a z axis (depth of tissue) into a two-dimensional plane. An ultrasound cell-tracking approach does not involve radiation exposure to the patient or stem cell and does not require genetic modification of the cell to enable detection. Because ultrasound technology does not involve ionizing radiation, repetitive imaging can be performed, allowing serial imaging of cell trafficking over time. Furthermore, the instrumentation involved is relatively simple (compared to magnetic resonance imaging [MRI] or positron emission tomography [PET]), and its portability confers the potential for widespread use in diverse clinical settings. Because ultrasound imaging provides both anatomic information and microbubble-specific detection, there is simultaneous coregistration of the underlying anatomy and stem cell signal, respectively.

As the gas within the microbubble is what confers acoustic activity, disruptions in microbubble shell integrity can result in rapid loss of the gas, particularly if the gas is highly soluble in the surrounding medium. It might be possible to manipulate the composition of the contrast agent shell such that shell integrity is sensitive to the intracellular milieu, with shell defects induced, and microbubble gas content released, when the cell undergoes death. This property would allow microbubble integrity, and hence the acoustic signal, to parallel cell viability. Such a link between the presence of an acoustic signal and the presence of cell viability has advantages over other modalities using stem cell labels that may persist or be internalized by macrophages, even after the demise of the stem cell [26].

## 16.6 Limitations to Ultrasound Cell Tracking

There are inherent limitations to an ultrasound approach to tracking stem cells. The major limitation is that ultrasound imaging is not a whole-body imaging method, and as such, it would be difficult to map the anatomic whole-body distribution of exogenously administered stem cells. The technique would be most useful when applied to the imaging of a target organ, such as the heart, in which the intracardiac/intramyocardial fate of locally or systemically delivered cells could be searched.

Current microbubble formulations have finite stability *in vivo*, which could limit the time frame available for meaningful cell imaging if the stem cell longevity exceeds microbubble survival. Intravascular injection leads to predominant cell loss within a week [27], which is likely to be within an achievable time frame for microbubble persistence. Localized injection of stem cells could lead to long-term survival and differentiation of a larger number of cells [28]; if this occurs microbubble labels capable of longer-term persistence remain to be developed.

## 16.7 Challenges in the Development of Ultrasound Cell-Tracking Methods

The use of microbubbles as probes to enable ultrasound imaging of stem cell fate is in the early stages of development. The advantages of ultrasound and microbubble technology as outlined are theoretical, and remain to be realized. The limited data available thus far have proven important concepts, including the fact that microbubbles can be internalized by stem cells; intracellular microbubbles remain acoustically active acutely; and antibody-targeted microbubbles can attach to endoluminally engrafted EPCs via engineered ligand-receptor interactions.

The ultimate clinical implementation of an acoustic approach to stem cell tracking will require several important advances. If the goal is to locate where the cells go, including extravascular migration, then intracellular incorporation of the microbubbles prior to injection would be preferable to intravenous injection of microbubbles targeted to a cell-specific epitope since the latter approach would be confined to the detection of endovascular engraftment only, given the exclusively intravascular location of the microbubbles. Importantly, an intracellular probe should not alter stem cell functions, so possible effects of microbubble internalization on cell activities must be evaluated. Furthermore, fundamental questions need to be answered, including how to modulate the number of microbubbles internalized by a cell; that is, what is the minimum number of bubbles necessary to render a cell acoustically active? The temporal course of acoustic activity needs to be defined, with the preferable time course paralleling stem cell viability. The mechanism of intracellular incorporation should be more fully understood to optimize approaches to ensuring microbubble "survival" within the cell. For example, if endocytosis is involved, a strategy for inducing lysosomal "escape" might be helpful for preventing microbubble destruction once internalized by the cell. Acoustic characterization of the microbubble-cell complexes using a combination of spectral analysis, ultrasound imaging, and ultra-high-speed microscopic imaging [29,30] under various ultrasound transmission parameters will be important for optimizing algorithms for detecting signals specific to labeled stem cells. Such acoustic parameters will ultimately need to be built into clinical scanning systems.

## 16.8 Conclusion

The determination of stem cell fate after exogenous delivery is important to the clinical implementation of cell-based therapies. Microbubble probes and ultrasound imaging may offer unique advantages for real-time, serial stem cell tracking in patients. Important proofs of concept have already been demonstrated. Further advances in microbubble chemistry and imaging technology along the lines described should facilitate clinical realization of this imaging approach and assist in the development of cell therapies for human populations.

## References

1. Kaul, S. 2008. Myocardial contrast echocardiography: A 25-year retrospective. *Circulation* 118:291–308.
2. Becher, H., Burns, P.N. 2000. *Handbook of Contrast Echocardiography*. New York: Springer-Verlag.
3. de Jong, N., Boukaz, A., Frinking, P. 2002. Basic acoustic properties of microbubbles. *Echocardiography* 12:229–40.
4. Burns, P. 2002. Instrumentation for contrast echocardiography. *Echocardiography* 19:241–88.
5. Hundley, W.G., Kizilbash, A.M., Afridi, I., et al. 1998. Administration of an intravenous perfluorocarbon contrast agent improves echocardiographic determination of left ventricular volumes and ejection fraction: Comparison with cine magnetic resonance imaging. *J Am Coll Cardiol* 32:1426–32.
6. Wei, K., Jayaweera, A.R., Firoozan, S., Linka, A., Skyba, D.M., Kaul, S. 1998. Quantification of myocardial blood flow with ultrasound-induced destruction of microbubbles administered as a constant venous infusion. *Circulation* 97:473–78.

7. Elhendy, A., O'Leary, E.L, Xie, F., McGrain, A.C., Anderson, J.R., Porter, T.R. 2004. Comparative accuracy of real time myocardial contrast perfusion imaging and wall motion analysis during dobutamine stress echocardiography for the diagnosis of coronary disease. *J Am Coll Cardiol* 44:2185–91.

8. Villanueva, F.S., Gertz, E.W., Csikari, M., Pulido, G., Fisher, D., Sklenar, J. 2001. Detection of coronary artery stenosis using power Doppler imaging. *Circulation* 103:2624–30.

9. Coggins, M.P., Sklenar, J., Le, E., Wei, K., Lindner, J.R., Kaul, S. 2001. Noninvasive prediction of ultimate infarct size at the time of acute coronary occlusion based on the extent and magnitude of collateral-derived myocardial blood flow. *Circulation* 104:2471–77.

10. Villanueva, F.S., Wagner, W.R. 2008. Ultrasound molecular imaging of cardiovascular disease. *Nature* 5 suppl 2:s26–s32.

11. Villanueva, F.S. 2008. Molecular imaging of cardiovascular disease using ultrasound. Bench to Bedside series. *J Nucl Cardiol* 15:576–86.

12. Wang, J., Kilic, S., Tom, E., et al. 2005. Vascular endothelial growth factor-conjugated ultrasound microbubbles adhere to angiogenic receptors. *Circulation* 112:II-502.

13. Villanueva, F.S., Lu, E., Bowry, S., et al. 2007. Myocardial ischemic memory imaging using molecular echocardiography. *Circulation* 115:345–52.

14. Weller, G.E.R., Wong, M.K.K., Modzelewski, R.A., et al. 2005. Ultrasonic imaging of tumor angiogenesis using contrast microbubbles targeted via the tumor-binding peptide RRL. *Cancer Res* 65:533–39.

15. Leong-Poi, H., Christiansen, J., Klibanov, A.L., Kaul, S., Lindner, J.R. 2003. Noninvasive assessment of angiogenesis by ultrasound and microbubbles targeted to $a_v$-integrins. *Circulation* 107:455–60.

16. Villanueva, F.S., Jankowski, R.J., Klibanov, S., et al. 1998. Microbubbles targeted to intercellular adhesion molecule-1 bind to activated coronary artery endothelial cells: A novel approach to assessing endothelial function using myocardial contrast echocardiography. *Circulation* 98:1–5.

17. Weller, G.E., Lu, E., Csikari, M.M., et al. 2003. Ultrasound imaging of acute cardiac transplant rejection with microbubbles targeted to intercellular adhesion molecule-1. *Circulation* 108:218–24.

18. Kaufmann, B.A., Sanders, J.M., Davis, C., et al. 2007. Molecular imaging of inflammation in atherosclerosis with targeted ultrasound detection of vascular cell adhesion molecule-1. *Circulation* 116:276–84.

19. Ellegala, D.B., Leong-Poi, H., Carpenter, J.E., et al. 2003. Imaging tumor angiogenesis with contrast ultrasound and microbubbles targeted to $a_v\beta_3$. *Circulation* 108:336–41.

20. Lindner, J.R., Dayton, P.A., Coggins, M.P., et al. 2000. Noninvasive imaging of inflammation by ultrasound detection of phagocytosed microbubbles. *Circulation* 102:531–38.

21. Cui, W., Wang, J., Chen, X., et al. 2008. Stem cell tracking using ultrasound contrast agents. *Circulation* 118:S642.

22. Leng, X., Wang, J., Fu, H., Fisher, A., Chen, X., Villanueva, F.S. In vivo stem cell tracking using ultrasound imaging. *Circulation* 120:S326.

23. Urbich, C., Dimmeler, S. 2004. Endothelial progenitor cells: Characterization and role in vascular biology. *Circ Res* 95:343–53.

24. Kuliszewski, M.A., Fugii, H., Liao, C., et al. 2009. Molecular imaging of endothelial progenitor cell engraftment using contrast-enhanced ultrasound and targeted microbubbles. *Cardiovasc Res* 83:653–62.

25. Sinusas, A.J., Bengel, F., Nahrendorf, M., et al. 2008. Multimodality cardiovascular molecular imaging. Part I. *Circ Cardiovasc Imaging* I:244–56.

26. Amsalem, Y., Mardor, Y., Feinberg M.S., et al. 2007. Iron-oxide labeling and outcome of transplanted mesenchymal stem cells in the infarcted myocardium. *Circulation*. 116(suppl I):I-38–I-45.

27. Toma, C., Wagner, W.R., Bowry, S., et al. 2009. Fate of culture-expanded mesenchymal stem cells in the microvasculature: *In vivo* observations of cell kinetics *Circ Res* 104(3):398–402.

28. Quevedo, H.C., Hatzistergos, K.E., Oskouei, B.N., et al. 2009. Allogeneic mesenchymal stem cells restore cardiac function in chronic ischemic cardiomyopathy via trilineage differentiating capacity. *Proc Natl Acad Sci USA* 106:14022–27.

29. Postema, M., van Wamel, A., ten Cate, F.J., de Jong, N. 2005. High-speed photography during ultrasound illustrates potential therapeutic applications of microbubbles. *Med Phys* 32:3707–11.

30. Chin, C.T., Lancee, C., Borsboom, J., Mastik, F., Frijlink, M.E., de Jong, N. 2003. Brandaris 128: A digital 25 million frames per second camera with 128 highly sensitive frames. *Rev Sci Instrum* 74:5026–34.

# 17

## X-Ray-Guided Delivery and Tracking of Cells

**Jeff W. M. Bulte and Aravind Arepally**

### CONTENTS

### 17.1 Introduction

Cellular therapeutics has recently emerged as a new way to treat or possibly cure myriad diseases. This includes the use of immune cells, including dendritic cell cancer vaccine for immunotherapy of cancer (Figdor et al. 2004), and progenitor and stem cells for repair of degenerative diseases (Fink 2009). To further guide these therapies into the clinic, the use of noninvasive imaging techniques is mandatory. These need to be applied not only to guide the cell injection itself but also to report on subsequent cell fate, including cell survival, cell movements, and cell differentiation (in the case of stem cells).

Clinically applicable imaging techniques that have been successfully used to monitor cell homing and cell trafficking are few, with labeling of cells with a radiotracer (e.g., [111]In-oxine) for nuclear scintigraphy (Kraitchman et al. 2005) as the only cell-tracking method approved by the Food and Drug Administration (FDA). The other two modalities that have only been recently introduced are positron emission tomography (PET) using the transgene enzyme herpes simplex virus (HSV) thymidine kinase (Yaghoubi et al. 2009) and magnetic resonance imaging (MRI) of cells labeled with superparamagnetic iron oxide (SPIO) particles (Bulte 2009). X-ray and computed tomographic (CT) imaging have not entered the arena yet, primarily due to the low sensitivity of radiopaque contrast agents and the potential cytotoxicity associated with them. However, for reasons stated in this chapter, it would be highly desirable to have a robust method that allows monitoring of radiopaque cells using X-ray. This chapter covers what has been accomplished so far in this field.

## 17.2 How Can Cell Therapy Benefit from Image-Guided Delivery?

For over three decades, physicians have used image guidance for delivery of various therapeutic agents. Initially described in the 1970s for catheter-directed delivery of vasoactive drugs to control emergent gastrointestinal bleeding, locoregional delivery of therapeutic agents has now exponentially grown and is utilized for a variety of conditions, including treatment of uterine fibroids, unresectable liver tumors, intracranial/extracranial neoplasm, treatment of vascular malformations, and catheter-directed thrombolysis. With the expansion in these clinical applications, the imaging technology has also grown in parallel and now provides unique capabilities for future targeted therapies. Finally, with this rapid growth in targeted delivery, vast resources of experience have been gained that can be parlayed for emerging applications, such as cellular therapies.

The main goal of local delivery is to increase the concentration of a specific therapeutic agent in a target tissue with minimal nontarget distribution. Compared to systemic therapy, local delivery provides a high level of therapeutic efficacy with minimal systemic effects. However, the success of any local therapy depends on the precision and accuracy of delivery to the region of interest. Currently, the majority of clinical delivery procedures, (such as transcatheter arterial chemoembolization [TACE] and arteriovenous malformation [AVM] embolizations) are performed predominantly with a transvascular approach. A transvascular approach for delivery not only provides the highest possible precision but also has several other advantages: (a) the local administration results in increased local tissue concentration, (b) the hemodynamics of the vascular bed can be altered through the administration of vasoactive compounds (i.e., vasodilation or embolization), and (c) the prolonged dwell time and slow uptake of the therapeutic agent results in greater efficacy (Llovet et al. 2002).

In an analogous manner, cellular therapies can take advantage of these techniques to enhance the delivery process. Although some cellular therapies are adequately managed with systemic delivery, certain organs and conditions require a targeted approach for maximum effect. In fact, due to the anatomical constraints of the vascularity of specific organs, systemic therapy is precluded, and image-guided techniques will be necessary to overcome some of these barriers. For example, the vasculature of the brain, liver, and pancreas have been shown to have unique anatomical boundaries that create a barrier to conventional therapies and therefore do not respond to conventional systemic therapies. In the brain, the blood-brain barrier acts as a physiologic barrier that prevents the migration and the transport of agents from the systemic vasculature. In both the liver and pancreas, there is a separate anatomic venous vascular supply, termed the *portomesenteric system*, that is completely isolated from systemic circulation. Therefore, with disorders involving these organs (i.e., stroke, cirrhosis, and diabetes) accessing these secluded organs will be critical to the success of such cellular therapeutics.

Finally, another relevant opportunity with image-guided therapy is the ability not only to deliver to target organs but also to administer only into injured tissues. As demonstrated by multiple studies, substantial mobilization of stem cells has been demonstrated after myocardial infarction and with liver injury. Clearly, the homing of cells to injured tissues is a major mechanism of regeneration, and there has been extensive work in utilizing image-guided therapy to help foster this process by providing targeted delivery into injured tissue.

## 17.3 X-Ray/CT Image-Guided Injections: The Prevalent Method of Image-Guided Injections

Image-guided delivery of therapeutic agents is now a routine clinical procedure. Often used for neoplasms in the liver and uterus and vascular anomalies, targeted drug delivery is a vital and compulsory component in the management of multiple clinical conditions. The current primary imaging modality for targeted delivery has been X-ray angiography. Initially, X-ray angiography was limited to simple fluoroscopy with limited imaging capability. However, advances in imaging technology, such as digital subtraction angiography, flat-panel detectors, and digital postprocessing, have created a powerful modality for targeted delivery.

However, despite these advances, a major limitation with this method is that anatomical information, which is derived from X-ray images, is two dimensional and has substantial anatomic ambiguity. Another major limitation is the inability to visualize the targeted tissues. Therefore, the ability to visualize the target tissue, monitor any nontarget distribution, and provide real-time validation of the treatment endpoint is not feasible with current X-ray-based imaging strategies.

Due to these inherent problems, CT guidance is also being explored as an alternative imaging modality for guiding and monitoring of drug therapy.

CT images provide a new and attractive alternative to conventional X-ray approaches for targeted drug delivery. Over the last decade, significant progress has been made in the use of CT to guide various interventions (Lederman 2005). In addition, interventional combined CT/X-ray (soft tissue imaging) systems have been implemented that have dual x-ray and CT capabilities in a single suite and allow for real-time interventional procedures to be performed in a clinical setting. In these suites, fluoroscopy is used to guide catheter placement; subsequently, adjunctive three-dimensional (3D) imaging is obtained immediately.

With recent advances in imaging, these systems now combine the advantages of 3D CT imaging with live X-ray imaging of all organs in one examination and on a single system. By rotating the C-arm of the fluoroscopy system at high speed around the patient, 3D CT-like images are produced. In this way, several hundred images are acquired and reconstructed as 3D volumes. If the acquisition is triggered via the patient ECG (electrocardiogram), time-dependent 3D volumes can be generated for visualization of the beating heart. Anatomical structure segments are overlaid with the live X-ray image, allowing the physician to navigate with the catheter quickly and confidently without the use of a contrast medium (Figure 17.1).

If drug or cellular delivery is required, these systems provide a significant advantage over conventional fluoroscopic systems. Previously, these combined suites were confined to a few academic centers, but currently multiple systems are being installed in many centers within North America.

## 17.4 Magnetic Resonance Image-Guided Injections: Pros and Cons

MR image-guided injections provide a new delivery alternative to conventional X-ray approaches. Over the last decade, significant progress has been made in the use of MRI to guide various vascular interventions. Initial clinical applications were seen for neurosurgery, where intraoperative MRI systems with movable magnets were used in the operating room (Maurer et al. 1998; Tronnier et al. 1997). With the advent of (3D) gradient echo techniques, further applications for vascular and nonvascular frontiers have appeared (Ladd and Debatin 2000). The advantage of MRI over X-ray is that it is a tomographic technique that allows one to view multiple planes rather than projections. The feasibility of aortic and iliac arterial angioplasty with MR tracking has been demonstrated in animal models and in limited clinical trials (Becker 2001; Godart et al. 2000; Manke et al. 2001; Serfaty et al. 2000). In addition, at our institution, advances have been made with multiple loopless antennas combined with surface coils to provide real-time visualization in an MRI environment (Arepally et al. 2004, 2005; Serfaty et al. 2000). Nonvascular

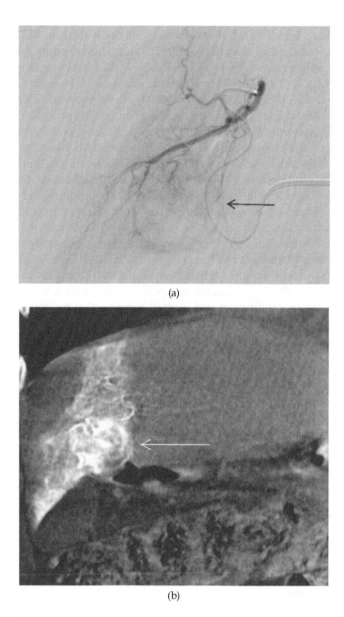

(a)

(b)

**FIGURE 17.1**
Hepatic angiogram for hepatocellular carcinoma during transarterial chemoembolization (TACE) with (a) conventional angiogram and (b) use of combined CT/X-ray unit. Arrow indicates tumor. Note the clear depiction of tumor and adjacent liver with combined CT/X-ray unit versus conventional angiogram.

applications have also been seen in the field of radio-frequency (RF) ablation, with MR guidance and monitoring of percutaneous treatment of liver, renal, and pancreatic tumors and, most recently, of bone ablation (Aschoff et al. 2002). Thus, even more than the ability to duplicate current X-ray-guided procedures, interventional MR will enable new therapies to emerge.

Because of the high spatial resolution, exquisite soft tissue anatomical detail, and the lack of ionizing radiation, MRI provides an ideal platform for applications such as cellular therapies. The capability to fuse previously acquired 3D anatomic MRI with interventional devices greatly amplifies the ability to replicate complex surgical procedures such as vascular anastomoses, vascular bypass grafts, and portal vein injections. Furthermore, with the rapid development of molecular imaging, catheter-based therapies for targeted injections of cellular therapeutics are now beginning to be explored. As shown over the last few years, the growth of catheter-based therapies for a variety of disorders, such as cancer, diabetes, and chronic limb ischemia, is closely tied to molecular imaging. For example, islet cell therapy for diabetes and stem cell implantation require catheter-directed delivery of these agents directly into the target organs (Hirshberg et al. 2002; Kraitchman et al. 2003). Most commonly, these procedures are performed under X-ray fluoroscopy, with poor anatomic definition and limited functional data. When combined with methods to MR-label cells, MR-guided islet therapy offers the additional benefit of direct confirmation of whether each injection successfully reaches the targeted organ.

A successful delivery procedure is defined by maximizing the concentration of the therapeutic agent in the target tissue with minimal nontarget distribution. As demonstrated, there has been significant improvement in both imaging and hardware design that now allows for highly targeted therapies under MR guidance. Furthermore, MRI provides multiple unique platforms with multiparametric imaging (anatomic, physiologic, and functional information) to now tailor delivery to an organ or tissues, which thus has significant advantages over other imaging modalities. Although most studies to date have been proof of principle, it is anticipated that MRI and guidance techniques will provide unique insights into the complex dynamics of cellular therapies and thus enhance these techniques for clinical implementation.

## 17.5 X-Ray Labeling of Cells: Promises and Pitfalls

As of the spring of 2011, to the best of our knowledge, only one study has been performed on direct prelabeling of cells followed by transplantation or transfusion using X-ray tracking. The study by Barnett et al. (2010) used perflurooctylbromide (PFOB) labeling of human pancreatic islets, containing about 1,000 cells to overcome this sensitivity problem. This is most likely due to the lack of sensitivity of CT contrast agents, as well as the potential toxicity of X-ray contrast agents. However, cells have been passively labeled *in vivo*, such as macrophages with iodinated nanoparticles for CT detection of atherosclerotic plaques in hyperlipidemic rabbits (Hyafil et al. 2007) and with bismuth sulfide nanoparticles for liver and lymph node imaging in mice (Rabin et al. 2006). This approach results in substantial accumulation of radiopaque agents in a focal area.

We have pursued a novel method of making cells X-ray visible, by encapsulating cells in semipermeable alginate capsules while loading the capsules with bismuth sulfide and barium sulfate during the polymerization of the alginate (Barnett et al. 2006). Encapsulated cell therapy has been widely pursued for the prevention of immunorejection of transplanted cells, while enabling incorporation of a high payload of contrast agent to allow detection of single capsules with MRI (Barnett et al. 2007) and X-ray imaging (Barnett et al. 2006). Microencapsulation creates a semipermeable membrane that prevents passage of antibodies and complement, thereby preventing graft rejection (Lim and Sun 1980; Orive et al. 2006). While antibodies are blocked, the selective permeability of the capsule allows for passage of therapeutic factors produced by encapsulated cells.

Our radiopaque capsules had a characteristic white appearance for barium sulfate capsules (Ba X-caps) or yellow for bismuth sulfide capsules (Bi X-caps) (Figure 17.2). As compared to unlabeled (nonradiopaque) microcapsules, we found both Ba X-caps and Bi X-caps had equal permeability to fluorescent lectins with different molecular weights. Capsules were permeable to fluorescent lectins smaller than 75 kDa but were impermeable to lectins larger than 120 kDa, that is, blocking antibodies while allowing penetration of smaller nutrients and secretion of insulin (MW ~ 5 kDa).

We have successfully imaged radiopaque capsules in mice and rabbits. Single Ba X-caps and Bi X-caps were clearly identified *in vivo* after transplantation into the peritoneal cavity of mice (Figure 17.3) and after intramuscular injection into the hind limb of rabbits (Figure 17.4). Two weeks after injection, both X-caps retained their radiopacity *in vivo* as compared to day 0 (Figure 17.4). Other recent approaches for making X-ray visible capsules include the use of PFOB (Barnett et al., 2011) and gold nanoparticles (Kim et al. 2011).

---

## 17.6 Summary and Conclusions

Cellular therapies are likely to become an integrated part of regenerative medicine, and image-guided injections, based on either CT/X-ray or MRI, will be needed to advance the clinical field. Studies on image-guided injections of cellular therapeutics have so far been limited to MRI using SPIO

**FIGURE 17.2**
Macroscopic (A) and (C) and fluoroscopic (B) and (D) images of (A) and (B) Bi X-caps and (C) and (D) Ba X-caps. Single capsules can be clearly visualized. (Reproduced from Barnett, B.P., D.L. Kraitchman, C. Lauzon, C.A. Magee, P. Walczak, W.D. Gilson, A. Arepally, and J.W. Bulte. 2006. *Mol Pharm* 3, no. 5: 531–38.)

labeling of cells. The advent of CT/X-ray imaging for similar targeted injections has yet to arise. Further improvement of labeling methods and development of radiopaque agents specifically tailored for CT/X-ray imaging of cells may lead to new applications. In the meantime, labeling capsules, which are housing therapeutic cells, with radiopaque agents represents an alternative method of creating a local high payload of contrast material without a potential harmful accumulation inside cells, allowing cells to be visible on X-ray and CT.

## Acknowledgment

Dr. Jeff W. M. Bulte is supported by NMSS RG3630, the TEDCO Maryland Stem Cell Fund ESC 06-29-01, RO1 NS045062, RO1 EB007825, EUREKA RO1 DA02699, and Roadmap R21 EB005252. He is a paid consultant for Surgivision Incorporated. This arrangement has been approved by the Johns Hopkins University in accordance with its conflict-of-interest policies. Dr.

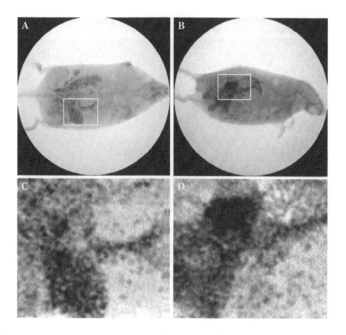

**FIGURE 17.3**

*In vivo* imaging of X-caps following transplantation in the peritoneal cavity of mice. (A) and (C) Fluoroscopic image following injection of 5,000 Ba X-caps. (B) and (D) Fluoroscopic image following injection of 5,000 Bi X-caps. (C) and (D) Enlargement of boxed areas in (A) and (B). (Reproduced from Barnett, B.P., D.L. Kraitchman, C. Lauzon, C.A. Magee, P. Walczak, W.D. Gilson, A. Arepally, and J.W. Bulte. 2006. *Mol Pharm* 3, no. 5: 531–38.)

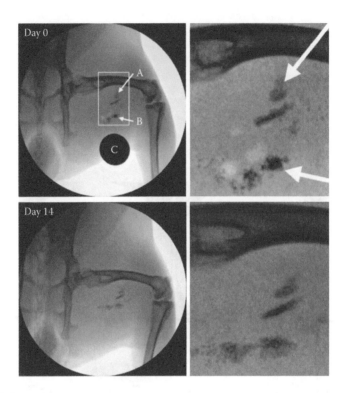

**FIGURE 17.4**

*In vivo* imaging of X-caps immediately (top row) and 2 weeks (bottom row) after intramuscular transplantation into a rabbit hind limb. A = 2,000 Ba X-caps; B = 2,000 Bi X-caps; C = quarter for reference of size and opacity. Magnification of fluoroscopic image is shown on right. (Reproduced from Barnett, B.P., D.L. Kraitchman, C. Lauzon, C.A. Magee, P. Walczak, W.D. Gilson, A. Arepally, and J.W. Bulte. 2006. *Mol Pharm* 3, no. 5: 531–38.)

A. Arepally is also a paid consultant for Surgivision LLC and founder of Surefire Medical Incorporated.

## References

Arepally, A., C. Georgiades, L.V. Hofmann, M. Choti, P. Thuluvath, and D.A. Bluemke. 2004. Hilar cholangiocarcinoma: Staging with intrabiliary MRI. *Am J Roentgenol* 183, no. 4: 1071–74.

Arepally, A., P.V. Karmarkar, C. Weiss, E.R. Rodriguez, R.J. Lederman, and E. Atalar. 2005. Magnetic resonance image-guided trans-septal puncture in a swine heart. *J Magn Reson Imaging* 21, no. 4: 463–67.

Aschoff, A.J., E.M. Merkle, S.N. Emancipator, C.A. Petersilge, J.L. Duerk, and J.S. Lewin. 2002. Femur: MR imaging-guided radio-frequency ablation in a porcine model feasibility study. *Radiology* 225, no. 2: 471–78.

Barnett, B.P., A. Arepally, P.V. Karmarkar, D. Qian, W.D. Gilson, P. Walczak, V. Howland, L. Lawler, C. Lauzon, M. Stuber, D.L. Kraitchman, and J.W. Bulte. 2007. Magnetic resonance-guided, real-time targeted delivery and imaging of magnetocapsules immunoprotecting pancreatic islet cells. *Nat Med* 13, no. 8: 986–91.

Barnett, B.P., D.L. Kraitchman, C. Lauzon, C.A. Magee, P. Walczak, W.D. Gilson, A. Arepally, and J.W. Bulte. 2006. Radiopaque alginate microcapsules for x-ray visualization and immunoprotection of cellular therapeutics. *Mol Pharm* 3, no. 5: 531–38.

Barnett, B.P., J. Ruiz-Cabello, P. Hota, R. Liddell, P. Walczak, V. Howland, V.P. Chacko, D.L. Kraitchman, A. Arepally, J.W.M. Bulte. Fluorocapsules for improved function, immunoprotection, and visualization of cellular therapeutics with MR, US, and CT imaging. Radiology, 258, 182–191, 2011.

Barnett, B.P., J. Ruiz-Cabello, P. Hota, M.J. Shamblott, C. Lauzon, P. Walczak, W.D. Gilson, V.P. Chacko, D.L. Kraitchman, A. Arepally, J.W.M. Bulte. Use of perfluorocarbon nanoparticles for non-invasive multimodal cell tracking of human pancreatic islets. Contrast Med. Mol. Imaging, 2010.

Becker, G.J. 2001. 2000 RSNA annual oration in diagnostic radiology: The future of interventional radiology. *Radiology* 220, no. 2: 281–92.

Bulte, J.W. 2009. In vivo MRI cell tracking: Clinical studies. *AJR Am J Roentgenol* 193, no. 2: 314–25.

Figdor, C.G., I.J. De Vries, W.J. Lesterhuis, and C.J. Melief. 2004. Dendritic cell immunotherapy: Mapping the way. *Nat Med* 10, no. 5: 475–80.

Fink, D.W., Jr. 2009. Fda regulation of stem cell-based products. *Science* 324, no. 5935: 1662–63.

Godart, F., J.P. Beregi, L. Nicol, B. Occelli, A. Vincentelli, V. Daanen, C. Rey, and J. Rousseau. 2000. Mr-guided balloon angioplasty of stenosed aorta: In vivo evaluation using near-standard instruments and a passive tracking technique. *J Magn Reson Imaging* 12, no. 4: 639–44.

Hirshberg, B., S. Montgomery, M.G. Wysoki, H. Xu, D. Tadaki, J. Lee, K. Hines, J. Gaglia, N. Patterson, J. Leconte, D. Hale, R. Chang, A.D. Kirk, and D.M. Harlan. 2002. Pancreatic islet transplantation using the nonhuman primate (rhesus) model predicts that the portal vein is superior to the celiac artery as the islet infusion site. *Diabetes* 51, no. 7: 2135–40.

Hyafil, F., J.C. Cornily, J.E. Feig, R. Gordon, E. Vucic, V. Amirbekian, E.A. Fisher, V. Fuster, L.J. Feldman, and Z.A. Fayad. 2007. Noninvasive detection of macrophages using a nanoparticulate contrast agent for computed tomography. *Nat Med* 13, no. 5: 636–41.

T. Kim, E. Momin, J. Choi, Q. Yuan, H. Zaidi, J. Kim, M. Park, N. Lee, M.T. McMahon, A. Quinones-Hinojosa, J.W.M. Bulte, T. Hyeon, A.A. Gilad. Mesoporous, silica-coated, hollow manganese oxide nanoparticles as positive T1 contrast agent for labeling and MR tracking of adipose-derived mesenchymal stem cells. JACS 133, 2955–2961 (2011).

Kraitchman, D.L., A.W. Heldman, E. Atalar, L.C. Amado, B.J. Martin, M.F. Pittenger, J.M. Hare, and J.W. Bulte. 2003. In vivo magnetic resonance imaging of mesenchymal stem cells in myocardial infarction. *Circulation* 107, no. 18: 2290–93.

Kraitchman, D.L., M. Tatsumi, W.D. Gilson, T. Ishimori, D. Kedziorek, P. Walczak, W.P. Segars, H.H. Chen, D. Fritzges, I. Izbudak, R.G. Young, M. Marcelino, M.F. Pittenger, M. Solaiyappan, R.C. Boston, B.M. Tsui, R.L. Wahl, and J.W. Bulte. 2005. Dynamic imaging of allogeneic mesenchymal stem cells trafficking to myocardial infarction. *Circulation* 112, no. 10: 1451–61.

Ladd, M.E. and J.F. Debatin. 2000. Interventional and intravascular MR angiography. *Herz* 25, no. 4: 440–51.

Lardo, A.C. 2000. Real-time magnetic resonance imaging: Diagnostic and interventional applications. *Pediatr Cardiol* 21, no. 1: 80–98.

Lederman, R.J. 2005. Cardiovascular interventional magnetic resonance imaging. *Circulation* 112, no. 19: 3009–17.

Lim, F. and A.M. Sun. 1980. Microencapsulated islets as bioartificial endocrine pancreas. *Science* 210, no. 4472: 908–10.

Llovet, J.M., M.I. Real, X. Montana, R. Planas, S. Coll, J. Aponte, C. Ayuso, M. Sala, J. Muchart, R. Sola, J. Rodes, and J. Bruix. 2002. Arterial embolisation or chemoembolisation versus symptomatic treatment in patients with unresectable hepatocellular carcinoma: A randomised controlled trial. *Lancet* 359, no. 9319: 1734–39.

Manke, C., W.R. Nitz, B. Djavidani, M. Strotzer, M. Lenhart, M. Volk, S. Feuerbach, and J. Link. 2001. MR imaging-guided stent placement in iliac arterial stenoses: A feasibility study. *Radiology* 219, no. 2: 527–34.

Maurer, C.R., Jr., D.L. Hill, A.J. Martin, H. Liu, M. Mccue, D. Rueckert, D. Lloret, W.A. Hall, R.E. Maxwell, D.J. Hawkes, and C.L. Truwit. 1998. Investigation of intraoperative brain deformation using a 1.5-T interventional MR system: Preliminary results. *IEEE Trans Med Imaging* 17, no. 5: 817–25.

Orive, G., S.K. Tam, J.L. Pedraz, and J.P. Halle. 2006. Biocompatibility of alginate-poly-l-lysine microcapsules for cell therapy. *Biomaterials* 27(20): 3691–700.

Rabin, O., J. Manuel Perez, J. Grimm, G. Wojtkiewicz, and R. Weissleder. 2006. An x-ray computed tomography imaging agent based on long-circulating bismuth sulphide nanoparticles. *Nat Mater* 5, no. 2: 118–22.

Serfaty, J.M., E. Atalar, J. Declerck, P. Karmakar, H.H. Quick, K.A. Shunk, A.W. Heldman, and X. Yang. 2000. Real-time projection MR angiography: Feasibility study. *Radiology* 217, no. 1: 290–95.

Tronnier, V.M., C.R. Wirtz, M. Knauth, G. Lenz, O. Pastyr, M.M. Bonsanto, F.K. Albert, R. Kuth, A. Staubert, W. Schlegel, K. Sartor, and S. Kunze. 1997. Intraoperative diagnostic and interventional magnetic resonance imaging in neurosurgery. *Neurosurgery* 40, no. 5: 891–900; discussion 900–2.

Yaghoubi, S.S., M.C. Jensen, N. Satyamurthy, S. Budhiraja, D. Paik, J. Czernin, and S.S. Gambhir. 2009. Noninvasive detection of therapeutic cytolytic T cells with 18F-FHBG PET in a patient with glioma. *Nat Clin Pract Oncol* 6, no. 1: 53–58.

# 18

## Multimodality Fusion Imaging: Toward Imaging of Structure and Function

Jason M. Criscione, Albert J. Sinusas, and Tarek M. Fahmy

### CONTENTS

### 18.1 Introduction

While cellular therapies have proven to be a promising approach for repairing complex physiological defects, assessing the success of these approaches *in vivo* has been challenging. This inherent difficulty is derived from the necessity to visualize the cellular therapies and functional responses of the affected tissues. In the case of stem cell therapy, this is of critical importance since the establishment of efficacy requires defining the trafficking of the cells *in vivo*, demonstrating their integration into the intended tissue, and determining their survival or fate. For stem cells expanded and differentiated *ex vivo*, it is imperative to assess whether the implanted cells retain the ability to migrate to the intended tissue and subsequently mature and func-

**TABLE 18.1**

Capabilities of Various Medical Imaging Modalities

| Imaging Modality | Energy | Spatial Resolution | Sensitivity |
|---|---|---|---|
| Magnetic resonance imaging (MRI) | Radio frequency | 50–250 μm | $10^{-3}$–$10^{-6}$ M |
| Computed tomography (CT) | X-ray | 25–150 μm | n/a |
| Positron emission tomography (PET) | γ-ray | 1–2 mm | $10^{-11}$–$10^{-12}$ M |
| Single-photon emission computed tomography (SPECT) | γ-ray | 0.3–1 mm | $10^{-10}$–$10^{-11}$ M |
| Optical imaging | Visible or near IR | 2–3 mm (*in vivo*) <1 μm (*in vitro*) | <$10^{-9}$ M |
| Ultrasound (US) | Ultrasound | 30–500 μm (frequency dependent) | $10^{-6}$–$10^{-9}$ M |

tion as intended. The ability to visualize cellular trafficking and function *in vivo* noninvasively will advance the field toward this end.

To achieve this goal, it is important to utilize imaging modalities that provide information regarding both anatomical structure and function. Imaging structure or anatomical features require imaging modalities manifesting high spatial resolution. In contrast, imaging function requires highly sensitive modalities that can report on dynamic metabolic processes or focal molecular events. The challenge has been integrating high spatial resolution with high sensitivity of detection since high-resolution modalities exhibit poor sensitivity and highly sensitive modalities exhibit low resolution (Table 18.1). Thus, observing structure and functional information on the cellular/tissue level is best achieved utilizing multiple modalities.

### 18.1.1 Multimodal Imaging

Multimodal imaging seeks to combine the high spatial resolution of one modality, for example, computed tomography (CT) or magnetic resonance imaging (MRI), with the high sensitivity of other modalities, like positron emission tomography (PET), single-photon emission computed tomography (SPECT), bioluminescence imaging (BLI), or fluorescence imaging. Currently, this multimodal approach is accomplished through utilization of either hybrid imaging systems or image fusion techniques (Judenhofer et al. 2008; Pichler et al. 2008a, 2008b, 2008c; Schäfers and Stegger 2008; Townsend 2008; Townsend and Cherry 2001). Both methods of multimodal imaging provide substantial information regarding anatomical structure and function; however, hybrid imaging systems offer the high spatial resolution of one modality and the high sensitivity of a second modality within a single instrumental setup, for example, SPECT/CT, PET/CT, or PET/MRI. Image fusion relies on software approaches to fuse complementary images acquired with several independent imaging modalities to meet the need for high spatial resolution and high sensitivity. These hybrid imaging platforms

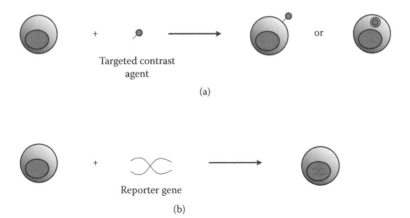

**FIGURE 18.1**
(a) Direct and (b) indirect methods of stem cell labeling.

have been available for preclinical applications and are now becoming part of routine clinical practice.

### 18.1.2 Cell Labeling by Direct and Indirect Methods

To visualize cellular trafficking adequately, it is essential that the cells of interest be either directly or indirectly labeled with an appropriate contrast agent to sufficiently enhance contrast and enable detection (Lee et al. 2008). Direct cell labeling requires that an exogenous contrast agent target the intended cell either to associate with the cell surface or to gain cell entry via endocytic pathways (Figure 18.1a). Contrast agents designed for direct cell labeling must generate sufficient signal or enhanced contrast, persist sufficiently with cell division, and be biocompatible to the extent that cellular functions, like migratory function or differentiation, are not altered (Ly et al. 2008) (Table 18.2). Indirect cell labeling (Figure 18.1b) requires efficient gene transfection without insertional mutagenesis and subsequent transcription to enable detection (Table 18.2). Thus, utilizing hybrid or fusion imaging for observation of cellular trafficking requires specific chemical and biological skills to design optimal multimodal contrast agents.

## 18.2 Hybrid Imaging Systems and Contrast Media

Hybrid imaging modalities, for example, PET/CT, SPECT/CT, and PET/MRI, are attractive because they offer the ability to generate images with both high spatial, anatomical resolution (~25–100 μm) and high sensitivity

**TABLE 18.2**

Examples of Probes Utilized for Direct and Indirect Stem Cell Labeling

| Cell Labeling | Probes | Cell Targeting | Imaging Modalities |
|---|---|---|---|
| Direct | Gd-DTPA dendrimer conjugates Perfluorocarbon nanoparticles Radioligands Iron oxide particles | Antibodies Peptides Aptamers Sugars Cationic molecules Membrane transport | PET/CT, SPECT/CT, US, MRI, optical imaging |
| Indirect | Reporter gene sequences encoding for Enzymes Receptors Fluorescent proteins | Lentivirus Adenovirus Electroporation | PET, SPECT, MRI, BLI, fluorescence imaging |

($<10^{-9}$ molar) using a single instrument (Oconnor and Kemp 2006; Pichler et al. 2008b; Schäfers and Stegger 2008; Townsend 2008). Although the design of the PET/MRI hybrid systems proved challenging (Judenhofer et al. 2008; Pichler et al. 2008a), the result may offer significant advantages over the PET/CT system (Table 18.3). First, MRI offers high spatial resolution and good soft tissue contrast without the imparted dose of radiation associated with the X-ray source of CT. Second, PET/MRI enables simultaneous acquisition of functional parameters and high-resolution anatomical structure; PET/CT requires sequential acquisition. Finally, PET/MRI offers the inherent ability to generate several types of functional information, as MRI affords the capabilities of spectroscopic analysis by magnetic resonance spectroscopy (MRS) or functional imaging by functional MRI (fMRI). MRS and fMRI offer the ability to explore biochemical and pathological processes (Judenhofer et al. 2008). Although MRI represents a seemingly versatile, high-resolution

**TABLE 18.3**

Advantages and Disadvantages of PET/CT and PET/MRI Hybrid Imaging Systems

| Imaging Modality | Advantages | Disadvantages |
|---|---|---|
| PET/CT | Preclinical and clinical availability Feasible PET attenuation correction Whole-body scans | X-ray radiation dose Sequential acquisition Motion correction and temporal registration |
| PET/MRI | Simultaneous acquisition Soft-tissue contrast MRS and fMRI offer additional functional information Motion correction Utilizes radio waves | Clinical availability Scans limited to the head and neck PET attenuation correction not straightforward |

modality, the image quantification is more robust and facile with the CT hybrid analogue. Specifically, CT offers a straightforward means to correct for both the tissue attenuation of the 511-keV photon in PET and associated partial volume errors. For PET/CT, attenuation correction is readily performed through correlating the tissue attenuations obtained by CT. Attenuation correction for PET/MRI is not as straightforward, as MRI only provides information regarding proton density, which does not necessarily correspond to attenuation ability (Pichler et al. 2008a). Attenuation correction with CT also facilitates the correction of partial volume errors that result in the underestimation of true regional radiolabel activity (Sinusas et al. 2008).

While MRI possesses specific advantages (Table 18.3), it is important to note that PET/MRI hybrid imaging systems are technically challenging to implement for clinical applications and may not be the best solution for all applications. For instance, the magnet bore diameter in a 3-T clinical scanner restricts the size of the PET insert and MR radio-frequency coil to only feasibly image the head and neck. Thus, a balance between practical availability and imaging capability needs to be considered for clinical applications.

### 18.2.1 PET/CT and SPECT/CT Systems

Both PET/CT and SPECT/CT molecular imaging systems utilize the high spatial resolution of CT to define anatomical structures and high sensitivity afforded by reporting directly and quantitatively from exogenous radionuclide labels, such as fluorine-18 ($^{18}$F) for PET and technetium-99m ($^{99m}$Tc) for SPECT, to generate "hot-spot" images. Although PET and SPECT offer high sensitivity, it is essential that the radiolabel administered possesses a reasonably long half-life and maintains sufficient signal with cell division. While $^{18}$F and $^{99m}$Tc have relatively short half-lives (110 min, Fowler and Ido 2002; and 6 hours, Eckelman 2009, respectively), other radionuclides offer longer half-lives and are more suitable for assessing the biodistribution of transplanted cells; for example, indium-111 ($^{111}$In) for SPECT and copper-64 ($^{64}$Cu) for PET possess half-lives of 2.8 days and 12.7 hours, respectively (Modo 2008). While PET/CT is routinely used to detect tumor structure and function, its utility for visualization of stem cells has been explored. For example, PET/CT has been effectively utilized to noninvasively observe tissue distribution of hematopoietic stem cells labeled with $^{18}$F-fluorodeoxyglucose ($^{18}$F-FDG) (Kang et al. 2006) (Figure 18.2).

### 18.2.2 PET/MRI Systems

The PET/MRI molecular imaging system was developed to utilize the high spatial resolution of proton MRI to define anatomical structures and the high sensitivity afforded by PET to generate hot-spot images. While PET/MRI has successfully been utilized for tumor and brain imaging (Judenhofer et al. 2008; Pichler et al. 2008b, 2008c), its utility for stem cell applications has not yet been

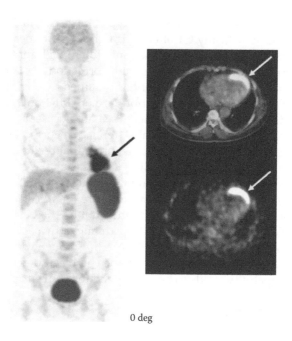

0 deg

**FIGURE 18.2 (See color insert.)**
PET/CT images of 65-year-old man with history of anterior wall infarction. After percutaneous intervention, [18]F-FDG-labeled stem cells were injected via intracoronary catheter. PET/CT images were obtained 2 hours after injection. Stem cell accumulation at myocardium is well visualized on transaxial views (arrow). Total amount of stem cells at myocardium was 2.1% of injected dose. (Adapted from Figure 2 of Kang, W.J., H.J. Kang, H.S. Kim, J.K. Chung, M.C. Lee, and D.S. Lee. 2006. *J Nucl Med* 47, no. 8: 1295–1301.)

explored. However, the success of PET/MRI image fusion in observing stem cell transplantation phenomena has effectively demonstrated its potential for future applications. For example, dual-modality imaging with PET and MRI enabled visualization of the effects of stem cell transplantation on cardiac function in a rat model of myocardial infarction (Chapon et al. 2009).

## 18.3  Image Fusion Platforms and Contrast Media

Image fusion presents an imaging methodology that utilizes the independent strengths of several modalities to acquire complementary sets of information for subsequent analysis and interpretation of complex physiological systems. The strength of fusion imaging is derived not only from the selection of imaging modalities but also from the design of the multimodal contrast agents (Figure 18.3) that enhance the imaging capabilities of the independent platforms (Cheon and Lee 2008).

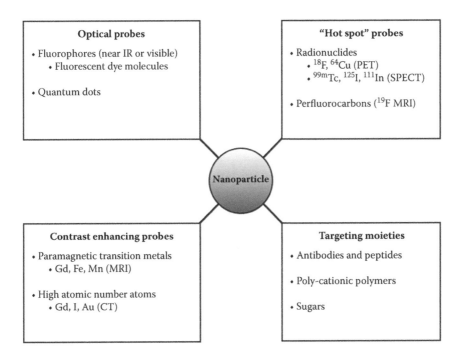

**FIGURE 18.3**
Components available for targeted, multimodal contrast agent design.

Multimodal contrast agents are typically fabricated from materials bearing surfaces comprising multiple functional groups. Multiple surface functional groups permit the attachment of the desired contrast medium and specific, high-affinity targeting moieties that are essential for effective direct cell labeling (Figure 18.1). The successful multimodal contrast agent should therefore generate sufficient signal or contrast enhancement for each modality utilized, persist sufficiently with cell division, and ensure complete biocompatibility (Ly et al. 2008). Typical materials suited for these applications include complex macromolecules, such as branched polymers (Bulte et al. 2001; Svenson and Tomalia 2005) or chain polymers with pendant functional groups (Maki et al. 2007), or nanoparticulate systems such as solid matrix (Hendryii et al. 2008) or vesicular nanoparticles (Partlow et al. 2007) formulated to bear functional groups (Figure 18.4).

### 18.3.1 MRI and Optical Imaging

Although the fusion of MRI and optical imaging creates the ability to provide complementary information regarding the migration and fate of labeled cells, its utility for noninvasive *in vivo* studies is limited by both the poor tissue depth penetration (<1 cm) and spatial resolution (2–3 mm) of optical methods (Cheon and Lee 2008), such as confocal and intravital microscopy.

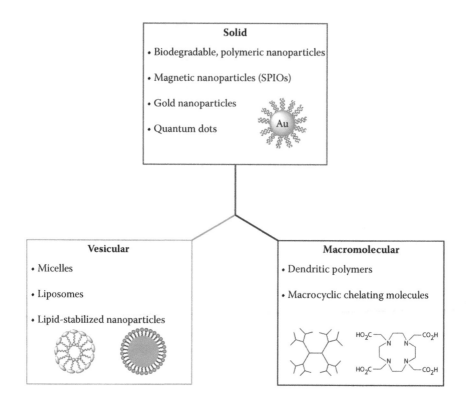

**FIGURE 18.4**

Examples of nanoparticles for therapeutic delivery and diagnostic imaging. SPIO, superparamagnetic iron oxide.

For this reason, multimodal contrast agents comprising paramagnetic and fluorescent components are routinely utilized to confirm direct cell labeling with MR active agents. Following confirmation of cell labeling by optical methods, MRI is utilized solely to image stem cell trafficking. For example, confocal microscopy has been utilized to assess iron-fluorophore particle (IFP) labeling of mesenchymal stem cells whose tissue integration and fate, following transplantation into the myocardium of normal and infarcted swine, was determined by MRI (Hill 2003).

### 18.3.2  ¹⁹F and ¹H MRI

Fluorine-based MRI (¹⁹F MRI) is an attractive, noninvasive imaging modality that offers a distinct advantage over contrast-based imaging platforms (Ahrens et al. 2005). Negligible fluorine background in tissue enables hot-spot imaging (Bulte 2005), in which selective reporting from the fluorinated probe ameliorates interpretation of the experiment. Since the hot-spot reports only on the detection of the fluorinated probe, collection of a spatially and geometrically identical proton ($^1$H) image is necessary to provide an appropriate anatomical map for fluorine localization. While the sensitivity of fluorine is comparable to that of proton, a dense localization of fluorine spins is required to ensure an adequate signal-to-noise ratio, a critical feature that only a few platforms possess. These platforms include lipid-stabilized perfluorocarbon nanoparticles (Ahrens et al. 2005; Caruthers et al. 2006; Janjic et al. 2008; Lanza et al. 2005; Partlow et al. 2007; Srinivas et al. 2007), liposomes encapsulating perfluorocarbons (Kimura et al. 2004), and chain polymers comprising fluorinated pendant groups (Maki et al. 2007). Among these platforms, lipid-stabilized perfluorocarbon nanoemulsions have proven useful for noninvasive visualization of stem cell tracking *in vivo* with $^{19}$F MRI (Partlow et al. 2007) (Figure 18.5).

## 18.4 Indirect Labeling Platforms

Indirect labeling (Figure 18.1) is an attractive means of stem cell labeling because it facilitates longitudinal studies following implantation. This is primarily due to the persistence of the reporter probe despite cell division since indirect labeling of stem cells relies on the expression of reporter genes, encoding for either enzymes or fluorescent proteins, transduced into the cells prior to transplantation. Visualization of these cells is facilitated through the fusion of several molecular imaging modalities. The combinations of imaging modalities that lend themselves to this form of fusion imaging are PET/CT/BLI (Love et al. 2007), MRI/BLI (Hendryii et al. 2008), or MRI/fluorescence imaging (Hill 2003).

Lentiviral or adenoviral vectors are traditionally employed to transfect double or triple fusion reporter gene sequences into stem cells to enable subsequent expression of fluorescent proteins, for example, green or red fluorescent protein (GFP or RFP, respectively), or enzymes, for example, firefly luciferase (fLuc) or herpes simplex virus type 1 thymidine kinase (HSV1-tk) (Duda et al. 2007). Expression of GFP (Duda et al. 2007) or RFP (Love et al. 2007) provides unique, effective reporter probes for optical imaging techniques. When expressed fLuc encounters exogenously delivered luciferin, detection of the cells is feasible with BLI (Lee et al. 2008; Love et al. 2007; Pomper et al. 2009). Similarly, when expressed HSV1-tk encounters its infused radiolabeled substrate, either 9-[4–$^{18}$F-fluoro-3-(hydroxymethyl) butyl]guanine ($^{18}$F-FHBG) (Love et al. 2007) or 1-(2'-deoxy-2'fluoro-β-D-arabinofuranosyl)-5-[$^{125}$I]-iodouracil ($^{125}$I-FIAU) (Pomper et al. 2009), detection of the cells is feasible with PET or SPECT, respectively. For example, the long-term fate and trafficking of transplanted mesenchymal stem cells transduced with the triple fusion reporter, fLuc-mRFP-tk, have been observed by PET/CT and BLI image fusion (Love et al. 2007) (Figure 18.6).

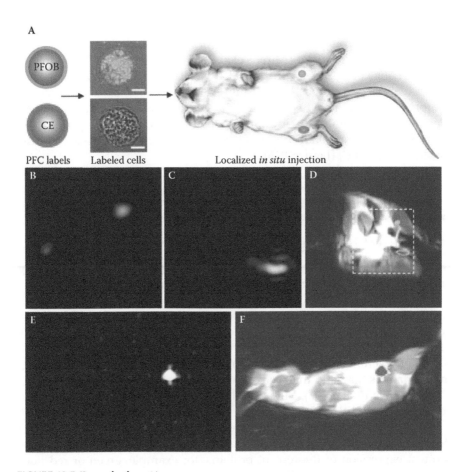

**FIGURE 18.5 (See color insert.)**
Localization of labeled cells after in situ injection by MRS/MRI. (A) To determine the utility for cell tracking stem/progenitor cells labeled with either perfluoro-octylbromide (PFOB, green) or crown ether (CE, red), nanoparticles were locally injected into mouse thigh skeletal muscle. (B)–(D) At 11.7 T, spectral discrimination permits imaging the fluorine signal attributable to about $1 \times 10^6$ PFOB-loaded (B) or CE-loaded cells (C) individually, which when overlaid onto a conventional $^1$H image of the site (D) reveals PFOB- and CE-labeled cells localized to the left and right leg, respectively (dashed line indicates $3 \times 3$ cm$^2$ field of view for $^{19}$F images). (E), (F) Similarly, at 1.5 T, $^{19}$F image of about $4 \times 10^6$ CE-loaded cells (E) locates to the mouse thigh in a $^1$H image of the mouse cross section (F). The absence of background signal in $^{19}$F images (B, C, E) enables unambiguous localization of perfluorocarbon-containing cells at both 11.7 and 1.5 T. (Adapted from Partlow, K., J. Chen, J. Brant, A. Neubauer, T. Meyerrose, M. Creer, J. Nolta, S. Caruthers, G. Lanza, and S. Wickline. 2007. *FASEB J* 21, no. 8: 1647–54.)

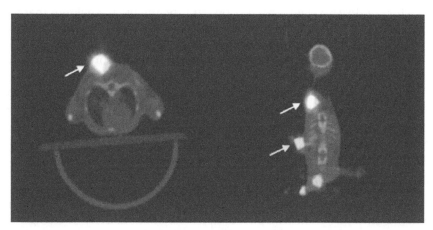

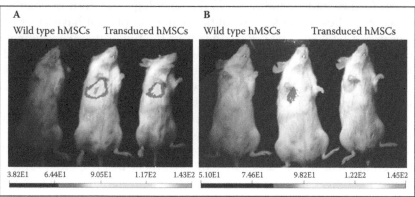

**FIGURE 18.6 (See color insert.)**
(Top) Overlay of PET (hot-metal pseudocolor) and CT (grayscale) images: transaxial (left) and coronal (right) views. Although top-row cubes (yellow arrows) had strong PET signals, second-row cubes (green arrow), loaded with mixture of empty-vector–transduced and reporter-transduced hMSCs (human mesenchymal stem cells), had visible signal despite 1:4 dilution. Cubes were also visible on CT images. (Bottom) BLI of tail vein-injected hMSCs 30 minutes after hMSC injection (A) and 24 hours after hMSC transplantation (B). One control animal injected with wild-type hMSCs is shown on left, and two animals injected with reporter-transduced hMSCs are shown on right. Scale bar for luminescence intensity is shown at bottom. (Adapted from Love, Z., F. Wang, J. Dennis, A. Awadallah, N. Salem, Y. Lin, A. Weisenberger, S. Majewski, S. Gerson, and Z. Lee. 2007. *J Nucl Med* 48, no. 12: 2011–20.)

## 18.5 Design of Intelligent Probes

Given that multiple imaging modalities are requisite for successful characterization of disease state, progression, and treatment, it is evident that the next generation of probes should incorporate functionalities of the intended modalities, while offering an effective means of therapeutic delivery. The design of intelligent probes would integrate these features into a biomimetic structure that possesses that ability to sense and respond to specific environmental cues. The durability of this concept is evident in the increasing number of reports that describe such systems for development of therapeutics. However, the utility of intelligent probes for imaging purposes is uncertain as development is in the early stages.

### 18.5.1 Probes Responsive to Microenvironment Stimuli

Probes designed to sense and respond to specific stimuli from the surrounding microenvironment, for example, enzymatic activity or changes in pH, can further understandings into the pathology and even mechanisms of disease. Probes responsive to subtle changes in pH can have great utility for exploring endocytic compartments or the tumor microenvironment. Inherent pH response is easily achieved through incorporation of a pH-sensitive polymer within the probe. For example, on exposure to low pH, poly(propylacrylic acid) (PPAA) has been demonstrated to elicit complete rupture of endosomal compartments (Kyriakides et al. 2002). Probes that respond to pH have enabled the development of on-off imaging regulation. For example, $H^+$ influx to polyamine-core, PEGylated nanogels elicit volume changes that facilitate the molecular motion of once-rigid internal $^{19}F$ MR active groups (Oishi et al. 2007). In addition, probes that respond to specific enzymatic activity can facilitate the development of similar on-off technologies. Lanthanide ($Ln^{3+}$) ion-chelating, paramagnetic chemical exchange saturation transfer (PARACEST) contrast agents possess this ability. An example of this phenomenon involves the enzymatic conversion of a pro-PARACEST agent to the active PARACEST agent (Chauvin et al. 2008).

### 18.5.2 Probes Responsive to External Triggers

Probes that are capable of responding to external triggers, for example, mechanical energy or X-rays, have implications for both imaging and therapy. For example, probes with echogenic properties (e.g., microbubbles) are susceptible to rupture following interaction with incident ultrasound (US) waves (Hernot and Klibanov 2008; Klibanov 2006; Liu et al. 2006). In concert with MRI/US fusion imaging, this technology enables image-guided, site-specific delivery of therapeutics. Under specific conditions, the interaction of the probe with the external trigger can elicit therapeutic effects. For example, while facilitating imaging, the interaction of gold nanoparticles with the X-ray source of CT generates sufficient heat to elicit cytotoxic effects toward tumor cells (Hainfeld et al. 2004).

### 18.5.3 Probes to Visualize Therapeutic Delivery

One of the most important goals of noninvasive imaging is the visualization of therapeutic delivery. Ongoing research is focusing on the design of nanoparticle-based, multimodal probes that possess the inherent ability to encapsulate and deliver a variety of therapeutics. Nanoparticles provide a versatile platform that enables site-specific or cell-specific targeting through receptor-ligand interactions. The targeting of specific endocytic receptors facilitates nanoparticle entry, a requisite event for numerous applications. Direct labeling of cells with nanoparticle-based, multimodal probes has the potential to enable visualization of both therapeutic delivery and cellular trafficking. For example, labeling of stem cells facilitates noninvasive determination of the efficacy of stem cell therapy (Hill 2003).

Designing these nanoparticle-based, multimodal probes requires sufficient understanding of biomaterials and their intended targets. Nanoparticles typically fall into the following categories: (a) solid, (b) vesicular, and (c) macromolecular (Figure 18.4). Biodegradable, polymeric nanoparticles and liposomes/lipid-stabilized nanoparticles are solid and vesicular biomimetic spheres, respectively, that enable encapsulation and release of therapeutics (Fahmy et al. 2005). Although their utility as molecular imaging probes has been limited, these systems hold great promise for future diagnostic applications. Macromolecules have seen widespread use in diagnostic and therapeutic applications. For example, dendrimers, a class of hyperbranched polymer whose multifunctional surface offers the ideal chemical structure for multimodal probe design, are routinely used as therapeutic delivery vehicles (Patri et al. 2005; Svenson and Tomalia 2005; Tomalia et al. 2007) and as paramagnetic contrast agents for MRI (Bulte et al. 2001; Jászberényi et al. 2007; Kobayashi and Brechbiel 2003, 2004, 2005; Regino et al. 2008; Talanov et al. 2006; Venditto et al. 2005; Wiener et al. 1994).

## 18.6 Conclusions

In addition to examining the native roles of stem cells *in vivo* and *ex vivo*, appropriate assessment of the efficacy of stem cell therapy must provide information regarding the fate of derived cells *in vivo*. For therapy to be realized, information pertaining to the central question of whether implanted stem cells mature as intended and home to their intended targets must be obtained. The synergy of fusion imaging and multimodal contrast agents offers the ability to provide the necessary insight and expand our understanding of stem cell therapies.

## References

Ahrens, E., R. Flores, H. Xu, and P. Morel. 2005. In vivo imaging platform for tracking immunotherapeutic cells. *Nat Biotechnol* 23, no. 8: 983–87.

Bulte, J.W. 2005. Hot spot MRI emerges from the background. *Nat Biotechnol* 23, no. 8: 945–46.

Bulte, J.W.M., T. Douglas, B. Witwer, S.C. Zhang, E. Strable, B.K. Lewis, H. Zywicke, B. Miller, P. Van Gelderen, B.M. Moskowitz, I.D. Duncan, and J.A. Frank. 2001. Magnetodendrimers allow endosomal magnetic labeling and in vivo tracking of stem cells. *Nat Biotechnol* 19, no. 12: 1141–47.

Caruthers, S.D., A.M. Neubauer, F.D. Hockett, R. Lamerichs, P.M. Winter, M.J. Scott, P.J. Gaffney, S.A. Wickline, and G.M. Lanza. 2006. In vitro demonstration using $^{19}F$ magnetic resonance to augment molecular imaging with paramagnetic per-fluorocarbon nanoparticles at 1.5 Tesla. *Invest Radiol* 41, no. 3: 305–12.

Chapon, C., J. Jackson, E. Aboagye, A. Herlihy, W. Jones, and K. Bhakoo. 2009. An in vivo multimodal imaging study using MRI and PET of stem cell transplantation after myocardial infarction in rats. *Mol Imaging Biol* 11, no. 1: 31–38.

Chauvin, T., P. Durand, M. Bernier, H. Meudal, B.-T. Doan, F. Noury, B. Badet, J.-C. Beloeil, and E. Toth. 2008. Detection of enzymatic activity by PARACEST MRI: a general approach to target a large variety of enzymes. *Angew Chem Int Ed Engl* 47: 4370–72.

Cheon, J. and J.H. Lee. 2008. Synergistically integrated nanoparticles as multimodal probes for nanobiotechnology. *Acc Chem Res* 41, no. 12: 1630–40.

Duda, J., M. Karimi, R.S. Negrin, and C.H. Contag. 2007. Methods for imaging cell fates in hematopoiesis. *Methods Mol Med* 134: 17–34.

Eckelman, W. 2009. Unparalleled contribution of technetium-99m to medicine over 5 decades. *JACC Cardiovasc Imaging* 2, no. 3: 364–68.

Fahmy, T., P. Fong, A. Goyal, and W. Saltzman. 2005. Targeted for drug delivery. *Mater Today* 8, no. 8: 18–26.

Fowler, J. and T. Ido. 2002. Initial and subsequent approach for the synthesis of $^{18}FDG^{+}$. *Semin Nucl Med* 32, no. 1: 6–12.

Hainfeld, J., D. Slatkin, and H. Smilowitz. 2004. The use of gold nanoparticles to enhance radiotherapy in mice. *Phys Med Biol* 49, no. 18: N309–15.

Hendry, S., K. Vanderbogt, A. Sheikh, T. Arai, S. Dylla, M. Drukker, M. McConnell, I. Kutschka, G. Hoyt, and F. Cao. 2008. Multimodal evaluation of in vivo magnetic resonance imaging of myocardial restoration by mouse embryonic stem cells. *J Thorac Cardiovasc Surg* 136, no. 4: 1028–37.

Hernot, S. and A. Klibanov. 2008. Microbubbles in ultrasound-triggered drug and gene delivery. *Adv Drug Delivery Rev* 60, no. 10: 1153–66.

Hill, J. 2003. Serial cardiac magnetic resonance imaging of injected mesenchymal stem cells. *Circulation* 108, no. 8: 1009–14.

Janjic, J.M., M. Srinivas, D.K.K. Kadayakkara, and E. Ahrens. 2008. Self-delivering nanoemulsions for dual fluorine-19 MRI and fluorescence detection. *J Am Chem Soc* 130, no. 9: 2832–41.

Jászberényi, Z., L. Moriggi, P. Schmidt, C. Weidensteiner, R. Kneuer, A. Merbach, L. Helm, and É. Tóth. 2007. Physicochemical and MRI characterization of Gd$^{3+}$-loaded polyamidoamine and hyperbranched dendrimers. *J Biol Inorg Chem* 12, no. 3: 406–20.

Judenhofer, M., H. Wehrl, D. Newport, C. Catana, S. Siegel, M. Becker, A. Thielscher, M. Kneilling, M. Lichy, M. Eichner, K. Klingel, G. Reischl, S. Widmaier, M. Röcken, R. Nutt, H. Machulla, K. Uludag, S. Cherry, C. Claussen, and B. Pichler. 2008. Simultaneous PET-MRI: A new approach for functional and morphological imaging. *Nat Med* 14, no. 4: 459–65.

Kang, W.J., H.J. Kang, H.S. Kim, J.K. Chung, M.C. Lee, and D.S. Lee. 2006. Tissue distribution of $^{18}$F-FDG-labeled peripheral hematopoietic stem cells after intracoronary administration in patients with myocardial infarction. *J Nucl Med* 47, no. 8: 1295–301.

Kimura, A., M. Narazaki, Y. Kanazawa, and H. Fujiwara. 2004. $^{19}$F magnetic resonance imaging of perfluorooctanoic acid encapsulated in liposome for biodistribution measurement. *Magn Reson Imaging* 22, no. 6: 855–60.

Klibanov, A.L. 2006. Microbubble contrast agents: targeted ultrasound imaging and ultrasound-assisted drug-delivery applications. *Invest Radiol* 41, no. 3: 354–62.

Kobayashi, H. and M. Brechbiel. 2005. Nano-sized MRI contrast agents with dendrimer cores. *Adv Drug Delivery Rev* 57, no. 15: 2271–86.

Kobayashi, H. and M.W. Brechbiel. 2003. Dendrimer-based macromolecular MRI contrast agents: characteristics and application. *Mol Imaging* 2, no. 1: 1–10.

Kobayashi, H. and M.W. Brechbiel. 2004. Dendrimer-based nanosized MRI contrast agents. *Curr Pharm Biotechnol* 5, no. 6: 539–49.

Kyriakides, T.R., C.Y. Cheung, N. Murthy, P. Bornstein, P.S. Stayton, and A.S. Hoffman. 2002. pH-sensitive polymers that enhance intracellular drug delivery in vivo. *J Control Release* 78, no. 1–3: 295–303.

Lanza, G., P. Winter, A. Neubauer, S. Caruthers, F. Hockett, and S. Wickline. 2005. $^1$H/$^{19}$F magnetic resonance molecular imaging with perfluorocarbon nanoparticles. *Curr Top Dev Biol* 70: 57–76.

Lee, Z., J. Dennis, and S. Gerson. 2008. Imaging stem cell implant for cellular-based therapies. *Exp Biol Med* 233, no. 8: 930–40.

Liu, Y., H. Miyoshi, and M. Nakamura. 2006. Encapsulated ultrasound microbubbles: therapeutic application in drug/gene delivery. *J Control Release* 114, no. 1: 89–99.

Love, Z., F. Wang, J. Dennis, A. Awadallah, N. Salem, Y. Lin, A. Weisenberger, S. Majewski, S. Gerson, and Z. Lee. 2007. Imaging of mesenchymal stem cell transplant by bioluminescence and PET. *J Nucl Med* 48, no. 12: 2011–20.

Ly, H., J. Frangioni, and R. Hajjar. 2008. Imaging in cardiac cell-based therapy: In vivo tracking of the biological fate of therapeutic cells. *Nat Clin Pract Cardiovasc Med* 5: S96–S102.

Maki, J., C. Masuda, S. Morikawa, M. Morita, T. Inubushi, Y. Matsusue, H. Taguchi, and I. Tooyama. 2007. The MR tracking of transplanted ATDC5 cells using fluorinated poly-l-lysine-cf3. *Biomaterials* 28, no. 3: 434–40.

Modo, M. 2008. Noninvasive imaging of transplanted cells. *Curr Opin Organ Transplant* 13, no. 6: 654–58.

O'Connor, M. and B. Kemp. 2006. Single-photon emission computed tomography/ computed tomography: basic instrumentation and innovations. *Semin Nucl Med* 36, no. 4: 258–66.

Oishi, M., S. Sumitani, and Y. Nagasaki. 2007. On-off regulation of $^{19}F$ magnetic resonance signals based on pH-sensitive pegylated nanogels for potential tumor-specific smart $^{19}F$ MRI probes. *Bioconjug Chem* 18, no. 5: 1379–82.

Partlow, K., J. Chen, J. Brant, A. Neubauer, T. Meyerrose, M. Creer, J. Nolta, S. Caruthers, G. Lanza, and S. Wickline. 2007. $^{19}F$ magnetic resonance imaging for stem/progenitor cell tracking with multiple unique perfluorocarbon nanobeacons. *FASEB J* 21, no. 8: 1647–54.

Patri, A.K., J.F. Kukowska-Latallo, and J.R. Baker, Jr. 2005. Targeted drug delivery with dendrimers: Comparison of the release kinetics of covalently conjugated drug and non-covalent drug inclusion complex. *Adv Drug Deliv Rev* 57, no. 15: 2203–14.

Pichler, B., M. Judenhofer, and C. Pfannenberg. 2008a. Multimodal imaging approaches: PET/CT and PET/MRI. *Handb Exp Pharmacol* no. 185 Pt 1: 109–32.

Pichler, B., M. Judenhofer, and H. Wehrl. 2008b. PET/MRI hybrid imaging: Devices and initial results. *Eur Radiol* 18, no. 6: 1077–86.

Pichler, B., H. Wehrl, A. Kolb, and M. Judenhofer. 2008c. Positron emission tomography/magnetic resonance imaging: The next generation of multimodality imaging? *Semin Nucl Med* 38, no. 3: 199–208.

Pomper, M., H. Hammond, X. Yu, Z. Ye, C. Foss, D. Lin, J. Fox, and L. Cheng. 2009. Serial imaging of human embryonic stem-cell engraftment and teratoma formation in live mouse models. *Cell Res* 19, no. 3: 370–79.

Regino, C.A., S. Walbridge, M. Bernardo, K.J. Wong, D. Johnson, R. Lonser, E.H. Oldfield, P.L. Choyke, and M.W. Brechbiel. 2008. A dual CT-MR dendrimer contrast agent as a surrogate marker for convection-enhanced delivery of intracerebral macromolecular therapeutic agents. *Contrast Media Mol Imaging* 3, no. 1: 2–8.

Schäfers, K. and L. Stegger. 2008. Combined imaging of molecular function and morphology with PET/CT and SPECT/CT: image fusion and motion correction. *Basic Res Cardiol* 103, no. 2: 191–99.

Sinusas, A., F. Bengel, M. Nahrendorf, F. Epstein, J. Wu, F. Villanueva, Z. Fayad, and R. Gropler. 2008. Multimodality cardiovascular molecular imaging, part I. *Circulation Cardiovasc Imaging* 1, no. 3: 244–56.

Srinivas, M., P. Morel, L. Ernst, D. Laidlaw, and E. Ahrens. 2007. Fluorine-19 MRI for visualization and quantification of cell migration in a diabetes model. *Magn Reson Med* 58, no. 4: 725–34.

Svenson, S. and D. Tomalia. 2005. Dendrimers in biomedical applications—reflections on the field. *Adv Drug Delivery Rev* 57, no. 15: 2106–29.

Talanov, V.S., C.A. Regino, H. Kobayashi, M. Bernardo, P.L. Choyke, and M.W. Brechbiel. 2006. Dendrimer-based nanoprobe for dual modality magnetic resonance and fluorescence imaging. *Nano Lett* 6, no. 7: 1459–63.

Tomalia, D.A., L.A. Reyna, and S. Svenson. 2007. Dendrimers as multi-purpose nanodevices for oncology drug delivery and diagnostic imaging. *Biochem Soc Trans* 35, Pt 1: 61–67.

Townsend, D. 2008. Dual-modality imaging: combining anatomy and function. *J Nucl Med* 49, no. 6: 938–55.

Townsend, D. and S. Cherry. 2001. Combining anatomy and function: the path to true image fusion. *Eur Radiol* 11, no. 10: 1968–74.

Venditto, V.J., C.A. Regino, and M.W. Brechbiel. 2005. PAMAM dendrimer based macromolecules as improved contrast agents. *Mol Pharm* 2, no. 4: 302–11.

Wiener, E.C., M.W. Brechbiel, H. Brothers, R.L. Magin, O.A. Gansow, D.A. Tomalia, and P.C. Lauterbur. 1994. Dendrimer-based metal chelates: A new class of magnetic resonance imaging contrast agents. *Magn Reson Med* 31, no. 1: 1–8.

# 19

## Other Non-Stem Cell Therapies for Cellular Tracking—Inflammatory Cell Tracking

Yijen Lin Wu, Qing Ye, Haosen Zhang, T. Kevin Hitchens, and Chien Ho

**CONTENTS**

### 19.1 Introduction

Inflammation plays crucial roles in a wide spectrum of disorders. In addition, cell-based immunotherapy has become more important in treating cancer and other diseases. Noninvasive longitudinal tracking of immune cells *in vivo* allows monitoring progression of a pathological condition, assessing therapeutic effectiveness, as well as serving as a surrogate determinant for clinical strategies. The scope of this chapter is to summarize our recent progress in noninvasive magnetic resonance imaging (MRI) of immune cells, using solid organ transplantation in rodents as a model system for imaging the immune response. The current gold standard for accessing allograft rejection after organ transplantation, biopsy, is not only invasive but also prone to sampling errors. Monitoring immune cell accumulation in the rejecting organs with MRI provides a noninvasive alternative to biopsy. In this chapter, we concentrate on our recent studies using renal and cardiac transplantation models to illustrate the unique benefits of cellular MRI.

To make the cells of interest detectable with MRI, a substantial differentiation in signal needs to be created between the target cells and the surrounding tissues. This can be achieved by labeling the target cells with MR contrast agents. MR contrast agents are usually metallic species capable of changing the nuclear relaxation properties (either T1 or T2) of the surrounding water proton spins. Thus, the cells containing contrast agent exhibit different $^1$H signal intensity than the surrounding tissue, and contrast is generated. The contrast agents are usually encapsulated with inert or biocompatible materials to shunt them from toxicity for biological applications. T2 or T2* contrast agents, such as iron oxide ($Fe_nO_m$) particles, create hypointensity in T2- or T2*-weighted MR images, whereas T1 contrast agents, such as gadolinium (Gd) chelates or manganese (Mn) derivatives, cause signal enhancement in T1-weighted imaging. This chapter focuses on iron oxide particles as contrast agents for cell labeling.

There are two main routes to label target cells with MR contrast agents. With the *ex vivo* (or *in vitro*) labeling scheme, the cells of interest from the host or provided by a cell line are loaded with MR contrast agents in culture, and the labeled cells are then introduced into the subject. Alternatively, with the *in vivo* (or in situ) labeling scheme, MR contrast agents are administered systemically, and the target cells take up the label in circulation or target tissues without cell isolation or culture. Phagocytic cells, like macrophages, can readily ingest a large quantity of iron oxide particles above the current detection threshold; thus, they are suitable for either of the labeling schemes. Nonphagocytic cells, such as T lymphocytes, usually require additional measures for sufficient labeling, such as electroporation (Walczak et al. 2005), HIV-TaT-peptide conjugative (Josephson et al. 1999; Lewin et al. 2000), coincubation with transfection agents (Frank et al. 2002), or other schemes for increasing incorporation. It is usually necessary to label nonphagocytic cells in culture unless the contrast agents can be conjugated with a sensitive and specific marker to target the cell type *in vivo*.

## 19.2 Iron Oxides as MR Contrast Agents for Cellular Imaging

Iron oxide ($Fe_nO_m$) is the most common MR contrast agent class for cell tracking. These agents are composed of an $Fe_2O_3$ and $Fe_3O_4$ core and are usually stabilized by different coating materials. Iron oxide particles exhibit superparamagnetism, causing rapid loss of proton spin coherence inside the iron oxide-containing cells and in the vicinity of these cells. As a result, an area of hypointensity or signal void can be observed with T2*-weighted MRI, such as gradient echo imaging. The presence of iron oxide particles can generate significant local magnetic field gradients, and it is possible for its effect on the surrounding water protons to propagate up to 50 times of the radius

(Lauterbur et al. 1986). Consequently, the areas with hypointensity or signal void on a T2*-weighted image can appear much larger than the actual sizes of the areas with iron oxide-containing cells. This phenomenon is known as the *blooming effect*. Although many might regard this phenomenon as "blooming artifacts," it can actually be advantageous because the effect of a single iron oxide-containing cell can be large enough to be detected with *in vivo* imaging, even though the actual size of the labeled cell (20 to 40 μm in diameter) is much smaller than an imaging voxel (at least 100 μm in diameter). This makes imaging single cells with MRI possible (Dodd et al. 1999; Foster-Gareau et al. 2003; Heyn et al. 2005, 2006; Shapiro et al., 2005, 2006a, 2006b; Wu et al. 2006). In general, at least 1 pg of iron per cell is needed to be detectable *in vivo* with current MRI instrumentation and techniques (Heyn et al. 2005, 2006).

There is a large variety of iron oxide particles with different sizes and coating materials, which affects their pharmacokinetics, labeling efficiency, biodistribution, and detectability. According to the sizes, iron oxide particles can be classified into three major categories: ultrasmall superparamagnetic iron oxide (USPIO), superparamagnetic iron oxide (SPIO), and micrometer-size particles (MPIO). The overall hydrodynamic sizes for USPIO, SPIO, and MPIO particles are 1 nm to tens of nanometers, up to a few hundreds of nanometers, and more than 1 nm to a few thousands of nanometers, respectively. Using different agents for cellular MRI can result in different contrast patterns, which are discussed elsewhere in this chapter.

Although having different hydrodynamic sizes, SPIO and USPIO particles generally have a similar core iron oxide architecture and comparable iron core sizes (5–8 nm and 4–5 nm, respectively). They also exhibit similar physicochemical properties. The effective relaxivity (effect on relaxation with unit concentration) for SPIO and USPIO particles in solution may not differ significantly; however, the efficacy of the iron oxide particle uptake into cells is greatly affected by the hydrodynamic size, both *in vitro* (in culture) and *in vivo*. Both SPIO and USPIO particles can be efficiently internalized by macrophages via active endocytosis processes, particularly those involved in scavenger receptor (SA-A) pathways (Raynal et al. 2004). Ferumoxide (SPIO) and ferumoxtran-10 (USPIO) are two iron oxide particles composed of the same coating material (both Dextran T-10), comparable iron core sizes (4.8 + 1.9 nm and 4.9 + 1.5 nm, respectively), but different overall hydrodynamic sizes (120–180 nm for ferumoxide and 15–30 nm for ferumoxtran-10). They exhibit similar relaxivity in solution at 0.47 T, with the same T2 relaxivity (both are 100 $s^{-1}mM^{-1}$) and comparable T1 relaxivity (19 $s^{-1}mM^{-1}$ for ferumoxide and 28 $s^{-1}mM^{-1}$ for ferumoxtran-10). In spite of this, they exhibit different iron uptake efficiencies *in vitro* (in culture), which is attributed to their overall hydrodynamic sizes (Raynal et al. 2004). In addition, also because of their overall size differences, USPIO and SPIO particles have very different blood clearance kinetics when administered systemically for *in vivo* labeling; this is due to their differences in affinity with the reticuloendothelial system

(RES). The larger SPIO particles, such as Feridex (Berlex Laboratories), which is approved by the Food and Drug Administration (FDA), have a shorter blood half-life and are cleared by the RES quickly; the smaller USPIO particles, on the other hand, have much longer blood half-life (24 to 36 hours in humans and about 2 hours in rats), which allows more accessibility to circulating monocytes and macrophages and deeper tissue penetration. Thus, for *in vivo* labeling schemes, SPIO particles are more suitable for liver and spleen imaging, whereas USPIO particles are better for deeper organs and tissues.

Historically, most of the USPIO and SPIO particles are manufactured with biocompatible and biodegradable coating materials, such as dextran or other polysaccharide derivatives. Consequently, both USPIO and SPIO particles that are commercially available share similar biodistribution and fate *in vivo*. Following intravenous injection, SPIO and USPIO particles are found in endosomes or lysosomes in labeled cells (Arbab et al. 2005) as well as in Kupffer cells in liver (Frank et al. 2002), followed by a transient increase in serum iron and ferritin levels in a few days (Frank et al. 2003). The iron is incorporated into hemoglobin in about a month (Arbab et al. 2003).

Much larger MPIOs have become attractive for MR cellular imaging, mainly due to the strong T2 relaxivity exhibited per particle. Currently available MPIO particles are composed of a magnetite core and an inert polymer coating, with sizes ranging from 1 μm to several micrometers. These particles can contain more than 1 pg of iron per particle, whereas it takes a few million (1 to $5 \times 10^6$) SPIO or USPIO particles to achieve 1 pg of iron. Because of the construction and the assembly of MPIO particles, more iron can be loaded into a smaller volume than the equivalent amount of USPIO or SPIO in a cell. In addition, even with the same iron concentration, MPIO particles have increased relaxivity over SPIO particles. T2* relaxivities ($s^{-1} mM^{-1}$) measured for SPIO and MPIO particles were $240 \pm 27$ and $356 \pm 21$, respectively, at 4.7 T and $498 \pm 19$ and $851 \pm 62$, respectively, at 11.7 T (Hinds et al. 2003) because of the vast iron packaging in a small space, resulting in a dominating relaxation mechanism shift from diffusion-sensitive to much larger static field relaxation (Bowen et al. 2002; Shapiro et al. 2006b; Weisskoff et al. 1994). Hence, it is possible to image single cells labeled with MPIO (Shapiro et al., 2005, 2006a, 2006b; Wu et al. 2006) or even single particles (Shapiro et al. 2004) by MRI.

MPIO particles currently available are constructed with an inert polymer matrix, such as polystyrene/divinyl benzene copolymers. Thus, they are not biodegradable at present. Despite a very short blood half-life (on the order of minutes in rodents), once ingested by cells, they can have a very long resident lifetime. Although this might be limiting for clinical application, for research it can be advantageous for allowing long-term longitudinal tracking of labeled cells for many days (Wu et al. 2006), weeks (Shapiro et al. 2006a; Sumner et al. 2007, 2009), or even months (Ye et al. 2008). It is conceivable that making MPIO particles with biodegradable coatings could take advantage of their vast relaxivity and enable clinical use.

## 19.3 Tracking Immune Cells in Renal Transplantation Models

In the acute renal allograft rejection model, the left kidney of the Brown Norway (BN) (RT1$^n$) recipient rat is replaced with a kidney from a Dark Agouti (DA) (RT1$^a$) donor rat. The right kidney of the recipient remains intact as the internal control and maintains renal function for the duration of the experiment. Without immunosuppression, the allograft kidney gradually develops mild acute rejection by postoperative day (POD) 4, and the rejection becomes severe by POD 5. Following intravenous infusion of USPIO particles (*in vivo* labeling scheme), monocytes and macrophages from circulation and in the rejecting kidney ingest USPIO particles and become MRI detectable (Ye et al. 2002).

Figure 19.1 shows USPIO-enhanced MRI for various allografts (Figure 19.1A–O) and an isograft (Figure 19.1P–R) with different doses of dextran-coated USPIO (0 to 12 mg Fe/kg) before contrast (left panels), immediately after USPIO administration on POD 4 (middle panels), and 1 day after USPIO administration on POD 5 (right panels) (Ye et al. 2002). The transplanted kidney (left kidney) appears on the right side of the images, whereas the native kidney (right kidney) appears on the left of the images.

In the absence of USPIO, a decrease in MRI intensity was seen in the allograft medulla regions on POD 5 (Figure 19.1C, left kidney), presumably due to hemorrhaging associated with renal graft rejection, while no significant change was observed in the cortex. Immediately after USPIO administration on POD 4, general darkening can be seen for both kidneys (Figure 19.1E, H, K, N, Q) for both allografts (Figure 19.1E, H, K, N) and the isograft control (Figure 19.1Q) due to the presence of USPIO particles in circulation. The circulating USPIO is cleared within a few hours. One day after USPIO administration on POD 5, allograft kidneys showed various degrees of hypointensity (Figure 19.1F, I, L, O, left kidneys), depending on the USPIO dosage given, but the hypointensity is not seen in the isograft control (Figure 19.1R, left kidney) or the native kidneys (Figure 19.1F, I, L, O, R, right kidneys). At the time of observation on POD 5, USPIO particles have been cleared from circulation, and the USPIO-induced hypointensity in the rejecting renal cortex observed is due to accumulation of USPIO-labeled macrophages in the rejecting grafts. This result was confirmed by histopathology. Pathological examination showed that the hypointense areas displayed high iron contents confirmed by Prussian blue staining, and the iron-containing cells were ED1$^+$ macrophages. This study in rats showed that approximately 6 mg Fe/kg (Figure 19.1L) appears to be the optimal dosage for the visualization of rejection-associated changes in the renal cortex by MRI. Similar hypointensities could be seen in patients after renal transplantation at 1.5 T (Hauger et al. 2007). The renal cortical signal drop was proportional to the number of macrophages infiltrated per unit area assessed by biopsy.

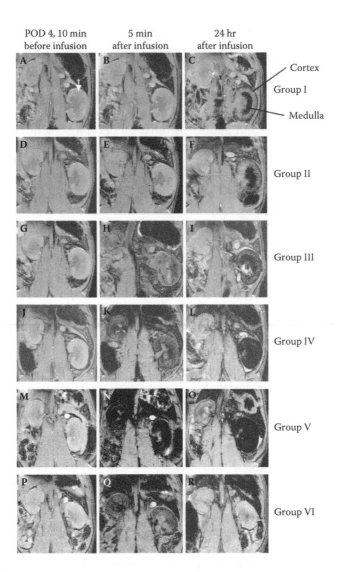

**FIGURE 19.1**

Representative gradient echo MR coronal images showing the effect of different doses of dextran-coated USPIO particles on transplanted rat kidneys. The transplanted kidneys appear on the right (arrow) and the native kidneys appear on the left in these panels. The MR images from immediately before, 5 minutes after, and 24 hours after infusion of dextran-coated USPIO particles are shown in the first, second, and third columns, respectively. Allograft images (A–C) are from group I without USPIO infusion (infused with PBS only); (D–F) are from group II infused with 1 mg Fe/kg; (G–I) are from group III infused with 3 mg Fe/kg; (J–L) are from group IV infused with 6 mg Fe/kg; and (M–O) are from group V infused with 12 mg Fe/kg. (P–R) Isograft MR images from group VI infused with 6 mg Fe/kg. The MRI experiments were carried out in a 4.7-T Bruker Avance DRX MR instrument equipped with a 40-cm horizontal bore superconducting solenoid. (Reprinted from Figure 3 of Ye, Q., D. Yang, M. Williams, D.S. Williams, C. Pluempitiwiriyawej, J.M. Moura, and C. Ho. 2002. *Kidney Int* 61, no. 3: 1124–35. With permission.)

With an *ex vivo* labeling scheme, T lymphocytes can be sufficiently loaded with SPIO particles by incubation with CL-SPIO-Tat-FITC (cross-linked supermagnetic iron oxide with HIV TaT peptide and fluorescein isothiocyanate) complexes and then introduced into transplant recipients. The FITC provided a fluorescent tag, and the SPIO-Tat-FITC-labeled T cells can be detected by both *in vivo* MRI and fluorescent microscopy of biopsy samples. Following administration of CL-SPIO-Tat-FITC-labeled T cells (Ho and Hitchens 2004), the MR signal in cortical areas of allotransplanted kidney (Figure 19.2a, left kidneys) gradually decreases, but not the native control kidney (Figure 19.2a, right kidneys). The areas with hypointensity after infusing *ex vivo* SPIO-labeled T-cells (Figure 19.2a) are compatible with those areas showing hypointensity after *in vivo* labeling of macrophages with USPIO (Figure 19.1). Fluorescence microscopy (Figure 19.2b) confirms that more labeled T cells are found in allograft kidneys.

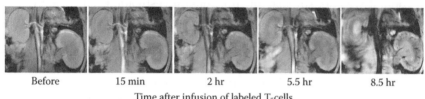

| Before | 15 min | 2 hr | 5.5 hr | 8.5 hr |

Time after infusion of labeled T-cells

(a) **MRI**

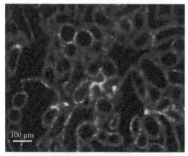

Transplanted kidney        Native kidney

(b) **Fluorescence**

**FIGURE 19.2 (See color insert.)**
(a) Serial gradient echo MR images of a rat with a transplanted kidney before and after infusion of CL-SPIO-Tat-FITC-labeled T cells on POD 4 where a BN rat was transplanted with a kidney from a DA rat; and (b) the corresponding fluorescence images of the allotransplanted and the native kidneys. A gradient echo sequence was used to obtain the coronal MR images with the following parameters: TR = 100 ms; TE = 7.3 ms; NEX = 32; matrix size = 256 × 256; FOV (field of view) = 6.4 × 6.4 cm; slice thickness = 1.4 mm with 0.1-mm slice gap for five slices. The MRI experiments were carried out in a 4.7-T Bruker Avance DRX MR instrument equipped with a 40-cm horizontal bore superconducting solenoid. (Reprinted from Figure 8 of Ho, C. and T.K. Hitchens. 2004. *Curr Pharm Biotechnol* 5: 551–66. With permission.)

## 19.4  Tracking Macrophages in Acute Cardiac Allograft Rejection Model

We have employed rodent heterotopic heart and lung transplantation models, using a transplantation pair of DA rat to BN rat, with transplantations of DA rat to DA rat and BN rat to BN rat as syngeneic controls (Kanno et al. 2000, 2001; Wu et al. 2004, 2006, 2009). The recipient receives an additional heart and lung in the abdomen, while the native organs support life, permitting investigation of the entire rejection process of the grafts without serious systemic physiological alteration or mortality of the recipient from severe rejection. All four chambers of the graft heart are intact. However, the transplanted hearts in the conventional single-anastomosis model used in our earlier work (Kanno et al. 2000, 2001) receive insufficient loading via regurgitation only. Although appropriate for studying immunological aspects of rejection, it was not suitable for functional investigation. The biventricular double-anastomosis working heart model used in our recent studies (Wu et al. 2006, 2009) preserves intact pulmonary circulation, physiological pressure, and volume loading. Although stroke volume may vary, ejection fraction of the graft hearts is near 100%, and isografts exhibit left ventricular strains similar to those of native hearts (Wu et al. 2009). Histological examination shows that allograft hearts exhibit grade II rejection on POD 4–5, grade III rejection on POD 5–6, and severe grade IV rejection on POD 7.

In the rodent heterotopic heart and lung transplantation models, 1 day after systemic administration of USPIO particles, hypointensities can be seen in the rejecting lung (Kanno et al. 2000) and heart (Kanno et al. 2001; Wu et al. 2006, 2009) with *in vivo* T2*-weighted gradient echo MRI at 4.7 T (Figure 19.3). Pathological examination showed that the hypointense areas on MRI were rich in iron-containing cells by Prussian Blue staining, and the iron-containing cells were ED1+ macrophages. Since the images were

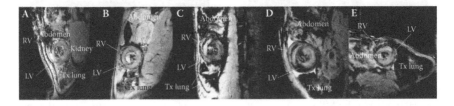

**FIGURE 19.3**
T2*-weighted image showing USPIO-labeled immune cell infiltration in transplanted hearts: (A), (B) isografts; (C), (D) grade III allografts; and (E) grade II allograft. Patches of hypointensity indicate areas with high numbers of USPIO-labeled immune cells. Imaged with 156-μm in-plane resolution at 4.7 T using a Bruker Biospec Avance-DBX MRI instrument. The upper left corner of the isograft in (A) has susceptibility artifact from the abdomen. LV, left ventricle; RV, right ventricle. (Reprinted from Figure 4 of Wu, Y.L., Q. Ye, K. Sato, L.M. Foley, T.K. Hitchens, and C. Ho. 2009. *JACC Cardiovasc Imaging* 2(6): 731–41. With permission.)

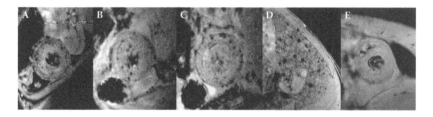

**FIGURE 19.4**
*In vivo* MRI of allograft hearts and lungs 1 day after intravenous injection of MPIO particles: (A) allograft heart on POD 5; (B), (C) allograft heart on POD 6; (D) allograft lung on POD 6; (E) isograft heart on POD 6 with 156-µm in-plane resolution at 4.7 T using a Bruker Biospec Avance-DBX MRI instrument. (Reprinted from Figure 1 of Wu, Y.L., Q. Ye, L.M. Foley, T.K. Hitchens, K. Sato, J.B. Williams, and C. Ho. 2006. *Proc Natl Acad Sci USA* 103, no. 6: 1852–57. With permission.)

collected 1 day post-USPIO administration, at the time when the circulation should be free of USPIO particles (blood half-life is about 2 hours for rats), the areas of hypointensity seen in the heart are a result of USPIO-labeled macrophages infiltrating the rejecting grafts. In addition, the degree of MR hypointensity correlates well with rejection status (Kanno et al. 2000).

MPIO particles, although having a very short blood half-life, can yield very high sensitivity for *in vivo* macrophage imaging because of the large amount of iron per particle. Following intravenous administration of MPIO particles, discrete specks of hypointensity can be readily seen throughout the rejecting allograft heart and lung (Figure 19.4). These features are absent in isograft control experiments. Fluorescence microscopy revealed that dragon green fluorescent MPIO particles are colocalized with ED1+ macrophages. Interestingly, with similar *in vivo* labeling procedures, USPIO and MPIO particles yield different cellular MRI contrast patterns (Figure 19.5). At the same rejection stage, infiltration of the iron-laden macrophages with USPIO labeling manifests itself over large continuous areas, whereas MPIO labeling yields a discrete punctate contrast pattern at a comparable location in the myocardium. In this particular experiment, the same dose of iron was

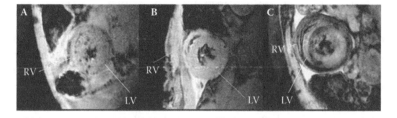

**FIGURE 19.5**
Contrast patterns with different contrast agent labeling: (A–C) *in vivo* MRI of macrophage accumulation on POD 6 of (A) an allograft heart with MPIO particle labeling; (B) an isograft heart with MPIO particle labeling; and (C) an allograft heart with USPIO labeling, with 156-µm in-plane resolution at 4.7 T. (Reprinted from Figure 3A–C of Wu, Y.L., Q. Ye, L.M. Foley, T.K. Hitchens, K. Sato, J.B. Williams, and C. Ho. 2006. *Proc Natl Acad Sci USA* 103, no. 6: 1852–57. With permission.)

given for both USPIO and MPIO *in vivo* labeling procedures. Since a single MPIO particle contains six orders of magnitude more iron oxide than one USPIO particle, with the same amount of iron given, more USPIO particles were administered. The USPIO particles stay in circulation much longer than MPIO particles, and as a result, *in vivo* labeling with USPIO in this case labels a larger population of macrophages, albeit with a smaller amount of iron oxide in each cell. In contrast, the MPIO particles only label few macrophages, yet each cell contains at least one order of magnitude more iron oxide. Therefore, although the actual macrophage infiltration pattern should be the same, *in vivo* labeling with USPIO particles gives continuous contrast over larger areas showing broad macrophage infiltration, whereas *in vivo* labeling with MPIO particles gives punctate contrast patterns identifying individual or small clusters of labeled cells.

In addition to differences in particle sizes, iron contents, and blood half-lives, MPIO particles contain an inert coating, whereas USPIO particles have a biodegradable dextran coating. Thus, once ingested into live macrophages, the USPIO label does not stay for long periods of time, regardless of the life span of the labeled cells, but the macrophage-ingested MPIO labels are stable and have a lifetime in live cells for at least several months. Consequently, longitudinal studies with USPIO or SPIO will require repetitive administration of iron oxide particles for each imaging session (Beckmann et al. 2003, 2006; Kanno et al. 2001), whereas a single administration of MPIO is sufficient for repetitive imaging to track macrophages for several months as observed in a model of chronic cardiac rejection (Ye et al. 2008). Moreover, after single injection of MPIO particles in the acute rejection model, MPIO-labeled monocytes and macrophages can remain systemic and the same pool of the MPIO-labeled cells can then be gradually recruited into the rejection sites over time as the rejection progresses without further contrast agent administration (Figure 19.6) (Wu et al. 2006).

Feridex, an FDA-approved SPIO particle for detecting liver lesions, could readily be translated into a clinical setting for cell tracking. One day after Feridex administration, some patches of hypointensity could be seen in the rejecting allograft hearts (Figure 19.7). Like other SPIO particles, Feridex has much shorter blood half-life than USPIO particles. So, these particles are cleared from the circulation much faster, resulting in lower *in vivo* labeling efficiency for macrophages. Thus, less SPIO-labeled macrophage infiltration can be seen with the same amount of Feridex labeling, compared to USPIO particles. A higher administration dosage can compensate for the lower *in vivo* labeling efficiency. On the other hand, the blood half-life of Feridex in human is at least 10 times longer than that of rats. Thus, a much lower administration concentration can be expected for human patients to achieve the same level of labeling *in vivo*.

Iron oxide particles exhibit particular magnetic dipole moments (Mills and Ahrens 2007). This dipole pattern can be used for computational automatic detection and quantification for iron oxide-labeled cells (Mills et al. 2008).

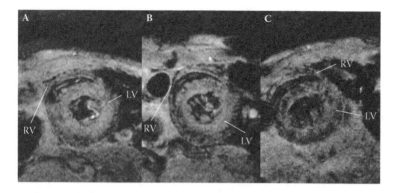

**FIGURE 19.6**
*In vivo* MRI of a rat allograft heart over time. MPIO particles were administered once on POD 3.5, and the same animal was imaged on POD 3.5 (A), 4.5 (B), and 5.5 (C), with 156-μm in-plane resolution at 4.7 T using a Bruker Biospec Avance-DBX MRI instrument. (Reprinted from Figure 2 of Wu, Y.L., Q. Ye, L.M. Foley, T.K. Hitchens, K. Sato, J.B. Williams, and C. Ho. 2006. *Proc Natl Acad Sci USA* 103, no. 6: 1852–57. With permission.)

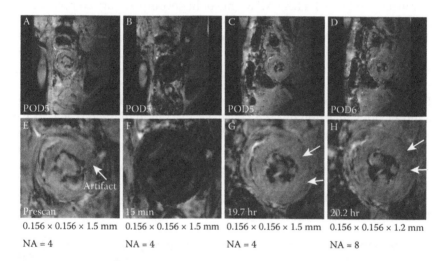

**FIGURE 19.7**
*In vivo* MRI of a rat allograft heart over time on POD 5 and 6 with Feridex labeling. The top row (A–D) is the original *in vivo* image of the transplanted heart in the abdominal region. The bottom row (E–H) is the corresponding magnified image of the heart. (A) and (E) The myocardial signal intensity is relatively uniform prior to Feridex injection. (B) and (F) The entire heart becomes dark 15 minutes after the intravenous injection of Feridex as the Feridex remains in the systemic circulation. (C) and (G) Focal hypointensity was observed 20 hours after the Feridex injection as the Feridex particles were incorporated into immune cells, mainly macrophages. (D) and (H) Thinner imaging slice with more signal averaging (to yield a higher signal-to-noise ratio) at similar time point shows better delineation of the focal hypointensities. Imaging parameters were as follows: TR around 1 s, TE = 8.1 ms, in-plane resolution is 156 × 156 μm, slice thickness was 1.5 or 1.2 mm, number of average = 4 or 8, total data acquisition time was 14 (average 4) or 28 minutes (average 8).

## 19.5 Tracking Macrophages in Chronic Cardiac Allograft Rejection Model

Chronic cardiac rejection, or chronic allograft vasculopathy (CAV), manifested years after transplantation, is the main cause of long-term graft loss and mortality. Despite well-controlled acute allograft rejection, patients still develop arteriosclerosis characterized by diffuse, concentric intimal thickening resulting in ultimate luminal occlusion of graft arteries. The severity and frequency of acute allograft rejection within the first year after heart transplantation is the best predictor for later development of the chronic rejection. Yet, immunosuppressive treatments cannot curtail the development of chronic rejection. How the machinery of acute allograft rejection, which is characterized by lymphocytic and monocytic inflammation, participates in the chronic rejection is still largely unknown. In addition, how immunosuppressive drugs contribute to the process cannot be easily studied in human patients. Shirwan et al. (2003) developed a transplantation rat pair that only differ in single class I, RT1.A$^u$, MHC (major histocompatibility complex) gene. Transplantation of the PVG.1U (RT1.A$^u$B$^u$D$^u$C$^u$) heart into a PVG.R8 (RT1. A$^a$B$^u$D$^u$C$^u$) recipient results in evidence of chronic cardiac rejection on POD 20 and most extensively by POD 100 in the absence of immunosuppressive treatment and acute rejection. Cellular MRI in combination with this unique transplantation model provides valuable information for chronic rejection.

Recipient macrophages were prelabeled 1 day prior to transplant surgery via intravenous injection of MPIO particles (0.9 μm in diameter) at a dose of 4.5 mg Fe/rat. The migration of labeled recipient cells in our CAV model can be assessed by serial *in vivo* MRI after a single MPIO injection (Figure 19.8) for at least up to 16 weeks. The location and distribution of labeled recipient cells were confirmed with MR microscopy (MRM) and histology (Figure 19.9). As the chronic rejection develops over time, more MPIO-labeled macrophages can be seen in allografts but not in isografts. Because of the inert coating of

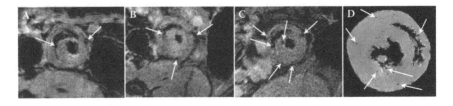

**FIGURE 19.8**
Repetitive T2*-weighted *in vivo* MRI with an in-plane resolution of 156 μm for an allograft on POD 7 (A), POD 14 (B), and POD 20 (C) after MPIO labeling. Few but very distinct hypointensities can be seen and were confirmed by MR microscopy (MRM) (D), indicated by arrows, which represent MPIO-labeled immune cells, mainly macrophages. (Reprinted from Figure 2 of Ye, Q., Y.L. Wu, L.M. Foley, T.K. Hitchens, D.F. Eytan, H. Shirwan, and C. Ho. 2008. *Circulation* 118, no. 2: 149–56. With permission.)

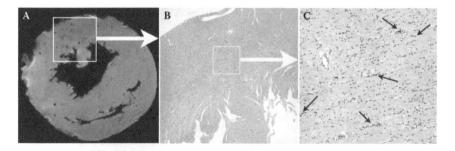

**FIGURE 19.9 (See color insert.)**
Correlation of MRM and iron staining of MPIO in a POD 94 allograft: Image from MRM shows the discrete and circular spots of hypointensity (A). These dark spots of hypointensity are due to the presence of MPIO particles, which was confirmed by the matching histological Perl's Prussian blue sections for iron (B, ×40 magnification) that correspond to the same area as the boxed region in MRM image (A). (C) The expansion of the boxed region in (B) (×200 magnification). (Reprinted from Figure 3 of Ye, Q., Y.L. Wu, L.M. Foley, T.K. Hitchens, D.F. Eytan, H. Shirwan, and C. Ho. 2008. *Circulation* 118, no. 2: 149–56. With permission.)

MPIO particles, they are stable once ingested by allowing macrophages, for long-term longitudinal imaging studies following a single administration of MPIO (Ye et al. 2008).

Because of the enormous amount of superparamagnetic iron packed into one MPIO particle, signal attenuation across a distance larger than one imaging pixel can be observed. Thus, single MPIO-labeled cells can be detected with *in vivo* MRI even with an imaging resolution larger than one cell. This enables *in vivo* imaging of sparse cell infiltration. In the early inflammation stage, very sparse macrophage infiltration in both allograft and isograft transplants can be detected *in vivo* with MPIO labeling (Figure 19.10). Similarly, sparse infiltration of macrophages in the brain after traumatic brain injury can be imaged *in vivo* by MPIO labeling (Foley et al. 2009). Each individual hypointensity represents a single macrophage, and the number of macrophages infiltrated can be quantified by directly counting of the number of hypointense or dark spots.

## 19.6 Beyond Mere Localization: Combining Cellular and Functional MRI

In addition to tracking cells *in vivo*, MRI, with good soft tissue contrast and penetration, has long been used for *in vivo* functional assessment. Renal perfusion (Sun et al. 2003; Wang et al. 1998; Yang et al. 2001), diffusion-weighted MRI for renal function (Yang et al. 2004), cerebral perfusion with arterial spin labeling (Foley et al. 2005, 2008), and cardiac regional wall motion with MRI tagging and strain analysis (Wu et al. 2009) can be evaluated simultaneously

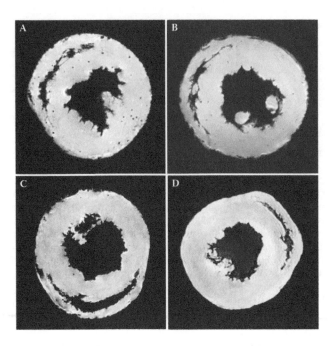

**FIGURE 19.10**
Representative MRM images of transplant and native hearts. Punctate regions of hypointensity can be clearly seen in the allograft heart harvested on POD 109 (A) after *in vivo* labeling with MPIO. A representative isograft heart harvested on POD 119 is shown in panel B. Two native hearts are also shown (C and D); (C) was harvested 7 days following MPIO injection and (D) was also harvested in 7 days, but did not receive any MPIO injection. (Reprinted from Figure 3 of Ye, Q., Y.L. Wu, L.M. Foley, T.K. Hitchens, D.F. Eytan, H. Shirwan, and C. Ho. 2008. *Circulation* 118(2): 149–56. With permission.)

alongside cellular tracking. This can provide cellular and functional information in the same subject for increased diagnostic accuracy.

## 19.7 Conclusion

*In vivo* cellular MRI for macrophages provides a noninvasive means for accurate assessment of a variety of disorders with inflammation. With the proper selection of contrast agents, one can tailor the cellular MRI to suit specific circumstances for each individual case. With the advancement of hardware and new agents, the detection sensitivity has markedly improved, making tracking even single cells or single particles possible. In addition to tracking cells *in vivo*, MRI has a great capacity to look beyond the mere cell location or temporal/spatial distribution of cells and simultaneously provide information on organ function and tissue composition.

## Acknowledgment

Our research is supported by research grants from the National Institutes of Health (R01EB-00318, R01HL-081349, and P41EB-001977).

## References

Arbab, A.S., L.A. Bashaw, B.R. Miller, E.K. Jordan, J.W. Bulte, and J.A. Frank. 2003. Intracytoplasmic tagging of cells with ferumoxides and transfection agent for cellular magnetic resonance imaging after cell transplantation: methods and techniques. *Transplantation* 76, no. 7: 1123–30.

Arbab, A.S., L.B. Wilson, P. Ashari, E.K. Jordan, B.K. Lewis, and J.A. Frank. 2005. A model of lysosomal metabolism of dextran coated superparamagnetic iron oxide (SPIO) nanoparticles: Implications for cellular magnetic resonance imaging. *NMR Biomed* 18, no. 6: 383–89.

Beckmann, N., C. Cannet, M. Fringeli-Tanner, D. Baumann, C. Pally, C. Bruns, H.G. Zerwes, E. Andriambeloson, and M. Bigaud. 2003. Macrophage labeling by SPIO as an early marker of allograft chronic rejection in a rat model of kidney transplantation. *Magn Reson Med* 49, no. 3: 459–67.

Beckmann, N., C. Cannet, S. Zurbruegg, R. Haberthur, J. Li, C. Pally, and C. Bruns. 2006. Macrophage infiltration detected at MR imaging in rat kidney allografts: Early marker of chronic rejection? *Radiology* 240, no. 3: 717–24.

Bowen, C.V., X. Zhang, G. Saab, P.J. Gareau, and B.K. Rutt. 2002. Application of the static dephasing regime theory to superparamagnetic iron-oxide loaded cells. *Magn Reson Med* 48, no. 1: 52–61.

Dodd, S.J., M. Williams, J.P. Suhan, D.S. Williams, A.P. Koretsky, and C. Ho. 1999. Detection of single mammalian cells by high-resolution magnetic resonance imaging. *Biophys J* 76 (1 Pt 1): 103–9.

Foley, L.M., T.K. Hitchens, C. Ho, K.L. Janesko, J.A. Melick, H. Bayir, and P.M. Kochanek. 2009. Magnetic resonance imaging assessment of macrophage accumulation in mouse brain after experimental traumatic brain injury. *J Neurotrauma* 26, no. 9: 1509–19.

Foley, L.M., T.K. Hitchens, P.M. Kochanek, J.A. Melick, E.K. Jackson, and C. Ho. 2005. Murine orthostatic response during prolonged vertical studies: effect on cerebral blood flow measured by arterial spin-labeled MRI. *Magn Reson Med* 54, no. 4: 798–806.

Foley, L.M., T.K. Hitchens, J.A. Melick, H. Bayir, C. Ho, and P.M. Kochanek. 2008. Effect of inducible nitric oxide synthase on cerebral blood flow after experimental traumatic brain injury in mice. *J Neurotrauma* 25, no. 4: 299–310.

Foster-Gareau, P., C. Heyn, A. Alejski, and B.K. Rutt. 2003. Imaging single mammalian cells with a 1.5 T clinical MRI scanner. *Magn Reson Med* 49, no. 5: 968–71.

Frank, J.A., B.R. Miller, A.S. Arbab, H.A. Zywicke, E.K. Jordan, B.K. Lewis, L.H. Bryant, Jr., and J.W. Bulte. 2003. Clinically applicable labeling of mammalian and stem cells by combining superparamagnetic iron oxides and transfection agents. *Radiology* 228, no. 2: 480–7.

Frank, J.A., H. Zywicke, E.K. Jordan, J. Mitchell, B.K. Lewis, B. Miller, L.H. Bryant, Jr., and J.W. Bulte. 2002. Magnetic intracellular labeling of mammalian cells by combining (FDA-approved) superparamagnetic iron oxide MR contrast agents and commonly used transfection agents. *Acad Radiol* 9 Suppl 2: S484–87.

Hauger, O., N. Grenier, C. Deminere, C. Lasseur, Y. Delmas, P. Merville, and C. Combe. 2007. USPIO-enhanced MR imaging of macrophage infiltration in native and transplanted kidneys: initial results in humans. *Eur Radiol* 17, no. 11: 2898–2907.

Heyn, C., C.V. Bowen, B.K. Rutt, and P.J. Foster. 2005. Detection threshold of single SPIO-labeled cells with fiesta. *Magn Reson Med* 53, no. 2: 312–20.

Heyn, C., J.A. Ronald, L.T. Mackenzie, I.C. Macdonald, A.F. Chambers, B.K. Rutt, and P.J. Foster. 2006. In vivo magnetic resonance imaging of single cells in mouse brain with optical validation. *Magn Reson Med* 55, no. 1: 23–29.

Hinds, K.A., J.M. Hill, E.M. Shapiro, M.O. Laukkanen, A.C. Silva, C.A. Combs, T.R. Varney, R.S. Balaban, A.P. Koretsky, and C.E. Dunbar. 2003. Highly efficient endosomal labeling of progenitor and stem cells with large magnetic particles allows magnetic resonance imaging of single cells. *Blood* 102, no. 3: 867–72.

Ho, C. and T.K. Hitchens. 2004. A non-invasive approach to detecting organ rejection by MRI: Monitoring the accumulation of immune cell cells at the transplanted organ. *Curr Pharm Biotechnol* 5: 551–66.

Josephson, L., C.H. Tung, A. Moore, and R. Weissleder. 1999. High-efficiency intracellular magnetic labeling with novel superparamagnetic-TaT peptide conjugates. *Bioconjug Chem* 10, no. 2: 186–91.

Kanno, S., P.C. Lee, S.J. Dodd, M. Williams, B.P. Griffith, and C. Ho. 2000. A novel approach with magnetic resonance imaging used for the detection of lung allograft rejection. *J Thorac Cardiovasc Surg* 120, no. 5: 923–34.

Kanno, S., Y.J. Wu, P.C. Lee, S.J. Dodd, M. Williams, B.P. Griffith, and C. Ho. 2001. Macrophage accumulation associated with rat cardiac allograft rejection detected by magnetic resonance imaging with ultrasmall superparamagnetic iron oxide particles. *Circulation* 104, no. 8: 934–38.

Lauterbur, P.C., J. Bernardo, M. L., M.H. Menonca Dias, and L.K. Hedges. 1986. Microscopic NMR imaging of the magnetic around magnetic particles. *Proc Soc Magn Reson Med*, 5th Annual Meeting, Montreal, Quebec, Canada: 229–30.

Lewin, M., N. Carlesso, C.H. Tung, X.W. Tang, D. Cory, D.T. Scadden, and R. Weissleder. 2000. TaT peptide-derivatized magnetic nanoparticles allow in vivo tracking and recovery of progenitor cells. *Nat Biotechnol* 18, no. 4: 410–14.

Mills, P.H. and E.T. Ahrens. 2007. Theoretical MRI contrast model for exogenous T2 agents. *Magn Reson Med* 57, no. 2: 442–47.

Mills, P.H., Y.J. Wu, C. Ho, and E.T. Ahrens. 2008. Sensitive and automated detection of iron-oxide-labeled cells using phase image cross-correlation analysis. *Magn Reson Imaging* 26, no. 5: 618–28.

Raynal, I., P. Prigent, S. Peyramaure, A. Najid, C. Rebuzzi, and C. Corot. 2004. Macrophage endocytosis of superparamagnetic iron oxide nanoparticles: Mechanisms and comparison of ferumoxides and ferumoxtran-10. *Invest Radiol* 39, no. 1: 56–63.

Shapiro, E.M., O. Gonzalez-Perez, J. Manuel Garcia-Verdugo, A. Alvarez-Buylla, and A.P. Koretsky. 2006a. Magnetic resonance imaging of the migration of neuronal precursors generated in the adult rodent brain. *Neuroimage* 32, no. 3: 1150–57.

Shapiro, E.M., K. Sharer, S. Skrtic, and A.P. Koretsky. 2006b. In vivo detection of single cells by MRI. *Magn Reson Med* 55, no. 2: 242–49.

Shapiro, E.M., S. Skrtic, and A.P. Koretsky. 2005. Sizing it up: Cellular MRI using micron-sized iron oxide particles. *Magn Reson Med* 53, no. 2: 329–38.

Shapiro, E.M., S. Skrtic, K. Sharer, J.M. Hill, C.E. Dunbar, and A.P. Koretsky. 2004. MRI detection of single particles for cellular imaging. *Proc Natl Acad Sci USA* 101, no. 30: 10901–6. Epub 2004 Jul 15.

Shirwan, H., A. Mhoyan, E.S. Yolcu, X. Que, and S. Ibrahim. 2003. Chronic cardiac allograft rejection in a rat model disparate for one single class I MHC molecule is associated with indirect recognition by CD4(+) T cells. *Transpl Immunol* 11, no. 2: 179–85.

Sumner, J.P., R. Conroy, E.M. Shapiro, J. Moreland, and A.P. Koretsky. 2007. Delivery of fluorescent probes using iron oxide particles as carriers enables in vivo labeling of migrating neural precursors for magnetic resonance imaging and optical imaging. *J Biomed Opt* 12, no. 5: 051504.

Sumner, J.P., E.M. Shapiro, D. Maric, R. Conroy, and A.P. Koretsky. 2009. In vivo labeling of adult neural progenitors for MRI with micron sized particles of iron oxide: Quantification of labeled cell phenotype. *Neuroimage* 44, no. 3: 671–78.

Sun, Y., D. Yang, Q. Ye, M. Williams, J.M. Moura, F. Boada, Z.P. Liang, and C. Ho. 2003. Improving spatiotemporal resolution of USPIO-enhanced dynamic imaging of rat kidneys. *Magn Reson Imaging* 21, no. 6: 593–98.

Walczak, P., D. Kedziorek, A.A. Gilad, S. Lin, and J.W. Bulte. 2005. Instant MR labeling of stem cells using magnetoelectroporation. *Magn Reson Med*, no. 54: 769–74.

Wang, J.J., K.S. Hendrich, E.K. Jackson, S.T. Ildstad, D.S. Williams, and C. Ho. 1998. Perfusion quantitation in transplanted rat kidney by MRI with arterial spin labeling. *Kidney Int* 53, no. 6: 1783–91.

Weisskoff, R.M., C.S. Zuo, J.L. Boxerman, and B.R. Rosen. 1994. Microscopic susceptibility variation and transverse relaxation: theory and experiment. *Magn Reson Med* 31, no. 6: 601–10.

Wu, Y.-J.L., K. Sato, Y. Qing, and C. Ho. 2004. MRI investigation of graft rejection following organ transplantation using rodent models. *Method Enzymol* 386: 73–105.

Wu, Y.L., Q. Ye, L.M. Foley, T.K. Hitchens, K. Sato, J.B. Williams, and C. Ho. 2006. In situ labeling of immune cells with iron oxide particles: An approach to detect organ rejection by cellular MRI. *Proc Natl Acad Sci USA* 103, no. 6: 1852–57.

Wu, Y.L., Q. Ye, K. Sato, L.M. Foley, T.K. Hitchens, and C. Ho. 2009. Non-invasive evaluation of cardiac allograft rejection by cellular and functional MRI. *JACC Cardiovasc Imaging* 2(6): 731–41.

Yang, D., Q. Ye, D.S. Williams, T.K. Hitchens, and C. Ho. 2004. Normal and transplanted rat kidneys: Diffusion MR imaging at 7 T. *Radiology* 231, no. 3: 702–9.

Yang, D., Q. Ye, M. Williams, Y. Sun, T.C. Hu, D.S. Williams, J.M. Moura, and C. Ho. 2001. USPIO-enhanced dynamic MRI: Evaluation of normal and transplanted rat kidneys. *Magn Reson Med* 46, no. 6: 1152–63.

Ye, Q., Y.L. Wu, L.M. Foley, T.K. Hitchens, D.F. Eytan, H. Shirwan, and C. Ho. 2008. Longitudinal tracking of recipient macrophages in a rat chronic cardiac allograft rejection model with noninvasive magnetic resonance imaging using micrometer-sized paramagnetic iron oxide particles. *Circulation* 118, no. 2: 149–56.

Ye, Q., D. Yang, M. Williams, D.S. Williams, C. Pluempitiwiriyawej, J.M. Moura, and C. Ho. 2002. In vivo detection of acute rat renal allograft rejection by MRI with USPIO particles. *Kidney Int* 61, no. 3: 1124–35.

# 20

## Bioluminescence, MRI, and PET Imaging Modalities of Stem Cell–Based Therapy for Neurological Disorders

Marcel M. Daadi, Raphael Guzman, and Gary K. Steinberg

### CONTENTS

## 20.1 Imaging in Stem Cell-Based Therapy for Neurological Disorders

Neurological disorders and central nervous system (CNS) trauma are associated with cellular loss or dysfunction. The discovery of self-renewable neural stem cells (NSCs) and progenitor cells in the adult brain brought hope that the brain could repair itself after disease or injury. When dissected out of the brain and given the appropriate growth factors, NSCs are able to generate a large number of progeny and differentiate into specialized CNS cell types, including neurons, astrocytes, and oligodendrocytes. However, contrary to tissues such as skin, gut, or blood, the endogenous CNS NSCs are unable to fully restore the cytoarchitecture and function of tissue damaged by neurodegeneration or injury. In acute CNS injuries, such as stroke or trauma, we predominantly see gliogenesis and the formation of glial scar at the site of injury. Consequently, cell loss leads to permanent functional deficits and characterizes most debilitating CNS diseases and injuries, including stroke, spinal cord injury, Parkinson's disease (PD), Huntington's disease, and others. Therefore, these neurodegenerative diseases and injuries are amenable to cell transplantation therapy. The rationale of neural transplantation therapy is to restore lost cells or to support dysfunctional cells through multiple mechanisms, including cell replacement, secretion of neurotrophic factors, anti-inflammatory actions, neurogenesis, and neovascularization (Park et al., 2002; Lee et al., 2007; Daadi and Steinberg, 2009). Preclinical studies in various animal models of neurological disorders have demonstrated the efficacy of stem cells derived from various sources in alleviating the behavioral deficits (Lindvall and Kokaia, 2010). However, the mechanisms mediating the functional integration of these exogenous grafted cells are not completely understood.

Recent advances in technology offer scientists the ability to study biological processes controlling graft-host interactions that govern integration and cellular differentiation of transplanted cells. One of these technologies is real-time *in vivo* imaging of grafted cells. Monitoring the cellular survival, delivery, migration, and function after transplantation is essential for successful development and optimization of cell transplantation protocols.

In this chapter, we review the application of a multimodal strategy employing bioluminescence imaging (BLI), magnetic resonance imaging (MRI), and positron emission tomography (PET) for imaging stem cell-based therapy for neurological disorders.

---

## 20.2 Bioluminescence Imaging

BLI involves the *ex vivo* genetic modification of stem cells, prior to transplantation, with a vector-carrying gene encoding for the firefly or *Renilla* luciferase (Luc) enzyme. An injection of the luciferase substrate D-luciferin or coalenterazine into the animal model results in the emission of photons that can be visualized and quantified in the precise region where Luc-expressing stem cells were transplanted. The simplicity combined with the accuracy and quantifiable measures make BLI powerful and the most utilized tool in preclinical imaging setup. BLI has been used for tracking survival, migration, immunogenicity, and tumorigenicity of transplanted therapeutic cells in animal models of stroke (Kim et al., 2004, 2006; Daadi et al., 2009), spinal cord injury (Okada et al., 2005), and intracranial brain tumor (Tang et al., 2003).

To image the migration of grafted neural progenitor cells (NPCs) to the ischemic infarct, C17.2 NPC lines were stably transfected with the firefly luciferase (fLuc) trangene under the cytomegalovirus (CMV) promoter (Kim et al., 2004). Longitudinal BLI of the grafted NPC demonstrated that by day 7 posttransplantation the cells migrated from the injection site, contralateral to the stroke, across the midline toward the infracted zone. BLI has efficiently visualized the fate of grafted human embryonic stem cell (hESC)-derived human neural stem cells (hNSCs) in stroke-damaged rat brain (Daadi et al., 2009). The hNSCs were genetically engineered with a lentiviral vector carrying a double-fusion (DF) reporter gene that stably expressed enhanced green fluorescence protein (eGFP) and fLuc reporter genes. The genetically

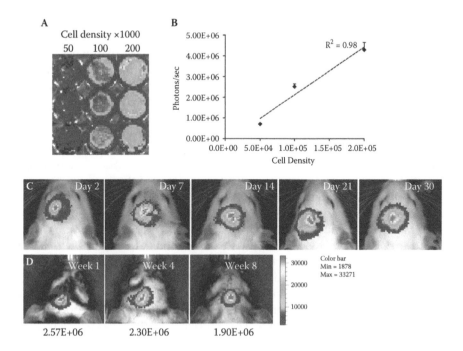

**FIGURE 20.1 (See color insert.)**

*In vitro* and *in vivo* bioluminescence imaging of the hNSCs. (A) *In vitro* imaging analysis of genetically engineered hNSCs show increasing fLuc activity with cell density and a linear correlation ($R^2$ = .98) (B). Data are representative of three independent experiments performed in triplicate. Representative BLI imaging of stroke-lesioned rats transplanted with the hNSCs and monitored for 4 weeks (C) and 8-week posttransplantation survival times (D). Quantitative analysis of the fLuc activity in these animals showed a stable BLI signal, which suggests the survival of the grafts and the nonproliferative property of the hNSCs. Color scale bar is in photon/second/square centimeter/sr. (Reprinted with permission from Daadi MM, Li Z, Arac A, Grueter BA, Sofilos M, Malenka RC, Wu JC, Steinberg GK. *Mol Ther* 2009 Jul;17(7):1282–91.)

engineered hNSCs were grafted into the MCAO rat model and monitored in real time for 2 months. Longitudinal quantitative analysis revealed no significant changes of the luciferase activity, demonstrating the nontumorigenic property of the hNSCs (Figure 20.1). The DF transgene did not alter the physiological properties of the cells. Grafted hNSCs migrated to the stroke-damaged area and differentiated into neurons, astrocytes, and oligodendrocytes (Daadi et al., 2009). Electron microscopy demonstrated that the hNSCs received synaptic contact with the host terminals. Electrophysiological recording of eGFP⁺ hNSCs in brain slices demonstrated that the grafted cells exhibited electrophysiological features of neurons with voltage-gated sodium currents. They also displayed excitatory postsynaptic currents, suggesting the establishment of synaptic input (Daadi et al., 2009).

Okada et al. (2005) used BLI to determine the optimal timing for transplanting NSCs into the animal model of spinal cord injury. NSCs expressing the fLuc were transplanted into a contusion injury model at the 10th thoracic vertebra level. The timing of the transplantation was either immediately or 9 days after injury. BLI was performed daily for 1 week, then weekly for 6 weeks. In both groups, BLI revealed an 80% decrease in fLuc activity that was proportional to the survival of grafted cells. However, histopathological analysis of the graft revealed that the acutely transplanted cells incorporated and contributed to the glial scar in the lesion epicenter. In the delayed transplant paradigm, grafted NSCs were found outside the scar area and differentiated mostly into neurons and oligodendrocytes.

BLI has also been used to accurately track migration of the subventricular zone (SVZ) precursors toward the olfactory brain and the *in vivo* actions of brain-derived neurotrophic factor (BDNF) on adult neurogenesis (Reumers et al., 2008). *In vivo* and *ex vivo* BLI of SVZ precursors transduced with the lentiviral vector encoding fLuc allowed for the kinetic analysis of migration toward the olfactory. This approach also demonstrated the negative effects of long-term BDNF overexpression on adult neurogenesis. Together, these studies and others demonstrated that BLI modality is a reliable method for long-term, noninvasive, real-time monitoring of biological processes. The major advantage of fLuc reporter gene imaging approach is the incorporation of the vector into the target cell DNA. This property ensures the specificity and the permanent expression of the reporter gene without dilution during cell divisions. In addition, BLI is fast, versatile, and cost-effective and offers high sensitivity with the ability to detect 100 to 1,000 cells in superficial anatomical areas (Contag et al., 2000; Wu et al., 2001; Ntziachristos et al., 2005; Walczak and Bulte, 2007). As such, variations in the cell density of grafted cells may be quantified, thus offering a simple way to monitor survival, immune reaction against the grafts, and tumorigenicity. Major disadvantages of BLI are the low spatial resolution and the dependence of light signal on tissue depth, which at present make it difficult for application to humans.

## 20.3 Cellular Magnetic Resonance Imaging

MRI is based on the distribution of hydrogen atoms in tissue. Therefore, this imaging technique essentially visualizes the relative distribution of water molecules in particular tissues. Each brain region exhibits different water content and has a unique composition of macromolecules, such as proteins and lipids that bind water in differential amounts. This water composition determines contrast between brain structures. MRI contrast agents alter the relaxivity of hydrogen atoms in their vicinity and therefore modify the MR signal. These contrast agents have been used for labeling stem cells *in vitro* and subsequently visualizing them at high resolution *in vivo* after transplantation into the brain. There are multiple contrast agents and strategies for labeling stem cells. A detailed description of labeling techniques of stem cells for MRI is reported in Chapter 15.

The iron oxide-based contrast agent superparamagnetic iron oxide (SPIO) provides a superior signal-to-noise ratio (Weissleder et al., 1990). The early approval of SPIO for clinical trials as an oral gastrointestinal contrast agent in MRI (Hahn et al., 1990) and the discovery of simple techniques for cell labeling (Frank et al., 2002; Arbab et al., 2003) led to its extensive use in cellular imaging and tracking studies. In fact, superparamagnetic contrast agents exist in three forms: ultrasmall SPIO, 10–40 nm in diameter; SPIO nanoparticles, 50–200 nm in diameter; and micron-size particles of iron oxide (MPIOs), with diameter around 0.76–1.63 μm. Given its size, the MPIOs are MR imaged as single-particle tracking a single labeled cell (Shapiro et al., 2004).

The efficiency of using SPIO to track NSCs *in vivo* has been reported in many experimental models of neurological disorders. In an animal model of multiple sclerosis (MS), demyelination is a major pathological component currently not adequately addressed by pharmacological means. Neural transplantation therapy of cells capable of myelinating axons may offer a single treatment with long-term relief. Myelinating cells such as NSC-derived, oligodendroglial progenitors, Schwann cells (SCs), or olfactory ensheathing cells (OECs) have been successfully prelabeled with SPIOs and imaged *in vivo* after engraftment in the spinal cord (Bulte et al., 1999; Franklin et al., 1999; Dunning et al., 2004). Given the high expression of transferring receptors, oligodendrocyte progenitors were labeled with magnetic nanoparticles 46L covalently linked to monoclonal antibody OX-26 raised against this receptor. This approach yielded high internalization of the contrast agent and labeling of the cells. After transplantation into the spinal cord of myelin-deficient rats, the oligodendrocyte progenitors retained their migratory and myelinating capacity *in vivo*. The grafted cells could also be easily visualized by MRI, which allowed the three-dimensional (3D) mapping of the extent of myelination (Bulte et al., 1999). SCs and OECs internalized SPIO from the culture medium by fluid phase pinocytosis. After transplantation into focal areas of demyelination in adult rat spinal cord, both transplanted SPIO-labeled SCs and OECs produced a signal reduction using T2-weighted MRI. The SPIO-labeled SCs and OECs were able to myelinate normally after transplantation into focal areas of demyelination (Dunning et al., 2004).

NSCs grown as neurospheres were labeled *in vitro* with SPIO coincubated with poly-L-lysine and transplanted into the experimental autoimmune encephalomyelitis (EAE) model of MS (Ben-Hur et al., 2007; Politi et al., 2007). MRI and quantitative analysis demonstrated the migration along the white matter and accumulation of grafted neural precursors at inflammatory lesions of the brain (Ben-Hur et al., 2007; Politi et al., 2007). Cellular MRI has been instrumental in demonstrating migration patterns of grafted stem cells in a stroke animal model (Hoehn et al., 2002; Modo et al., 2002; Zhang et al., 2003; Guzman et al., 2007; Walczak et al., 2008; Daadi et al., 2009; Song et al., 2009).

Early studies used the contrast agent gadolinium rhodamine dextran (GRID) to track grafted NSCs in stroke-lesioned brain, demonstrating the feasibility of the cellular MRI approach in monitoring cell migration *in vivo*. Interestingly, in a recent study (Modo et al., 2009) the authors reported the deleterious effects that gadolinium-based contrast agents have on the long-term *in vivo* functional properties of the grafted NSCs in the stroke model. MRI visualization of cross-callosal migration toward the stroke-damaged striatum was demonstrated (Hoehn et al., 2002). A total of 60,000 undifferentiated ESCs labeled with USPIO were grafted into the contralateral cortex-corpus callosum border. Three weeks later, these cells migrated transhemispherically along the corpus callosum to the stroke-damaged striatum.

Other innovative ways of cell labeling, delivery, and measurement of MRI parameters to assess poststroke recovery have been reported (Zhang et al., 2003; Jiang et al., 2005). SVZ-derived neural precursors were labeled with SPIO precoated with lipophilic dye DiI using a biolistic device "gene gun." This technique is reported to label over 95% of cultured cells (Jiang et al., 2005). MRI tracking of SPIO-labeled neurospheres injected intracisternally demonstrated selective migration toward the ischemic parenchyma at a mean speed of $65 \pm 14.6$ µm/hour in living rats. Quantitative MRI measurements of cerebral blood flow, cerebral blood volume, and blood-to-brain transfer constantly predicted graft-induced angiogenesis in the ischemic tissue (Jiang et al., 2005).

We used SPIO-labeled hNSCs derived from fetal brain (Guzman et al., 2007) and from hESCs (Daadi et al., 2009) to visualize migration and integration of hNSCs in the developing brain and in a stroke animal model. Using T2-weighted spin echo and 3D gradient echo MRI, the migration of the SPIO-labeled hNSCs was monitored during the first 3 weeks from the injection site in the lateral ventricle to the lateral and fourth ventricles. At the 9-, 12-, and 18-week posttransplant periods, the hNSCs were MR imaged migrating through the rostral migratory stream (Figure 20.2) to the olfactory bulb (Guzman et al., 2007).

We investigated the fate of grafted SPIO-labeled hNSCs derived from hESCs in stroke environment over time and the fidelity of MRI to detect dose-dependent effects of the grafts on the lesion (Daadi et al., 2009). Cultures of hNSCs were treated with an SPIO-poly-L-lysine mixture for 3 days *in vitro*, harvested, and transplanted medially to the stroke at an increasing cell density of 50,000, 200,000, and 400,000 cell doses. MRI analysis clearly detected the grafts as hypointense areas in the striatum and the stroke zone as a hyperintense region in the striatum and cortex on T2-weighted images (Figure 20.3). The graft size for each analysis in the serial MRI scans demonstrated a linear correlation between the injected cell dose and the MRI size of the transplant. Three-dimensional reconstructions of the grafts and stroke by surface rendering from the MR scans allowed for an accurate representation of both the graft and the stroke sizes and visualized the relationship between small and large grafts to stroke region. Immunohistological

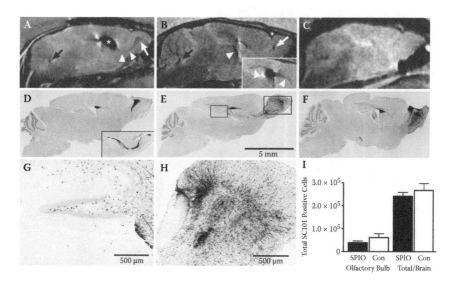

**FIGURE 20.2**

Migration and integration of SPIO-labeled human central nervous system stem cells (hCNS-SCns). MRI detects widespread migration of SPIO-labeled hCNS-SCns after intraventricular injection in the P0/P1 NOD-SCID mouse brain. (A) Three weeks after transplantation, sagittal MRI shows hypointensities representing SPIO-labeled hCNS-SCns in the lateral ventricle (asterisk), along the RMS (arrowheads), toward the OB (arrow), and in the fourth ventricle (black arrow). (D) Corresponding section stained with the human-specific cytoplasmic marker SC121. (Inset) Shows RMS in adjacent section. (B) Sagittal MRI 18 weeks after transplantation showing that hCNS-SCns have integrated in the ventricular wall (arrowhead), in the core of the OB (white arrow), and in the fourth ventricle (black arrow). (Inset) Demonstrates migration along the corpus callosum (arrowheads). (E) Corresponding histological section. (C) Sagittal MRI of a control animal transplanted with unlabeled hCNS-SCns 18 weeks after transplantation shows no cell signal. (F) Corresponding histological section. (G) and (H) Higher-magnification sagittal images (areas boxed in Figure 20.2E) show hCNS-SCns in the CA1, CA3, and dentate gyrus of the hippocampus (G) and in the OB (H) of the NOD-SCID mouse. (I) There was robust cell survival at 18 weeks after transplantation without any statistically significant difference between SPIO-labeled ($n$ = 4 animals) and unlabeled ($n$ = 3 animals) cells. Results are mean ± SEM (standard error of the mean). OB, olfactory bulb; RMS, rostral migratory stream; SVZ, subventricular zone. (Reprinted with permission from Guzman R, Uchida N, Bliss TM, He D, Christopherson KK, Stellwagen D, Capela A, Greve J, Malenka RC, Moseley ME, Palmer TD, Steinberg GK. *Proc Natl Acad Sci USA* 2007 Jun 12;104(24):10211–16.)

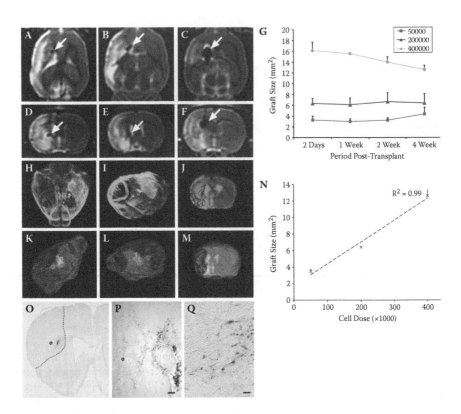

**FIGURE 20.3**
MRI analysis of the hNSC grafts in experimental stroke model. MRI horizontal (A, B, C) and frontal (D, E, F) scans show dose-dependent size of the SPIO-labeled hNSCs grafts as hypointense areas in the striatum (arrow) medially in the penumbral zone of the stroke distinguished as strongly hyperintense areas on T2-weighted images. The cell doses are 50,000 cells (A, D); 200,000 cells (B, E); and 400,000 cells (C, F). (G) Quantitative analysis of graft size, in consecutive coronal MRI scans, 600 μm spaced in the three animal groups ($n = 15$) over the posttransplant survival time confirmed the BLI imaging data and showed a stable graft size, demonstrating survival of the graft. Three-dimensional surface rendering reconstruction of grafted rat brain from high-resolution T2 MRI illustrates the grafts (green) and stroke (pink, red) in a representative animal from the low-dose (H, I, J) and intermediate-dose (K, L, M) groups. The MRI-measured graft size shows a strong correlation ($R^2 = .99$) with the cell dose transplanted (N). (O–Q) Histological analysis using Prussian blue staining for SPIO particles demonstrated cytosolic deposition of blue crystals in the grafted hNSCs and migration toward stroke area (asterisk in O, P). Interrupted line in (O) shows the boundary of stroke zone. Bars: (P) 50 μm; (Q) 20 μm. (Reprinted with permission from Daadi MM, Li Z, Arac A, Grueter BA, Sofilos M, Malenka RC, Wu JC, Steinberg GK. *Mol Ther* 2009 Jul;17(7):1282–91.)

analysis of SPIO-labeled grafts performed with confocal and Prussian blue staining confirmed the MRI data and demonstrated the survival of the grafts and migration toward the stroke-damaged areas (Daadi et al., 2009).

Cellular MRI is also useful in cell therapeutic approaches for brain tumor as a sensitive method of assessing the efficacy of NSCs to target and eliminate tumors *in vivo*. It has been reported that NSCs home toward brain tumors (Arboody et al., 2000; Benedetti et al., 2000; Ehtesham et al., 2002; Shah et al., 2005). As such, they could be used as a vehicle to deliver therapeutic genes to the tumor. To generate animal models of brain tumor, GFP-labeled gliomas were transplanted into the forebrain (Zhang et al., 2004). One week later, SPIO-labeled NSCs were injected into the cisterna magna and tail vein. MRI demonstrated the ability of the NSCs to migrate and infiltrate the tumor mass. Immunocytochemistry and Prussian blue staining analysis confirmed the tumor infiltration and the overlapping of Prussian blue-stained NSCs over GFP⁺ glioma cells.

Clinical studies have demonstrated the feasibility of MRI of grafted neural cells labeled with ferumoxides into patients with traumatic brain injury in China (Zhu et al., 2006). A second study, performed in Brazil (Callera and de Melo, 2007) used magnetic bead-labeled bone marrow cells transplanted into patients with spinal cord injury.

## 20.4 Positron Emission Tomography

In the past two decades, PET imaging has been a rapidly advancing technology in molecular imaging of neurological disorders. Due to accuracy and noninvasiveness, PET is now routinely used for diagnosis and post-therapeutic monitoring of patients. The most commonly used radiolabeled probe to image brain activity is $^{18}$F-fluorodeoxyglucose ($^{18}$F-FDG). The biological parameter estimated in this molecular imaging modality is the rate of regional glucose utilization. The regional concentration of the radioactivity may be accurately measured and normalized to a reference region of the brain. FDG-PET is approved by the U.S. Food and Drug Administration and is currently the most commonly performed modality in the clinical arena.

FDG-PET has been used to image the transplantation of neurons into stroke patients (Kondziolka et al., 2000). Patients were injected with 7 mCi of $^{18}$F-FDG, and PET imaging was performed 40 minutes later. PET images were normalized, and the $^{18}$F-FDG uptake in the infarct and peri-infarct zones was expressed as percentage of baseline. PET scans performed 6 months after surgery revealed a 15% increase in $^{18}$F-FDG uptake at the transplant site or in the ipsilateral adjacent parenchyma. This was observed in 6 of 11 patients and was relative to pretransplant scans. Interestingly, 4 of these 6 patients demonstrated improvement in European Stroke Scale, suggesting correlation with a PET-increased graft function.

Neurological functional recovery after cell transplantation therapy may be assessed with PET imaging radiotracers binding to receptors, neurotransmitter, or neurotransmitter transporters. For instance, carbon-11-labeled 2β-carbomethoxy-3β-(4-fluorophenyl)tropane ($^{11}$C-CFT) was used to visualize and quantify striatal presynaptic dopamine (DA) transporters (DATs) in patients with PD and PD animal models. PET imaging demonstrated that, in the 6-OHDA rat model, DAT binding was reduced in the lesioned side by 15–35% (Bjorklund et al., 2002). Interestingly, functional recovery in this model occurred gradually after transplantation of mouse ESC-derived DA neurons and became significant when the $^{11}$C-CFT binding ratio reached 75–90% of the intact side (Bjorklund et al., 2002). Thus, PET imaging of metabolic function, receptor-ligand interactions, neurotransmitter and neurotransmitter transporter binding is critical in stem cell transplantation therapy for monitoring graft fate and function, as well as potential side effects, such as dyskinesias in patients with PD, for instance (Ma et al., 2002).

PET use of reporter genes is a novel and elegant approach, originally developed for detecting and killing cancer cells and currently applied to stem cell-based therapy and neurological functional imaging (Gambhir et al., 1998; Tjuvajev et al., 1999; MacLaren et al., 2000). This concept is based on imaging gene expression *in vivo* with PET (Gambhir et al., 2000b). Gambhir et al. employed a PET reporter-probe imaging approach using the herpes simplex virus type 1 thymidine kinase (HSV1-tk) genes (Gambhir et al., 1998). The PET reporter gene is introduced to stem cells or directly to a subject via viral vectors. The reporter gene product is a protein, in this case an enzyme, that phosphorylates a radiolabeled substrate transported into the cells. As a result, the phosphorylated substrate is trapped and imaged only in cells carrying the HSV1-tk gene. HSV-sr139tk, a mutated form of the thymidine kinase gene, was more efficient than the wild type in converting the radiotracer (Gambhir et al., 2000a). This approach has been successfully used for imaging hESC-based therapy for the heart (Sun et al., 2009).

One study used the HSV1-tk reporter gene approach to image the C17.2 murine NPCs after implantation in an intracranial glioma animal model (Waerzeggers et al., 2008). The reporter probe 9-[4-[18F]fluoro-3-hydroxymethyl)butyl]guanine ($^{18}$F-FHBG) was used to visualize the migratory pattern of the C17.2 cells. However, PET imaging was possible only in regions where the blood-brain barrier was disrupted. In addition, an aberrant migratory pattern of the cells was detected. Given the triple imaging modality, complexity of the model, and route of cell delivery, further studies are necessary to reconcile discrepancies with previous reports on the migratory patterns of the cells and PET imaging.

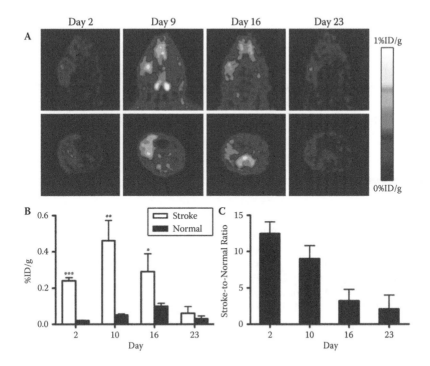

**FIGURE 20.4**

$^{64}$Cu-DOTA-VEGF$_{121}$ PET of the dMCAo rats. (A) Axial and coronal PET images of the dMCAo rat brain at 2 hours postinjection of 1 mCi of $^{64}$Cu-DOTA-VEGF$_{121}$. (B) $^{64}$Cu-DOTA-VEGF$_{121}$ uptake in the stroke area and the contralateral brain. (C) The ratio of $^{64}$Cu-DOTA-VEGF$_{121}$ uptake in the stroke area versus the contralateral brain. $n = 4$; *$P$ .05; **$P$ .01; ***$P$ .001. (Reprinted with permission from Cai W, Guzman R, Hsu AR, Wang H, Chen K, Sun G, Gera A, Choi R, Bliss T, He L, Li ZB, Maag AL, Hori N, Zhao H, Moseley M, Steinberg GK, Chen X. *Stroke* 2009 Jan;40(1):270–77.)

PET may also be used to visualize and quantify the repair mechanisms induced by NSC grafts. Cai et al. (2009) used radiolabeled vascular endothelial growth factor (VEGF) $^{64}$Cu-DOTA-VEGF to image VEGF receptors as indicators of poststroke angiogenic repair. This study demonstrated a transient increase in $^{64}$Cu-DOTA-VEGF uptake in the stroke area, at a maximum 10 days poststroke and back to background level by 3 weeks (Figure 20.4).

Development of new tracers to target specific biological properties or signaling pathways in the cells could be extended to image other biological processes, such as neurogenesis, neuroinflammation, axonal sprouting, and neovascularization, known to mediate stem cell-induced functional recovery after stroke (reviewed in Daadi and Steinberg, 2009). This will be a powerful approach to monitor efficacy and dosing of the therapeutic approach at the cellular level.

## Acknowledgments

This work was supported in part by Russell and Elizabeth Siegelman, Bernard and Ronni Lacroute, the William Randolph Hearst Foundation, Edward G. Hills Fund, and NIH NINDS grants RO1 NS27292, P01 NS37520, and R01 NS058784.

## References

Arbab AS, Bashaw LA, Miller BR, Jordan EK, Lewis BK, Kalish H, Frank JA. (2003) Characterization of biophysical and metabolic properties of cells labeled with superparamagnetic iron oxide nanoparticles and transfection agent for cellular MR imaging. *Radiology* 229:838–846.

Arboody KS, Brown A, Rainov NG, Bower KA, Liu S, Yang W, Small JE, Herrlinger U, Ourednik V, Black PM, Breakefield XO, Snyder EY. (2000) Neural stem cells display extensive tropism for pathology in adult brain: Evidence from intracranial gliomas. *Proc Natl Acad Sci USA* 97:12846–12851.

Benedetti S, Pirola B, Pollo B, Magrassi L, Bruzzone MG, Rigamonti D, Galli R, Selleri S, Di Meco F, De Fraja C, Vescovi A, Cattaneo E, Finocchiaro G. (2000) Gene therapy of experimental brain tumors using neural progenitor cells. *Nat Med* 6:447–450.

Ben-Hur T, van Heeswijk RB, Einstein O, Aharonowiz M, Xue R, Frost EE, Mori S, Reubinoff BE, Bulte JW. (2007) Serial in vivo MR tracking of magnetically labeled neural spheres transplanted in chronic EAE mice. *Magn Reson Med* 57:164–171.

Bjorklund LM, Sanchez-Pernaute R, Chung S, Andersson T, Chen IY, McNaught KS, Brownell AL, Jenkins BG, Wahlestedt C, Kim KS, Isacson O. (2002) Embryonic stem cells develop into functional dopaminergic neurons after transplantation in a Parkinson rat model. *Proc Natl Acad Sci USA* 99:2344–2349.

Bulte JW, Zhang S, van Gelderen P, Herynek V, Jordan EK, Duncan ID, Frank JA. (1999) Neurotransplantation of magnetically labeled oligodendrocyte progenitors: magnetic resonance tracking of cell migration and myelination. *Proc Natl Acad Sci USA* 96:15256–15261.

Cai W, Guzman R, Hsu AR, Wang H, Chen K, Sun G, Gera A, Choi R, Bliss T, He L, Li ZB, Maag AL, Hori N, Zhao H, Moseley M, Steinberg GK, Chen X. (2009) Positron emission tomography imaging of poststroke angiogenesis. *Stroke* 40:270–277.

Callera F, de Melo CM. (2007) Magnetic resonance tracking of magnetically labeled autologous bone marrow CD34+ cells transplanted into the spinal cord via lumbar puncture technique in patients with chronic spinal cord injury: CD34+ cells' migration into the injured site. *Stem Cells Dev* 16:461–466.

Contag CH, Jenkins D, Contag PR, Negrin RS. (2000) Use of reporter genes for optical measurements of neoplastic disease in vivo. *Neoplasia* 2:41–52.

Daadi MM, Li Z, Arac A, Grueter BA, Sofilos M, Malenka RC, Wu JC, Steinberg GK. (2009) Molecular and magnetic resonance imaging of human embryonic stem cell-derived neural stem cell grafts in ischemic rat brain. *Mol Ther* 17:1282–1291.

Daadi MM, Steinberg GK. (2009) Manufacturing neurons from human embryonic stem cells: Biological and regulatory aspects to develop a safe cellular product for stroke cell therapy. *Regen Med* 4:251–263.

Dunning MD, Lakatos A, Loizou L, Kettunen M, ffrench-Constant C, Brindle KM, Franklin RJ. (2004) Superparamagnetic iron oxide-labeled Schwann cells and olfactory ensheathing cells can be traced in vivo by magnetic resonance imaging and retain functional properties after transplantation into the CNS. *J Neurosci* 24:9799–9810.

Ehtesham M, Kabos P, Gutierrez MA, Chung NH, Griffith TS, Black KL, Yu JS. (2002) Induction of glioblastoma apoptosis using neural stem cell-mediated delivery of tumor necrosis factor-related apoptosis-inducing ligand. *Cancer Res* 62:7170–7174.

Frank JA, Zywicke H, Jordan EK, Mitchell J, Lewis BK, Miller B, Bryant LH, Jr., Bulte JW. (2002) Magnetic intracellular labeling of mammalian cells by combining (FDA-approved) superparamagnetic iron oxide MR contrast agents and commonly used transfection agents. *Acad Radiol* 9 Suppl 2:S484–S487.

Franklin RJ, Blaschuk KL, Bearchell MC, Prestoz LL, Setzu A, Brindle KM, Ffrench-Constant C. (1999) Magnetic resonance imaging of transplanted oligodendrocyte precursors in the rat brain. *Neuroreport* 10:3961–3965.

Gambhir SS, Barrio JR, Wu L, Iyer M, Namavari M, Satyamurthy N, Bauer E, Parrish C, MacLaren DC, Borghei AR, Green LA, Sharfstein S, Berk AJ, Cherry SR, Phelps ME, Herschman HR. (1998) Imaging of adenoviral-directed herpes simplex virus type 1 thymidine kinase reporter gene expression in mice with radiolabeled ganciclovir. *J Nucl Med* 39:2003–2011.

Gambhir SS, Bauer E, Black ME, Liang Q, Kokoris MS, Barrio JR, Iyer M, Namavari M, Phelps ME, Herschman HR. (2000a) A mutant herpes simplex virus type 1 thymidine kinase reporter gene shows improved sensitivity for imaging reporter gene expression with positron emission tomography. *Proc Natl Acad Sci USA* 97:2785–2790.

Gambhir SS, Herschman HR, Cherry SR, Barrio JR, Satyamurthy N, Toyokuni T, Phelps ME, Larson SM, Balatoni J, Finn R, Sadelain M, Tjuvajev J, Blasberg R. (2000b) Imaging transgene expression with radionuclide imaging technologies. *Neoplasia* 2:118–138.

Guzman R, Uchida N, Bliss TM, He D, Christopherson KK, Stellwagen D, Capela A, Greve J, Malenka RC, Moseley ME, Palmer TD, Steinberg GK. (2007) Long-term monitoring of transplanted human neural stem cells in developmental and pathological contexts with MRI. *Proc Natl Acad Sci USA* 104:10211–10216.

Hahn PF, Stark DD, Lewis JM, Saini S, Elizondo G, Weissleder R, Fretz CJ, Ferrucci JT. (1990) First clinical trial of a new superparamagnetic iron oxide for use as an oral gastrointestinal contrast agent in MR imaging. *Radiology* 175:695–700.

Hoehn M, Kustermann E, Blunk J, Wiedermann D, Trapp T, Wecker S, Focking M, Arnold H, Hescheler J, Fleischmann BK, Schwidt W, Buhrle C. (2002) Monitoring of implanted stem cell migration in vivo: A highly resolved in vivo magnetic resonance imaging investigation of experimental stroke in rat. *Proc Natl Acad Sci USA* 99:16267–16272.

Jiang Q, Zhang ZG, Ding GL, Zhang L, Ewing JR, Wang L, Zhang R, Li L, Lu M, Meng H, Arbab AS, Hu J, Li QJ, Pourabdollah Nejad DS, Athiraman H, Chopp M. (2005) Investigation of neural progenitor cell induced angiogenesis after embolic stroke in rat using MRI. *Neuroimage* 28:698–707.

Kim DE, Schellingerhout D, Ishii K, Shah K, Weissleder R. (2004) Imaging of stem cell recruitment to ischemic infarcts in a murine model. *Stroke* 35:952–957.

Kim DE, Tsuji K, Kim YR, Mueller FJ, Eom HS, Snyder EY, Lo EH, Weissleder R, Schellingerhout D. (2006) Neural stem cell transplant survival in brains of mice: assessing the effect of immunity and ischemia by using real-time bioluminescent imaging. *Radiology* 241:822–830.

Kondziolka D, Wechsler L, Goldstein S, Meltzer C, Thulborn KR, Gebel J, Jannetta P, DeCesare S, Elder EM, McGrogan M, Reitman MA, Bynum L. (2000) Transplantation of cultured human neuronal cells for patients with stroke. *Neurology* 55:565–569.

Lee JP, Lee JP, Jeyakumar M, Gonzalez R, Takahashi H, Lee PJ, Baek RC, Clark D, Rose H, Fu G, Clarke J, McKercher S, Meerloo J,Muller FJ, Park KI, Butters TD, Dwek RA, Schwartz P, Tong G, Wenger D, Lipton SA, Seyfried TN, Platt FM, Snyder EY. (2007) Stem cells act through multiple mechanisms to benefit mice with neurodegenerative metabolic disease. *Nat Med* 13:439–447.

Lindvall O, Kokaia Z. (2010) Stem cells in human neurodegenerative disorders—time for clinical translation? *J Clin Invest* 120:29–40.

Ma Y, Feigin A, Dhawan V, Fukuda M, Shi Q, Greene P, Breeze R, Fahn S, Freed C, Eidelberg D. (2002) Dyskinesia after fetal cell transplantation for parkinsonism: A PET study. *Ann Neurol* 52:628–634.

MacLaren DC, Toyokuni T, Cherry SR, Barrio JR, Phelps ME, Herschman HR, Gambhir SS. (2000) PET imaging of transgene expression. *Biol Psychiatry* 48:337–348.

Modo M, Beech JS, Meade TJ, Williams SC, Price J. (2009) A chronic 1 year assessment of MRI contrast agent-labelled neural stem cell transplants in stroke. *Neuroimage* 47 Suppl 2:T133–T142.

Modo M, Cash D, Mellodew K, Williams SC, Fraser SE, Meade TJ, Price J, Hodges H. (2002) Tracking transplanted stem cell migration using bifunctional, contrast agent-enhanced, magnetic resonance imaging. *Neuroimage* 17:803–811.

Ntziachristos V, Ripoll J, Wang LV, Weissleder R. (2005) Looking and listening to light: The evolution of whole-body photonic imaging. *Nat Biotechnol* 23:313–320.

Okada S, Ishii K, Yamane J, Iwanami A, Ikegami T, Katoh H, Iwamoto Y, Nakamura M, Miyoshi H, Okano HJ, Contag CH, Toyama Y, Okano H. (2005) In vivo imaging of engrafted neural stem cells: its application in evaluating the optimal timing of transplantation for spinal cord injury. *FASEB J* 19:1839–1841.

Park KI, Teng YD, Snyder EY. (2002) The injured brain interacts reciprocally with neural stem cells supported by scaffolds to reconstitute lost tissue. *Nat Biotechnol* 20:1111–1117.

Politi LS, Bacigaluppi M, Brambilla E, Cadioli M, Falini A, Comi G, Scotti G, Martino G, Pluchino S. (2007) Magnetic-resonance-based tracking and quantification of intravenously injected neural stem cell accumulation in the brains of mice with experimental multiple sclerosis. *Stem Cells* 25:2583–2592.

Reumers V, Deroose CM, Krylyshkina O, Nuyts J, Geraerts M, Mortelmans L, Gijsbers R, Van den Haute C, Debyser Z, Baekelandt V. (2008) Noninvasive and quantitative monitoring of adult neuronal stem cell migration in mouse brain using bioluminescence imaging. *Stem Cells* 26:2382–2390.

Shah K, Bureau E, Kim DE, Yang K, Tang Y, Weissleder R, Breakefield XO. (2005) Glioma therapy and real-time imaging of neural precursor cell migration and tumor regression. *Ann Neurol* 57:34–41.

Shapiro EM, Skrtic S, Sharer K, Hill JM, Dunbar CE, Koretsky AP. (2004) MRI detection of single particles for cellular imaging. *Proc Natl Acad Sci USA* 101:10901–10906.

Song M, Kim Y, Kim Y, Ryu S, Song I, Kim SU, Yoon BW. (2009) MRI tracking of intravenously transplanted human neural stem cells in rat focal ischemia model. *Neurosci Res* 64:235–239.

Sun N, Lee A, Wu JC. (2009) Long term non-invasive imaging of embryonic stem cells using reporter genes. *Nat Protoc* 4:1192–1201.

Tang Y, Shah K, Messerli SM, Snyder E, Breakefield X, Weissleder R. (2003) In vivo tracking of neural progenitor cell migration to glioblastomas. *Hum Gene Ther* 14:1247–1254.

Tjuvajev JG, Chen SH, Joshi A, Joshi R, Guo ZS, Balatoni J, Ballon D, Koutcher J, Finn R, Woo SL, Blasberg RG. (1999) Imaging adenoviral-mediated herpes virus thymidine kinase gene transfer and expression in vivo. *Cancer Res* 59:5186–5193.

Waerzeggers Y, Klein M, Miletic H, Himmelreich U, Li H, Monfared P, Herrlinger U, Hoehn M, Coenen HH, Weller M, Winkeler A, Jacobs AH. (2008) Multimodal imaging of neural progenitor cell fate in rodents. *Mol Imaging* 7:77–91.

Walczak P, Bulte JW. (2007) The role of noninvasive cellular imaging in developing cell-based therapies for neurodegenerative disorders. *Neurodegener Dis* 4:306–313.

Walczak P, Zhang J, Gilad AA, Kedziorek DA, Ruiz-Cabello J, Young RG, Pittenger MF, van Zijl PC, Huang J, Bulte JW. (2008) Dual-modality monitoring of targeted intraarterial delivery of mesenchymal stem cells after transient ischemia. *Stroke* 39:1569–1574.

Weissleder R, Elizondo G, Wittenberg J, Rabito CA, Bengele HH, Josephson L. (1990) Ultrasmall superparamagnetic iron oxide: Characterization of a new class of contrast agents for MR imaging. *Radiology* 175:489–493.

Wu JC, Sundaresan G, Iyer M, Gambhir SS. (2001) Noninvasive optical imaging of firefly luciferase reporter gene expression in skeletal muscles of living mice. *Mol Ther* 4:297–306.

Zhang Z, Jiang Q, Jiang F, Ding G, Zhang R, Wang L, Zhang L, Robin AM, Katakowski M, Chopp M. (2004) In vivo magnetic resonance imaging tracks adult neural progenitor cell targeting of brain tumor. *Neuroimage* 23:281–287.

Zhang ZG, Jiang Q, Zhang R, Zhang L, Wang L, Zhang L, Arniego P, Ho KL, Chopp M. (2003) Magnetic resonance imaging and neurosphere therapy of stroke in rat. *Ann Neurol* 53:259–263.

Zhu J, Zhou L, XingWu F. (2006) Tracking neural stem cells in patients with brain trauma. *N Engl J Med* 355:2376–2378.

# 21

## Clinical Application of Noninvasive Molecular Imaging in Cancer Cell Therapy: The First Reporter Gene–Based Imaging Clinical Trial

**Shahriar S. Yaghoubi and Sanjiv S. Gambhir**

### CONTENTS

## 21.1 Introduction: Whole-Body Therapeutic Cell Pharmacokinetic Analysis through Noninvasive Molecular Imaging

There are three currently conceived techniques for noninvasively imaging the pharmacokinetics of therapeutic cells (TCs) in humans (Table 21.1). Pharmacokinetics includes the locations, quantity at a specific location, viability of the TCs, proliferation, and changes in the characteristics of the

369

**TABLE 21.1**

Three Techniques for Noninvasive Imaging of Therapeutic Cell Pharmacokinetics in Humans

| Technique | *Ex Vivo* Labeling with Imaging Probes (Mainly Radioisotope Labeled or MRI Probes) Prior to Administration | Incorporation of Radionuclide or MRI-Based Reporter Genes into Therapeutic Cells; Imaging Pharmacokinetics of Administered Cells by Reporter Gene Expression | Image Cells Using a Probe that Can Detect a Specific Target Associated with the Therapeutic Cell |
|---|---|---|---|
| Advantages | 1. Relatively easy to implement<br>2. Relatively higher sensitivity due to absence of probe background if probe is not released<br>3. Relatively lower radiation exposure when using radioactive probes<br>4. Generalizable | 1. Potential for imaging every aspect of therapeutic cell pharmacokinetics<br>2. Generalizable<br>3. No false positives: dead cells cannot express the reporter gene<br>4. Capability for long-term monitoring | 1. Long-term monitoring technique<br>2. Lack of false positives |
| Disadvantages | 1. Limited pharmacokinetic information: cannot obtain information about cell status<br>2. False positive images when cells die in vivo<br>3. Probe dilution when cells divide<br>4. Radioactive decay of radionuclide probes only allows short-term imaging<br>5. Toxicity caused by inserted probe is possible | 1. Necessary to incorporate a reporter transgene into therapeutic cells<br>2. Lower sensitivity than direct labeling method due to background noise | 1. Not generalizable: a new probe may have to be developed for each therapeutic cell of interest<br>2. Lower sensitivity than direct labeling method due to background noise |

administered TC, through time. Whole-body imaging is especially important in human studies because it is the only practical way of obtaining the
pharmacokinetic information for clinical applications. The pharmacokinetic information is necessary for early assessment of efficacy or potential
adverse effects.

The most conventional method that can allow imaging the location
and quantity of TCs after administration is direct labeling of cells *ex vivo*
with radionuclide or magnetic resonance imaging (MRI) probes, such as
indium-111 ($^{111}$In) oxine or ferumoxides, respectively. Trafficking of human
dendritic cells has been imaged in human patients by *ex vivo* direct labeling of the cells by superparamagnetic iron oxide particles and $^{111}$In-oxine.[1,2]
Probes of other imaging modalities, such as fluorescent nanoparticles, have
limited application for imaging TCs in patients due to tissue attenuation of
signal; hence, whole-body imaging is currently not possible. The advantage
of the direct probe labeling technique is easy implementation and reduced
whole-body radiation exposure when radionuclide probes are used. The
signal-to-background ratio will also be low if cells do not release (efflux) the
probe once inside the patient's body. However, the direct labeling technique
may yield false-positive images about cell locations and viability if the probe
is released (such as when cells die). Furthermore, probes are diluted after
cell division, and the activity of radionuclide probes will decay; hence, this is
primarily a shorter-term monitoring technique.

Another approach is to use a probe that detects a specific receptor found
only on the surface of the administered TCs or an intracellular protein target.
This method is challenging to implement as it is not a general method and
requires development of specific imaging probes for potentially every type
of TC. In addition, the sensitivity may be low. However, this is a long-term
monitoring technique and can potentially also allow imaging TC survival
and proliferation. To our knowledge, this technique has not yet been used to
image TCs in humans. However, there are reports of studies imaging T cells
in mice using radiolabeled probes.[3–5]

The third technique is the reporter gene- (RG-) based cell imaging method.
This technique can potentially allow analyzing all aspects of the pharmacokinetics of a TC (stated in previous paragraphs). To image whole-body TC
location and determine survival and proliferation ideally by quantifying
TCs at any location inside the patient's body, using the RG technique, TCs
are first genetically engineered to constitutively express the RG in a stable
manner. This can be achieved by making a genetic construct containing
the RG under control of a constitutive promoter, such as ubiquitin or cytomegalovirus (CMV) promoters. The genetic construct is then delivered into
the TC either using a viral vector that stably integrates the construct into
the cell chromosome or by antibiotic selection after stable transfection or
electroporation/nucleofection. Ideally, the promoter is "strong," causing a
high level of RG expression, it is not silenced, and the level of RG expression

is constant to allow accurate quantification. To accurately quantify using a radionuclide-based RG technique, tracer kinetic modeling will be necessary. As far as RG delivery, ideally transgene construct chromosomal integration is site specific, and the site of integration is known to be nonmutagenic. Otherwise, constant episomal expression would be fine when the TC does not proliferate, avoiding the risk of insertional mutagenesis.

The RG-based technique is also useful for imaging changes in TC status. For example, it would be important to detect therapeutic stem cell differentiation, adult TC dedifferentiation, or general health of the TCs once administered into a patient. Obtaining this information may be possible by genetically engineering TCs to have a transgene construct containing a RG regulated by a promoter that is conditionally activated. For example, to image differentiation one may have the RG regulated by a tissue-specific promoter, which is activated when the stem cell differentiates into that specific adult tissue cell.

Therefore, the main advantage of the RG-based technique is that it can potentially allow obtaining TC pharmacokinetic information through noninvasive whole-body imaging. Radionuclide-based and MRI RGs are currently the only modalities foreseen for such applications in humans or large-animal models. Positron emission tomography (PET) and single-photon emission computed tomography (SPECT) RGs (PRGs and SRGs, respectively) and reporter probes (RPs) are currently the most advanced for these applications. In fact, the first demonstration of RG-based TC imaging reported in humans is with the PET/SPECT RG, herpes simplex virus type 1 thymidine kinase (*HSV1-tk*) and the PET RP (PRP) [18]F-radiolabeled 9-[4-fluoro-3-(hydroxymethyl)butyl]guanine ([18]F-FHBG) (see Section 21.2). Another advantage of the RG-based technique is potential long-term whole-body pharmacokinetic analysis through noninvasive imaging. Similar to the direct-labeling technique, the RG-based technique is also generalizable.

The main challenge for this technique is genetic modification of TCs without significantly affecting the characteristics of the cell or causing insertional mutagenesis. Another potential problem is immune reaction to the reporter protein when using a nonhuman RG. The sensitivity of this technique is determined by the pharmacokinetics of the RP and the level of RG expression per cell. PET and SPECT RG (PRG and SRG) systems are more sensitive than MRI RG systems, and their probes are administered in trace doses, practically eliminating risk of pharmacological toxicity. MRI, however, has a better spatial resolution than PET or SPECT. Detailed discussion of RGs and RPs and imaging their expression is beyond the scope of this book chapter; hence, the reader is referred to a book on this topic by Gambhir and Yaghoubi.[6] Readers can also refer to reviews by Gilad et al. on MRI RGs,[7,8] Greer et al.,[9] and Contag et al.[10] on bioluminescent RGs and Gambhir et al.[11–13] on radionuclide-based RGs.

## 21.2 Imaging the Survival of Therapeutic Cells: The First Reporter Gene-Based Therapeutic Cell Imaging Clinical Trial

In this section, we describe the first reported clinical trial using the PRG *HSV1-tk* and the PRP ¹⁸F-FHBG, successfully imaging the survival and locations of autologous cytolytic T cells (CTLs) administered to treat patients with glioma after resection of their recurrent tumors. The first case of this study was published by Yaghoubi et al. in *Nature Clinical Practice Oncology* and much of the information contained in this section is from that case report.[14] To our knowledge, at the time this chapter was written, there were no reports of another RG-based TC imaging clinical trial, although we and others are currently performing several new trials.

### 21.2.1 Herpes Simplex Virus Type 1 Thymidine Kinase and ¹⁸F-FHBG

*HSV1-tk* is a PRG, an SRG, and a suicide/safety gene. As a PRG or SRG, *HSV1-tk* and its mutants have been used for a variety of preclinical and clinical molecular imaging purposes, including imaging therapeutic transgene expression, cell trafficking, the location of implanted cells, and endogenous gene expression.[15-23] ¹⁸F-FHBG is a side-chain fluorine-18 radiolabeled analog of the antiherpes virus drug penciclovir. When ¹⁸F-FHBG is taken up by cells expressing *HSV1-tk* and some of its mutants (such as *HSV1-sr39tk*),[24] this PRP is phosphorylated and trapped within the cells (Figure 21.1). Otherwise, ¹⁸F-FHBG effluxes from cells not expressing *HSV1-tk* and mutant PRGs and is cleared from the body of a human, although cells that do not express *HSV1-tk* or its mutants can still trap a few *HSV1-tk* probes due to inefficient phosphorylation by the mammalian thymidine kinase. In addition, competition from thymidine can affect the level of phosphorylation of *HSV1-tk* probes as it is also a substrate for the *HSV1-tk* enzyme. The protocols for preclinical and clinical imaging of ¹⁸F-FHBG and detection of the gene expression of *HSV1-tk* or a mutant have been previously published by Yaghoubi et al.[25,26]

Pharmacological safety of cold FHBG at 100X tracer dose has also been demonstrated in rats and rabbits.[27] We have also analyzed the safety, dosimetry, and pharmacokinetics of ¹⁸F-FHBG in healthy human volunteers.[28] ¹⁸F-FHBG is currently approved by the U.S. Food and Drug Administration (FDA) as an investigational new drug (IND 61,880). Figure 21.2 illustrates biodistribution of ¹⁸F-FHBG in a human 2 hours after injecting it intravenously. As shown in the image, rapid clearance of ¹⁸F-FHBG results in low PRP background, except in the liver, kidneys, bladder, and intestines, which are involved in its clearance. ¹⁸F-FHBG does not cross the blood-brain barrier (BBB), and imaging within the central nervous system is only possible when

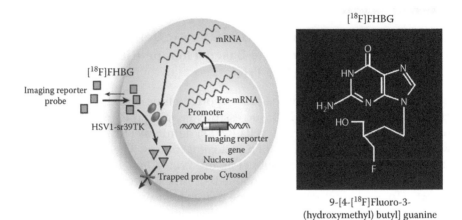

9-[4-[$^{18}$F]Fluoro-3-
(hydroxymethyl) butyl] guanine

**FIGURE 21.1**
Diagram illustrating the mechanism of imaging an enzyme-based PRG. *HSV1-sr39tk* is a mutant of herpes simplex virus type 1 thymidine kinase (*HSV1-tk*). The PRG *HSV1-sr39tk* encodes the enzyme *HSV1-sr39TK*, which can phosphorylate acycloguanosine and uracil nucleoside analogs. $^{18}$F-FHBG, a side-chain fluorine-18 radiolabeled analog of the drug penciclovir, is an acycloguanosine analog, most efficiently phosphorylated through *HSV1-sr39TK* enzyme catalysis. On phosphorylation, $^{18}$F-FHBG cannot cross the membrane; hence, cells expressing *HSV1-sr39tk* entrap $^{18}$F-FHBG. Above background accumulation of $^{18}$F-FHBG in the tissues of living mammals is an indication of the presence of *HSV1-tk-* or *HSV1-sr39tk*-expressing cells within the tissues.

the BBB is compromised, as is the case when a brain tumor is present or is surgically resected. Other probes are available, and new probes are being developed for imaging the expression of *HSV1-tk* and its mutants,[22] but their use for imaging TCs in humans has not been reported yet.

## 21.2.2 The Patients

Currently, six patients diagnosed with glioblastomas have had whole-body $^{18}$F-FHBG PET scans. All of the patients were administered less than 7 mCi or 2 µg of $^{18}$F-FHBG intravenously. All of the patients had the following safety monitoring:

- Blood chemistry and complete blood count (CBC) at baseline and then 1 day and 1 week after injection of $^{18}$F-FHBG
- Two hours of temperature, heart rate, breathing rate, blood pressure, pulse oximetry, and EKG (electrocardiogram) monitoring after $^{18}$F-FHBG injection
- Mini-Mental Status exam at baseline and 1 week after $^{18}$F-FHBG injection

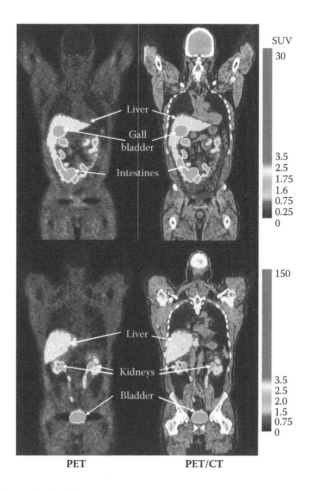

**FIGURE 21.2 (See color insert.)**
Whole-body PET and PET/CT (computed tomographic) images of ¹⁸F-FHBG biodistribution in a human 2 hours after its intravenous injection. Two coronal slices are shown to illustrate activity within the liver, gall bladder, intestines, kidneys, and bladder, which are organs involved with the clearance of ¹⁸F-FHBG from the body. Background activity in all other tissues is relatively low due to the absence of cells expressing *HSV1-tk* or *HSV1-sr39tk* within the body of this human volunteer.

Four of the patients, who served as controls, did not receive infusions of any CTLs. From these patients, 1 mL of blood was collected at seven different time points after the injection of ¹⁸F-FHBG for blood time-activity analysis (refer to Figure 21.3 for the blood time-activity curve). Also, urine specimens were collected from these patients to analyze stability of ¹⁸F-FHBG with high-performance liquid chromatography (HPLC) (refer to Figure 21.4 for a graph illustrating ¹⁸F-FHBG stability). Of these four patients, three had surgically resected tumors in the rear cerebrum or frontal lobe, and one had intact

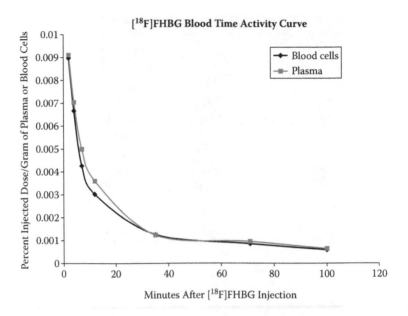

**FIGURE 21.3**
Blood time-activity of ¹⁸F-FHBG in a female patient volunteer with glioma. Half of the radiolabeled probe is cleared from plasma and blood cells in less than 10 minutes, and only about 10% of the original activity is present after 40 minutes. Radioactivity measurements were decay corrected back to the injection time.

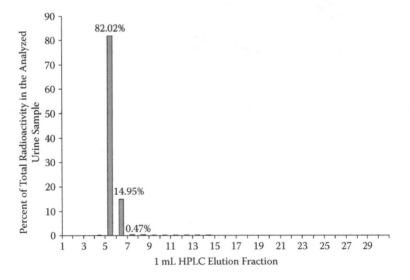

**FIGURE 21.4**
HPLC analysis of ¹⁸F-FHBG stability. Urine was collected approximately 2 hours after intravenous bolus injection of ¹⁸F-FHBG from a female patient volunteer with glioma. At least 82% of radioactivity detected in the urine sample was associated with intact ¹⁸F-FHBG.

tumors in the rear of his cerebrum and near ventricles. Two of these patients were females, and two were males.

Two of the six patients with glioma receiving [18]F-FHBG PET scans were also enrolled in another trial, receiving genetically engineered autologous CTLs (described in Section 21.2.3). The first patient (CTL patient 1) was a 57-year-old Caucasian man diagnosed with grade IV glioblastoma multiforme, 1 year prior to the CTL infusions and the [18]F-FHBG PET scan. This patient had received 12 sessions of intracranial CTL infusions after resection of his recurrent tumor in the right parietal occipital lobe of his brain, directly into the tumor resection site. He received a single [18]F-FHBG tracer injection and had a single whole-body PET scan 3 days after completion of the CTL infusions. At the time the first patient was enrolled, the FDA only allowed a single [18]F-FHBG PET scan per patient. The second patient (CTL patient 2) was a 41-year-old Caucasian woman who had been diagnosed with glioblastoma in the right parieto-occipital lobe of her brain 4.5 years prior to receiving CTL infusions. This patient had a recurrent tumor in the same area in November 2008. The patient had a [18]F-FHBG PET scan once prior to CTL infusions and a second time 3 days after completion of the 12 sessions of CTL infusions. CTL patients 1 and 2 met the inclusion and exclusion criteria discussed next prior to enrollment in the cell therapy trial.

### 21.2.2.1 Inclusion Criteria for Study Enrollment

Histological verification of grade III or IV malignant glioma at original diagnosis

Male or female subjects between 18 and 70 years, inclusive

Unifocal site of original disease in the cerebral cortex

Primary therapy completed and steroid independent no less than 4 weeks

If patient is taking adjunct cytotoxic chemotherapy, must be at least 2 weeks from finishing most recent course and recovered from all acute side effects

Adequate renal function as evidenced by creatinine below 1.6.

Adequate bone marrow function as evidenced by white blood cells (WBCs) 2,000/dL or higher (or Absolute Neutrophil Count [ANC] > 1,000) and platelets 100,000/dL or higher unsupported by transfusion or growth factor

Normal liver function as evidenced by bilirubin below 1.5 and SGOT (aspartate aminotransferase) and SGPT (alanine aminotransferase) more than twice upper limits of normal

Female patients of childbearing potential must not be pregnant as evidenced by a serum β-hCG (human chorionic gonadotropin) pregnancy test obtained within 7 days of enrollment

Participants having reproductive potential must agree to use effective contraception during participation on this protocol

### 12.2.2.2 Exclusion Criteria for Study Enrollment

Prior reresection for recurrent/progressive disease

Communication of resection cavity with ventricles/deep cerebrospinal fluid (CSF) pathways

Survival expectation less than 3 months

Karnofsky Performance Score (KPS) below 70

Organ function:

Pulmonary: Requirement for supplemental oxygen use that is not expected to resolve within 2 weeks

Cardiac: Uncontrolled cardiac arrhythmia, hypotension requiring pressor support

Renal: Dialysis dependent

Neurologic: Refractory seizure disorder, clinically evident progressive encephalopathy

Patients with any nonmalignant intercurrent illness that is either poorly controlled with currently available treatment or is of such severity that the investigators deem it unwise to enter the patient on protocol shall be ineligible

Patients being treated for severe infection or who are recovering from major surgery are ineligible until recovery is deemed complete by the investigator

Failure to understand the basic elements of the protocol or the risks/benefits of participating in this pilot study

History of ganciclovir or Prohance contrast allergy or intolerance

Positive serology for HIV based on testing performed within 3 months prior to protocol enrollment

### 21.2.3 Preparation and Administration of Genetically Engineered Autologous Cytolytic T Cells for Cellular Immunotherapy of Glioma

The CTL patients 1 and 2 had enrolled in an FDA-authorized (BB-IND 10109) adoptive cellular gene immunotherapy (ACGIT) clinical trial after their initial diagnosis and glioma tumor resection and consented and donated the blood that was used to prepare their own CTLs. CD8$^+$ T cells had been isolated from each patient's peripheral blood mononuclear cells and electroporated, delivering a plasmid DNA construct encoding the interleukin 13 (IL-13) zetakine and the hygromycin/*HSV1-tk* genes under the transcriptional control of a modified human elongation factor 1α (EF1α) promoter and the CMV immediate/early promoter, respectively, in a Good Manufacturing Practices (GMP) cell production facility at City of Hope National Medical Center in

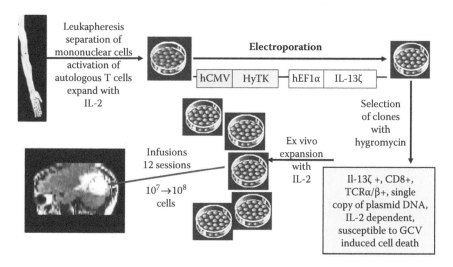

**FIGURE 21.5**

Description of the procedures involved in the preparation of genetically engineered CTLs and their infusion into the recurrent glioma tumor resection site.

Duarte, California (refer to Figure 21.5 for a diagram of CTL preparation). Hygromycin-resistant CTLs were cloned in limiting dilution then expanded using the Rapid Expansion Method (REM) method to numbers in excess of $10^9$ and cryopreserved. Following the diagnosis of relapse, the cryopreserved cells were thawed, expanded, and formulated for intracranial infusion in 2 mL of preservative-free normal saline (PFNS). These cells were infused over a period of 5 weeks on Mondays, Wednesdays, and Fridays, with a break on week 3 (refer to Figure 21.6 for a timeline of events). The recurrent tumor was resected, and a Rickham reservoir was inserted to allow infusion of genetically engineered autologous $CD8^+$ CTLs. The patients started with a cell dose of $1 \times 10^7$. Both patients tolerated that dose well, and their cell infusion dose was escalated to $1 \times 10^8$ per day. By the end of the CTL infusions, the patients had received approximately $1 \times 10^9$ genetically engineered autologous CTLs (refer to Table 21.2 for quality assurance analysis of infused CTLs).

### 21.2.4 Discussion of Positron Emission Tomography Image Results

As discussed in Section 21.2.2, six glioblastoma patients were enrolled in the $^{18}$F-FHBG whole-body PET scan clinical trials, of whom four did not have any CTL infusions and two (CTL patients 1 and 2) had CTL infusions as described in Section 21.2.3. Of the four control patients, three only had resected tumor sites in the brain, and one had an intact side and an intact center tumor. CTL patient 1 had surgical resection of a side recurrent tumor and a new recurrent intact tumor in the center of his brain, discovered after the $^{18}$F-FHBG PET scan. CTL patient 2 had a surgically resected side tumor,

**TABLE 21.2**

Quality Assurance of Infused CTLs

| Test for | Release Criteria | Testing Method |
|---|---|---|
| Viability of clinical preparation | >75% | Trypan blue exclusion |
| Cell-surface phenotype | Uniformly TCRα/β+, CD4-,CD8+, IL-13+ | Flow cytometric evaluation with isotype controls. |
| Vector rearrangement | Single band | Southern Blot with HyTK-Specific Probe |
| IL-13 zetakine expression | Presence of chimeric zeta and endogenous zeta bands | Western blot with human zeta-specific primary antibody |
| IL-13Rα2-specific cytolytic activity | >25% specific lysis at E:T ratio of 5:1 against IL-13Rα2+ Daudi. Lysis of Daudi parental < 2X percent lysis of parental Daudi at same E:T. | Four-hour chromium release assay |
| Sensitivity to ganciclovir | <15% cell viability after 14-days of coculture in 5 μM ganciclovir. | Trypan blue-exclusion cell enumeration |
| Sterility | All screening bacterial/ fungal cultures negative for > 14 days. Mycoplasma negative at time of cyropreservation and within 72 hours of each treatment cycle Endotoxin level < 5 EI/kg in washed cell preparation. Gram stain of cell culture medium negative on day of reinfusion. | Bacterial/fungal by routine clinical specimen culture Mycoplasma by gene-probe radioimmunoassay (RIA) Endotoxin by enzyme-linked immunosorbent assay (ELISA) Gram stain by clinical microbiology lab |

and she was scanned twice for $^{18}$F-FHBG PET, once 2 weeks prior to the infusion of CTLs and the second time 3 days after completion of the entire CTL regimen.

Figure 21.7 illustrates the MRI and $^{18}$F-FHBG PET-over-MRI superimposed brain images of CTL patient 1. Higher than brain background $^{18}$F-FHBG activity can be seen in site 1, which is where the CTLs were infused after surgical removal of the recurrent tumor in the right parietal lobe. Interestingly, we observed another region of $^{18}$F-FHBG activity near the corpus callosum of this patient, which was later detected by MRI to be the site of an intact recurrent tumor. When we quantified the $^{18}$F-FHBG activity in this patient and the control patients, $^{18}$F-FHBG activity was 2.6 times higher in the resection site of CTL patient 1 versus the average activity in the resection sites of three control patients. $^{18}$F-FHBG activity was 2.8 times higher in the intact corpus callosum tumor of CTL patient 1 versus the intact tumor of a control patient. Table 21.3

**Timeline of Events**

Patient enrollment
& leukapheresis

Isolation and cryopreservation of
IL-13 zetakine+ HYTK+ CD8+ CTL clones

Radiographic evidence of tumor progression/recurrence
re-biopsy, re-resection and reservoir placement

Cell Infusions

| Cycle: | Week: | Day: | Intracavitary cell dose: |
|--------|-------|------|--------------------------|
| 1 | 1 | 1 | $10^7$ |
|   |   | 3 | $5 \times 10^7$ |
|   |   | 5 | $10^8$ |
| 2 | 2 | 1 | $10^8$ |
|   |   | 3 | $10^8$ |
|   |   | 5 | $10^8$ |
|   | 3 | Rest/restaging | |
| 3 | 4 | 1 | $10^8$ |
|   |   | 3 | $10^8$ |
|   |   | 5 | $10^8$ |
| 4 | 5 | 1 | $10^8$ |
|   |   | 3 | $10^8$ |
|   |   | 5 | $10^8$ |

6        1    $^{18}$F-FHBG PET scan
Rest/restaging & follow-up safety monitoring

**FIGURE 21.6**
Timeline of CTL therapy and $^{18}$F-FHBG imaging.

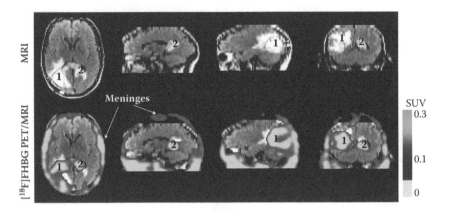

**FIGURE 21.7**
MRI and PET-over-MRI superimposed brain images of CTL patient 1. Images were acquired approximately 2 hours after $^{18}$F-FHBG injection. The patient had a surgically resected tumor (1) in the right parietal lobe and a new nonresected tumor in the center (2), near the corpus callosum of his brain. The infused cells had localized at the site of tumor 1 and trafficked to tumor 2. $^{18}$F-FHBG activity is higher than the brain background at both sites. Background $^{18}$F-FHBG activity is low within the central nervous system due to its inability to cross the blood-brain barrier. Background activity is relatively higher in all other tissues. Activity can also be observed in the meninges. The tumor 1/meninges and tumor 2/meninges $^{18}$F-FHBG activity ratios in this patient were 1.75 and 1.57, respectively, whereas the average resected tumor site/meninges and intact tumor site-to-meninges $^{18}$F-FHBG activity ratios in control patients were 0.86 and 0.44, respectively.

includes the average activity ratios of the two CTL patients and the four control patients. Including CTL patient 2, the results are still similar.

Figure 21.8 illustrates the pre- and post-CTL $^{18}$F-FHBG head PET images of CTL patient 2. Higher $^{18}$F-FHBG activity can be seen at the infusion site post-CTL infusions, and unlike pre-CTL infusions, for which the center showed very low activity, the center of the resection site has the higher-activity post-CTL infusions. Quantifying the data, we determined there was a 61% increase in tumor/brain background ratio after CTL infusions (see Figure 21.9 and Table 21.4). Biopsy further confirmed the presence of CTLs after the PET scans.

## 21.3 Future Directions

In this chapter, we described the three methods currently envisioned for studying the pharmacokinetics of TCs through noninvasive imaging. Currently, none of these techniques is widely used in clinical trials despite their importance in cell therapy. Possible reasons are the additional costs, additional regulatory efforts, and lack of widespread knowledge of the

**TABLE 21.3**

Mean Region of Interest $^{18}$F-FHBG Activity Ratios of Tumors to Other Tissues

| Ratios | Patients with Resected Side Tumors after Infusion of CTLs ($n$ = 2) (Average + SEM) | Patient with Intact Side Tumor without CTL Infusions ($n$ = 1) | Patients with Resected Side Tumors without CTL Infusions ($n$ = 3) (AVG + SEM) |
|---|---|---|---|
| Side tumor/brain BKGD | 4.89 ± 0.07 | 1.54 | 1.82 ± 0.07 |
| Side tumor/meninges | 2.22 ± 0.47 | 0.44 | 0.86 ± 0.14 |
| Side tumor/heart | 0.85 ± 0.03 | 0.57 | 0.38 ± 0.08 |
| Side tumor/liver | 0.14 ± 0.08 | 0.08 | 0.08 ± 0.02 |

| Ratios | Patient with Intact Center Tumor after Infusion of CTLs ($n$ = 1) | Patient with Intact Center Tumor without CTL Infusions ($n$ = 1) |
|---|---|---|
| Center tumor/brain BKGD | 4.33 | 1.55 |
| Center tumor/ meninges | 1.57 | 0.44 |
| Center tumor/heart | 0.74 | 0.57 |
| Center tumor/liver | 0.13 | 0.08 |

*Note:* SEM, standard error of the mean; BKGD, background.

existence of these technologies. In addition, these imaging technologies need to be further developed to answer the pharmacokinetics questions they can potentially answer. For example, image-based quantification of TCs has yet to be addressed, and without quantification the information obtained will be incomplete.

The study described in this chapter was made possible by early collaboration of molecular imaging and cell therapy investigators. Even though when first planned, the CTL therapy protocol did not include $^{18}$F-FHBG PET imaging, starting the collaboration 2 years prior to study initiation permitted the inclusion of imaging in the cell therapy trial. In future studies that have been planned from the beginning, we will incorporate the mutant PRG *HSV1-sr39tk*, which will increase sensitivity several-fold. Furthermore, all future patients will have an $^{18}$F-FHBG PET scan before and after CTL infusions, providing the best control, the patient him- or herself. Early collaboration will ensure reduction of regulatory efforts, implementation of the most appropriate clinical trial protocol, and efficient acquisition of the necessary preclinical data.

PRG-based clinical imaging technologies are not limited to monitoring cell trafficking and survival in adoptive cellular gene therapy of cancer. These technologies can also be used to study pharmacokinetics of other TCs, such

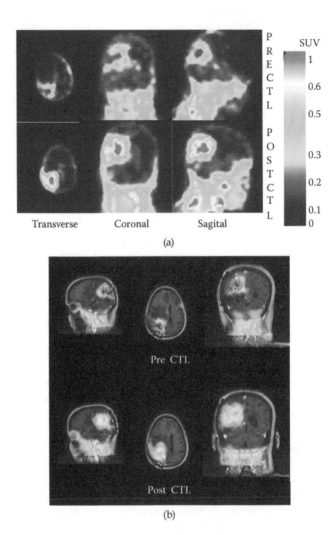

**FIGURE 21.8**

(a) 18F-FHBG PET images of the head of CTL patient 2 prior to initiation of CTL infusions and 3 days after completion of all CTL infusions. (b) 18F-FHBG head PET images superimposed over corresponding MRI images of CTL patient 2 illustrating increased 18F-FHBG accumulation after CTL infusions at the recurrent tumor resection site. Images acquired approximately 2 hours after bolus intravenous 18F-FHBG injection.

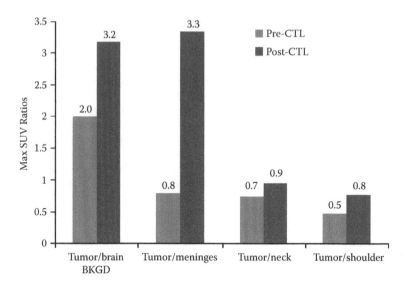

**FIGURE 21.9**
Ratios of $^{18}$F-FHBG signal intensity in tumor resection site over a brain background site, the surrounding meninges, neck, and shoulder. All of the ratios increased after CTL infusions.

**TABLE 21.4**

Percentage Change in $^{18}$F-FHBG Accumulation and Tumor/BKGD $^{18}$F-FHBG Post-CTL Infusion

| FHBG Accumulation | | Tumor/Tissue Ratios | |
|---|---|---|---|
| Tumor | 58% | | |
| Brain BKGD | −2% | Tumor/brain BKGD | 61% |
| Meninges | −63% | Tumor/meninges | 423% |
| Neck | 23% | Tumor/neck | 28% |
| Shoulder | −3% | Tumor/shoulder | 62% |

as stem cells, pancreatic islets, and dendritic cells, aimed at curing other ailments, such as neurodegenerative, cardiovascular, and autoimmune diseases. In all of these therapeutic procedures, it will be important to identify TC locations, determine quantity at all locations, assess survival and proliferation, and monitor changes in TC characteristics through time.

# References

1. De Vries IJM, Lesterhuis WJ, Barentsz JO, Verdijk P, Van Krieken JH, Boerman OC, et al. Magnetic resonance tracking of dendritic cells in melanoma patients for monitoring of cellular therapy. *Nat Biotechnol* 2005; 23: 1407–1413.

2. Morse MA, Coleman RE, Akabani G, Niehaus N, Coleman D, Lyerly HK. Migration of human dendritic cells after injection in patients with metastatic malignancies. *Cancer Res* 1999; 59: 56–58.

3. Malviya G, D'Alessandria C, Bonanno E, Vexler V, Massari R, Trotta C, et al. Radiolabeled humanized anti-CD3 monoclonal antibody Visilizumab for imaging human T-lymphocytes. *J Nucl Med* 2009; 50: 1683–1691.

4. Matsui K, Wang Z, McCarthy TJ, Allen PM, Reichert DE. Quantitation and visualization of tumor-specific T cells in the secondary lymphoid organs during and after tumor elimination by PET. *Nucl Med Biol* 2004; 31: 1021–1031.

5. Annovazzi A, D'Alessandria C, Bonanno E, Mather SJ, Cornelissen B, Van de Wiele C, et al. Synthesis of $^{99m}$Tc-HYNIC-interleukin-12, a new specific radiopharmaceutical for imaging T lymphocytes. *Eur J Nucl Med Mol Imaging* 2006; 33: 474–482.

6. Gambhir SS, Yaghoubi SS (Eds.). *Molecular Imaging with Reporter Genes.* Cambridge University Press, Cambridge, UK, 2010.

7. Gilad AA, Winnard PT, Jr, Van Zijl PCM, Bulte JWM. Developing MR reporter genes: Promises and pitfalls. *NMR Biomed* 2007; 20: 275–290.

8. Gilad AA, Ziv K, McMahon MT, Van Zijl PCM, Neeman M, Bulte JWM. MRI reporter genes. *J Nucl Med* 2008; 49: 1905–1908.

9. Greer LF, III, Szalay AA. Imaging of light emission from the expression of luciferases in living cells and organisms: A review. *Luminescence* 2002; 17: 43–74.

10. Contag CH, Ross BD. It's not just about anatomy: In vivo bioluminescence imaging as an eyepiece into biology. *J Magn Reson Imaging* 2002; 16: 378–387.

11. Gambhir SS, Barrio JR, Herschman HR, Phelps ME. Assays for noninvasive imaging of reporter gene expression. *Nucl Med Biol* 1999; 26: 481–490.

12. Gambhir SS, Barrio JR, Herschman HR, Phelps ME. Imaging gene expression: Principles and assays. *J Nucl Cardiol* 1999; 6: 219–233.

13. Gambhir SS, Herschman HR, Cherry SR, Barrio JR, Satyamurthy N, Toyokuni T, et al. Imaging transgene expression with radionuclide imaging technologies. *Neoplasia* 2000; 2: 118–138.

14. Yaghoubi SS, Jensen MC, Satyamurthy N, Budhiraja S, Paik D, Czernin J, et al. Noninvasive detection of therapeutic cytolytic T cells with $^{18}$F-FHBG in a patient with glioma. *Nat Clin Pract Oncol* 2009; 6: 53–58.

15. Min J, Gambhir SS. Molecular imaging of PET reporter gene expression. In: Semmler W, Schwaiger M (Eds.), *Molecular Imaging II*, vol. 185/2. Springer-Verlag: Berlin, 2008, pp. 277–303.

16. Peñuelas I, Yaghoubi SS, Prósper F, Gambhir SS. Clinical applications of reporter gene technology. In: Yaghoubi SS, Gambhir SS (eds.), *Molecular Imaging with Reporter Genes*. Cambridge University Press, Cambridge, UK, 2010, pp. 299–316.

17. Penuelas I, Haberkorn U, Yaghoubi S, Gambhir S. Gene therapy imaging in patients for oncological applications. *Eur J Nucl Med Mol Imaging* 2005; 32: S384–S403.

18. Ray P, Gambhir SS. Multimodality imaging of reporter genes. In: Yaghoubi SS, Gambhir SS (Eds.), *Molecular Imaging with Reporter Genes*. Cambridge University Press, Cambridge, UK, 2010, pp. 113–126.

19. Yaghoubi SS, Gambhir SS. Gene therapy and imaging of transgene expression in living subjects. In: Yaghoubi SS, Gambhir SS (Eds.), *Molecular Imaging with Reporter Genes*. Cambridge University Press, Cambridge, UK, 2010, pp. 227–238.

20. Serganova I, Blasberg R. Reporter gene imaging: potential impact on therapy. *Nucl Med Biol* 2005; 32: 763–780.

21. Blasberg RG, Gelovani-Tjuvajev J. In vivo molecular-genetic imaging. *J Cell Biochem* 2002; 39: 172–183.

22. Chung J, Kang JH, Kang KW. Reporter gene imaging with PET/SPECT. In: Gambhir SS, Yaghoubi SS (eds.), *Molecular Imaging with Reporter Genes*. Cambridge University Press, Cambridge, UK, 2010, pp. 70–87.

23. Rodriguez-Porcel MG, Gambhir SS. Imaging of reporter genes and stem cells. In: Gambhir SS, Yaghoubi SS (Eds.), *Molecular Imaging with Reporter Genes*. Cambridge University Press, Cambridge, UK, 2010, pp. 277–298.

24. Gambhir SS, Bauer E, Black ME, Liang Q, Kokoris MS, Barrio JR, et al. A mutant herpes simplex virus type 1 thymidine kinase reporter gene shows improved sensitivity for imaging reporter gene expression with positron emission tomography. *Proc Natl Acad Sci USA* 2000; 97: 2785–2790.

25. Yaghoubi SS, Gambhir SS. Measuring herpes simplex virus thymidine kinase reporter gene expression in vitro. *Nat Protoc* 2006; 1: 2137–2142.

26. Yaghoubi SS, Gambhir SS. PET imaging of herpes simplex virus type 1 thymidine kinase (*HSV1-tk*) or mutant *HSV1-sr39tk* reporter gene expression in mice and humans using [$^{18}$F]FHBG. *Nat Protoc* 2007; 1: 3069–3075.

27. Yaghoubi SS, Couto MA, Chen C, Polavaram L, Cui G, Sen L, et al. Preclinical safety evaluation of $^{18}$F-FHBG: A PET reporter probe for imaging herpes simplex virus type 1 thymidine kinase (*HSV1-tk*) or mutant *HSV1-sr39tk*'s expression. *J Nucl Med* 2006; 47: 706–715.

28. Yaghoubi SS, Barrio JR, Dahlbom M, Iyer M, Namavari M, Satyamurthy N, et al. Human pharmacokinetic and dosimetry studies of [$^{18}$F]FHBG: a reporter probe for imaging herpes simplex virus type-1 thymidine kinase reporter gene expression. *J Nucl Med* 2001; 42: 1225–1234.

# 22

# Clinical Cardiology Stem Cell Applications

Anthony J. White, Rachel Ruckdeschel Smith,
Raj Makkar, and Eduardo Marbán

## CONTENTS

## 22.1 Introduction

Until recently, traditional thinking held that no new cardiomyocytes are produced in postnatal life. Because human myocardial infarction heals mostly by collagenous scar formation rather than regeneration of contractile myocardium, there seemed little reason to question this paradigm during the twentieth century. However, the early years of this century have brought definitive demonstrations of dividing cardiomyocytes (or precardiomyocytes) after myocardial damage in animals[1] and humans.[2] In addition, based on [14]C labeling of human DNA, the concept of postnatal endogenous regeneration of the adult heart is now firmly established.[3] Rather than being born with all the cardiomyocytes that will ever exist, the normal adult human heart actually has a small rate of homeostatic cellular turnover, with about 1–2% of cardiomyocytes replaced per annum.[3] Furthermore, populations of endogenous resident cardiac stem cells have been described,[4–9] and in some cases these cells have been shown to have clonal capacity and demonstrated an ability to differentiate into smooth muscle cells, endothelial cells, or cardiomyocytes.[4,8–10]

These recent insights into basic cardiac biology have allowed for the rational consideration of the concept of myocardial regeneration and potential clinical applications. Some suggested cardiac clinical applications of stem cell therapy include regeneration of contractile myocardium, cellular revascularization for cases of ischemic cardiomyopathy in which percutaneous or surgical revascularization are not possible, and cellular therapy for bradyarrhythmias—a so-called biological pacemaker.[11]

As with any nascent field, there are many challenges to overcome and questions to answer, including the following:

- What is the best type of cell to use for any given clinical indication?
- What is the optimal route of administration for any given clinical indication?
- Is cardiac stem cell therapy safe? In particular, what are the risks of unintended proarrhythmia, teratoma, or cancer? With allogeneic cell therapy, will there be immune rejection?
- Do the administered cells engraft and survive within the heart? Can they be imaged or quantified *in vivo*?
- What is the mechanism of action of the cells? Are they fusing with existing cells, differentiating into effector cells themselves, or exerting their biological effect by paracrine or nonspecific physical mechanisms?
- Do we need to administer cells? Can the same effect be achieved by administration of small molecules or proteins to simulate the paracrine effect of cells or to activate the endogenous stem cell population?
- How can imaging modalities best be utilized to monitor safety, efficacy, and mechanism of action?

In this chapter, we seek to review the status of this rapidly evolving field and shed some light on these questions.

## 22.2 Myocardial Regeneration

Multiple candidate cell types have been investigated in *preclinical* models for their ability to regenerate cardiac function, whether directly through differentiation to form new tissue or indirectly via paracrine effects. Cells that have been investigated for such effects in the preclinical arena include embryonic stem cells,[12–14] induced pluripotent stem cells,[15] neonatal cardiomyocytes,[16,17] skeletal muscle myoblasts (SKMMs),[18–20] endothelial progenitor cells,[21] bone marrow mononuclear cells (BMMCs),[1] mesenchymal stem cells (MSCs),[22–27] and most recently, resident cardiac stem cells[4,5,7,28] and cardiosphere-derived cells (CDCs).[8,9]

Clinical testing of various cells types has also begun (Table 22.1)—albeit not always with the benefit of a convincing preclinical development program. Clinical trials examining the safety and efficacy of circulating or bone

**TABLE 22.1**

Clinical Trials of Various Cell Types for Cells for the Purpose of Augmenting Function of Damaged Heart

| Cell Type | Trials |
|---|---|
| Mononuclear cells (circulating or bone marrow derived) | BOOST[31, 32] |
| | ASTAMI[33] |
| | Janssens et al.[36] |
| | REPAIR-AMI[30] |
| | TOPCARE-AMI[29, 35] |
| | TOPCARE-CHD[39] |
| Skeletal myoblasts | MAGIC[41] |
| Cardiosphere-derived cells | CADUCEUS[55] |
| Mesenchymal cells | PROMETHEUS[53] |

marrow-derived mononuclear cells,[29–39] SKMMs,[40–42] and bone marrow-derived mesenchymal cells[43] in patients with acute myocardial infarction (MI) have established a variable track record for cell therapy thus far.

## 22.3 Bone Marrow Mononuclear Cells

Testing of BMMCs in humans[34] began on the basis of an early report showing that implanted BMMCs are capable of *in vivo* cardiogenic differentiation in mice.[1] Although this claim has subsequently been challenged,[44,45] in the meantime clinical trials of BMMCs forged ahead. Delivery of autologous bone marrow or circulating mononuclear cells was associated with favorable left ventricular (LV) remodeling over a 1-year follow-up period in the TOPCARE-AMI clinical trial.[29,35] Subsequent placebo-controlled, prospective, randomized studies of BMMCs have yielded varied results. The REPAIR-AMI trial showed a modest improvement in ejection fraction (EF) at 1 year,[30] while the similarly designed ASTAMI trial showed no functional benefit,[33] a difference that has since been attributed to the quality of the cells as measured by an *in vitro* cell migration assay.[46] Similarly, a randomized controlled Belgian clinical study[36] found no functional benefit from coronary administration of BMMCs. A separate study (BOOST), using routine care controls, examined patients after 18 months and concluded that the functional benefit seen at 6 months[31] was not sustained.[32] The BMMC trials described so far were performed in patients with acute MI. Promising results have also been reported with BMMCs in patients with chronic ischemic cardiomyopathy in a small nonrandomized trial[37,38] and a randomized study.[39] A meta-analysis of the pooled trials of BMMC therapy suggested an approximately 3.7% beneficial effect on absolute EF derived from BMMC administration after MI.[47]

A number of generalizations can be made from the BMMC acute MI studies, which by far are the most numerous to date: (a) the safety profile has been favorable; (b) efficacy, as gauged by an increase in EF, has been inconsistent and, overall, modest; and (c) the patient population was not very ill at baseline, most having suffered their first MI with prompt reperfusion and mean EFs of about 50% pretherapy. The legacy of these studies has left the field with cautious optimism, while motivating a search for better cell types, more effective delivery methods, and more suitable patient populations. In addition, the somewhat mixed results of the clinical trials of cardiac administration of bone marrow-derived cells emphasizes the importance of *in vivo* monitoring of the injected cells for meaningful evaluation of the efficacy and mechanism of any potential cardiac cellular therapy.

## 22.4 Skeletal Muscle Myoblasts

Skeletal muscle myoblasts (SKKMs) are derived from a biopsy of skeletal muscle, typically from the thigh. SKKMs have been investigated as potential replacement cardiac cells in preclinical models for over a decade and found to have the ability to engraft in the damaged heart and to improve pump function.[20] On this basis, several small phase 1 trials were conducted delivering SKMMs by direct intramyocardial injection during coronary artery bypass surgery into patients with ischemic cardiomyopathy.[40,42,48] These studies indicated a long-term functional benefit of SKKMs, albeit with a high incidence of ventricular arrhythmia, a problem that thus far has not been associated with delivery of blood or bone marrow-derived cells.[49] However, the first prospective, randomized, placebo-controlled SKMM trial—the so called MAGIC trial[41]—was discontinued for lack of efficacy. The premature termination of the randomized trial, together with the perception that these cells may produce a proarrhythmic effect, probably means that SKKMs will not see mainstream clinical use in the near future.

## 22.5 Mesenchymal Stem Cells

Mesenchymal stem cells (MSCs) are derived from the bone marrow, where they fulfill a stromal support function. MSCs are thought to be immunoprivileged by virtue of low expression of major histocompatibility complex (MHC) class II molecule and secretion of immunosuppressive cytokines.[50,51] These hypoimmunogenic features, together with preclinical demonstrations of efficacy for the treatment of cardiac dyfunction,[22-27] have led to speculation that it may be possible to use "off-the-shelf" allogenic MSCs to treat individuals with LV dysfunction. A safety study of human intramyocardial injection of MSCs during coronary artery bypass graft surgery is in progress, known as the PROMETHEUS (Prospective Randomized Assessment of Mesenchymal Stem Cell Therapy in Patients Undergoing Surgery) study.[52,53]

## 22.6 Cardiosphere-Derived Cells

Cardiosphere-derived cells (CDCs) are derived from a cardiac biopsy specimen and amplified *ex vivo*. Following expansion, the cells can be delivered back into the heart to areas of injury, where there is extensive evidence to indicate that they engraft and augment functional recovery of the heart.[8,54] These cells represent a logical candidate for cardiac cell therapy because they are resident in the heart, can be expanded by routine cell culture methods, and are preprogrammed with the capability to reconstitute all cardiac lineages.[8,10] Based on these preclinical data, we have begun the first clinical study of autologous CDCs, known as CADUCEUS (CArdiosphere-Derived aUtologous Stem CElls to Reverse ventricUlar dySfunction).[55] The CADUCEUS study assesses intracoronary delivery of CDCs in patients with ischemic cardiac damage and has safety as the primary endpoint.

## 22.7 Imaging of Injected Cells

In animal models, a number of techniques are available to track the location, longevity, and viability of injected cells[56] (Table 22.2). As is discussed elsewhere in this book, these techniques are often unsuitable for use in human studies, at least at their current levels of development.

It is possible to load cells with superparamagnetic iron oxide (SPIO) or ferumoxides, which can then be detected *in vivo* with magnetic resonance imaging (MRI). While this concept has been demonstrated in animals,[57–59] a major concern with this technique is the deposition of the iron label in tissues if the injected cells die,[60,61] such that a positive signal on MRI may represent the label scavenged by macrophages rather than viable injected cells (false-positive signal). Nevertheless, the method may be of short-term utility in the assessment of acute cell retention and biodistribution.

Genetic labeling of cells with a bioluminescent label can allow for direct detection in small animals.[62] However, this is less likely to be a successful technique in human imaging due to attenuation of the signal by the longer distances of tissue penetration required in humans compared to rodents.

**TABLE 22.2**

Techniques for Imaging of Injected Cells in Preclinical Models

| Technique/Description and Example References | Allows Localization of Cells? | Provides Information About Viability? |
|---|---|---|
| Genetic labeling of cells with luminescent proteins for detection by optical imaging; "bioluminescence"[62] | Yes | Yes |
| Genetic labeling of cells with fluorescent proteins[63, 64] | Yes | Yes |
| Derivation of cells from transgenic animals that constitutively express luciferase or fluorescent proteins[73] | Yes | Yes |
| Superparamagnetic iron oxide (SPIO) detected with MRI[57–59] | Yes | No—iron signal remains even if the cells die. |
| Passive nuclear scintigraphic imaging or PET with a preloaded radiolabel[68] | Yes | No—isotope decays before long term viability can be assessed |
| Genetic labeling of cells to take up or accumulate a radiolabel for detection by SPECT or PET[65–67] | Yes | Yes |
| Labeling of cells by fluorescent semiconductor nanoparticles[74,75] | Yes | No |

Similarly, injected cells can be labeled with genes encoding fluorescent proteins to enable later detection.[63,64]

In a related approach, cells can be genetically induced to take up or activate a tracer for subsequent detection by MRI, positron emission tomography (PET), or single-photon emission computed tomography (SPECT). There are several variations on this theme. In one example, transfection of cells with the sodium iodide symporter induces the cells to take up the tracer technicium-99m, which can later be detected by SPECT, or another tracer (iodine-124) for detection by PET.[65,66] In another example, cells may be induced to express the herpes simplex enzyme thymidine kinase (TK). Presence of TK causes cellular accumulation of the tracer $^{18}$F-fluoro-3-hydroxymethylbutyl guanine, allowing for detection of the labeled cells by PET.[67] A clinical limitation of lentiviral or retroviral labeling of cells with all of these reporter genes in humans involves potential safety concerns over the somatic gene transfer vector used to introduce the symporter, enzyme, or luminescent/fluorescent label.

Finally, cells can be radiolabeled to allow for *in vivo* detection. Cells labeled with indium-111[68] or technicium-99m[69] may be detected by SPECT, or cells may be preloaded with $^{18}$F-fluorodeoxyglucose (FDG) and detected by PET.[70] Of the various techniques outlined here and in Table 22.2, only such cells preloaded with radiolabels have been used in humans, as discussed in greater detail next.

## 22.8 Clinical Imaging of Injected Cells

Imaging has a dual role in the evaluation of potential cardiac cell therapies. First, imaging is critical in the measurement of cardiac function and perfusion, which assess the efficacy of any cell-based treatment under evaluation. These methodologies are not discussed in detail here as these are techniques in routine clinical use.

Second, imaging has a potential role in the assessment of engraftment, localization, viability, and possibly the differentiation of injected cells in future clinical trials of cell therapy.

In contrast to the large number of techniques that have been tried for tracking of cell viability and engraftment in animal models that have been outlined, in humans only radiolabeling of cells has been successfully used. Both of the published examples of radiolabeling of injected cells were from Germany. In 2005, the Drexler lab performed radiolabeling of autologous human bone marrow cells with $^{18}$F-FDG by injecting the labeled cells via intracoronary infusion and imaging their subsequent fate by three-dimensional PET.[71] They demonstrated that cell uptake in the myocardium corresponded to the distribution territory of the injected coronary artery (Figure 22.1). In 2008, Schächinger et al. from Zeiher's lab similarly radiolabeled bone marrow mononuclear cells, this time with $^{111}$In oxine, to localize and quantify their engraftment.[72] In Figure 22.2, the images from that study are shown.[72]

These proof-of-concept studies established that radiolabeling of cells by preloading is feasible to image cells injected into myocardium. However, the utility of the method is limited by the lifetime of the radiolabel. It must be stated that clinical imaging of injected cells is in its infancy. Future advances in this field of imaging would be a great step toward facilitating more accurate clinical assessment of emerging novel cell therapies. Once perfected, successful clinical cell imaging has the exciting potential to provide, for the first time, vital information about cell engraftment, viability, longevity following injection, and mechanism of action.

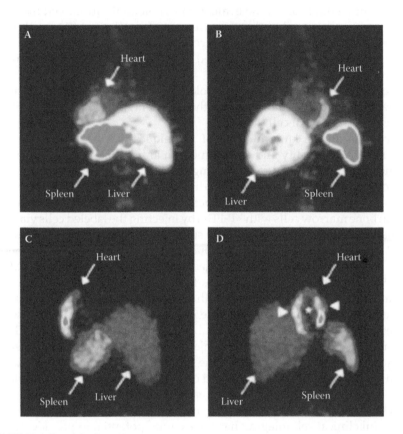

**FIGURE 22.1 (See color insert.)**
Clinical imaging of injected bone marrow cells (BMCs) by radiolabeling with
[18]F-fluorodeoxyglucose (FDG) and subsequent imaging by positron emission tomography
(PET). Left posterior oblique (A) and left anterior oblique (B) views are shown of the chest and
upper abdomen of patient 2 taken 65 minutes after transfer of [18]F-FDG-labeled, unselected
BMCs into left circumflex. BMC homing is detectable in the lateral wall of the heart (infarct
center and border zone), liver, and spleen. Left posterior oblique (C) and left anterior oblique
(D) views of chest and upper abdomen of patient 7 taken 70 minutes after transfer of [18]F-FDG-
labeled, CD34-enriched BMCs into left anterior descending coronary artery. Homing of CD34-
enriched cells is detectable in the anteroseptal wall of the heart, liver, and spleen. CD34 cell
homing is most prominent in infarct border zone (arrowheads) but not infarct center (asterisk).
(From Hofmann M, Wollert KC, Meyer GP, Menke A, Arseniev L, Hertenstein B, Ganser A,
Knapp WH, and Drexler H. *Circulation* 2005;111(17): 2198–2202. With permission.)

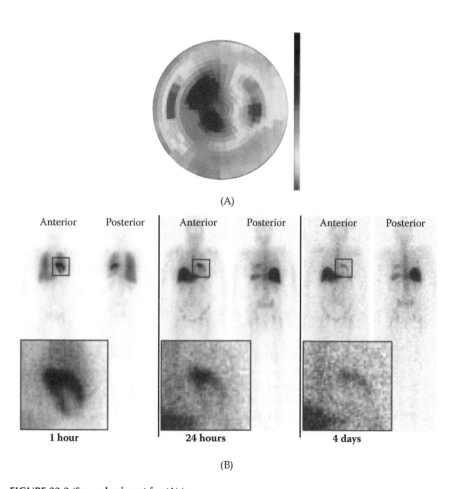

(A)

Anterior   Posterior   Anterior   Posterior   Anterior   Posterior

1 hour          24 hours          4 days

(B)

**FIGURE 22.2 (See color insert for (A).)**
Clinical imaging of injected circulating progenitor cells (CPCs) radiolabeled with indium[111]. (A) FDG-PET imaging: Dark scale indicates low viability. (B) Imaging of [111]In distribution with gamma camera at 1 hour, 24 hours, and 4 days after infusion of [111]In-oxine-labeled CPCs. Anterior and posterior whole-body scans were acquired. The inserts show the heart at a higher magnification. Cardiac [111]In activity was highest immediately after infusion and progressively diminished in the lung and heart during the next days. (From Schachinger V, Aicher A, Dobert N, Rover R, Diener J, Fichtlscherer S, Assmus B, Seeger FH, Menzel C, Brenner W, Dimmeler S, and Zeiher AM. *Circulation* 2008;118(14): 1425–32. With permission.)

## References

1. Orlic D, Kajstura J, Chimenti S, Jakoniuk I, Anderson SM, Li B, Pickel J, McKay R, Nadal-Ginard B, Bodine DM, Leri A, and Anversa P. Bone marrow cells regenerate infarcted myocardium. *Nature*, 2001; 410(6829): 701–5.
2. Beltrami AP, Urbanek K, Kajstura J, Yan SM, Finato N, Bussani R, Nadal-Ginard B, Silvestri F, Leri A, Beltrami CA, and Anversa P. Evidence that human cardiac myocytes divide after myocardial infarction. *N Engl J Med*, 2001; 344(23): 1750–57.
3. Bergmann O, Bhardwaj RD, Bernard S, Zdunek S, Barnabe-Heider F, Walsh S, Zupicich J, Alkass K, Buchholz BA, Druid H, Jovinge S, and Frisen J. Evidence for cardiomyocyte renewal in humans. *Science*, 2009; 324(5923): 98–102.
4. Beltrami AP. Adult cardiac stem cells are multipotent and support myocardial regeneration. *Cell*, 2003; 114(6): 763–76.
5. Oh H, Bradfute SB, Gallardo TD, Nakamura T, Gaussin V, Mishina Y, Pocius J, Michael LH, Behringer RR, Garry DJ, Entman ML, and Schneider MD. Cardiac progenitor cells from adult myocardium: Homing, differentiation, and fusion after infarction. *Proc Natl Acad Sci USA*, 2003; 100(21): 12313–18.
6. Pfister O, Mouquet F, Jain M, Summer R, Helmes M, Fine A, Colucci WS, and Liao R. CD31⁻ but not CD31⁺ cardiac side population cells exhibit functional cardiomyogenic differentiation. *Circ Res*, 2005; 97(1): 52–61.
7. Ott HC, Matthiesen TS, Brechtken J, Grindle S, Goh SK, Nelson W, and Taylor DA. The adult human heart as a source for stem cells: Repair strategies with embryonic-like progenitor cells. *Nat Clin Pract Cardiovasc Med*, 2007; 4 Suppl 1: S27–S39.
8. Smith RR, Barile L, Cho HC, Leppo MK, Hare JM, Messina E, Giacomello A, Abraham MR, and Marban E. Regenerative potential of cardiosphere-derived cells expanded from percutaneous endomyocardial biopsy specimens. *Circulation*, 2007; 115(7): 896–908.
9. Messina E, De Angelis L, Frati G, Morrone S, Chimenti S, Fiordaliso F, Salio M, Battaglia M, Latronico MV, Coletta M, Vivarelli E, Frati L, Cossu G, and Giacomello A. Isolation and expansion of adult cardiac stem cells from human and murine heart. *Circ Res*, 2004; 95(9): 911–21.
10. Davis DR, Zhang Y, Smith RR, Cheng K, Terrovitis J, Malliaras K, Li TS, White A, Makkar R, and Marbán E. Validation of the cardiosphere method to culture cardiac progenitor cells from myocardial tissue. *PLoS one*, 2009; 4(9): e7195.
11. Miake J, Marban E, and Nuss HB. Biological pacemaker created by gene transfer. *Nature*, 2002; 419(6903): 132–33.
12. Laflamme MA, Chen KY, Naumova AV, Muskheli V, Fugate JA, Dupras SK, Reinecke H, Xu C, Hassanipour M, Police S, O'Sullivan C, Collins L, Chen Y, Minami E, Gill EA, Ueno S, Yuan C, Gold J, and Murry CE. Cardiomyocytes derived from human embryonic stem cells in pro-survival factors enhance function of infarcted rat hearts. *Nat Biotechnol*, 2007; 25(9): 1015–24.
13. Caspi O, Huber I, Kehat I, Habib M, Arbel G, Gepstein A, Yankelson L, Aronson D, Beyar R, and Gepstein L. Transplantation of human embryonic stem cell-derived cardiomyocytes improves myocardial performance in infarcted rat hearts. *J Am Coll Cardiol*, 2007; 50(19): 1884–93.

14. Dai W, Field LJ, Rubart M, Reuter S, Hale SL, Zweigerdt R, Graichen RE, Kay GL, Jyrala AJ, Colman A, Davidson BP, Pera M, and Kloner RA. Survival and maturation of human embryonic stem cell-derived cardiomyocytes in rat hearts. *J Mol Cell Cardiol*, 2007; 43(4): 504–16.
15. Nelson TJ, Martinez-Fernandez A, Yamada S, Perez-Terzic C, Ikeda Y, and Terzic A. Repair of acute myocardial infarction by human stemness factors induced pluripotent stem cells. *Circulation*, 2009; 120(5): 408–16.
16. Muller-Ehmsen J, Peterson KL, Kedes L, Whittaker P, Dow JS, Long TI, Laird PW, and Kloner RA. Rebuilding a damaged heart: Long-term survival of transplanted neonatal rat cardiomyocytes after myocardial infarction and effect on cardiac function. *Circulation*, 2002; 105(14): 1720–26.
17. Reffelmann T, Dow JS, Dai W, Hale SL, Simkhovich BZ, and Kloner RA. Transplantation of neonatal cardiomyocytes after permanent coronary artery occlusion increases regional blood flow of infarcted myocardium. *J Mol Cell Cardiol*, 2003; 35(6): 607–13.
18. Jain M, DerSimonian H, Brenner DA, Ngoy S, Teller P, Edge AS, Zawadzka A, Wetzel K, Sawyer DB, Colucci WS, Apstein CS, and Liao R. Cell therapy attenuates deleterious ventricular remodeling and improves cardiac performance after myocardial infarction. *Circulation*, 2001; 103(14): 1920–27.
19. He KL, Yi GH, Sherman W, Zhou H, Zhang GP, Gu A, Kao R, Haimes HB, Harvey J, Roos E, White D, Taylor DA, Wang J, and Burkhoff D. Autologous skeletal myoblast transplantation improved hemodynamics and left ventricular function in chronic heart failure dogs. *J Heart Lung Transplant*, 2005; 24(11): 1940–49.
20. Murry CE, Wiseman RW, Schwartz SM, and Hauschka SD. Skeletal myoblast transplantation for repair of myocardial necrosis. *J Clin Invest*, 1996; 98(11): 2512–23.
21. Kawamoto A, Gwon HC, Iwaguro H, Yamaguchi JI, Uchida S, Masuda H, Silver M, Ma H, Kearney M, Isner JM, and Asahara T. Therapeutic potential of ex vivo expanded endothelial progenitor cells for myocardial ischemia. *Circulation*, 2001; 103(5): 634–37.
22. Amado LC, Saliaris AP, Schuleri KH, St John M, Xie JS, Cattaneo S, Durand DJ, Fitton T, Kuang JQ, Stewart G, Lehrke S, Baumgartner WW, Martin BJ, Heldman AW, and Hare JM. Cardiac repair with intramyocardial injection of allogeneic mesenchymal stem cells after myocardial infarction. *Proc Natl Acad Sci USA*, 2005; 102(32): 11474–79.
23. Nagaya N, Fujii T, Iwase T, Ohgushi H, Itoh T, Uematsu M, Yamagishi M, Mori H, Kangawa K, and Kitamura S. Intravenous administration of mesenchymal stem cells improves cardiac function in rats with acute myocardial infarction through angiogenesis and myogenesis. *Am J Physiol Heart Circ Physiol*, 2004; 287(6): H2670–76.
24. Dai W, Hale SL, Martin BJ, Kuang JQ, Dow JS, Wold LE, and Kloner RA. Allogeneic mesenchymal stem cell transplantation in postinfarcted rat myocardium: Short- and long-term effects. *Circulation*, 2005; 112(2): 214–23.
25. Schuleri KH, Feigenbaum GS, Centola M, Weiss ES, Zimmet JM, Turney J, Kellner J, Zviman MM, Hatzistergos KE, Detrick B, Conte JV, McNiece I, Steenbergen C, Lardo AC, and Hare JM. Autologous mesenchymal stem cells produce reverse remodelling in chronic ischaemic cardiomyopathy. *Eur Heart J*, 2009.

26. Quevedo HC, Hatzistergos KE, Oskouei BN, Feigenbaum GS, Rodriguez JE, Valdes D, Pattany PM, Zambrano JP, Hu Q, McNiece I, Heldman AW, and Hare JM. Allogeneic mesenchymal stem cells restore cardiac function in chronic ischemic cardiomyopathy via trilineage differentiating capacity. *Proc Natl Acad Sci USA*, 2009; 106(33): 14022–27.

27. Makkar RR, Price MJ, Lill M, Frantzen M, Takizawa K, Kleisli T, Zheng J, Kar S, McClelan R, Miyamota T, Bick-Forrester J, Fishbein MC, Shah PK, Forrester JS, Sharifi B, Chen PS, and Qayyum M. Intramyocardial injection of allogenic bone marrow-derived mesenchymal stem cells without immunosuppression preserves cardiac function in a porcine model of myocardial infarction. *J Cardiovasc Pharmacol Ther*, 2005; 10(4): 225–33.

28. Martin CM, Meeson AP, Robertson SM, Hawke TJ, Richardson JA, Bates S, Goetsch SC, Gallardo TD, and Garry DJ. Persistent expression of the ATP-binding cassette transporter, Abcg2, identifies cardiac SP cells in the developing and adult heart. *Dev Biol*, 2004; 265(1): 262–75.

29. Schachinger V, Assmus B, Britten MB, Honold J, Lehmann R, Teupe C, Abolmaali ND, Vogl TJ, Hofmann WK, Martin H, Dimmeler S, and Zeiher AM. Transplantation of progenitor cells and regeneration enhancement in acute myocardial infarction: final one-year results of the TOPCARE-AMI Trial. *J Am Coll Cardiol*, 2004; 44(8): 1690–99.

30. Schachinger V, Erbs S, Elsasser A, Haberbosch W, Hambrecht R, Holschermann H, Yu J, Corti R, Mathey DG, Hamm CW, Suselbeck T, Werner N, Haase J, Neuzner J, Germing A, Mark B, Assmus B, Tonn T, Dimmeler S, and Zeiher AM. Improved clinical outcome after intracoronary administration of bone-marrow-derived progenitor cells in acute myocardial infarction: Final 1-year results of the REPAIR-AMI trial. *Eur Heart J*, 2006; 27(23): 2775–83.

31. Wollert KC, Meyer GP, Lotz J, Ringes-Lichtenberg S, Lippolt P, Breidenbach C, Fichtner S, Korte T, Hornig B, Messinger D, Arseniev L, Hertenstein B, Ganser A, and Drexler H. Intracoronary autologous bone-marrow cell transfer after myocardial infarction: The BOOST randomised controlled clinical trial. *Lancet*, 2004; 364(9429): 141–48.

32. Meyer GP, Wollert KC, Lotz J, Steffens J, Lippolt P, Fichtner S, Hecker H, Schaefer A, Arseniev L, Hertenstein B, Ganser A, and Drexler H. Intracoronary bone marrow cell transfer after myocardial infarction: Eighteen months' follow-up data from the randomized, controlled BOOST (BOne marrOw transfer to enhance ST-elevation infarct regeneration) trial. *Circulation,* 2006; 113(10): 1287–94.

33. Lunde K, Solheim S, Aakhus S, Arnesen H, Abdelnoor M, Egeland T, Endresen K, Ilebekk A, Mangschau A, Fjeld JG, Smith HJ, Taraldsrud E, Grogaard HK, Bjornerheim R, Brekke M, Muller C, Hopp E, Ragnarsson A, Brinchmann JE, and Forfang K. Intracoronary injection of mononuclear bone marrow cells in acute myocardial infarction. *N Engl J Med*, 2006; 355(12): 1199–1209.

34. Strauer BE, Brehm M, Zeus T, Kostering M, Hernandez A, Sorg RV, Kogler G, and Wernet P. Repair of infarcted myocardium by autologous intracoronary mononuclear bone marrow cell transplantation in humans. *Circulation*, 2002; 106(15): 1913–18.

35. Assmus B, Schachinger V, Teupe C, Britten M, Lehmann R, Dobert N, Grunwald F, Aicher A, Urbich C, Martin H, Hoelzer D, Dimmeler S, and Zeiher AM. Transplantation of Progenitor Cells And Regeneration Enhancement in Acute Myocardial Infarction (TOPCARE-AMI): *Circulation*, 2002; 106(24): 3009–17.

36. Janssens S, Dubois C, Bogaert J, Theunissen K, Deroose C, Desmet W, Kalantzi M, Herbots L, Sinnaeve P, Dens J, Maertens J, Rademakers F, Dymarkowski S, Gheysens O, Van Cleemput J, Bormans G, Nuyts J, Belmans A, Mortelmans L, Boogaerts M, and Van de Werf F. Autologous bone marrow-derived stem-cell transfer in patients with ST-segment elevation myocardial infarction: Double-blind, randomised controlled trial. *Lancet*, 2006; 367(9505): 113–21.

37. Perin EC, Dohmann HF, Borojevic R, Silva SA, Sousa AL, Mesquita CT, Rossi MI, Carvalho AC, Dutra HS, Dohmann HJ, Silva GV, Belem L, Vivacqua R, Rangel FO, Esporcatte R, Geng YJ, Vaughn WK, Assad JA, Mesquita ET, and Willerson JT. Transendocardial, autologous bone marrow cell transplantation for severe, chronic ischemic heart failure. *Circulation,* 2003; 107(18): 2294–2302.

38. Perin EC, Dohmann HF, Borojevic R, Silva SA, Sousa AL, Silva GV, Mesquita CT, Belem L, Vaughn WK, Rangel FO, Assad JA, Carvalho AC, Branco RV, Rossi MI, Dohmann HJ, and Willerson JT. Improved exercise capacity and ischemia 6 and 12 months after transendocardial injection of autologous bone marrow mononuclear cells for ischemic cardiomyopathy. *Circulation*, 2004; 110(11 Suppl 1): II213–18.

39. Assmus B, Honold J, Schachinger V, Britten MB, Fischer-Rasokat U, Lehmann R, Teupe C, Pistorius K, Martin H, Abolmaali ND, Tonn T, Dimmeler S, and Zeiher AM. Transcoronary transplantation of progenitor cells after myocardial infarction. *N Engl J Med*, 2006; 355(12): 1222–32.

40. Menasche P, Hagege AA, Vilquin JT, Desnos M, Abergel E, Pouzet B, Bel A, Sarateanu S, Scorsin M, Schwartz K, Bruneval P, Benbunan M, Marolleau JP, and Duboc D. Autologous skeletal myoblast transplantation for severe postinfarction left ventricular dysfunction. *J Am Coll Cardiol*, 2003; 41(7): 1078–83.

41. Menasche P, Alfieri O, Janssens S, McKenna W, Reichenspurner H, Trinquart L, Vilquin JT, Marolleau JP, Seymour B, Larghero J, Lake S, Chatellier G, Solomon S, Desnos M, and Hagege AA. The Myoblast Autologous Grafting in Ischemic Cardiomyopathy (MAGIC) trial: First randomized placebo-controlled study of myoblast transplantation. *Circulation*, 2008; 117(9): 1189–1200.

42. Hagege AA, Marolleau JP, Vilquin JT, Alheritiere A, Peyrard S, Duboc D, Abergel E, Messas E, Mousseaux E, Schwartz K, Desnos M, and Menasche P. Skeletal myoblast transplantation in ischemic heart failure: Long-term follow-up of the first phase I cohort of patients. *Circulation,* 2006; 114(1 Suppl): I108–13.

43. Chen SL, Fang WW, Ye F, Liu YH, Qian J, Shan SJ, Zhang JJ, Chunhua RZ, Liao LM, Lin S, and Sun JP. Effect on left ventricular function of intracoronary transplantation of autologous bone marrow mesenchymal stem cell in patients with acute myocardial infarction. *Am J Cardiol*, 2004; 94(1): 92–95.

44. Murry CE, Soonpaa MH, Reinecke H, Nakajima H, Nakajima HO, Rubart M, Pasumarthi KB, Virag JI, Bartelmez SH, Poppa V, Bradford G, Dowell JD, Williams DA, and Field LJ. Haematopoietic stem cells do not transdifferentiate into cardiac myocytes in myocardial infarcts. *Nature*, 2004; 428(6983): 664–68.

45. Balsam LB, Wagers AJ, Christensen JL, Kofidis T, Weissman IL, and Robbins RC. Haematopoietic stem cells adopt mature haematopoietic fates in ischaemic myocardium. *Nature*, 2004; 428(6983): 668–73.

46. Seeger FH, Tonn T, Krzossok N, Zeiher AM, and Dimmeler S. Cell isolation procedures matter: A comparison of different isolation protocols of bone marrow mononuclear cells used for cell therapy in patients with acute myocardial infarction. *Eur Heart J*, 2007; 28(6): 766–72.

47. Abdel-Latif A, Bolli R, Tleyjeh IM, Montori VM, Perin EC, Hornung CA, Zuba-Surma EK, Al-Mallah M, and Dawn B. Adult bone marrow-derived cells for cardiac repair: A systematic review and meta-analysis. *Arch Intern Med*, 2007; 167(10): 989–97.

48. Siminiak T, Kalawski R, Fiszer D, Jerzykowska O, Rzezniczak J, Rozwadowska N, and Kurpisz M. Autologous skeletal myoblast transplantation for the treatment of postinfarction myocardial injury: Phase I clinical study with 12 months of follow-up. *Am Heart J*, 2004; 148(3): 531–37.

49. Katritsis DG, Sotiropoulou P, Giazitzoglou E, Karvouni E, and Papamichail M. Electrophysiological effects of intracoronary transplantation of autologous mesenchymal and endothelial progenitor cells. *Europace*, 2007; 9(3): 167–71.

50. Beyth S, Borovsky Z, Mevorach D, Liebergall M, Gazit Z, Aslan H, Galun E, and Rachmilewitz J. Human mesenchymal stem cells alter antigen-presenting cell maturation and induce T-cell unresponsiveness. *Blood*, 2005; 105(5): 2214–19.

51. Aggarwal S and Pittenger MF. Human mesenchymal stem cells modulate allogeneic immune cell responses. *Blood*, 2005; 105(4): 1815–22.

52. Hare JM. Translational development of mesenchymal stem cell therapy for cardiovascular diseases. *Tex Heart Inst J*, 2009; 36(2): 145–47.

53. ClinicalTrials.gov. Identifier NCT00587990. PROMETHEUS—Prospective Randomized assessment Of MEsenchymal stem cell THErapy in patients Undergoing Surgery. URL http://www.clinicaltrials.gov/ct2/results?term=NCT00587990. Date accessed: 4/14/2011.

54. Johnston PV, Sasano T, Mills K, Evers R, Lee ST, Smith RR, Lardo AC, Lai S, Steenbergen C, Gerstenblith G, Lange R, and Marban E. Engraftment, differentiation, and functional benefits of autologous cardiosphere-derived cells in porcine ischemic cardiomyopathy. *Circulation*, 2009; 120(12): 1075–83, 7 p. following 1083.

55. ClinicalTrials.gov. Identifier NCT00893360. CADUCEUS—CArdiosphere-Derived aUtologous Stem CElls to Reverse ventricUlar dySfunction. URL http://www.clinicaltrials.gov/ct2/show/NCT00893360?term=NCT00893360 &rank=1. Date accessed: 4/14/2011.

56. Zhang SJ and Wu JC. Comparison of imaging techniques for tracking cardiac stem cell therapy. *J Nucl Med*, 2007; 48(12): 1916–19.

57. Kraitchman DL, Heldman AW, Atalar E, Amado LC, Martin BJ, Pittenger MF, Hare JM, and Bulte JW. In vivo magnetic resonance imaging of mesenchymal stem cells in myocardial infarction. *Circulation*, 2003; 107(18): 2290–93.

58. Hill JM, Dick AJ, Raman VK, Thompson RB, Yu ZX, Hinds KA, Pessanha BS, Guttman MA, Varney TR, Martin BJ, Dunbar CE, McVeigh ER, and Lederman RJ. Serial cardiac magnetic resonance imaging of injected mesenchymal stem cells. *Circulation*, 2003; 108(8): 1009–14.

59. Rogers WJ, Meyer CH, and Kramer CM. Technology insight: In vivo cell tracking by use of MRI. *Nat Clin Pract Cardiovasc Med*, 2006; 3: 554–62.

60. Terrovitis J, Stuber M, Youssef A, Preece S, Leppo M, Kizana E, Schar M, Gerstenblith G, Weiss RG, Marban E, and Abraham MR. Magnetic resonance imaging overestimates ferumoxide-labeled stem cell survival after transplantation in the heart. *Circulation*, 2008; 117(12): 1555–62.

61. Amsalem Y, Mardor Y, Feinberg MS, Landa N, Miller L, Daniels D, Ocherashvilli A, Holbova R, Yosef O, Barbash IM, and Leor J. Iron-oxide labeling and outcome of transplanted mesenchymal stem cells in the infarcted myocardium. *Circulation*, 2007; 116(11 Suppl): I38–I45.

62. Cao F, Lin S, Xie X, Ray P, Patel M, Zhang X, Drukker M, Dylla SJ, Connolly AJ, Chen X, Weissman IL, Gambhir SS, and Wu JC. In vivo visualization of embryonic stem cell survival, proliferation, and migration after cardiac delivery. *Circulation*, 2006; 113(7): 1005–14.

63. Shi PA, Hematti P, von Kalle C, and Dunbar CE. Genetic marking as an approach to studying in vivo hematopoiesis: Progress in the non-human primate model. *Oncogene*, 2002; 21(21): 3274–83.

64. Huhn RD, Tisdale JF, Agricola B, Metzger ME, Donahue RE, and Dunbar CE. Retroviral marking and transplantation of rhesus hematopoietic cells by nonmyeloablative conditioning. *Hum Gene Ther*, 1999; 10(11): 1783–90.

65. Miyagawa M, Beyer M, Wagner B, Anton M, Spitzweg C, Gansbacher B, Schwaiger M, and Bengel FM. Cardiac reporter gene imaging using the human sodium/iodide symporter gene. *Cardiovasc Res*, 2005; 65(1): 195–202.

66. Terrovitis J, Kwok KF, Lautamaki R, Engles JM, Barth AS, Kizana E, Miake J, Leppo MK, Fox J, Seidel J, Pomper M, Wahl RL, Tsui B, Bengel F, Marban E, and Abraham MR. Ectopic expression of the sodium-iodide symporter enables imaging of transplanted cardiac stem cells in vivo by single-photon emission computed tomography or positron emission tomography. *J Am Coll Cardiol*, 2008; 52(20): 1652–60.

67. Wu JC, Inubushi M, Sundaresan G, Schelbert HR, and Gambhir SS. Positron emission tomography imaging of cardiac reporter gene expression in living rats. *Circulation*, 2002; 106(2): 180–83.

68. Aicher A, Brenner W, Zuhayra M, Badorff C, Massoudi S, Assmus B, Eckey T, Henze E, Zeiher AM, and Dimmeler S. Assessment of the tissue distribution of transplanted human endothelial progenitor cells by radioactive labeling. *Circulation*, 2003; 107(16): 2134–39.

69. Barbash IM, Chouraqui P, Baron J, Feinberg MS, Etzion S, Tessone A, Miller L, Guetta E, Zipori D, Kedes LH, Kloner RA, and Leor J. Systemic delivery of bone marrow-derived mesenchymal stem cells to the infarcted myocardium: Feasibility, cell migration, and body distribution. *Circulation*, 2003; 108(7): 863–68.

70. Doyle B, Kemp BJ, Chareonthaitawee P, Reed C, Schmeckpeper J, Sorajja P, Russell S, Araoz P, Riederer SJ, and Caplice NM. Dynamic tracking during intracoronary injection of 18F-FDG-labeled progenitor cell therapy for acute myocardial infarction. *J Nucl Med*, 2007; 48(10): 1708–14.

71. Hofmann M, Wollert KC, Meyer GP, Menke A, Arseniev L, Hertenstein B, Ganser A, Knapp WH, and Drexler H. Monitoring of bone marrow cell homing into the infarcted human myocardium. *Circulation*, 2005; 111(17): 2198–2202.

72. Schachinger V, Aicher A, Dobert N, Rover R, Diener J, Fichtlscherer S, Assmus B, Seeger FH, Menzel C, Brenner W, Dimmeler S, and Zeiher AM. Pilot trial on determinants of progenitor cell recruitment to the infarcted human myocardium. *Circulation*, 2008; 118(14): 1425–32.

73. Li Z, Lee A, Huang M, Chun H, Chung J, Chu P, Hoyt G, Yang P, Rosenberg J, Robbins RC, and Wu JC. Imaging survival and function of transplanted cardiac resident stem cells. *J Am Coll Cardiol*, 2009; 53(14): 1229–40.
74. Dubertret B, Skourides P, Norris DJ, Noireaux V, Brivanlou AH, and Libchaber A. In vivo imaging of quantum dots encapsulated in phospholipid micelles. *Science*, 2002; 298(5599): 1759–62.
75. Jaiswal JK, Mattoussi H, Mauro JM, and Simon SM. Long-term multiple color imaging of live cells using quantum dot bioconjugates. *Nat Biotechnol*, 2003; 21(1): 47–51.

# 23

# Musculoskeletal Clinical Applications of Stem Cells

Antônio J. Machado S., John A. Carrino, and Lew C. Schon

## CONTENTS

## 23.1 Introduction

Current techniques for the treatment of musculoskeletal disorders can be limited by the availability, quality, and quantity of materials, such as grafts to perform surgical repair. This has led to the exploration and development of novel methods of intervention based on tissue engineering and regenerative medicine. Stem cells, due to their ability to differentiate toward various lineages, may serve as an important tool in regenerative medicine and tissue engineering, supported by experimental and clinical data. This chapter reviews the applications for stem cells in the musculoskeletal system (bones, joints, and soft tissues).

## 23.2 Mesenchymal Stem Cells

Treatments using stem cells in musculoskeletal medicine most often employ the mesenchymal lineage. Mesenchymal stem cells (MSCs) are multipotential and fibroblast-like cells of embryonic origin, or from an adult individual, capable of multiplying and differentiating into tissues such as muscle, bone, collagen, and cartilage. The International Society for Cytotherapy put forward qualifying criteria for MSCs (Abdallah et al. 2005; Abdallah and Kassem 2008; Dominici et al. 2006; Foster et al. 2005), which include possessing tripotential mesodermal differentiation capability into osteoblasts, chondrocytes, and adipocytes (Chen and Tuan 2008; Delorme et al. 2006). The origins of MSCs include muscle-derived satellite cells (cells that reside beneath the basal lamina of skeletal muscle fibers), bone marrow-derived MSCs, muscle or bone marrow side population cells, circulating CD133[+] cells, and cells derived from blood vessel walls, such as mesoangioblasts or pericytes, adipose tissue, liver, and lung (Bhagavati 2008; Delorme et al. 2006; In 't Anker et al. 2003; Jackson et al. 2007). Some studies have considered that cell sex may have an important role in determining the *in vivo* outcome of stem cell therapies (Ishida and Heersche 1997; Schwartz et al. 1994).

## 23.3 Bone

The cells important in bone biology are osteoblasts, osteocytes, and osteoclasts. Osteoblasts are the primary cells responsible for bone formation (osteogenesis) and mineralization, while osteoclasts are primarily responsible for bone resorption. Osteoblasts and osteocytes are derived from MSCs, while osteoclasts are derived from hematopoietic stem cells and are related to monocytes/macrophages (Safadi et al. 2009).

### 23.3.1 Fracture Nonunion

Despite advances in orthopedic surgery, a subset of fractures occurs with a deficiency in bone repair and culminates in nonunion, which is defined as failure to heal in 6 to 8 months (Hernigou et al. 2005a; Undale et al. 2009). Bone has an uncommon capacity to regenerate without the development of a fibrous scar and hence heal without compromising its mechanical properties (Kanczler and Oreffo 2008). During normal fracture healing, undifferentiated MSCs recruited by the inflammatory response and regulatory cytokines, with the aid of bone morphogenetic proteins (BMPs), proliferate and differentiate into chondrocytes and osteoblasts to form bone, thereby repairing the injury (Tseng et al. 2008).

The causes of bone fracture nonunion may be the type of fracture, host factors (i.e., diabetes, smoking, etc.), or the surgical procedure. The incidence of nonunions varies by fracture site and can be as high as 5% to 20% (Undale et al. 2009). Nonunions have traditionally been classified as hypertrophic, oligotrophic, and atrophic, and this is related to the degree of biologic viability and metabolic activity of the bone ends. These differences drive the surgical and clinical treatment decisions (Tseng et al. 2008).

MSCs have been utilized by several investigators to reinforce the osteogenic properties of allografts, xenografts, and composite grafts by mixing bone marrow removed during the operation with bone grafts based on scaffolds to improve clinical outcome (Hernigou et al. 2005b). The chemical composition of the scaffold is crucial for the osteoconductive and resorptive properties of the material. Synthetic polymer matrices are contamination free and versatile as three-dimensional (3D) porous moldable structures to fit the anatomical defects but lack osteoconductivity. Porous bioceramic scaffold polymers (i.e., coral scaffolds) present internal structures permissive for vascular invasion, possess adequate mechanical properties, and neither induce an immune/inflammatory reaction nor present a cytotoxicity response. The enhanced engineered scaffolds that liberate growth factors promoting angiogenesis beyond the evolution of stem cells are considered an important approach to succeed in treatment of bone nonunion (Kanczler and Oreffo 2008; Mastrogiacomo et al. 2005; Petite et al. 2000; Porada et al. 2006).

Animal studies of stem cells for nonunion repair have succeeded in both large and small animals (Arinzeh et al. 2003; Bruder et al. 1998; Kadiyala et al. 1997; Kon et al. 2000), and these accomplishments encouraged the pursuit of clinical research applications (Figure 23.1). Clinical use of culture-expanded, autologous, bone marrow-derived MSCs in conjunction with hydroxyapatite macroporous ceramic scaffolds has been reported in small numbers of patients (Marcacci et al. 2007; Quarto et al. 2001). Bone formation was most prominent over the external surface and within the inner channel of the implants, which could be related to a higher density of loaded cells within the outermost portions of the bioceramics. With time, the scaffold implants revealed progressive appearance of cracks and fissures indicative of bioceramic disintegration, whereas bone formation progressed, and finally the implants were completely integrated into the existing bone (Undale et al. 2009). Percutaneous autologous bone marrow grafting is an effective and safe method for treating atrophic tibial diaphyseal nonunion, and the success appears to be linked to the number and concentration of stem cells available for injection (Hernigou et al. 2006). Aspirate cell concentrate (ACC) can also be used for treatment of nonunion or delayed unions (Schon et al.), and high-risk bone-healing situations show a substantially improved rate of radiographic union (Figure 23.2).

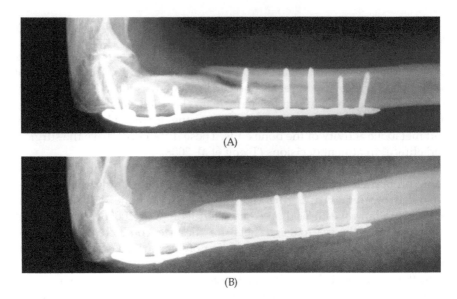

(A)

(B)

**FIGURE 23.1**
Fracture nonunion: (A) 38-year-old white female, smoker with a 9-month-old nonunion who had failed a trial using a bone stimulator. MSCs were implanted under X-ray fluoroscopy into the fracture lucency. (B) Radiograph 5 weeks after MSC transplant shows improved bone consolidation. (Courtesy of Regenerative Sciences, Inc.)

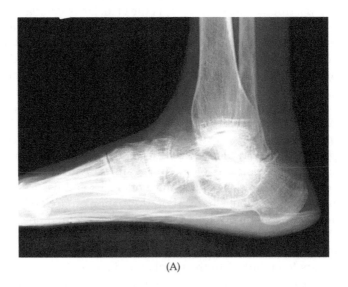

(A)

**FIGURE 23.2**
Fracture nonunion: (A) 66-year-old woman with hypothyroidism who sustained a calcaneus fracture 2 years ago. She had severe ankle and hindfoot pain, no subtalar motion, and dysfunction in activities of daily living. X-ray showed subsidence of the talus into the crushed calcaneus with loss of height and impingement of the talar neck at the ankle joint.        *(Continued)*

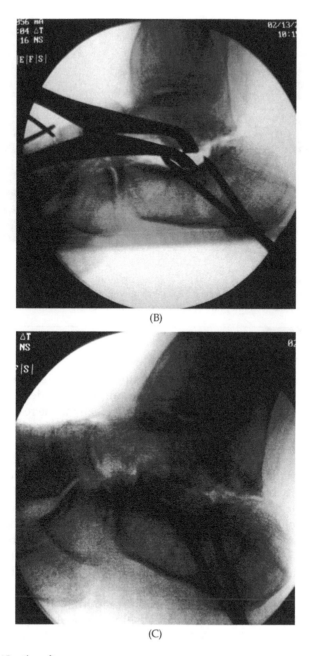

(B)

(C)

**FIGURE 23.2 (*Continued*)**
Fracture nonunion: (B) and (C) Intraoperative radiograph showing two pins in the calcaneus and allograft bone block inserted between calcaneus and talus. The allograft bone is soaked in concentrated bone marrow aspirate obtained from the iliac crest. The aspirate contains a high concentration of growth factors, hematopoietic cells, mesenchymal cells, and osteoprogenitors. (*Continued*)

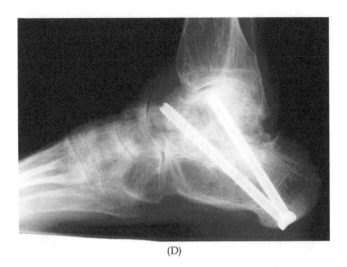

(D)

**FIGURE 23.2 (*Continued*)**
Fracture nonunion: (D) Lateral X-ray at 12 weeks showing complete incorporation of the marrow concentrate-enhanced allograft. (Courtesy of Dr. Lew Schon.)

## 23.4 Cartilage and Osteoarthritis

Osteoarthritis (OA) is the most common form of joint disease and is characterized by cartilage degeneration (thinning or focal defects), marginal osteophytosis, and subarticular bone marrow reaction. The results of current therapies suggest that tissue damage in progressive OA might be related to the depletion or functional alteration of MSC populations. The proliferative, chondrogenic, and adipogenic capacities of MSCs obtained from patients with OA are reduced (Murphy et al. 2002). Also, there is evidence for age-dependent reductions in the differentiation capability of MSCs (Im et al. 2006). However, some have found sufficient MSCs with adequate chondrogenic differentiation potential from patients with OA, independent of etiology of disease or their age (Im et al. 2006; Scharstuhl et al. 2007). Nonetheless, when considering the potential application of MSCs in OA treatment, researchers should ascertain whether MSCs obtained from OA patients differ functionally from those of healthy individuals.

The method of delivery seems to be an important requirement for the success of OA treatment using stem cells. Intra-articular injection of MSCs (Figure 23.3) and matrix-guided application of MSCs with native MSCs and genetically modified MSCs can be an effective delivery mechanism for certain lesions (Noth et al. 2008). Compared with direct intra-articular injection, MSC application to eroded cartilage surfaces via a scaffold offers more control. Another approach for OA treatment is the use of MSCs as vehicles for

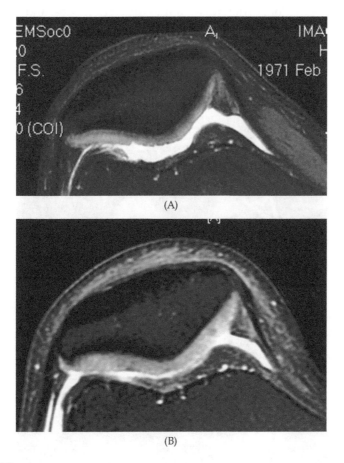

(A)

(B)

**FIGURE 23.3**
Cartilage: A 3.0-T axial proton density fat saturation MRI taken before and after three injections of autologous mesenchymal stem cells (MSCs) into the medial trochlear groove in a 37-year-old white male. Prior to MSC treatment (A), note the appearance of damaged cartilage, which shows ragged appearance with breaks on the patellar surface and the medial trochlear groove surface. Three months posttreatment (B), the matching axial slice shows improved contour of the cartilage with filling in of the defects. (Courtesy of Regenerative Sciences, Inc.)

gene delivery. MSCs seem to be receptive to transduction with various viral vectors. This approach involves isolation of MSCs, *ex vivo* genetic modification of the MSCs, and transplantation of the modified cells into the diseased joint (Evans et al. 2006). For cartilage repair and regeneration, in general, stem cells have become an attractive cell source due to the ability to sustain their own population, differentiation into different tissues, and response to stress (Figure 23.4).

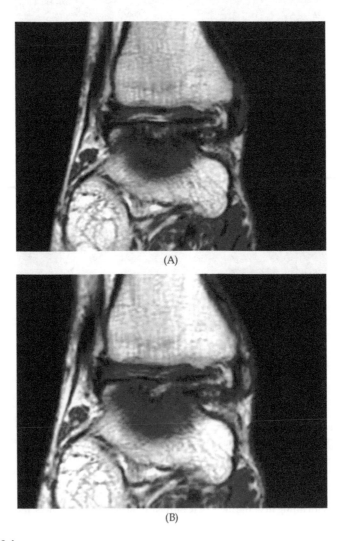

(A)

(B)

**FIGURE 23.4**
Cartilage: (A) An MRI of a 30-year-old white male with a congenital absence of the fibula and a dysplastic foot with chronic pain in the talar dome region prior to MSC treatment. (B) The preinjection T1-weighted MRI shows an area of high signal that resolved 6 months after the autologous MSC injection treatment. (Courtesy of Regenerative Sciences, Inc.)

## 23.5 Intervertebral Disc

Some studies estimated a prevalence of up to 84% for the presence of low back pain at some point in a patient's lifetime, and intervertebral disc degeneration accounts for approximately 20% of this type of pain (Sakai 2008; Walker 2000). Surgery limits movements, and fusion procedures translate stress to adjacent levels. In addition, adverse outcomes, like pseudoarthrosis or infection, may occur in a substantial proportion of cases. No effective therapy has proven to reverse or retard disc degeneration, and this is the current focus of treatment strategies. These therapies include induction of cytokines and growth factors, tissue engineering (e.g., reimplantation of nucleus pulposus cells expanded in culture), and cell transplantation (e.g., autologous MSCs) to aid in the regeneration of the intervertebral extracellular matrix. Cell-based therapies, like gene therapies, have the potential advantage of overcoming the limited time duration of molecular treatments (Figure 23.5). MSC therapy is being pursued both as an *in vitro* method for inducing the differentiation of stem cells into nucleus pulposus cells and as an *in vivo* demonstration for the feasibility of exogenous cell delivery, retention, and survival in the pressurized disc space (Crevensten et al. 2004) and with transplanted autologous MSCs embedded in collagen gel into a rabbit model for disc degeneration (Sakai 2008).

## 23.6 Tendon and Enthesis

Tendon repairs are commonly performed operations and vary depending on the site and chronicity, with three main options for grafts. Autografts are favored since there are no issues with immunogenicity or disease transmission. However, their availability is limited, and there is some donor site compromise and morbidity. Synthetics have not proven to be a viable option as they have been shown to break down and cause failure of the reconstruction, foreign-body reaction, or synovitis (Indelicato et al. 1989). Allografts are used and fare better than the synthetics. Again, availability, immunogenicity, and worries of disease transmission are limiting factors to wide adoption.

Although there are no current treatments approved by the Food and Drug Administration (FDA) related to the use of stem cells (Andres and Murrell 2008), this technique together with tissue engineering has been seen as an alternative to improve the treatment of surgical tendon repairs—seeking to enhance biologic activity by delivering cells with a biologic scaffold to a repair site in an attempt to augment the healing response (Figure 23.6). A biologic scaffold has also been used (Liu et al. 2008) to help the delivery of MSCs as a combined silk scaffold with microporous silk sponges to form

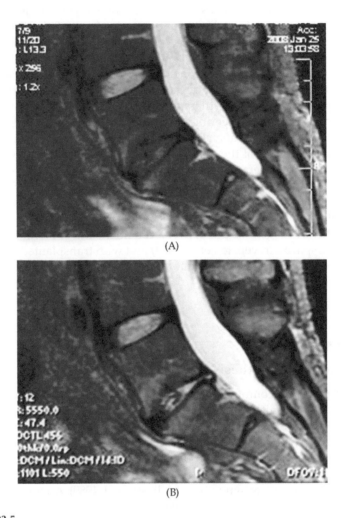

**FIGURE 23.5**
Intervertebral disc disease: MRIs before (A) and 1-year after (B) MSC injection in a 39-year-old white female. A short tau inversion recovery (STIR) sequence demonstrates a brighter signal in nucleus pulposus of L5–S1 discs that closely matches the brightness of L4–L5 discs after treatment.

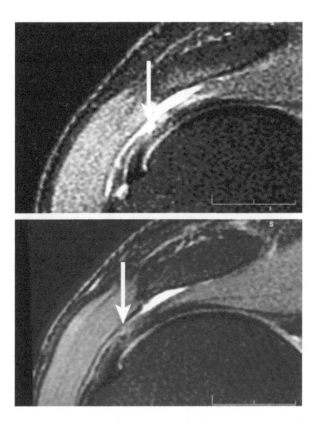

**FIGURE 23.6**
Tendons: A 34-year-old white male who injured his shoulder in a car accident. There was an acromioclavicular joint injury as well as a suspected labral tear with impingement. He failed physical therapy and subsequently underwent a subacromial decompression. The surgery failed, and a new rotator cuff tear (top arrow showing white gap in the tendon) was identified in more advanced postoperative 3.0-T MRI (top). The patient received MSC injection treatment, and 4 months later the MRI showed improvement (bottom). (Courtesy of Regenerative Sciences, Inc.)

a hybrid mesh tested *in vitro* and *in vivo*. Histologically, the scaffold-MSC mesh resembled the original structure of the tendon-bone insertion, and measurements of tensile strength on regenerated ligament were compatible with requirements of daily activities. Investigation was made of the effect on tendon-bone healing in rats, and it was found that the implantation of synovial MSCs into bone tunnel accelerated early remodeling of enthesial (tendon-bone) healing (Ju et al. 2008).

Bone marrow-derived MSCs in a fibrin carrier increased the modulus and improved collagen organization compared with control tendons in a rabbit model (Chong et al. 2007). Multiple research teams have shown MSC scaffolds designed to support and continuously improve tendon repair under a functional tissue engineering paradigm.

An attempt to overcome the problem of inconsistent repair quality was performed by designing a new cell-gel-sponge repair system to replace the suture and maintain axial strains on the gel (Butler et al. 2008). Although possessing a significantly higher stiffness and failure force than for natural healing, the initial grafts were weaker when compared to normal tendon. This new approach has shown better performance in repaired tendons when stem cells are appropriately applied to the collagen fibers and subjected to mechanical stimulus postsurgery.

The effect of coating tendon grafts with MSCs on the rate and quality of graft osteointegration in ACL (anterior cruciate ligament) reconstruction has been tested (Lim et al. 2004) and showed improved healing, with the presence of a zone of cartilage resembling the chondral enthesis of normal ACL insertions rather than the noncoated ones, whose healing process formed collagen fibers and scar tissue.

The senior author (L.C.S.) uses ACC for supplementing tendon repairs and reconstructions of the lower extremity in patients, with clinically good-to-excellent outcomes.

## 23.7 Ligaments

Acute ligament damage frequently heals poorly, and chronic recalcitrant injuries occur because of multiple reinjury or catastrophic injuries that are ineffectively treated. The healing process depends on different biomechanical variables, such as the usual stress to which the injured ligament is submitted, the size of gap, and the characterization of a partial or complete tear. Diverse strategies have been used to reestablish the original properties and functions of a ligament. They comprise controlled joint motion, tissue engineering, gene therapy, biochemical modulation, surgical repair, and grafting with or without the use of MSCs. The proliferation rate and collagen excretion of bone marrow-derived MSCs are higher than for ligament fibroblasts, and MSCs can survive for at least 6 weeks in knee joints, constituting a potentially better cell source than ligament fibroblasts for tissue engineering (Ge et al. 2005). The intra-articular injection of bone marrow-derived stem cells was shown to accelerate the healing and the postsurgical performance of a partially torn ACL in an animal model of Sprague-Dawley rats (Kanaya et al. 2007) and thus could be a viable option for humans (Figure 23.7). MSCs isolated from ligaments demonstrated that this type of cell shows potential to form bone, fat, and cartilage in addition to an increased potential to form ligament fibroblasts when compared to bone marrow-derived MSCs (Huang et al. 2006).

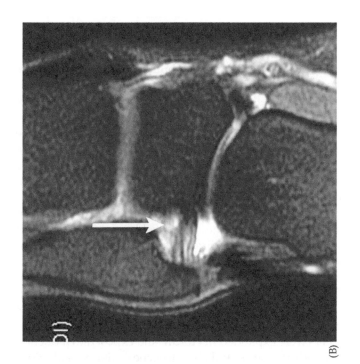

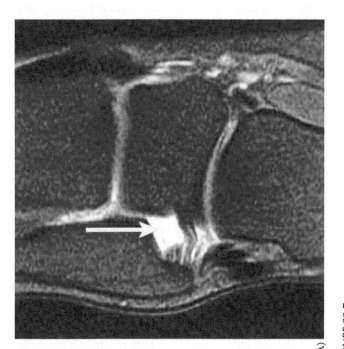

**FIGURE 23.7**

Ligaments: A 32-year-old white female with a several year history of significant ankle pain from a fall. Before (A) and 3 months after (B) MSC transplantation, coronal proton density fat suppression 3.0-T MRI of the lateral ankle ligaments and talar dome. The patient had failed arthroscopic debridement, steroids, prolotherapy, and physical therapy and still had chronic ankle pain. Note the partially disrupted talo-fibular ligament (arrow) on the preoperative image. The same ligament in the post-MSC administration MRI shows that the upper portion of the ligament has been repaired and that the "crimped" appearance of the subfailure stretch injury present in the left image has returned to the more normal morphology. The patient had complete resolution of lateral ankle pain, and this procedure required no immobilization. (Courtesy of Regenerative Sciences, Inc.)

## 23.8 Muscle

Muscle cells are multinucleated (beyond several hundreds of nuclei per cubic centimeter of fiber volume), and these nuclei arise from undifferentiated myogenic cells, called satellite cells, located underneath the basal lamina of muscle fibers. These cells are normally quiescent in adult muscle but act as a reserve population of cells, able to proliferate in response to injury and give rise to regenerated muscle and to more satellite cells (Morgan and Partridge 2003). This satellite cell pool can be heterogeneous and contain self-renewing satellite "stem" cells and myogenic precursors with limited replicative potential in the same anatomical location (Zammit 2008).

Studies have demonstrated that muscle regeneration derived from stem cells has potential to become a reality in the clinical field in the future. Most of the studies are related to identification of different potential lineage precursor cells that best suits the challenge of muscle regeneration. Local injection of autologous bone marrow-derived cells can improve functional muscle regeneration after blunt skeletal muscle injury, and after treatment, muscle contractile force improved in the bone marrow-derived delivery group (Matziolis et al. 2006). Isolated mesenchymal multiprogenitor cells from debrided muscle of posttraumatic soldiers were expanded in culture and exposed to induction media for osteogenesis, adipogenesis, and chondrogenesis (Nesti et al. 2008). The results suggest that discarded tissue contains cellular building blocks that might be useful in future treatment and tissue regeneration strategies. In another investigation (Jackson et al. 2009), mesenchymal progenitor cells (MPCs) derived from traumatized muscle demonstrated the ability to differentiate into osteoblasts, adipocytes, and chondrocytes even after relatively late population doubling *in vitro* (eight population doublings). Thus, MPCs could provide an alternative to bone marrow-derived MSCs with comparable morphology, proliferation, and differentiation capacities.

## 23.9 Mesenchymal Stem Cell Imaging

The development of *in vivo* imaging tracking of transplanted MSCs constitutes an important component for monitoring successful therapeutic cell engraftment or diagnosing early treatment failure. Ideally, these techniques should be noninvasive, reproducible, reliable, and sensitive to changes in the cellular environment. The modalities most likely to meet these requirements are magnetic resonance imaging (MRI) and optical imaging (OI).

MRI is the most commonly used imaging modality for *in vivo* tracking of labeled stem cells. MRI provides good contrast resolution (especially for soft tissues) and relatively high spatial resolution and allows dynamic assessment of cell migration into target tissues for nanomolar-to-micromolar concentrations of labeling agents. The most commonly used labeling agents are represented by nanoparticles with an iron oxide core of magnetite ($Fe_3O_4$) or maghemite ($\gamma Fe_2O_3$), a ferrimagnetic cubic form of Fe(III) oxide encased in polysaccharide, synthetic polymer, or monomer coatings (Arbab et al. 2003; Frank et al. 2003; Stark et al. 1988). These agents decrease the relaxation time in MRI T2* measurements due to a disturbance in magnetic field in their surroundings, making them appear as signal voids. Thus, their usefulness may be limited in certain applications since the hypointensities may be inconspicuous or confounded on present musculoskeletal MRI protocols. In the case of muscle regeneration, hypointensities caused by magnetically labeled MSCs in the MRIs represent an important concern for *in vivo* visualization of cells since structures of normal tissue, such as blood vessels, tendons, and boundary layers between the muscles, also show hypointense MRI signals, which may limit the ability to distinguish treatment effect (Winkler et al. 2008). Protocols to increase conspicuity, such as SWI (susceptibility-weighted imaging), could be employed, but the blooming artifact may obscure the structures of interest. One potential drawback to using iron oxide-labeled MSCs is an observation that the magnetic labeling procedure may affect the differentiation (Kostura et al. 2004) of labeled MSCs, thus inhibiting the originally intended function. For some applications, routine MRI may be useful to document specific morphological features (e.g., cartilage defect filling, tendon/ligament continuity, muscle reconstitution) or biochemical status (e.g., GLycosaminoglycan concentration).

OI has several advantages for tracking stem cells as it is rapid, inexpensive, and noninvasive; does not involve ionizing radiation; and provides single-cell sensitivity. A major disadvantage is related to the low detectable depth penetration of the near-infrared (NIR) radiation, which is not suitable in many clinical scenarios. However, OI has been applied to monitor MSC localization in arthritic joints (Sutton et al. 2009). For applications such as fracture healing, radiography and computed tomography (CT) are often adequate to show osseous union but cannot directly detect whether MSCs are located at the fracture site or have migrated to remote areas. Therefore, imaging in some capacity will play an important role for monitoring of treatment effect for patients undergoing stem cell therapy.

## 23.10 Conclusion

Stem cell therapy has the potential to provide treatment for several degenerative diseases, traumatic orthopedic problems, and rheumatologic conditions with the ability to improve the quality of life for this segment of the population that suffers from the morbidity of these disorders. MSCs are the most commonly developed, studied, and used instrument for stem cell therapy in musculoskeletal medicine. MSCs may be applied in isolation, as injectables on scaffolds, or in combination with reconstructive surgeries. The feasibility and reliability of MSC treatments are an area of intensive *in vitro* research. There is potential to create functional load-bearing repairs that might bring substantial progress to surgical reconstruction after bone, cartilage, tendon, and ligament injury. The use of MSCs may facilitate faster healing and, thus bring benefit by improving postsurgical recovery and rehabilitation. Encouraging recent positive reports of clinical case series promise a pathway for the use of stem cells *in vivo*. However, the effectiveness of cellular therapies in randomized clinical trials will need to be assessed as the next step. Short- and long-term safety should also be addressed.

## References

Abdallah, B.M., M. Haack-Sorensen, J.S. Burns, B. Elsnab, F. Jakob, P. Hokland, and M. Kassem. 2005. Maintenance of differentiation potential of human bone marrow mesenchymal stem cells immortalized by human telomerase reverse transcriptase gene despite [corrected] extensive proliferation. *Biochemical and Biophysical Research Communications* 326, no. 3: 527–38.

Abdallah, B.M. and M. Kassem. 2008. Human mesenchymal stem cells: from basic biology to clinical applications. *Gene Therapy* 15, no. 2: 109–16.

Andres, B.M. and G.A. Murrell. 2008. Treatment of tendinopathy: what works, what does not, and what is on the horizon. *Clinical Orthopaedics and Related Research* 466, no. 7: 1539–54.

Arbab, A.S., L.A. Bashaw, B.R. Miller, E.K. Jordan, J.W. Bulte, and J.A. Frank. 2003. Intracytoplasmic tagging of cells with ferumoxides and transfection agent for cellular magnetic resonance imaging after cell transplantation: methods and techniques. *Transplantation* 76, no. 7: 1123–30.

Arinzeh, T.L., S.J. Peter, M.P. Archambault, C. Van Den Bos, S. Gordon, K. Kraus, A. Smith, and S. Kadiyala. 2003. Allogeneic mesenchymal stem cells regenerate bone in a critical-sized canine segmental defect. *Journal of Bone and Joint Surgery. American Volume* 85-A, no. 10: 1927–35.

Bhagavati, S. 2008. Stem cell based therapy for skeletal muscle diseases. *Current Stem Cell Research and Therapy* 3, no. 3: 219–28.

Bruder, S.P., K.H. Kraus, V.M. Goldberg, and S. Kadiyala. 1998. The effect of implants loaded with autologous mesenchymal stem cells on the healing of canine segmental bone defects. *Journal of Bone and Joint Surgery. American Volume* 80, no. 7: 985–96.

Butler, D.L., N. Juncosa-Melvin, G.P. Boivin, M.T. Galloway, J.T. Shearn, C. Gooch, and H. Awad. 2008. Functional tissue engineering for tendon repair: a multidisciplinary strategy using mesenchymal stem cells, bioscaffolds, and mechanical stimulation. *Journal of Orthopaedic Research: Official Publication of the Orthopaedic Research Society* 26, no. 1: 1–9.

Chen, F.H. and R.S. Tuan. 2008. Mesenchymal stem cells in arthritic diseases. *Arthritis Research and Therapy* 10, no. 5: 223.

Chong, A.K., A.D. Ang, J.C. Goh, J.H. Hui, A.Y. Lim, E.H. Lee, and B.H. Lim. 2007. Bone marrow-derived mesenchymal stem cells influence early tendon-healing in a rabbit achilles tendon model. *Journal of Bone and Joint Surgery. American Volume* 89, no. 1: 74–81.

Crevensten, G., A.J. Walsh, D. Ananthakrishnan, P. Page, G.M. Wahba, J.C. Lotz, and S. Berven. 2004. Intervertebral disc cell therapy for regeneration: mesenchymal stem cell implantation in rat intervertebral discs. *Annals of Biomedical Engineering* 32, no. 3: 430–34.

Delorme, B., S. Chateauvieux, and P. Charbord. 2006. The concept of mesenchymal stem cells. *Regenerative Medicine* 1, no. 4: 497–509.

Dominici, M., K. Le Blanc, I. Mueller, I. Slaper-Cortenbach, F. Marini, D. Krause, R. Deans, A. Keating, D. Prockop, and E. Horwitz. 2006. Minimal criteria for defining multipotent mesenchymal stromal cells. The International Society for Cellular Therapy position statement. *Cytotherapy* 8, no. 4: 315–17.

Evans, C.H., S.C. Ghivizzani, and P.D. Robbins. 2006. Will arthritis gene therapy become a clinical reality? *Nature Clinical Practice. Rheumatology* 2, no. 7: 344–45.

Foster, L.J., P.A. Zeemann, C. Li, M. Mann, O.N. Jensen, and M. Kassem. 2005. Differential expression profiling of membrane proteins by quantitative proteomics in a human mesenchymal stem cell line undergoing osteoblast differentiation. *Stem Cells (Dayton, Ohio)* 23, no. 9: 1367–77.

Frank, J.A., B.R. Miller, A.S. Arbab, H.A. Zywicke, E.K. Jordan, B.K. Lewis, L.H. Bryant, Jr., and J.W.M. Bulte. 2003. Clinically applicable labeling of mammalian and stem cells by combining superparamagnetic iron oxides and transfection agents. *Radiology* 228: 480–87.

Ge, Z., J.C. Goh, and E.H. Lee. 2005. Selection of cell source for ligament tissue engineering. *Cell Transplantation* 14, no. 8: 573–83.

Hernigou, P., G. Mathieu, A. Poignard, O. Manicom, F. Beaujean, and H. Rouard. 2006. Percutaneous autologous bone-marrow grafting for nonunions. Surgical technique. *Journal of Bone and Joint Surgery. American Volume* 88 Suppl 1 Pt 2: 322–27.

Hernigou, P., A. Poignard, F. Beaujean, and H. Rouard. 2005a. Percutaneous autologous bone-marrow grafting for nonunions. Influence of the number and concentration of progenitor cells. *Journal of Bone and Joint Surgery. American Volume* 87, no. 7: 1430–37.

Hernigou, P., A. Poignard, O. Manicom, G. Mathieu, and H. Rouard. 2005b. The use of percutaneous autologous bone marrow transplantation in nonunion and avascular necrosis of bone. *Journal of Bone and Joint Surgery. British volume* 87, no. 7: 896–902.

Huang, J.I., M.M. Durbhakula, P. Angele, B. Johnstone, and J.U. Yoo. 2006. Lunate arthroplasty with autologous mesenchymal stem cells in a rabbit model. *Journal of Bone and Joint Surgery. American Volume* 88, no. 4: 744–52.

Im, G.I., N.H. Jung, and S.K. Tae. 2006. Chondrogenic differentiation of mesenchymal stem cells isolated from patients in late adulthood: the optimal conditions of growth factors. *Tissue Engineering* 12, no. 3: 527–36.

Indelicato, P.A., M.S. Pascale, and M.O. Huegel. 1989. Early experience with the Goretex polytetrafluoroethylene anterior cruciate ligament prosthesis. *American Journal of Sports Medicine* 17, no. 1: 55–52.

In 'T Anker, P.S., W.A. Noort, S.A. Scherjon, C. Kleijburg-Van Der Keur, A.B. Kruisselbrink, R.L. Van Bezooijen, W. Beekhuizen, R. Willemze, H.H. Kanhai, and W.E. Fibbe. 2003. Mesenchymal stem cells in human second-trimester bone marrow, liver, lung, and spleen exhibit a similar immunophenotype but a heterogeneous multilineage differentiation potential. *Haematologica* 88, no. 8: 845–52.

Ishida, Y. and J.N. Heersche. 1997. Progesterone stimulates proliferation and differentiation of osteoprogenitor cells in bone cell populations derived from adult female but not from adult male rats. *Bone* 20, no. 1: 17–25.

Jackson, L., D.R. Jones, P. Scotting, and V. Sottile. 2007. Adult mesenchymal stem cells: differentiation potential and therapeutic applications. *Journal of Postgraduate Medicine* 53, no. 2: 121–27.

Jackson, W.M., A.B. Aragon, F. Djouad, Y. Song, S.M. Koehler, L.J. Nesti, and R.S. Tuan. 2009. Mesenchymal progenitor cells derived from traumatized human muscle. *Journal of Tissue Engineering and Regenerative Medicine* 3, no. 2: 129–38.

Ju, Y.J., T. Muneta, H. Yoshimura, H. Koga, and I. Sekiya. 2008. Synovial mesenchymal stem cells accelerate early remodeling of tendon-bone healing. *Cell and Tissue Research* 332, no. 3: 469–78.

Kadiyala, S., N. Jaiswal, and S.P. Bruder. 1997. Culture-expanded, bone marrow-derived mesenchymal stem cells can regenerate a critical-sized segmental bone defect. *Tissue Engineering* 3, no. 2: 173–85.

Kanaya, A., M. Deie, N. Adachi, M. Nishimori, S. Yanada, and M. Ochi. 2007. Intra-articular injection of mesenchymal stromal cells in partially torn anterior cruciate ligaments in a rat model. *Arthroscopy: The Journal of Arthroscopic and Related Surgery: Official Publication of the Arthroscopy Association of North America and the International Arthroscopy Association* 23, no. 6: 610–17.

Kanczler, J.M. and R.O. Oreffo. 2008. Osteogenesis and angiogenesis: the potential for engineering bone. *European Cells and Materials* 15: 100–14.

Kon, E., A. Muraglia, A. Corsi, P. Bianco, M. Marcacci, I. Martin, A. Boyde, I. Ruspantini, P. Chistolini, M. Rocca, R. Giardino, R. Cancedda, and R. Quarto. 2000. Autologous bone marrow stromal cells loaded onto porous hydroxyapatite ceramic accelerate bone repair in critical-size defects of sheep long bones. *Journal of Biomedical Materials Research* 49, no. 3: 328–37.

Kostura, L., D.L. Kraitchman, A.M. Mackay, M.F. Pittenger, and J.W. Bulte. 2004. Feridex labeling of mesenchymal stem cells inhibits chondrogenesis but not adipogenesis or osteogenesis. *NMR Biomed* 17, no. 7: 513–17.

Lim, J.K., J. Hui, L. Li, A. Thambyah, J. Goh, and E.H. Lee. 2004. Enhancement of tendon graft osteointegration using mesenchymal stem cells in a rabbit model of anterior cruciate ligament reconstruction. *Arthroscopy: The Journal of Arthroscopic and Related Surgery: Official Publication of the Arthroscopy Association of North America and the International Arthroscopy Association* 20, no. 9: 899–910.

Liu, H., H. Fan, E.J.W. Wong, S. Lok Toh, and J.C.H. Goh. 2008. Silk-based scaffold for ligament tissue engineering. In *14th Nordic-Baltic conference on biomedical engineering and medical physics*, eds. A. Katashev, Y. Dekhtyar, and J. Spigulis, 34–37. Springer, Berlin.

Marcacci, M., E. Kon, V. Moukhachev, A. Lavroukov, S. Kutepov, R. Quarto, M. Mastrogiacomo, and R. Cancedda. 2007. Stem cells associated with macroporous bioceramics for long bone repair: 6- to 7-year outcome of a pilot clinical study. *Tissue Engineering* 13, no. 5: 947–55.

Mastrogiacomo, M., A. Muraglia, V. Komlev, F. Peyrin, F. Rustichelli, A. Crovace, and R. Cancedda. 2005. Tissue engineering of bone: search for a better scaffold. *Orthodontics and Craniofacial Research* 8, no. 4: 277–84.

Matziolis, G., T. Winkler, K. Schaser, M. Wiemann, D. Krocker, J. Tuischer, C. Perka, and G.N. Duda. 2006. Autologous bone marrow-derived cells enhance muscle strength following skeletal muscle crush injury in rats. *Tissue Engineering* 12, no. 2: 361–67.

Morgan, J.E. and T.A. Partridge. 2003. Muscle satellite cells. *International Journal of Biochemistry and Cell Biology* 35, no. 8: 1151–56.

Murphy, J.M., K. Dixon, S. Beck, D. Fabian, A. Feldman, and F. Barry. 2002. Reduced chondrogenic and adipogenic activity of mesenchymal stem cells from patients with advanced osteoarthritis. *Arthritis and Rheumatism* 46, no. 3: 704–13.

Nesti, L.J., W.M. Jackson, R.M. Shanti, S.M. Koehler, A.B. Aragon, J.R. Bailey, M.K. Sracic, B.A. Freedman, J.R. Giuliani, and R.S. Tuan. 2008. Differentiation potential of multipotent progenitor cells derived from war-traumatized muscle tissue. *Journal of Bone and Joint Surgery. American Volume* 90, no. 11: 2390–98.

Noth, U., A.F. Steinert, and R.S. Tuan. 2008. Technology insight: adult mesenchymal stem cells for osteoarthritis therapy. *Nature Clinical Practice. Rheumatology* 4, no. 7: 371–80.

Petite, H., V. Viateau, W. Bensaid, A. Meunier, C. De Pollak, M. Bourguignon, K. Oudina, L. Sedel, and G. Guillemin. 2000. Tissue-engineered bone regeneration. *Nature Biotechnology* 18, no. 9: 959–63.

Porada, C.D., E.D. Zanjani, and G. Almeida-Porad. 2006. Adult mesenchymal stem cells: a pluripotent population with multiple applications. *Current Stem Cell Research and Therapy* 1, no. 3: 365–69.

Quarto, R., M. Mastrogiacomo, R. Cancedda, S.M. Kutepov, V. Mukhachev, A. Lavroukov, E. Kon, and M. Marcacci. 2001. Repair of large bone defects with the use of autologous bone marrow stromal cells. *New England Journal of Medicine* 344, no. 5: 385–86.

Safadi, F.F., M.F. Barbe, S.M. Abdelmagid, M.C. Rico, R.A. Aswad, J. Litvin, and S.N. Popoff. 2009. Bone structure, development and bone biology. In *Bone pathology*, ed. J.S. Khurana, 1–50. Humana Press, Totowa, NJ.

Sakai, D. 2008. Future perspectives of cell-based therapy for intervertebral disc disease. *European Spine Journal: Official Publication of the European Spine Society, the European Spinal Deformity Society, and the European Section of the Cervical Spine Research Society* 17 Suppl 4: 452–58.

Scharstuhl, A., B. Schewe, K. Benz, C. Gaissmaier, H.J. Buhring, and R. Stoop. 2007. Chondrogenic potential of human adult mesenchymal stem cells is independent of age or osteoarthritis etiology. *Stem Cells (Dayton, Ohio)* 25, no. 12: 3244–51.

Schon, L.C., R. Dunn and L. Gamez. 2010. Cell concentrate from autologous bone marrow augments bone grafting in the lower extremity. AOFAS annual meeting. Washington, D.C.

Schwartz, Z., E. Nasatzky, A. Ornoy, B.P. Brooks, W.A. Soskolne, and B.D. Boyan. 1994. Gender-specific, maturation-dependent effects of testosterone on chondrocytes in culture. *Endocrinology* 134, no. 4: 1640–47.

Stark, D.D., R. Weissleder, G. Elizondo, P.F. Hahn, S. Saini, L.E. Todd, J. Wittenberg, and J.T. Ferrucci. 1988. Superparamagnetic iron oxide: clinical application as a contrast agent for MR imaging of the liver. *Radiology* 168: 297–301.

Sutton, E.J., S.E. Boddington, A.J. Nedopil, T.D. Henning, S.G. Demos, R. Baehner, B. Sennino, Y. Lu, and H.E. Daldrup-Link. 2009. An optical imaging method to monitor stem cell migration in a model of immune-mediated arthritis. *Optics Express* 17, no. 26: 24403–13.

Tseng, S.S., M.A. Lee, and A.H. Reddi. 2008. Nonunions and the potential of stem cells in fracture-healing. *Journal of Bone and Joint Surgery. American Volume* 90 Suppl 1: 92–98.

Undale, A.H., J.J. Westendorf, M.J. Yaszemski, and S. Khosla. 2009. Mesenchymal stem cells for bone repair and metabolic bone diseases. *Mayo Clinic Proceedings. Mayo Clinic* 84, no. 10: 893–902.

Walker, B.F. 2000. The prevalence of low back pain: a systematic review of the literature from 1966 to 1998. *Journal of Spinal Disorders* 13, no. 3: 205–17.

Winkler, T., P. von Roth, M. Schuman, K. Sieland, G. Stoltenburg-Didinger, M. Taupitz, C. Perka, G. Duda, and G. Matziolis. 2008. In vivo visualization of locally transplanted mesenchymal stem cells in the severly injured muscle in rats. *Tissue Engineering, Part A* 14: 1149–60.

Zammit, P.S. 2008. All muscle satellite cells are equal, but are some more equal than others? *Journal of Cell Science* 121, Pt 18: 2975–82.

# 24

## Regulatory Hurdles to Translation

Adrian D. Nunn

### CONTENTS

### 24.1 Introduction

Labeled cells of various lineages have been used clinically for some years, but the capacity of stem cells for self-renewal, their robust proliferative potential, and their ability to differentiate into varied, disparate tissue phenotypes in response to appropriate biologic cues set them apart. This chapter summarizes the regulatory framework under which labeled stem cells might be used. The opinions are somewhat speculative as although the regulatory authorities have published extensively on the regulation of somatic cell therapies, and some (mainly hematopoietic) stem cell therapies are of considerable vintage, it does not appear that an imageable stem cell Investigational New Drug (IND) application has been formally presented to the Food and Drug Administration (FDA) or other authorities to test the system. As such, a more cautious approach has been taken. Most emphasis is on the regulatory environment in the United States; but as harmonization progresses, there is less difference between the various jurisdictions. Along with the regulatory aspects, this chapter expands on some of the pharmacology/toxicology issues that should be considered.

## 24.2 Regulation of Stem Cell Studies in Humans

The FDA regulates stem cell therapies under the tissue action plan framework as human cells, tissues, or cellular or tissue-based products (HCT/Ps) under 21 CFR 1271 [1]. The degree of regulation anticipated is dependent on the source of the cells, the manipulation required prior to administration, the addition of noncell components (other than excipients), and their intended use [2]. Although the agency considers cell selection (e.g., selection of stem cells from among lymphocytes and mature cells of other lineages) to be minimal manipulation (and therefore subject to less regulation), the addition of some means of imaging the cells *in vivo* (addition of noncell components) would appear to be subject to more comprehensive regulatory controls than unlabeled stem cells (Table 24.1) [3]. Thus, except in limited circumstances, the FDA intends to regulate stem cells such that an IND application is required prior to first in man.

A guidance document has been issued by the FDA for somatic cell therapy that identifies combination products that contain a human somatic cell therapy biological product in combination with a drug or device as part of the final product, and specific mention is made of diagnostic applications involving cells that are modified by radiolabeling [4]. A labeling agent added to stem cells to enable *in vivo* tracking would appear to fall under such a description. Similarly, incorporation of genetic material to provide a means of generating an (imageable) signal *in vivo* is also addressed as stated,

**TABLE 24.1**

Concern: Clinical Safety or Clinical Effectiveness (Including Use-Specific Concerns)

(a)  More than minimal manipulation
(b)  Nonhomologous use
(c)  Combination with noncell/nontissue
(d)  Metabolic use (other than reproductive except when used autologously or in a close family member)

| Characteristic | Action Required | Regulatory Submission |
|---|---|---|
| 1. Without any of factors a, b, c, or d | None | No FDA submission. |
| 2. For local, structural reconstruction or repair and has factors a, b, or c | Would have to gather clinical safety and effectiveness data | Studies would have to be done under IND or IDE |
| 3. For reproductive or metabolic use with factors a, b, c, or d | Would have to gather clinical safety and effectiveness data | Studies would have to be done under IND |

*Note:*   IDE = Investigational Device Exemption.
*Source:*   Adapted from Food and Drug Administration. Proposed Approach to the Regulation of Cellular and Tissue-Based Products. Table 1(c). February 28, 1997. Available at http://www.fda.gov/CBER/gdlns/celltissue.pdf (accessed March 17, 2009).

"Cells may be modified *ex vivo* for subsequent administration to humans. ... The genetic manipulation ... may provide a way of marking cells for later identification" [4].

The FDA issued a guidance document describing the information that should be considered when submitting an IND application for human somatic cell therapy [5]. In this, they described reagents used in manufacturing the product, which are those materials used in the preparation that are not intended to be part of the final product. Similarly, they described excipients, such as human serum albumin or dimethyl sulfoxide, that support the cells in their final formulation. Clearly, incorporation of an imaging agent is somewhat different.

## 24.3 Characterization of a Stem Cell Preparation Intended for First in Man

Many of the controls needed to ensure that a cell preparation destined for clinical use under an IND is both representative of previous preparations used to justify first in man and suitability for the intended testing are similar to those required of drugs and biologics. These relate to a defined and controlled method of preparation leading to a final form with characteristics within specifications. As with drugs, there is some leeway in the rigor of the procedures for early clinical trials. There are, however, some areas, such as the characterization of the cell source and the viability of the cells, that are quite different. Most of these topics are discussed in depth in the guidance document [5]. There are two areas related to imageable stem cells that are much less straightforward than for normal drugs. These relate to the assessment of potential toxicity and of efficacy. The two general methods of incorporating an imageable signal into stem cells—as an imaging agent itself (see the discussion of labeling agent in the next section) such as radioactivity (single-photon emission computed tomography/positron emission tomography, SPECT/PET) or iron oxide particles (magnetic resonance imaging, MRI) or by genetic means (e.g., enzyme, receptor, or transporter expression—present different toxicity and efficacy issues.

## 24.4 Regulation of a Labeling Agent

The administration of labeled cells requires appropriate testing even if the labeling agent has already been introduced into humans as in many cases the administration route has changed or the cell is different such that the risk/

benefit equation has changed. Thus, the use of In-111 oxine (an approved drug for labeling cells), indocyanine green (approved for intravenous administration), or magnetic iron particles (some of which are approved for intravenous administration) each represents a different environment for administration that requires reexamination of the equation.

Traditional phase 1 trials explore the safety profile of the drug in relation to the dose-response profile determined in animals and in particular seek to determine the maximum tolerated dose (MTD). The phase 0, E (Exploratory)-IND or microdosing concept has been introduced by the regulatory authorities in an effort to allow easier testing of new chemical entities (NCEs) in a manner that minimizes the risk but still allows testing and collection of the necessary data. Microdosing trials, unlike the traditional phase 1, do not seek to establish the MTD in humans but instead use low doses to establish proof of concept or pharmacokinetic characteristics of a single or multiple related compounds.

The general requirements for filing an IND application in the United States are delineated in 21 CFR 312 [6], and specific rules for radiopharmaceuticals (only) can be found in 21 CFR 315 [7]. A discussion of the requirements that need to be fulfilled before the regulatory authorities will allow first-in-man trials of an NCE and the evolution of these requirements to accommodate microdosing concepts has been published [8]. More recently, the major regulatory authorities have come to agreement under the International Conference on Harmonization of Technical Requirements for Registration of Pharmaceuticals for Human Use (ICH) to further elaborate on the microdosing concept and to incorporate their thoughts into the overall requirements for safety studies [9].

In the United States, there is an additional route that might be considered as a means of performing first-in-man labeled cell studies. This is the Radioactive Drug Research Committee (RDRC), and as its name implies, it is limited to radioactive compounds [10]. This permits basic research using radioactive drugs in humans without an IND application when the drug is administered for the purpose of advancing scientific knowledge. It is intended to obtain basic information regarding the metabolism (including kinetics, distribution, dosimetry, and localization) of a radioactive drug or regarding human physiology, pathophysiology, or biochemistry. It is not intended that this be the first-in-man use of the radioactive drug. If attempting to use this approach, an argument will need to be made that one is following the different pharmacokinetics (PK) of the radioactivity because it is incorporated into the cells or following physiology using radioactivity. But, this may not be a tenable argument, and it would be wise to approach the FDA as well as the RDRC before too many resources are expended. In addition, the benefit of this approach would only accrue if it changed the category to level C1 as described in References 2 and 3 and Table 24.1 and thus lead to reduced regulation. This approach, of course, also requires that there be a suitable radioactive drug available, and the utility is limited by the (short) physical half-life of the radionuclide used. Thus, it may only be useful for the initial distribution phase of administered cells.

## 24.5 Regulation of a Reporter Gene

The insertion of reporter genes is the alternative approach either through incorporation of light-generating systems, such as luciferase (leaving aside the practicality of clinical optical imaging in this case), or engineered to express enzymes, such as a mutant herpes simplex virus type 1 thymidine kinase (HSV1-sr39tk), which can be selectively imaged by administration of [18F]FHBG (9-[4–18F-fluoro-3-(hydroxymethyl)butyl]guanine) and others [11] or other radioactive compounds. Such approaches have been covered by FDA guidances (as discussed here) [4] at least from the genetic point of view.

## 24.6 Toxicity Assessment

No matter which imaging mechanism is employed, the most important concept that must be justified before first-in-man use is the risk/benefit ratio. How does one establish the long-term toxic effects, if any, of labeling such cells? These are issues that are not touched on in the published regulatory advice.

The mass of labeling agent administered with the cells is necessarily small and is limited by the volume of cells. One might think that this minimizes the risk and places the labeling agent in the microdosing category. However, the "standard" methods of assessing the toxicity of a small molecule or biologic are inappropriate in that they focus on systemic toxicity. The use of cells labeled with iron particles is a case in point. Iron oxide-labeled dendritic cells have been administered to melanoma patients and imaged by MRI [12]. The authors elected to use an approved iron particle and the normal phagocytic capacity of the immature dendritic cells. The particles in question are approved in Europe and the United States but not for this indication. The authors administered $7.5 \times 10^6$ cells containing 10–30 pg of iron per cell or a total of 1–2 µg, which lies well within the mass amount limits of microdosing. The study was performed in stage III melanoma patients, so a higher starting dose (10% of the MTD) of iron may well be acceptable because the risk/benefit ratio for oncology patients is different [13]. However, although the safety profile of the iron particles after intravenous administration is established, the toxicity to these particular cells is not, so what is the relevance of a (whole-body) MTD in this case? The authors performed functional assays *in vitro* and observed the anticipated behavior *in vivo*, which suggested that the cells were not grossly altered by their high iron loading. This is in contrast to the results of Kostura et al. [14], who found some alteration in behavior *in vitro* when mesenchymal stem cells were labeled with iron particles. An additional factor in the study by de Vries [12] is that the cells were also loaded with tumor-derived antigenic peptides.

Cardiac stem cells have been labeled using $^{18}$F-FDG (fluorodeoxyglucose) prior to administration and followed for as long as the physical half-life allowed [15]. Viability was assessed and deemed acceptable, but of course, after decay of the radionuclide the behavior of the cells is unknown. The cells were exposed to very low masses of FDG, but 2-deoxyglucose derivatives are metabolic poisons, and it would be desirable to understand not only their effect on the viability of the stem cells but also their subsequent behavior.

When it comes to reporter genes, the assessment of toxicity is still not straightforward. There is not only the genetic load to consider but also the substrate. Thus, for animal imaging using the luciferase system, 375 mg/kg body weight D-luciferin is administered intraperitoneally [11], which translates to many grams of material in an average patient—hardly a microdose. An alternative is to use an enzyme, transporter, or receptor-type system and then administer a radioactive substrate. In this case, the mass of substrate injected is small, but one must also assess the risk of periodically irradiating stem cells due to decay of internalized radioactivity.

Certainly, patients with end-stage cancer are an appropriate choice to start, but the transition to patients with greater life expectancy or worse risk/benefit ratio will require discussion with the regulatory authorities and probably recourse to the traditional IND application. Some of the reporter genes (HSV1-*tk*, herpes simplex virus type 1 thymidine kinase) have been administered to humans and imaged, so there is some clinical experience [15].

## 24.7 Efficacy

The FDA stated that products with a "metabolic" mode of action (e.g., pancreatic islet cells, pituitary cells, stem cells) are sensitive to perturbations and may not retain normal function after the transplantation process. Thus, minimally manipulated cellular and tissue-based products with metabolic function raise greater clinical safety and effectiveness concerns than do products with structural or reproductive function. If there is frank *in vitro* cell toxicity due to the addition of imaging capability, the results are relatively easy to interpret, especially if there is a reasonable dose-response curve, but *in vivo* cell death could be due to other factors as well. Far more difficult to interpret is a change in the behavior of the cells either *in vitro* or *in vivo*.

The understanding of the "normal" behavior of these cells is incomplete, and they also have many degrees of freedom. Another issue is that they or their daughters can be long lived, so it is conceivable that behavioral changes induced by labeling are not manifest until a considerable time after administration. To date, animal studies have been performed for over 6 months using a gene-based system with the appropriate substrate [15].

## 24.8 Summary

The rules under which the clinical use of stem cells are regulated are relatively clear, and the issued guidance documents describe in some detail the requirements necessary to perform first-in-man clinical trials. The requirements have not been tested to determine their feasibility or appropriateness when imaging capability is added to enable *in vivo* tracking. The biggest issue is not how to determine frank toxicity, due to the incorporation of imaging ability by the means mentioned, but how to identify changes in behavior due to the labeling method—not only safety, but also efficacy.

## References

1. Fink, D.W., Jr., Embryonic stem cell-based therapies: U.S. FDA regulatory expectations. 2007. Available at http://www.fda.gov/cber/genetherapy/stemcell012907df.pdf (accessed March 17, 2009).
2. Food and Drug Administration. Proposed approach to the regulation of cellular and tissue-based products. February 28, 1997. Available at http://www.fda.gov/CBER/gdlns/celltissue.pdf (accessed March 17, 2009).
3. Food and Drug Administration. Proposed approach to the regulation of cellular and tissue-based products. Section 2c and Table 1, row C. February 28, 1997. Available at http://www.fda.gov/CBER/gdlns/celltissue.pdf (accessed March 17, 2009).
4. Food and Drug Administration. Guidance for human somatic cell therapy and gene therapy. March 1998. p. 3. Available at http://www.fda.gov/CBER/gdlns/somgene.pdf (accessed March 17, 2009).
5. Food and Drug Administration. Guidance for FDA reviewers and sponsors. Content and review of chemistry, manufacturing, and control (CMC) information for human somatic cell therapy Investigational New Drug applications (INDs). April 2008. Available at http://www.fda.gov/cber/gdlns/cmcsomcell.htm (accessed March 17, 2009).
6. Investigational new drug application. 21 CFR 312. Available at http://ecfr.gpoaccess.gov/cgi/t/text/text-idx?c=ecfr&tpl=/ecfrbrowse/Title21/21cfr312_main_02.tpl (accessed March 28, 2006).

7. Regulations for in vivo radiopharmaceuticals used for diagnosis and monitoring. 21 CFR 315. Available at http://ecfr.gpoaccess.gov/cgi/t/text/text-idx?c=ecfr&tpl=/ecfrbrowse/Title21/21cfr315_main_02.tpl (accessed March 28, 2006).

8. Nunn, A.D. Translating promising experimental approaches to clinical trials. In *Molecular and Cellular MR Imaging*, pp. 395–403. Eds. M. Modo and J.F.W. Bulte, CRC Press, New York, 2007.

9. ICH M3(R2) Guidance on nonclinical safety studies for the conduct of human clinical trials and marketing authorization for pharmaceuticals. July 15, 2008. http://www.ema.europa.eu/docs/en-GB/document_library/scientific_guideline/2009/09/WC500002720.pdf.

10. Radioactive Drug Research Committee (RDRC). 21 CFR 361.1. April 2007. Available at http://www.fda.gov/CDER/regulatory/RDRC/default.htm (accessed March 17, 2009).

11. Wilson, K., Yu, J., Lee, A., and Wu, J.C. In vitro and in vivo bioluminescence reporter gene imaging of human embryonic stem cells. 2008. *JoVE* 14. Available at http://www.jove.com/index/Details.stp?ID=740, doi: 10.3791/740 (accessed March 17, 2009).

12. de Vries, I.J., Lesterhuis, W.J., Barentsz, J.O., Verdijk, P., van Krieken, J.H., Boerman, O.C., Oyen, W.J., Bonenkamp, J.J., Boezeman, J.B., Adema, G.J., Bulte, J.W., Scheenen, T.W., Punt, C.J., Heerschap, A., and Figdor, C.G. Magnetic resonance tracking of dendritic cells in melanoma patients for monitoring of cellular therapy. *Nat Biotechnol* 23(11), 1407–13, 2005.

13. ICH topic S9. Nonclinical evaluation for anticancer pharmaceuticals. December 2008. European Medicines Agency, London. Available at http://www.emea. europa.eu/pdfs/human/ich/64610708en.pdf (accessed March 17, 2009).

14. Kostura, L., Kraitchman, D.L., Mackay, A.M., Pittenger, M.F., and Bulte, J.W.M. Feridex labeling of mesenchymal stem cells inhibits chondrogenesis but not adipogenesis or osteogenesis. *NMR Biomed* 17, 513–17, 2004.

15. Zhang, Y., Ruel, M., Beanlands, R.S., deKemp, R.A., Suuronen, E.J., and DaSilva, J.N. Tracking stem cell therapy in the myocardium: Applications of positron emission tomography. *Curr Pharm Des* 14, 3835–53, 2008.

# Index

Printed and bound by CPI Group (UK) Ltd, Croydon, CR0 4YY

01/11/2024

01782618-0006